# Electrical Methods
# in Geophysical Prospecting

*by*

## GEORGE V. KELLER

Colorado School of Mines,
Golden, Colorado, U.S.A.

*and*

## FRANK C. FRISCHKNECHT

U.S. Geological Survey,
Denver, Colorado, U.S.A.

D1338050

## PERGAMON PRESS

OXFORD · NEW YORK · TORONTO
SYDNEY · PARIS · FRANKFURT

| U.K. | Pergamon Press Ltd., Headington Hill Hall, Oxford OX3 0BW, England |
|---|---|
| U.S.A. | Pergamon Press Inc., Maxwell House, Fairview Park, Elmsford, New York 10523, U.S.A. |
| CANADA | Pergamon of Canada Ltd., 75 The East Mall, Toronto, Ontario, Canada |
| AUSTRALIA | Pergamon Press (Aust.) Pty. Ltd., 19a Boundary Street, Rushcutters Bay, N.S.W. 2011, Australia |
| FRANCE | Pergamon Press SARL, 24 rue des Ecoles, 75240 Paris, Cedex 05, France |
| WEST GERMANY | Pergamon Press GmbH, 6242 Kronberg-Taunus, Pferdstrasse 1, West Germany |

First edition 1966

Reprinted 1970, 1977

Library of Congress Catalog Card No. 66–12653

*Printed in Great Britain by Biddles Ltd., Guildford, Surrey*

ISBN 0 08 011525 X

# CONTENTS

# PREFACE

We have written this text in an effort to share our conviction that electrical methods in geophysical prospecting have sound foundations in theory and are useful in application. While there is an extensive literature on electrical prospecting in the journals of the various professional societies which deal with geophysical exploration, there has been no textbook treatment in English since the publication of such texts as Jakosky's *Exploration Geophysics* and Heiland's *Geophysical Exploration* nearly a quarter of a century ago.

The mathematical theory on which electrical exploration is based is sufficiently difficult that a text such as this could be written on as restrictive a level as one might choose. While we have included a share of theory which requires something more than senior-level preparation as a background, we have also used a parallel approach in which nothing more than a background in sophomoric physics is required. Therefore, we advise the reader not to be discouraged by the more mathematical sections, which are by no means required in the application of electrical prospecting methods, but which are included for the more mathematically inclined reader who wishes to pursue the theoretical foundations more deeply.

Little of the material in this text is original. The reader who is familiar with the literature in this field is bound to recognize a similarity between the first part of Chapter III and E. N. Kalenov's book, *Interpretatzia Krivix Vertikalnovo Elektricheskovo Zondirovania*; between Chapters IV and V and M. N. Berdichevskii's book, *Elektricheskaia Razvedka Metodom Telluricheskikh Tokov*, and between portions of Chapter VI and numerous papers by J. R. Wait. However, outside of the primary references, we have provided little guidance to the literature. For this, we suggest the serious student refer to the very complete bibliography given by Volker Fritsch in *Grundzüge der angewandten Geoelektrik*, published by Manzsche Verlags- und Universitätsbuchhandlung, Vienna 1949. For more recent literature references, one cannot go wrong in using *Geophysical Abstracts*, published by the U.S. Geological Survey.

We are deeply indebted to the U.S. Geological Survey for having provided us with the opportunity to become interested in electrical prospecting over the past decade. This text to a large degree represents our experience in prospecting with electrical methods while with the U.S.G.S. One of us (G.V.K) is indebted also to the Colorado School of Mines for providing the opportunity of trying this text on groups of students over the past several years. These students have proved to be most enthusiastic in seeking out errors in logic and presentation, and if any persist, it is not their fault.

# CHAPTER 1

# ELECTRICAL PROPERTIES OF EARTH MATERIALS

ELECTRICAL conduction in minerals may take place by electronic or ionic processes. Solid conductors may be divided into three classes depending on the mechanism of current conduction: metals, electron semiconductors and solid electrolytes. Examples of each type of conduction, as well as examples of combinations of several types of conduction can be found readily in a list of common rock-forming and economically important minerals. The native metals, such as gold and copper, are examples of metallic conductors. Most of the sulfide ore minerals belong to the semiconductor class, while the silicate rock-forming minerals are solid electrolytes.

Rocks are not formed of minerals alone, and the electrical properties of a rock are not necessarily determined by the properties of the mineral constituents alone. All rocks at the earth's surface are porous. Highly porous rocks such as pumice or poorly compacted mudstones may have pore volumes as great as 50 per cent or more of the total rock volume. Igneous rocks, evaporites and dense carbonate rocks have far lower porosities, perhaps less than 1 per cent of the total volume, but no surface rock is completely non-porous. Under any reasonable circumstances, these pores are partly or completely filled with water. This water usually carries some salt in solution, so that the water content of a rock has a far greater capacity for carrying current than does the solid matrix of the rock, unless highly conducting minerals are present. Solid conduction mechanisms can be expected to be important in comparison with electrolytic conduction through pore water in three cases: in a rock containing a high percentage of conducting minerals, in a completely frozen rock and in a rock which is far enough below the surface of the earth that all pore spaces are closed by overburden pressure.

The relative abilities of materials to conduct electricity when a voltage is applied are expressed as *conductivities*; conversely, the resistance offered by a material to current flow is expressed in terms of *resistivity*. The resistivity of a material is defined by the mathematical expression of Ohm's law, a law which states that the electric field strength at a point in a material is proportional to the current density passing that point:

$$E = \varrho j \tag{1}$$

where $E$ is the electric field strength, expressed in volts per meter and $j$ is the current density in amperes per square meter. This definition implies the use of a zero frequency or direct current. The mks unit for resistivity is the ohm-meter. The ohm-meter is a unit of convenient size for expressing the resistivity of earth materials and is preferable to the cgs unit, the ohm-centimeter, which also is used frequently. The conversion between the two units is straight-forward:

1 ohm-m = 100 ohm-cm
1 ohm-cm = 0·01 ohm-m

It is sometimes convenient to discuss the *conductivity* of a rock, rather than the resistivity. Since conductivity is the reciprocal of resistivity, the dimensions of the unit for expressing conductivity are either the mho per meter (mks) or mho per centimeter (cgs).

The ohm-meter may also be defined in terms of a hypothetical experiment in which the electrical resistance through a cube of material with dimensions of one meter on a side is measured. The resistance between opposite faces of such a cube is equal numerically to the resistivity of the material, providing the cube is completely uniform.

The resistivity of a medium is one of three physical properties which determine the behavior of electromagnetic fields in the medium, the other two properties being the dielectric constant and the magnetic permeability. Resistivity is usually the most important of the three properties in determining electrical current flow. Dielectric constant and magnetic permeability will be discussed briefly later in this chapter.

## 1. CONDUCTIVITY OF METALS

Native metals occur infrequently, but such occurences are of some economic importance. Two of the more important native metals are copper and gold. Such metals as platinum, iridium, osmium and iron occur in the elemental form but extremely rarely. Carbon occurs commonly in the form of graphite, which is a metallic conductor with some peculiar properties.

The structure of a metal and the nature of bonding between atoms is difficult to explain, but on an elementary level, a metal appears to be an orderly packing of metal ions surrounded by an atmosphere of valence electrons. The energy required to move a valence electron from one atom to the next in the structure is very small, so that the valence electrons are not closely identified with any particular atom. The high conductivity of metals, as well as the other metallic properties, is due to the large number of mobile electrons, a number which is approximately equal to the number of atoms in the metal.

If the atoms in a metal are perfectly ordered, there should be almost no resistance to the motion of electrons through the metal when an electric field is applied. In practice, imperfections in the crystal structure are always present, and such imperfections interfere with the free movement of the valence electrons. The greater the number of such imperfections, the higher will be the resistivity of a metal.

Thermal motion of the metal ions about their equilibrium positions also interferes with the freedom of motion of the conduction electrons. The higher the temperature, the greater will be the amplitude of the oscillations of the ions about their rest positions and the greater will be the probability that any particular ion will interfere with the freedom of motion of the electron atmosphere. Thus, metals are characterized by a positive temperature coefficient of resistivity: resistivity in many cases increasing by about $\frac{1}{2}$ per cent for each degree centigrade rise in temperature.

The resistivities of refined metals and of several naturally-occurring metallic minerals are listed in Table 1.

## 2. CONDUCTIVITY OF ELECTRONIC SEMICONDUCTORS

Semiconductors are non-metallic materials in which conduction is by electron motion, but in which the conductivity is lower than that in true metals. The lower conductivity is not a result of lower mobility of the electrons, but rather due to a lesser number of conduction electrons. Relatively few electrons are free to move through the crystal lattice in a semiconductor, in comparison with a number of electrons about equal to the number of atoms in a metal.

Semiconductors differ from metals in that the energy level of the conduction electrons must be raised by an appreciable amount before the electrons are free to wander through the crystal lattice, as in the case of a metal. This energy is most commonly provided in the form of heat, so that in semiconductors, the number of conduction electrons is found to increase with temperature according to the law:

$$n_e \propto e^{-E/kT} \tag{2}$$

where $n_e$ is the number of conduction electrons, $E$ is the energy required to raise the energy level of the conduction electrons to the point where they are free to move through the lattice, $k$ is Boltzman's constant and $T$ is the absolute temperature. The activation energy, $E$, is a characteristic of the material. In materials containing several elements, the activation energy for electrons from different elements will differ, but ordinarily, only the lowest activation energy need be considered.

All materials which are not true metals are electronic semiconductors to a greater or lesser extent. If a material has electrons available with a relatively low activation energy, the material may be nearly as conductive as a metal. This is the case with many of the sulfide ore minerals, which have resistivities in the range from $10^{-6}$ to $10^{-3}$ ohm-m. If the activation energy is large, a material may be nearly a perfect insulator. Most of the silicate minerals require large activation energies to provide conduction electrons, and under normal

TABLE 1
## Resistivities of metals and metallic minerals (zero frequency)

*Refined metals at 0°C*

| | |
|---|---|
| Lithium | $8.5 \times 10^{-8}$ ohm-m |
| Beryllium | $5.5 \times 10^{-8}$ |
| Sodium | $4.3 \times 10^{-8}$ |
| Magnesium | $4.0 \times 10^{-8}$ |
| Aluminium | $2.5 \times 10^{-8}$ |
| Potassium | $6.3 \times 10^{-8}$ |
| Calcium | $4.2 \times 10^{-8}$ |
| Titanium | $83 \times 10^{-8}$ |
| Chromium | $15.3 \times 10^{-8}$ |
| Iron | $9.0 \times 10^{-8}$ |
| Cobalt | $6.3 \times 10^{-8}$ |
| Nickel | $6.3 \times 10^{-8}$ |
| Copper | $1.6 \times 10^{-8}$ |
| Zinc | $5.5 \times 10^{-8}$ |
| Gallium | $41 \times 10^{-8}$ |
| Arsenic | $35 \times 10^{-8}$ |
| Rubidium | $11.6 \times 10^{-8}$ |
| Strontium | $33 \times 10^{-8}$ |
| Zirconium | $42 \times 10^{-8}$ |
| Molybdenum | $4.3 \times 10^{-8}$ |
| Ruthenium | $11.7 \times 10^{-8}$ |
| Rhenium | $4.5 \times 10^{-8}$ |
| Palladium | $10.0 \times 10^{-8}$ |
| Silver | $1.5 \times 10^{-8}$ |
| Cadmium | $6.7 \times 10^{-8}$ |
| Indium | $8.5 \times 10^{-8}$ |
| Tin | $10.0 \times 10^{-8}$ |
| Antimony | $36 \times 10^{-8}$ |
| Cesium | $18 \times 10^{-8}$ |
| Barium | $59 \times 10^{-8}$ |
| Lanthanum | $59 \times 10^{-8}$ |
| Cerium | $71 \times 10^{-8}$ |
| Praseodymium | $62 \times 10^{-8}$ |
| Hafnium | $29 \times 10^{-8}$ |
| Tantalum | $14 \times 10^{-8}$ |
| Tungsten | $5.0 \times 10^{-8}$ |
| Osmium | $9.1 \times 10^{-8}$ |
| Iridium | $5.0 \times 10^{-8}$ |
| Platinum | $9.8 \times 10^{-8}$ |
| Gold | $2.0 \times 10^{-8}$ |
| Tellurium | $14 \times 10^{-8}$ |
| Lead | $19 \times 10^{-8}$ |
| Bismuth | $100 \times 10^{-8}$ |

*Metallic minerals*

| | |
|---|---|
| Native copper | $1.2$ to $30 \times 10^{-8}$ ohm-m |
| Graphite (carbon) | $36$ to $100 \times 10^{-8}$ (current flow parallel to cleavage) |
| | $28$ to $9900 \times 10^{-6}$ (current flow across cleavage) |
| Ulmanite, NiSbS | $9.0$ to $120 \times 10^{-8}$ |
| Breithauptite, NiSb | $3.0$ to $50 \times 10^{-8}$ |

circumstances, electron conduction in silicates is negligible in comparison with ion conduction.

Since the number of electrons available for conduction in a semiconductor increases with temperature, such materials usually have a negative temperature coefficient of resistivity; resistivity decreasing with increasing temperature. This is not always the case, though, since increasing temperature decreases the mobility of the electrons, as it does in metals, and this effect may more than offset the effect of an increased number of conduction electrons. In such a case, semiconductors may have a positive coefficient for a portion of the temperature range, and a negative thermal coefficient of resistivity for other portions of the range.

The important minerals which are semiconductors are the various metal–sulfur and related compounds and some of the metal oxides. Resistivities for these minerals are listed in Table 2.

TABLE 2

**Resistivities of semiconducting minerals (zero frequency)**

*Native elements*

| | |
|---|---|
| Diamond (C) | $2 \cdot 7$ ohm-m |

*Sulfides*

| | |
|---|---|
| Argentite, $Ag_2S$ | $1 \cdot 5$ to $2 \cdot 0 \times 10^{-3}$ |
| Bismuthinite, $Bi_2S_3$ | 3 to 570 |
| Bornite, $Fe_2S_3 . n\,Cu_2S$ | $1 \cdot 6$ to $6000 \times 10^{-6}$ |
| Chalcocite, $Cu_2S$ | 80 to $100 \times 10^{-6}$ |
| Chalcopyrite, $Fe_2S_3 . Cu_2S$ | 150 to $9000 \times 10^{-6}$ |
| Covellite, $CuS$ | $0 \cdot 30$ to $83 \times 10^{-6}$ |
| Galena, $PbS$ | $6 \cdot 8 \times 10^{-6}$ to $9 \cdot 0 \times 10^{-2}$ |
| Haverite, $MnS_2$ | 10 to 20 |
| Marcasite, $FeS_2$ | 1 to $150 \times 10^{-3}$ |
| Metacinnabarite, $4\,HgS$ | $2 \times 10^{-6}$ to $1 \times 10^{-3}$ |
| Millerite, $NiS$ | 2 to $4 \times 10^{-7}$ |
| Molybdenite, $MoS_2$ | $0 \cdot 12$ to $7 \cdot 5$ |
| Pentlandite, $(Fe, Ni)_9S_8$ | 1 to $11 \times 10^{-6}$ |
| Pyrrhotite, $Fe_7S_8$ | 2 to $160 \times 10^{-6}$ |
| Pyrite, $FeS_2$ | $1 \cdot 2$ to $600 \times 10^{-3}$ |
| Sphalerite, $ZnS$ | $2 \cdot 7 \times 10^{-3}$ to $1 \cdot 2 \times 10^4$ |

*Antimony–sulfur compounds*

| | |
|---|---|
| Berthierite, $FeSb_2S_4$ | $0 \cdot 0083$ to $2 \cdot 0$ |
| Boulangerite, $Pb_5Sb_4S_{11}$ | $2 \times 10^3$ to $4 \times 10^4$ |
| Cylindrite, $Pb_3Sn_4Sb_2S_{14}$ | $2 \cdot 5$ to 60 |
| Franckeite, $Pb_5Sn_3Sb_2S_{14}$ | $1 \cdot 2$ to 4 |
| Hauchecornite, $Ni_9(Bi, Sb)_2S_8$ | 1 to $83 \times 10^{-6}$ |
| Jamesonite, $Pb_4FeSb_6S_{14}$ | $0 \cdot 020$ to $0 \cdot 15$ |
| Tetrahedrite, $Cu_3SbS_3$ | $0 \cdot 30$ to 30,000 |

*Arsenic–sulfur compounds*

| | |
|---|---|
| Arsenopyrite, FeAsS | 20 to 300 $\times$ $10^{-6}$ |
| Cobaltite, CoAsS | 6·5 to 130 $\times$ $10^{-3}$ |
| Enargite, $Cu_3AsS_4$ | 0·2 to 40 $\times$ $10^{-3}$ |
| Gersdorffite, NiAsS | 1 to 160 $\times$ $10^{-6}$ |
| Glaucodote, (Co, Fe)AsS | 5 to 100 $\times$ $10^{-6}$ |

*Antimonide*

| | |
|---|---|
| Dyscrasite, $Ag_3Sb$ | 0·12 to 1·2 $\times$ $10^{-6}$ |

*Arsenides*

| | |
|---|---|
| Allemonite, $SbAs_3$ | 70 to 60,000 |
| Lollingite, $FeAs_2$ | 2 to 270 $\times$ $10^{-6}$ |
| Nicollite, NiAs | 0·1 to 2 $\times$ $10^{-6}$ |
| Skutterudite, $CoAs_3$ | 1 to 400 $\times$ $10^{-6}$ |
| Smaltite, $CoAs_2$ | 1 to 12 $\times$ $10^{-6}$ |

*Tellurides*

| | |
|---|---|
| Altaite, PbTe | 20 to 200 $\times$ $10^{-6}$ |
| Calavarite, $AuTe_2$ | 6 to 12 $\times$ $10^{-6}$ |
| Coloradoite, HgTe | 4 to 100 $\times$ $10^{-6}$ |
| Hessite, $Ag_2Te$ | 4 to 100 $\times$ $10^{-6}$ |
| Nagyagite, $Pb_6Au(S, Te)_{14}$ | 20 to 80 $\times$ $10^{-6}$ |
| Sylvanite, $AgAuTe_4$ | 4 to 20 $\times$ $10^{-6}$ |

*Oxides*

| | |
|---|---|
| Braunite, $Mn_2O_3$ | 0·16 to 1·0 |
| Cassiterite, $SnO_2$ | 4·5 $\times$ $10^{-4}$ to 10,000 |
| Cuprite, $Cu_2O$ | 10 to 50 |
| Hollandite, (Ba, Na, K)$Mn_8O_{16}$ | 2 to 100 $\times$ $10^{-3}$ |
| Ilmenite, $FeTiO_3$ | 0·001 to 4 |
| Magnetite, $Fe_3O_4$ | 52 $\times$ $10^{-6}$ |
| Manganite, MnO . OH | 0·018 to 0·5 |
| Melaconite, CuO | 6000 |
| Psilomelane, $KMnO . MnO_2 . nH_2O$ | 0·04 to 6000 |
| Pyrolusite, $MnO_2$ | 0·007 to 30 |
| Rutile, $TiO_2$ | 29 to 910 |
| Uraninite, UO | 1·5 to 200 |

The data contained in Table 2 are taken primarily from measurements published by Harvey (1928) and by Parasnis (1956). A wide range of resistivities is reported for most minerals. The narrow ranges reported for a few minerals do not represent more reliable measurements necessarily, but rather, a smaller number of measurements. The large range in values may be in part real, since many of the semiconducting minerals listed in Table 2 have indefinite composition, and the electrical properties may be altered considerably by minor changes in composition. However, Harvey points out that in many cases, erroneously high values of resistivity may be measured on samples which have hairline cracks. In this case, the lower resistivities reported for a mineral are probably more correct than the higher values.

Despite the wide ranges of values included for each mineral, the data in Table 2 suggest several generalizations. With a few exceptions, the metal sulfides and arsenides are nearly as conductive as true metals. The exceptions are galena, haverite and molybdenite, which have moderately high resistivities. The arsenides and tellurides also have very low resistivities, but these compounds may be metallic alloys, rather than semiconducting compounds. The antimony compounds and the metal oxides, except for magnetite, have relatively high resistivities.

## 3. SOLID ELECTROLYTES

Most rock-forming minerals can be considered solid electrolytes, so far as conduction of current is concerned. Electrolytic conduction can take place in ionic bonded crystals. In ionic bonding, metal ions give up their valence electrons to complete the outer electron shell of the other element forming the compound. For example, in the compound NaCl, the sodium atom surrenders its lone valence electron to fill the valence shell of the chlorine atom, which ordinarily has seven of the eight electron orbits in its outer shell filled. After this transfer of valence electrons, the chlorine and sodium ions are held together by coulomb attraction forces, since the ions are oppositely charged.

When an electric field is applied to an ionic bonded crystal structure, the force exerted on each ion by the field is small compared to the strength of the binding forces, so that in an ideal crystal, one would not expect any electrolysis, or conduction by ion movement, to take place. However, crystals are not perfect and electrolysis does take place when an electric field is applied. There are two important types of imperfection in crystals which aid in current conduction; inherent imperfections in the crystal lattice (Schottky defects), and thermally induced imperfections (Frenkel's defects.)

All crystals have inherent imperfections in the form of impurity ions of the wrong valence substituted in the lattice, or in the form of ions missing from the lattice. All the ions in the lattice vibrate about their rest positions, with the amplitude of vibration increasing with increasing temperature. Occasionally, an ion will oscillate far enough that it jumps into such a vacancy. Such jumps occur continually, but in random directions, even when no external electric field is applied to the crystal. When a field is applied, jumps in the direction of the field are favored statistically over jumps in other directions, and a net transfer of ions, or current, takes place. The frequency with which jumps take place depends on temperature, according to the law:

$$n_j \propto e^{-U/kT} \qquad (3)$$

where $n_j$ is the number of jumps per unit time, $U$ is the the height of the potential energy barrier across which the jump takes place, $k$ is Boltzman's constant ($8 \cdot 62 \times 10^{-3}$ eV/degC or $1 \cdot 38 \times 10^{-16}$ erg/degC) and $T$ is the absolute temperature.

According to Frenkel, an ion may be displaced occasionally from its equilibrium position in the crystal lattice to an interstitial position by thermal agitation. The ion will ultimately return to a normal lattice position as it moves about in the interstices of the lattice. As in the case of inherent defects, the motion will not be random when an external electric field is applied to the crystal, but motion along the field direction will take preference to motion in other directions.

The resistivity of a solid electrolyte should be inversely proportional to the number of charge carriers available, with the constant of proportionality being determined by the mobility of the ions. The mobility of the ions will depend on the relative size of the moving ions and the interstices in the crystal lattice. Small ions may move readily through a lattice composed of large immobile ions. Approximate ionic radii are listed in Table 3.

TABLE 3

**Approximate radii of ions**

| | |
|---|---|
| $O^=$ | 1·40 Å |
| $Cl^-$ | 1·81 |
| $H^-$ | 2·08 |
| $Na^+$ | 0·95 |
| $Mg^{++}$ | 0·65 |
| $Al^{+++}$ | 0·50 |
| $Si^{++++}$ | 0·41 |
| $K^+$ | 1·33 |
| $Ca^{++}$ | 0·99 |
| $Fe^{++}$ | 0·76 |
| $Fe^{+++}$ | 0·64 |
| $Mn^{++}$ | 0·66 |
| $Mn^{++++}$ | 0·54 |

These radii pertain only to single-nucleus ions. In rocks, ions consisting of several atoms bonded together with covalent bonds are common, particularly in the form of the oxides of silicon and aluminium. In rocks in which the stable lattice consists of these super-ions, the small ions which are free to move through the interstices are those of sodium, magnesium and iron. In any event, anionic dimensions are much larger than the cationic dimensions, and conduction must take place largely by the transfer of cations.

Noritomi (1958) has studied the material deposited at the cathode after the electrolysis of silicate rock samples at high temperature, and found that iron, aluminium, calcium and sodium had been deposited.

It has been observed that generally there are two main regions in the curve relating conductivity to temperature for ionic conductors such that in each region, the logarithm of conductivity is proportional to the inverse absolute temperature, as predicted by Eq. (3). At high temperatures (above 750°K), the relationship between conductivity and temperature is an intrinsic property of a material, varying little from sample to sample. The low-temperature portion of the relationship is sensitive to the structure of a sample and its thermal history, and so does not reproduce well. Conductivity in ionic crystals may usually be approximated with the equation:

$$\frac{1}{\varrho} = \sigma = A_1 e^{-U_1/kT} + A_2 e^{-U_2/kT} \tag{4}$$

where the parameters $A_1$ and $A_2$ are determined by the numbers of ions available for conduction and their mobility through the lattice, and $U_1$ and $U_2$ are the activation energies required to liberate these ions. Generally, $U_1$ is only about half as large as $U_2$, and $A_1$ is about five orders of magnitude smaller than $A_2$. The low-temperature conductivity is called *extrinsic* or *structure-sensitive* conductivity, and is due to weakly-bonded impurities or defects in the crystal. The high-temperature conductivity is called *intrinsic* conductivity, and is due to ions from the regular lattice which have been displaced by thermal vibrations. Several examples of the relation between conductivity and temperature are shown graphically in Fig. 1. Values for the parameters in Eq. (4) are given for a variety of rocks in Table 4.

TABLE 4

**Parameters defining the temperature dependence of resistivity in solid electrolytes**

| Rock | $A_1$ | $A_2$ | $U_1$ | $U_2$ |
|---|---|---|---|---|
| Granite | $5 \times 10^{-4}$ mho/cm | $10^5$ mho/cm | 0·62 eV | 2·5 eV |
| Gabbro | $7 \times 10^{-3}$ | $10^5$ | 0·70 | 2·2 |
| Basalt | $7 \times 10^{-3}$ | $10^5$ | 0·57 | 2·0 |
| Peridotite | $4 \times 10^{-2}$ | $10^5$ | 0·81 | 2·3 |
| Andesite | $6 \times 10^{-3}$ | | 0·7 | 1·6 |

The resistivity of a conductor or a semiconductor is the same whether it is measured with direct current or with a high-frequency alternating current. This is not true for solid electrolytes, with resistivity usually depending to some extent on the frequency of the current used in measurements. A typical set of curves showing the relation between resistivity and the frequency of the current used in determining it is plotted in Fig. 2. At low temperatures, the resistivity is nearly inversely proportional to the frequency. At high temperatures, the resistivity is much lower, and nearly constant. Extrinsic conductivity, the type which is dominant in solid electrolytes at low temperatures, is frequency-dependent at all frequencies above a few cycles per second, whereas intrinsic conductivity is not frequency dependent, at least up to frequencies of some megacycles per second.

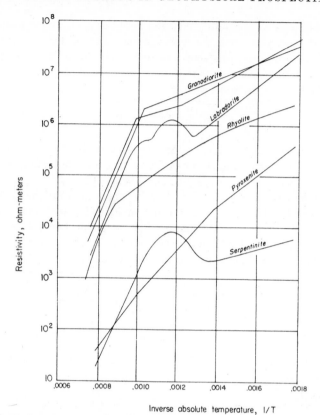

FIG. 1. Resistivity as a function of $1/T$ for several rocks. The right-hand segments of the curves represent extrinsic conduction by impurities and crystal defects, the left-hand, or high-temperature portions, represent intrinsic conduction.

The extrinsic conductivity is frequency dependent since impurity ions and defects are not free to move far through a crystal lattice. Each impurity ion is held weakly to a vacancy in the crystal lattice. When high-frequency currents are used in measuring resistivity, the mobile ions are not required to move very far during one cycle, and so, these ions can move with the current. When low-frequency currents are used, ions may move to the limit of their freedom before each cycle is completed, and so, contribute less efficiently to the conduction of current. Thus extrinsic conduction is lower at low frequencies than at high frequencies.

At high temperatures, enough energy is available that conduction ions are not held near the vacancy left in the crystal lattice, and these ions are free to move large distances through the lattice. Thus, intrinsic conduction is not frequency dependent.

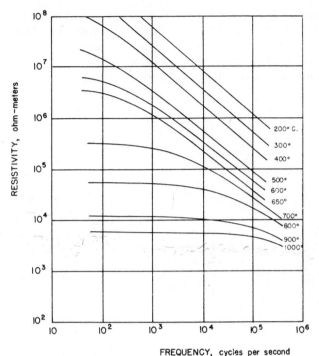

FIG. 2. The resistivity of a sample of granodiorite measured as a function of frequency at various temperatures. The low temperature, frequency-dependent resistivity is extrinsic; the high temperature, non-variant conduction is intrinsic.

## 4. SOLID CONDUCTION IN ROCKS

For most rocks near the earth's surface, conduction of electricity is entirely through ground water contained in the pores of the rocks. Conduction through minerals takes place in some metallic ore minerals in rocks, providing these minerals occur in high enough concentrations. Many conducting minerals are listed in Table 2, but few of them occur commonly enough or in sufficient quantity to change appreciably the electrical properties of the rock in which they occur. The conducting minerals which do occasionally occur in sufficient quantities to render large volumes of rock conductive are magnetite, specular hematite, carbon, graphite, pyrite and pyrrhotite.

The habit of a mineral, the form in which it usually occurs, is as important as the amount in determining the effect on bulk resistivity. The minerals listed above commonly occur in dendritic patterns, so that a small amount of conductive mineral will decrease the bulk resistivity of a rock tremendously.

As an example, Fig. 3 shows the effect chalcopyrite has on the resistivity of a gabbroic rock which occurs in Maine. Each point plotted on this graph is an average of several measurements; otherwise, the scatter of the plotted points would be much greater. Eight per cent chalcopyrite (by volume) is all that is required to increase the conductivity of the rock manyfold.

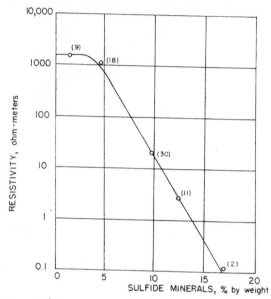

Fig. 3. Observed relationship between resistivity and the amount of pyrrhotite and pentlandite present in gabbro. Samples were taken from an ore body in southern Maine. The numbers in parentheses indicate the number of measurements averaged and plotted as a single point.

Magnetite and specular hematite commonly render large volumes of rock conductive in the iron ranges of the north central United States. Five per cent magnetite or specular hematite can reduce the resistivity of a rock to $0 \cdot 1$ ohm-m or less. Such concentrations of magnetic and specular hematite occur in many places in the Precambrian chert beds around Lake Superior. Figure 4 shows graphically some resistivity measurements made with equipment lowered in a drill hole penetrating such ferrugenous cherts in the Gogebic Iron Range in northern Wisconsin. The hole also penetrates silicate chert and slate, the slate having a resistivity of a few ohm-meters, owing to its water content, and the silicate chert having a high resistivity, several thousands of ohm-meters, where it does not contain magnetite. At depths between 205 and 460 ft in this drill hole, the chert is known to contain from 5 to 29 per cent magnetite, by weight. Over a portion of this interval, from 310 to 420 ft, the magnetite occurs in such a manner that the resistivity of the chert is reduced from a normal value

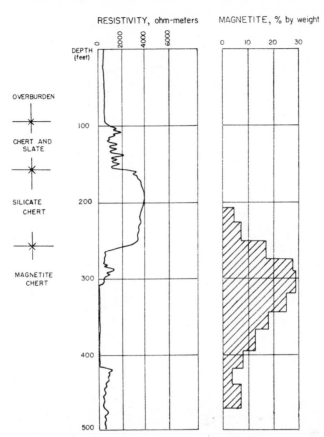

Fig. 4. Borehole resistivity measurements from a drill hole penetrating magnetic chert in the Gogebic Iron Range, northern Wisconsin.

of some thousands of ohm-meters to less than 1 ohm-m. The highest magnetite content, 49 per cent, does not have as great an effect as lesser magnetite contents in the lower zones. It appears the effect of magnetite on the conductivity in this case depends more on the distribution in the rock than on the amount.

Graphite (or carbon) and pyrite are probably the minerals which most commonly cause rock to be conductive. Carbon, graphite and pyrite are especially common in slates, but it is uncertain which mineral contributes to the conduction observed in such rocks. It is possible that pyritic slates conduct because of their carbon content, which usually goes unnoticed due to the small grain size. Figure 5 shows resistivity measurements made in a drill hole penetrating pyritic and graphitic Precambrian slates in northern Michigan. This hole penetrated cherts and slates with resistivities averaging about 1000 ohm-m.

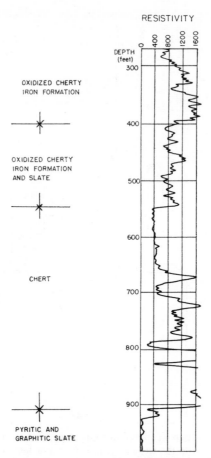

FIG. 5. Borehole resistivity measurements from Precambrian rocks in Michigan. The rocks below 930 ft are rendered conductive by the presence of pyrite and graphite.

The ferrugenous slates in this case are not conductive because the iron is present as goethite, rather than as magnetite and hematite, as in the example in Fig. 4. Below 900 ft, the slate contains a few per cent pyrite and graphite or carbon. In this zone, the resistivity of the rock has been reduced from 1000 to less than 1 ohm-m.

Although a few per cent conductive mineral will render a rock conductive as a whole when the mineral is properly distributed; in other cases, 10 to 15 per cent of an ore mineral will have no detectable effect. This is the case if ore minerals occur widely disseminated in the host rock, as in the porphyry ore deposits in the western states. The same is true for vein-type deposits in general, if the ore minerals grow as single crystals separated from one another by non-conducting gangue minerals. The Mission Prospect in southern Arizona is an example of an ore deposit containing fairly large concentrations of conductive minerals such as pyrite and bornite, and yet which is not conductive. Resistivity measurements made in a drill hole penetrating the ore zone in the Mission Prospect are shown in Fig. 6, along with a bar graph of the sulfide

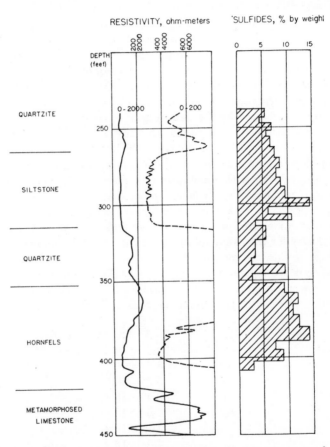

FIG. 6. Borehole resistivity measurements made in a porphyry ore deposit in southern Arizona. The rock resistivity is high despite the presence of conducting sulfide minerals.

mineral content in the rock. The sulfides do not lower the resistivity of the rock, and in fact, the resistivity in the ore zone is about 100 times greater than that in non-mineralized rock of the same type, since deposition of non-conducting gangue minerals in the pore structure of the rock has reduced the normal conduction in the rock to a low value.

## 5. CONDUCTION IN WATER-BEARING ROCKS

For most rocks near the earth's surface, conduction will be electrolytic, the conducting medium being an aqueous solution of common salts, distributed in a complicated manner through the pore structure of a rock. The resistivity of a water-bearing rock will depend on the amount of water present, the salinity of this water and the way in which the water is distributed in the rock. The electrical properties of a water-bearing rock should be describable in the same terms as the electrical properties of an electrolyte.

When a salt goes into solution in water, the constituent ions in the solid salt separate, and these ions are free to move around independently in the solution. When an electric field is applied across an electrolytic solution, cations will be accelerated toward the negative pole and anions to the positive pole. Once the ions have been accelerated by the electric field, a viscous drag force limits the velocity which is attained. In aqueous solutions, the time required for an ion to reach this terminal velocity is considerably less than a microsecond. This terminal velocity is defined as the *mobility* of an ion. Specifically, the mobility of an ion is the velocity with which it moves, measured in meters per second, when a voltage gradient of one volt per meter is applied.

Mobility depends on both temperature and concentration. Increasing the temperature of an electrolyte decreases the viscosity, permitting a higher terminal velocity for the same voltage gradient. If a solution contains a high concentration of ions, the motion of one ion will be influenced by the motion of ions close to it, reducing the terminal velocity. For these reasons, the temperature and concentration for which an ion mobility applies must be specified. The mobilities of a few of the commoner ions at 25°C in dilute solution are:

$$
\begin{array}{lll}
H^+ & 36 \cdot 2 \times 10^{-8}\ \mathrm{m^2/sec\ V} \\
OH^- & 20 \cdot 5 \times 10^{-8} \\
SO_4^= & 8 \cdot 3 \times 10^{-8} \\
Na^+ & 5 \cdot 2 \times 10^{-8} \\
Cl^- & 7 \cdot 9 \times 10^{-8} \\
K^+ & 7 \cdot 6 \times 10^{-8} \\
NO_3^- & 7 \cdot 4 \times 10^{-8} \\
Li^+ & 4 \cdot 0 \times 10^{-8} \\
HCO_3^- & 4 \cdot 6 \times 10^{-8}
\end{array}
$$

When an electric field is applied to an electrolyte, the amount of current which flows is found by multiplying the number of ions present (concentration) by the velocity with which they move. Since one gram equivalent weight of salt in solution contains 96,500 coulombs of charge (Faraday's number, $F$), the current flowing through an electrolyte per one volt per meter applied field is:

$$ I = A F(c_1 v_1 + c_2 v_2 + c_3 v_3 \ldots) \tag{5} $$

where the $c_n$ and $v_n$ represent the concentration and mobility for each species of ion in solution, and $A$ is the cross-sectional area through which current flows. The resistivity of the solution may be determined readily by considering the current flow through a cross-sectional area of one square meter at a voltage gradient of one volt per meter:

$$\frac{1}{\varrho} = F(c_1 m_1 + c_2 m_2 + c_3 m_3 \ldots) \tag{6}$$

where $m_n$ are the mobilities measured at the proper temperatures and concentrations.

Many hundreds of ground-water analyses have been grouped and averaged on the basis of the geologic age of the rock, with the given results in Table 5.

TABLE 5

Average values of connate water resistivity (in part from Chebotarev, 1955)

| Nature of samples | Number of samples averaged | Average resistivity (ohm-m) |
|---|---|---|
| Igneous rocks, Europe | 314 | 7·6 |
| Igneous rocks, South Africa | 175 | 11·0 |
| Metamorphic rocks, South Africa | 88 | 7·6 |
| Metamorphic rocks, Australian Precambrian | 31 | 3·6 |
| Pleistocene to recent sedimentary rocks from Europe | 610 | 3·9 |
| Pleistocene to recent sedimentary rocks from Australia | 323 | 3·2 |
| Tertiary sedimentary rocks from Europe | 993 | 1·4 |
| Tertiary sedimentary rocks from Australia | 240 | 3·2 |
| Mesozoic sedimentary rocks from Europe | 105 | 2·5 |
| Paleozoic sedimentary rocks from Europe | 161 | 0·93 |
| Chloride waters from oil fields | 967 | 0·16 |
| Sulfate waters from oil fields | 256 | 1·2 |
| Bicarbonate waters from oil fields | 630 | 0·98 |
| Morrison formation (Jurassic), Colorado-Utah | 240 | 1·8 |
| Cisco series, North Texas | 85 | 0·061 |
| Des Moines series (Pennsylvania), Central Oklahoma | 94 | 0·062 |

The concept of an *equivalent salinity* is frequently used in discussing the resistivity of ground water. Since ground water may have a variety of salts in solution, it is not always easy to compute the resistivity of the water from a chemical analysis. For this reason, the equivalent salinity of a solution is defined as the salinity of a sodium chloride solution which would have the same resistivity as that of the particular solution for which the equivalent salinity is being expressed. The equivalent salinity should be fairly close to the true salinity since mobilities of ions do not vary widely.

The use of equivalent salinity has the advantage that only tables (or graphs) for a single salt are needed to determine the resistivity of a solution. The resistivity of solutions of sodium chloride at 10°C is given in Table 6. Curves

TABLE 6

## Resistivity in ohm-meters of sodium chloride electrolytes (from Dakhnov, 1962)

| Temperature (°C) | NaCl g/l | | | | | |
|---|---|---|---|---|---|---|
| | 58·45* | 29·23 | 5·845 | 2·933 | 0·5845 | 0·2923 |
| 0 | 0·211 | 0·386 | 1·73 | 3·36 | 15·82 | 31·2 |
| 2 | 0·200 | 0·368 | 1·65 | 3·19 | 15·1 | 29·6 |
| 4 | 0·190 | 0·352 | 1·57 | 3·02 | 14·3 | 28·1 |
| 6 | 0·182 | 0·336 | 1·49 | 2·86 | 13·7 | 26·7 |
| 8 | 0·174 | 0·320 | 1·42 | 2·73 | 12·9 | 25·0 |
| 10 | 0·165 | 0·304 | 1·35 | 2·57 | 12·3 | 23·9 |
| 12 | 0·157 | 0·288 | 1·28 | 2·43 | 11·7 | 22·6 |
| 14 | 0·149 | 0·274 | 1·21 | 2·31 | 11·1 | 21·4 |
| 16 | 0·142 | 0·260 | 1·15 | 2·19 | 10·5 | 20·3 |
| 18 | 0·135 | 0·248 | 1·09 | 2·09 | 9·8 | 19·3 |
| 20 | 0·129 | 0·238 | 1·04 | 2·00 | 9·5 | 18·4 |
| 22 | 0·123 | 0·228 | 1·00 | 1·92 | 9·0 | 17·6 |
| 24 | 0·117 | 0·219 | 0·96 | 1 84 | 8·6 | 16·8 |
| 26 | | 0·210 | 0·93 | 6 | 8·2 | 16·2 |
| 28 | | 0·200 | 0·87 | 8 | 7·9 | 15·6 |
| 30 | | 0·191 | 0·84 | 1·61 | 7·5 | 14·9 |
| 32 | | 0·183 | 0·80 | 1·55 | 7·2 | 14·3 |
| 34 | | 0·176 | 0·77 | 1·49 | 6·9 | 13·7 |

* Normal solution.

showing the relationship between the resistivity and the salinity of sodium chloride solutions at various temperatures are presented in Fig. 7. The temperatures range from 0 to 140°C, the higher temperatures being accompanied by sufficiently high pressures to keep water in a liquid state. The conductivity of a sodium chloride solution increases about sevenfold as temperature is increased from 0 to 140°. The conductivity of a brine-saturated rock should behave in the same way.

Figure 8 shows a series of curves for the relation between salinity and solution resistivity for solutions of a variety of salts, all at 18°C.

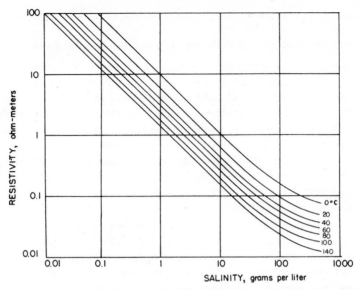

FIG. 7. Resistivity of solutions of sodium chloride as a function of concentration and temperature.

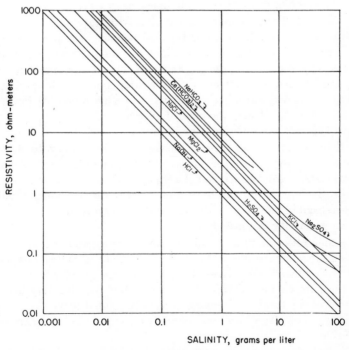

FIG. 8. Relationship between resistivity and concentration for various salt solutions at a temperature of 18°C.

## 5a. Relationship between resistivity, porosity and the texture of a rock

The resistivity of a water-bearing rock decreases with increasing water content. In fully saturated rocks, water content may be equated with porosity, but in partially desaturated rocks, the effect of desaturation on resistivity must be considered. The texture of a rock also has some effect on the resistivity.

If a simple pore geometry is assumed, the relationship between resistivity and water content may be calculated exactly. Calculations of the resistivity of a matrix consisting of uniform spheres have been reported in the literature. In consolidated rocks, the pore geometry is not so simple that it can be described with simple equations, so the relationship between resistivity and pore volume must be determined empirically.

Rocks may be grouped into three general categories on the basis of their pore geometries. In some rocks, such as consolidated sedimentary rocks, volcanic ash beds, and so on, porosity is *intergranular* in nature, consisting of the space left over after the rock grains were compacted. In other rocks, and particularly in igneous rocks, porosity occurs primarily in the form of *joints*. A third form of porosity, common in limestones and in some volcanic rocks, is *vugular* porosity, consisting of large, irregular cavities formed either by solution, as in limestones, or by large gas bubbles, as in volcanic rocks.

Pore spaces must be interconnected and filled with water in order that a rock may conduct electricity. In all three types of porosity, the pore volume may consist of two parts; the larger voids, which are called *storage pores*, and the finer *connecting pores*. Most of the resistance to current flow is met in the connecting pores. A rock with a high ratio of storage pores to connecting pores, such as a rock with vugular porosity, will have a higher resistivity for the same porosity than a rock in which the opposite is true.

All three types of porosity are usually present in any rock in varying proportions. Sedimentary sandstones and shales may have from 5 to 60 per cent intergranular porosity, with the lower end of the range corresponding to Paleozoic quartzitic sandstones and the upper end to more recent unconsolidated muds. Such rocks will also have some joint porosity, though usually the amount is negligible compared to the amount of intergranular porosity, so that its effect on resistivity can usually be ignored.

Most igneous rocks contain a small amount of intergranular porosity; space left between the mineral grains as the rock cooled and contracted, but generally the total pore volume contributed by intergranular porosity is in the same range as the pore volume contained in joints. Joint porosity may be as large as 2 per cent of the total rock volume in heavily jointed rocks.

The ranges in porosity which may normally be expected in rocks are summarized in Table 7.

A great deal of work has been done in correlating resistivity with water content for petroleum-bearing rocks. For these rocks, which are primarily

TABLE 7

**Normal ranges in porosity for rocks**

| Rock type | Intergranular porosity (%) | Joint porosity (%) | Vugular porosity* (%) |
|---|---|---|---|
| Paleozoic sandstones and shale | 5–30 | 0–1 | 0 |
| Paleozoic limestones | 2–10 | 0–2 | 0 |
| Paleozoic clastic volcanics | 5–30 | 0–2 | 0 |
| Post-Paleozoic sandstones and shale | 10–40 | 0 | 0 |
| Post-Paleozoic limestones | 4–20 | 0–2 | 0 |
| Post-Paleozoic clastic volcanics | 10–60 | 0 | 0 |
| Precambrian sediments and low-rank metamorphosed sediments | 1–8 | 0–2 | 0 |
| Precambrian igneous rocks and high-rank metamorphic rocks | 0–2 | 0–2 | 0 |
| More recent igneous rocks | 0–10 | 0–2 | 0 |

\* Vugular porosity accounts for an appreciable total porosity only in rare cases.

porous sandstones and limestones, it has been observed that resistivity varies approximately as the inverse square of the porosity when the rock is fully saturated with water. This observation has led to the widespread use of an empirical function relating resistivity and porosity which is known as *Archie's law*:

$$\varrho = a\varrho_w\varphi^{-m} \tag{7}$$

where $\varrho$ is the bulk resistivity of the rock, $\varrho_w$ is the resistivity of the water contained in the pore structure, $\varphi$ is the porosity expressed as a fraction per unit volume of rock and $a$ and $m$ parameters whose values are assigned arbitrarily to make the equation fit a particular group of measurements. The value for the parameter $a$ varies from slightly less than 1 for rocks with intergranular porosity to slightly more than 1 for rocks with joint porosity. The exponent $m$ is somewhat larger than 2 for cemented and well-sorted granular rocks and somewhat less than 2 for poorly sorted and poorly comented granular rocks.

Equations of the form given in (7) that have been reported in the literature are tabulated in Table 8.

It is necessary to make a large number of measurements of both porosity and resistivity in order to determine the values of $a$ and $m$ with a good degree of reliability. Ordinarily, this is not practical, and for a first approximation, a value of 1 may be assumed for $a$ and a value of 2 for $m$. This simple inverse square relationship between resistivity and porosity will provide about the same answer as the more exact equations in the normal porosity range, 10 to 30 per cent. This may be seen by comparing the various curves plotted accord-

TABLE 8

**Expressions for Archie's law which have been reported in the literature**

| Formations for which equations were developed | Porosity range | Number of measurements | Equation |
|---|---|---|---|
| Frio sandstone (Oligocene) Bradford sandstone (Devonian) Woodbine sand (Cretaceous) Wilcox sand (Eocene) | 0·15–0·37 | 30 | $F = 0·62\,\varphi^{-2·15}$ |
| Pennsylvanian sandstone, Oklahoma | 0·08–0·20 | 97 | $F = 0·65\,\varphi^{-1·91}$ |
| Morrison sandstone (Jurassic), Colorado | 0·14–0·23 | 243 | $F = 0·62\,\varphi^{-2·10}$ |
| Clean Miocene sandstone, Weeks Island, Louisiana | 0·11–0·26 | 35 | $F = 0·78\,\varphi^{-1·92}$ |
| Clean Cretaceous sandstone, Paluxy sand, Texas | 0·08–0·25 | 50 | $F = 0·47\,\varphi^{-2·23}$ |
| Clean Ordovician sandstone, Simpson sand, Oklahoma | 0·07–0·15 | 44 | $F = 1·3\,\varphi^{-1·71}$ |
| Shaley sandstone (Eocene), Wilcox formation, Texas | 0·09–0·22 | 72 | $F = 1·8\,\varphi^{-1·64}$ |
| Shaley sandstone (Oligocene), Frio sands, Texas | 0·07–0·26 | 63 | $F = 1·7\,\varphi^{-1·65}$ |
| Shaley sandstone (Cretaceous), Taylor sand, Texas | 0·07–0·31 | 36 | $F = 1·7\,\varphi^{-1·80}$ |
| Oolitic limestone (Cretaceous), Pettit limestone, Texas | 0·07–0·19 | 13 | $F = 2·3\,\varphi^{-1·64}$ |
| Oolitic limestone (Jurassic), Smackover limestone, Ark. | 0·09–0·26 | 42 | $F = 0·73\,\varphi^{-2·10}$ |
| Siliceous limestone (Devonian), Texas | 0·07–0·30 | 58 | $F = 1·2\,\varphi^{-1·88}$ |
| Limestone (Cretaceous), Rodessa limestone, Texas | 0·08–0·30 | 37 | $F = 2·2\,\varphi^{-1·65}$ |

ing to Archie's law in Fig. 9. In practice the uncertainty in knowledge of the proper numerical values for $a$ and $m$ is less serious than the uncertainty in the choice of the proper value for $\varrho_w$, the resistivity of the water in the pores. Unless water samples are available from wells, vague relationships such as those presented earlier for the dependence of water salinity on geologic age and environment must be used. Even when water samples can be obtained, it is not always certain that the conductivity of the water is the same after the water is removed from the rock as before.

## 5b. Interaction between electrolytic solutions and the rock framework

In using Eq. (7) to relate the bulk resistivity of a rock to the porosity and the resistivity of the water in the pore space, the appropriate value for the water resistivity is not always the same as that which would be measured on a sample extracted from the pore space. Equation (7) indicates that the ratio of bulk resistivity to water resistivity should be a constant for a given porosity; that it should not depend on the resistivity of the water in the rocks. This ratio is called the *formation factor*. However, it is usually found that this ratio is less

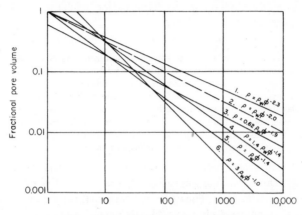

Ratio of bulk resistivity to water resistivity

FIG. 9. Graphical presentation of the various forms of Archie's law. The dashed curve is a simple inverse square relationship between resistivity and porosity which is usually a good approximation to the more exact expressions. Curve 6 applies to a model consisting of three sets of straight tubular pores, each set of pores being at right angles to the directions of the other two sets.

when a rock is saturated with a dilute solution than when the rock is saturated with a highly-saline solution. This may be explained by considering that the conductivity of the water distributed through the pore space is usually increased by two phenomena: ionization of clay minerals and surface conductance.

Clay minerals, such as kaolinite, halloysite, montmorillonite, vermiculite, illite, chlorite and others, have the property of sorbing certain anions and cations and retaining these in an *exchangeable state* (see Grim, 1953). The common exhangeable ions adsorbed on clay are Ca, Mg, H, K, Na and $NH_3$, in order of decreasing abundance. The quantity of exchangeable ions attached to a clay is usually expressed in terms of the weight of ions in milliequivalents absorbed per 100 g of clay. The exchange capacities of some common clays are:

| | |
|---|---|
| Kaolinite | 3 to 15 m-equiv/100 g |
| Halloysite . $2H_2O$ | 5 to 10 |
| Halloysite . $4H_2O$ | 40 to 50 |
| Montmorillonite | 80 to 150 |
| Illite | 10 to 40 |
| Vermiculite | 100 to 150 |
| Chlorite | 10 to 40 |
| Attapulgite | 20 to 30 |

Clay minerals are not the only materials exhibiting cation exchange capacity. All fine-grained minerals including quartz have an appreciable cation exchange capacity resulting from unsatisfied crystal bonds along the edges of grains. Exchange capacity is larger for finer grained particles. Zeolite minerals which are common in some volcanic rocks have cation exchange capacities of the order of 100 to 300 m-equiv per 100 g.

There are two principle causes for cation exchange properties in clays:

1. *Broken bonds* around the edges of the silica–alumina units in the crystal lattice contain unsatisfied ionic charges which are balanced by adsorbed ions. A single cation may be adsorbed on a clay mineral with a wide range of bonding energies.

2. Trivalent aluminium may *substitute* for quadrivalent silicon in the tetrahedral sheet structure of a clay mineral, leaving an unbalanced charge in the crystal. Ions of lower valence, such as magnesium, may substitute for aluminium in the lattice with the same effect. Exchangeable ions adsorbed by the unbalanced charges resulting from such substitution are usually found on the cleavage surfaces of the clay mineral.

In clay–water mixtures where there is more water than needed to make the clay plastic, the exchange ions may separate from the clay mineral in a process resembling ionization. The desorbed cations form a mobile cloud around the now negatively-charged clay particle, or *micelle*, as it is sometimes called. Only a portion of the adsorbed ions are likely to be desorbed, with the percentage desorbed depending on the particular clay mineral, the concentration of clay in water, the particular cation involved in the desorption and the concentration of ions already in the solution.

Since all rocks possess some exchange capacity, the conductivity of an electrolyte in a pore structure will always be increased by ions supplied by desorption. As an example, if a sandstone with 20 per cent pore volume contains 0·1 per cent by weight of sodium-charged montmorillonite clay with an exchange capacity of 100 m-equiv per 100 g, the resistivity of the pore water would be determined as follows if the rock were initially saturated with distilled water:

1. Determination of the weight of clay per cm³ of rock:

$$w = 0 \cdot 001 \times (1 - \varphi) \, \delta_m$$
$$= 0 \cdot 001 \times (1 - 0 \cdot 2) \times 2 \cdot 76$$
$$= 0 \cdot 0022 \text{ g}$$

where $\varphi$ is the porosity and $\delta_m$ is the density of the minerals forming the rock framework.

2. Determination of the quantity of exchangeable ions:

$$q = cw$$
$$= \frac{100 \times 0 \cdot 0022}{100 \text{ g}}$$
$$= 0 \cdot 0022 \text{ m-equiv}$$

where $c$ is the exchange capacity. Since one milliequivalent of sodium ion weighs 23·0 mg, the weight of sodium ion present is 0·508 mg.

3. Determination of the concentration of sodium ion in the pore water:

$$q/\varphi = 0·0508/0·2 = 0·254 \text{ m-equiv/ml}$$

4. Conversion of sodium ion content to an equivalent concentration of sodium chloride, using data on ion mobilities:

$$\frac{m_{Na}}{m_{Na} + m_{Cl}} \times \frac{q}{\varphi} = \frac{4·35 \times 10^{-8}}{4·35 \times 10^{-8} + 6·55 \times 10^{-8}} \times 0·254$$

$$= 0·102 \text{ mg per ml equivalent NaCl concentration}$$

Referring to Fig. 7, the resistivity of this equivalent solution is found to be about 6·0 ohm-m at a temperature of 18°C.

Not all the exchangeable cations would necessarily be desorbed from the clay, so that in fact, the conductivity of the pore water might be less than the value computed in this example. Moreover, the mobility of the desorbed ions is less than the mobility of the same ions in a free solution since they remain close to the negatively charged clay particle. The example does serve to show how little exchange capacity is required to lower the pore-water resistivity significantly. Even in rocks with very low exchange capacities, pore water resistivity rarely exceeds 10 ohm-m. In fine-grained rocks, such as shale, apparent pore water resistivities are always much lower than would be expected on the basis of a chemical analysis of water extracted from the rock. The added salinity is relatively unimportant if the normal salinity of the pore water is high, but if the pore water is dilute, it is practically impossible to predict the resistivity of the water in the pores unless the cation exchange capacity of the rock is known.

The nature of surface conduction is less well known, but the phenomenon is important in water-bearing rocks. Rock-forming minerals usually fracture in such a way that one species of ion in the crystal is commonly closer to the surface than others. In silicates, the oxygen ions are usually closest to the surface (see Fig. 10). When an electrolyte is in contact with such a surface, it will seem to the ions in solution that the surface is charged negatively. Cations will be attracted to the surface by coulomb forces and adsorbed, while anions will be repelled. A similar effect takes place with water molecules, which are *polar*. The water molecule is not symmetrically constructed from an oxygen atom and two hydrogen atoms; rather, there is an angle of 105° between the bonds from the oxygen to the two hydrogen atoms, as shown schematically in Fig. 11. As a result, the center of mass of the positive charges (the hydrogen atoms) does not coincide with the center of mass of the negative charges (the oxygen atom). When viewed from the oxygen side, the water molecule appears to carry a negative charge; when viewed from the hydrogen side, it appears to carry a positive charge.

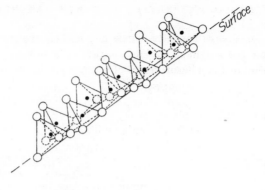

○  Oxygen

●  Silicon

Fig. 10.

Fig. 11.

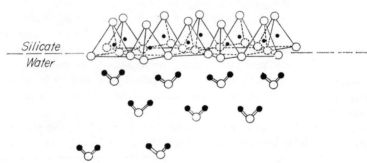

Fig. 12.

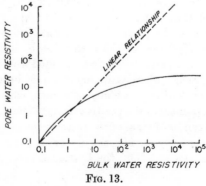

Fig. 13.

When an electrolyte is in contact with a surface such as that shown in Fig. 10, several layers of water molecules will become adsorbed to the surface, as shown in Fig. 12. This layer may be several molecules thick, if there are relatively few ions in the electrolyte, since each layer of oriented water molecules will absorb other water molecules. A single adsorbed cation will neutralize a surface charge which otherwise would hold several water molecules in a chain.

The conductivity of water in this oriented adsorbed phase is higher than the conductivity of free water, and so contributes to the overall conductivity of a rock. However, the increased pressure in the adsorbed layers increases the viscosity of the water and decreases the mobility of ions. If many ions are adsorbed, the conductivity of the electrolyte may be significantly reduced.

If all these factors that affect the conductivity of water in the pores of a rock are considered, the conductivity might behave as indicated in Fig. 13. The dotted curve shows the relationship which would hold between rock resistivity and water resistivity if there were no interaction between the water and the rock framework; the bulk resistivity of the rock should be directly proportional to the resistivity of the water placed in the pores. The solid curve shows the more probable relation between rock resistivity and water resistivity when interaction between the water and the solid minerals takes place; at high salinities (low resistivity), the resistivity of the electrolyte in the pores is increased slightly by the greater viscosity of the water adsorbed on the grains; at lower salinities (higher resistivities), the resistivity of the pore water approaches a maximum value rather than increasing indefinitely. The resistivity of the pore water cannot exceed some fairly low value, determined by the amount of interaction. Both the increase in resistivity at high salinities and the decrease in resistivity at low salinities is more pronounced in fine-grained rocks than in coarse-grained rocks. A limiting value of 10 ohm-m would be characteristic of a rock with a low cation exchange capacity, while the limiting value in a clay-rich rock may be as low as 0·1 ohm-m.

## 5 c. Resistivity of rocks only partially saturated with water

The pore space of a rock need not necessarily be filled with an electrolyte; in oil reservoir rocks, some of the pore space may be filled with oil or petroleum gas, while in near-surface rocks, part of the pore space may be filled with air. The second case is more common than the first.

Although the term *water table* has no precise scientific definition, it is usually used to indicate the depth at which the pore spaces remain fully saturated. The expression and its usage are misleading in that it infers that there exists a depth below which water is present and above which water is absent. In fact, the transition from complete saturation of the pore space to partial saturation above the water table is gradual and difficult to pinpoint. There is a transition from a surficial zone (which may be thousands of feet in arid regions) through which meteoric waters (those arriving from the atmosphere in the form of rain or snowfall) percolate in an erratic way to a region in which water content is static, or moving very slowly.

Circulation of water above the static water level involves several steps: (1) the infiltration of rain water from the surface into the soil or rock immediately beneath the surface during rainy periods; (2) the downward or lateral movement of this water through an aerated or partially saturated zone above the water table; and (3) return of the water to the atmosphere during dry periods or transpiration through plants. It is evident that if the permanent water table is stable (no water is being withdrawn from wells, etc.), all the water delivered to the zone of aeration during the rainy periods must be returned to the atmosphere during dry periods. There is nowhere else for it to go. This circulation of water in the zone of aeration is an important factor in determining near-surface resistivities.

The amount of water held in the zone of aeration changes slowly with time— evaporation or drainage of water through most soils is not rapid. Pore structures in soil may be very fine, so that capillary forces holding water in the fine pores of a soil are much larger than the gravitational force causing the moisture to drain downward. Frequently, it is found that the water content of the rock or soil above the water table depends more on grain size than the distance above the water table. Fine-grained zones hold more water than coarse-grained zones. In areas where the permanent water table is at a considerable depth, as it is in the southwestern United States, this leads to the phenomenon of a *perched* water table, or zones of complete saturation above the permanent water table. An example of the effect of a perched water table on near-surface resistivities is shown by the electric log in Fig. 14. The static water table in the area where these measurements were made is at least 200 ft deep. However, the electric log shows alternating zones of high and low resistivity above the water table. Sandstone lenses are partially desaturated and have a high resistivity while shaley zones retain full water saturation by capillarity, and so, have a low resistivity.

A quantitative relationship between resistivity and the extent of desaturation in a rock has been observed to hold:

$$\frac{\varrho}{\varrho_{100}} = S_w{}^{-n_1}; \quad S_w > S_{wc} \tag{8}$$

where $\varrho$ is the bulk resistivity of a partially desaturated rock, $\varrho_{100}$ is the resistivity of the same rock when completely saturated with the same electrolyte, $S_w$ is the fraction of the total pore volume filled with electrolyte, and $n_1$ is a parameter determined experimentally, and which usually has a value of approximately 2. Equation (8) holds providing the water content is greater than some critical value, $S_{wc}$, which depends on the texture of the rock. This critical saturation represents the least saturation for which there is a continuous film

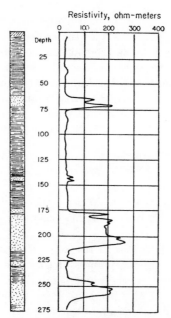

FIG. 14. Resistivity measurements which were made in a well penetrating a sequence of sandy and shaley beds lying above the permanent water table. The shaley beds are fully saturated with water while the sandstone is partially desaturated.

FIG. 15.

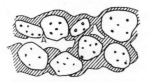

FIG. 16.

of water over all the surfaces in a rock. At high saturations, desaturation proceeds through the removal of small amounts of water from the centers of large pores, as shown in Fig. 15. This increases the overall resistance to current flow only moderately, since the resistance through connecting pores (where most of the resistance is met) is not affected. Once the critical saturation has been reached, further desaturation will break the continuous film of water over the grains, as at point A in Fig. 16, meaning that a small loss of water is accompanied by a large increase in resistance. At saturations below the critical saturation, the equation relating resistivity to the degree of saturation is:

$$\frac{\varrho}{\varrho_{100}} = aS_w^{-n_2}; \quad S_w < S_{uc} \tag{9a}$$

where $a$ and $n_2$ are parameters determined experimentally. The multiplying parameter, $a$, varies from about 0·05 for sandstones to about 0·5 for igneous rocks, while the exponent, $n_2$, has a value between 4 and 5.

The critical water saturation is about 25 per cent of total pore space for sandstones and similar permeable rocks, but may be as large as 70 to 80 per cent in igneous rocks.

In areas of much rainfall, the water in the zone of aeration may have a very low conductivity since over long periods of time the circulating water will leach the exchange ions from the clay minerals, leaving behind ions which do not readily dissociate from the clay particles. In arid climates, the amount of evaporation from the ground surface may be greater than the recharge from rainfall, the extra water coming from migration of water from the permanent water table. This migration of water upwards results in an increase in the salinity of the near surface water and in extreme cases, in the formation of caliche zones. In such areas, the increase in salinity of the water above the water table may more than offset the effect of partial desaturation, with the result that soils in arid areas are moderately conductive.

This leads to a paradox: in arid areas, soils tend to be more conductive than soils from humid areas. A typical soil from the southwestern United States will usually have a resistivity of 20 to 100 ohm-m, while soils from the north-central and Atlantic Coast areas will have resistivities from 200 to 1000 ohm-m.

## 5 d. Effect of temperature on the resistivity of water-bearing rocks

Extreme ranges in temperature may affect the resistivity of a water-bearing rock markedly, particularly if the temperature is high enough to drive water from the rock as steam or low enough to freeze the water in the pores of a rock. At moderate temperatures, a change in temperature changes the conductivity of a rock only in so far as the conductivity of the electrolyte in the rock is changed. The conductivity of aqueous electrolytes increases with increasing temperature since the viscosity of the water is decreased, in turn increasing the mobility of the ions. The dependence of resistivity on temperature for either an electrolyte or a rock saturated with an electrolyte is given by the equation:

$$\varrho_t = \frac{\varrho_{18°}}{1 + \alpha_t(t - 18°)} \tag{9b}$$

where $\varrho_{18°}$ is the resistivity measured at a reference temperature of 18°C (any other reference temperature may be used), $t$ is the ambient temperature and $\alpha$ is the temperature coefficient of resistivity, which has a value of about 0·025 per degree centigrade for most electrolytes.

Temperatures within the first few miles of the earth's crust rise gradually with depth at a rate of about 0·5°C per 100 ft in sedimentary rocks and about 0·2°C per 100 ft in igneous rocks. Temperature at a depth of 8000 ft in sedimentary rocks will be about 40° higher than the temperature at the surface, and this difference in temperatures means the rocks at a depth of 8000 ft have a resistivity only half as large as that which the same rock would have at the surface. At depths of 15,000 to 20,000 ft in sedimentary rocks, the temperature may be 100 to 150°C, with rock resistivity being a quarter or less of the value at surface temperatures.

The behavior of resistivity in rocks below the freezing point is important since about one-seventh of the earth's surface is frozen or permanently covered with ice or snow. One might expect that the resistivity of a frozen rock would be very high since the resistivity of ice is high. Curves relating the resistivity of ice to temperature and frequency are shown in Fig. 17. The low frequency

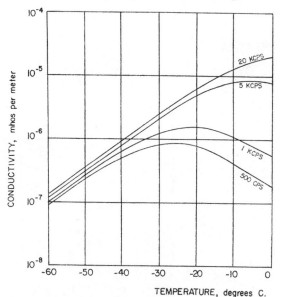

FIG. 17. The electrical conductivity of pure ice as a function of temperature and frequency.

resistivity of ice is very high, being about $10^1$ ohm-m, so ice should contribute no more to conduction in a rock than many of the common rock-forming minerals.

Measurements of rock resistivity at sub-freezing temperatures have shown that at $-12°C$, the resistivity of a rock is about 10 to 100 times larger than the resistivity measured at $18°C$. The fact that freezing has only a moderate effect on resistivity is explained by two factors:

1. Most ground water is moderately saline, and the presence of salts in solution lowers the freezing point. Moreover, if electrolytes are frozen slowly, they do not freeze uniformly. Salt ions migrate from the solidifying phase to the still liquid phase, increasing the salinity of this fraction and further lowering its freezing point. As a result, pockets of liquid brine remain in ice at temperatures down to $-60°C$. Freezing can be considered then as merely a reduction in the fraction of pore space which is saturated with water.

2. Pressure also lowers the freezing point of water. Water adsorbed on grain surfaces is under great pressure and will not reorient into ice crystals until the temperature is considerably below the normal freezing point. Also, as some water does freeze, it attempts to occupy a greater volume than it has in the liquid state, further increasing the pressure on the unfrozen water. Surface adsorption pressures would be expected to be greatest in fine-grained rocks with large surface areas exposed in the pore structure.

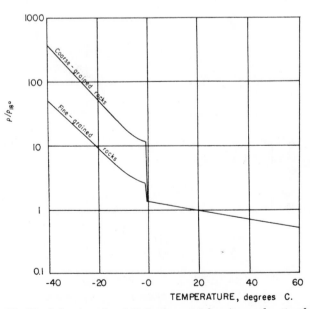

Fig. 18. The behavior of resistivity in water-bearing rocks at sub-freezing temperatures.

These two factors cause the freezing process in rocks to take place over an extended temperature range rather than at a single temperature. The higher the salinity of the pore water, the lower will be the temperature at which freezing first starts to take place, and the finer the grain size, the broader will be the temperature range through which freezing continues. Figure 18 shows the behavior of resistivity in granular rocks as the temperature is lowered below the freezing point.

It is unlikely that the water in any rock will be completely converted to ice at the temperatures prevailing on earth.

## 6. DESCRIPTION OF A GEOELECTRIC SECTION

So far, the electrical properties of single minerals and simple aggregates of minerals which form uniform rocks have been discussed. In geoelectric exploration, it is necessary to consider also the way in which electrical properties should be averaged over a large volume of rock which may not necessarily be homogeneous. Some rocks may have remarkably uniform properties through thousands of feet of section, while other rocks may consist of alternating layers with differing resistivities, with each layer being only a few inches thick. This poses the problem of how the average resistivity of a collection of heterogeneous rocks should be defined. It turns out that there are two important cases to be considered; that of a layered sequence of rocks such as a sedimentary section, and that of a faulted and jointed mass of dense rock, such as an igneous intrusion.

In discussing the electrical properties of a sequence of layered rocks, a distinction must be made between the *geoelectric section* and the *geologic section*. The geoelectric section differs from the geologic section in that boundaries between layers are determined by resistivity contrasts rather than by the combination of factors used by the geologist in establishing the boundaries between beds. The geologist attaches importance to such things as fossils in establishing formational boundaries, as well as on texture. Many of these things, particularly fossils, have no effect on the electrical properties of a rock, since the electrical character is determined primarily by texture and water content. Boundaries in the geoelectric section coincide with boundaries in the geologic section only when there is a pronounced change in texture at such a boundary. A single named geologic unit may consist of several rocks with quite different textures, and so, correspond to several units in the geoelectric section. The converse situation is also common; rocks covering a long period geologically may be uniform electrically, and all can be combined into a single unit in the geoelectric section.

The average electrical properties of each unit in a layered geoelectric section may be described with five parameters: the average resistivity along the bedding planes, $\varrho_l$; the total conductance in the direction of the bedding planes through a column 1 m², $S$; the average resistivity across the bedding planes, $\varrho_t$;

the total resistance through a 1 m² column cut perpendicular to the bedding planes, $T$; and a coefficient of anisotropy, $\lambda$. These geoelectric parameters may be defined in terms of a column of rock 1 m² cut from the geoelectric unit, as shown in Fig. 19. This column consists of $m$ horizontal beds, each with its own characteristic resistivity, $\varrho_i$, and thickness, $h_i$. The total thickness is $H$,

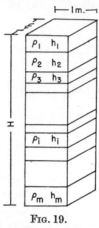

Fig. 19.

the thickness of the geoelectric unit. The total resistance presented to current flowing vertically through such a column is found simply by adding in series the resistance contributed by each individual layer. The resistance of a single layer, the $i$th layer, is found from the definition of resistivity to be:

$$R_i = \varrho_i \frac{l}{A} \tag{10}$$

where $l$ is the distance current flows in the $i$th layes or the thickness, $h_i$, and $A$ is the cross-sectional area presented for current flow. Since the column is 1 m², Eq. (10) reduces to:

$$R_i = \varrho_i h_i \tag{11}$$

The sum of resistances contributed by all the beds in the section is the transverse resistance, $T$:

$$T = \sum_{i=1}^{m} \varrho_i h_i \tag{12}$$

This is one parameter describing the geoelectric unit. A second parameter, the average resistivity to current flowing across the bedding planes (the transverse resistivity, $\varrho_t$) is found by dividing the transverse resistance, $T$, by the total thickness of the unit, $H$:

$$\varrho_t = T/H = \frac{\sum\limits_{i=1}^{m} \varrho_i h_i}{\sum h_i} \tag{13}$$

The conductance for current flowing horizontally through the column of rock shown in Fig. 19 is found by summing the conductances through each individual layer. For the $i$th layer, the conductance is:

$$1/R_i = \frac{1}{\varrho_i} \frac{A}{l}$$
$$= h_i/\varrho_i \qquad (14)$$

Designating the total conductance as $S$, it is found to be:

$$S = \sum_{i=1}^{m} \frac{h_i}{\varrho_i} \qquad (15)$$

The average conductivity for horizontal current flow is determined by dividing the total conductance by the height of the column:

$$\sigma_l = S/H \qquad (16)$$

The reciprocal of the average conductivity is the average longitudinal resistivity, $\varrho_l$:

$$\varrho_l = \frac{H}{S} = \frac{\sum h_i}{\sum h_i/\varrho_i} \qquad (17)$$

The longitudinal resistivity is always smaller than the transverse resistivity. This dependence of resistivity on the direction of current flow is *anisotropy*. A coefficient of anisotropy may be defined by taking the square root of the ratio of resistivities measured in the two principle directions, along the bedding planes and across the bedding planes:

$$\lambda = \sqrt{\frac{\varrho_t}{\varrho_l}} = \sqrt{\frac{ST}{H^2}} = \sqrt{\frac{\sum h_i\varrho_i \cdot \sum \dfrac{h_i}{\varrho_i}}{(\sum h_i)^2}} \qquad (18)$$

Electrical well logs provide virtually the only method for determining the electrical parameters describing a geoelectric section. In well logging, resistivity measurements are made while moving an array of electrodes through the borehole. Resistivities may be averaged over vertical distances ranging from several inches to some tens of feet with the electric logging equipment in common usage. If the individual beds comprising an anisotropic rock are thick enough to be shown on an electric log, the rock is considered to be anisotropic on a macroscopic scale, or *macroanisotropic*. Rocks may also be anisotropic on a microscopic scale, either consisting of thin layers of isotropic rock or being inherently anisotropic because of some preferential orientation of texture. The distinction between macroanisotropy and microanisotropy is vague, and there is no direct way to determine either the amount of microanisotropy or the total anisotropy of a layered rock.

Normal ranges in macroanisotropy are listed in Table 9.

TABLE 9

### Coefficients of anisotropy for layered rock

| Rock type | Coefficient of anisotropy |
|---|---|
| Volcanic tuff, Eocene and younger, from Nevada | 1·10–1·20 |
| Alluvium, thick sections from the southwestern United States | 1·02–1·10 |
| Interbedded limestones and limey shales from northeastern Colorado | 2·0 –3·0 |
| Interbedded anhydrite and shale, northeastern Colorado | 4·0 –7·5 |
| Massive shale beds | 1·02–1·05 |
| Interbedded shale and sandstone | 1·05–1·15 |
| Baked shale or low-rank slate | 1·10–1·60 |
| Slates | 1·40–2·25 |
| Bitumenous coal and mudstone | 1·7 –2·6 |
| Anthracite coal and associated rocks | 2·0 –2·6 |
| Graphitic slate | 2·0 –2·8 |

Consider an example of the computation of the various parameters describing a geoelectric unit. Assume that the unit consists of an alternating series of beds with a total thickness of 100 m, the individual beds being isotropic, one meter thick and with resistivities alternating between 50 and 200 ohm-m.

The transverse resistance of the unit is:

$$T = \sum \varrho_i h_i = 50 \times 50 + 200 \times 50 = 12{,}500 \text{ ohm-m}.$$

The average transverse resistivity is:

$$\varrho_t = T/H = 12{,}500/100 = 125 \text{ ohm-m}.$$

The longitudinal conductance is:

$$S = \sum \sigma_i h_i = 50 \times \frac{1}{50} + 50 \times \frac{1}{200} = 1 \cdot 25 \text{ mhos}.$$

The average longitudinal resistivity is:

$$\varrho_l = H/S = 100/1 \cdot 25 = 80 \text{ ohm-m}.$$

The coefficient of macroanisotropy is:

$$\lambda = \sqrt{\frac{\varrho_t}{\varrho_l}} = \sqrt{\frac{125}{80}} = 1 \cdot 25$$

Many igneous and metamorphic rocks may show a layered or zoned electrical structure similar to the electrical layering found in sedimentary rocks. Volcanic rocks frequently are layered, whereas intrusive rocks such as granite and gabbro less commonly have zoned textures to the degree that they may be thought of as layered. Metamorphic rocks may be electrically layered. In any of these cases, the rock may be described with the various geoelectric parameters defined above. If a rock is not layered, but has random variations in resistivity, the parameters defined for a layered sequence are of no value, and in fact, there is no way to represent such a rock. Random resistivity variations are not common, and usually when a rock is not layered, it is a dense rock whose bulk properties are determined largely by a set of faults or joints. Such a rock may be described in terms of average resistivities and patterns of anisotropy if the directions by faulting and jointing can be described.

Consider a simple model of a jointed rock in which the conduction takes place through a randomly oriented set of fractures. Random orientation is assured if in counting the density of fractures with traces falling in some range of angles $\Delta\alpha$, the density of fracture traces is the same for any angular interval throughout a full circle. If current is assumed to flow in the direction indicated in Fig. 20, the difference in direction of current flow and any particular fracture trace can be designated by two angles: $\beta$ indicating the angle between the projection of the fracture trace on the plane of the paper in Fig. 20 and the direc-

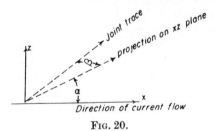

FIG. 20.

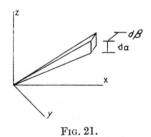

FIG. 21.

tion of current flow, and $\alpha$ indicating the angle between the fracture trace and its projection on the paper. In a small sector with dimensions $\Delta\alpha$, $\Delta\beta$, and radius $1/\cos\alpha\cos\beta$, as shown in Fig. 21, the fraction of the total porosity which is present is:

$$\Delta\varphi = \frac{\Delta\beta \cdot \Delta\alpha}{\pi^2}\,\varphi\,\cos\alpha\,\cos\beta \tag{19}$$

Assuming the fractures are filled with a material having a resistivity, $\varrho_w$, the conductance for the fractures in the segment shown in Fig. 21 is:

$$\varDelta\left(\frac{1}{R}\right) = \frac{\varphi \cos\beta \, \cos\alpha \, d\alpha \, d\beta}{\pi^2 \varrho_w} \tag{20}$$

The conductance along joints oriented in all conceivable directions may be found by integrating this expression with angles $\alpha$ and $\beta$ varying from $-\pi/2$ to $\pi/2$:

$$\frac{1}{R} = \int\limits_{-\pi/2}^{\pi/2}\!\!\int \frac{\varphi \cos\beta \, \cos\alpha \, d\alpha \, d\beta}{\pi^2 \varrho_w}$$

$$R = \frac{\pi^2}{4} \varrho_w \varphi^{-1}$$

$$= 2 \cdot 46 \, \varrho_w \varphi^{-1} \tag{21}$$

In this simple model, the electrical properties are isotropic inasmuch as it was assumed that the fracture traces were randomly oriented. Usually, a rock fracture system has several preferred directions. This can be taken into account in Eq. (19) by considering that the fraction of the total porosity present in any particular sector depends on the radial directions of the sector boundaries. For example, consider a model in which there are three joint systems, all mutually perpendicular, and each system accounting for one third of the total porosity in the rock. If the direction of current flow is along one of the joint directions, the other two systems of joints will not contribute to electrical conduction. The resistivity measured in a direction parallel to any one of the three joint directions will be:

$$\varrho = 3 \, \varrho_w \varphi^{-1} \tag{22}$$

If the direction of current flow is at an angle of 45° to two of the joint directions, but still perpendicular to the third, the resistivity will be:

$$\varrho = 2 \cdot 1 \, \varrho_w \varphi^{-1} \tag{23}$$

A minimum resistivity results when an equal amount of current flows through all three sets of joints. A pattern of anisotropy may be constructed by plotting resistivity as a function of direction as shown in Fig. 22. Figure 22(a) shows a pattern for an isotropic rock. The pattern, shown in Fig. 22(a), is a circle since the resistivity does not depend on the direction in which it is measured. The pattern for a layered rock is shown in Fig. 22(b). It is an ellipse, as will be proven in Chapter III, with a major axis equal in length to the numerical value of the transverse resistivity, and with a minor axis equal in length to the numerical value for the longitudinal resistivity. The pattern for a rock with three orthogonal fracture systems is shown in Fig. 22(c) for the case in which the current flow is always at right angles to one of the fracture systems. This pattern shows a maximum resistivity each time the direction of current flow is normal to one of the fracture directions. Sets of fracture directions do not need to be orthogonal in actual rocks, so the anisotropy patterns need not be symmetrical as the one shown in Fig. 22(c). However, it is reasonable to expect that in jointed rocks a maximum resistivity will be observed whenever current flow is directed normal to one of the joint systems, so that the porosity contained in that system does not contribute to conduction.

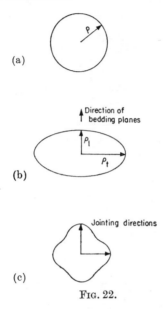

Fig. 22.

## 7. RESISTIVITY SUMMARY

Conduction in near-surface rocks is almost entirely through the water filling pore spaces in rocks. Conduction through mineral grains is important only in rare cases in surface rocks, such as when large concentrations of minerals such as magnetite, graphite or pyrrhotite are found, or at depth within the earth, where pore structures in the rock are closed by overburden pressure.

In water-bearing rocks, there is an indirect relationship between resistivity and lithology or geologic age, since these two factors tend to control the porosity or water storage capacity of a rock, and to a lesser extent, the salinity of the water contained in a rock. The dependence of resistivity on age and lithology is indicated in a general way in Table 10. The vertical columns are arranged in decreasing order of porosity from left to right, starting with marine sedimentary rocks, which may have porosities as high as 40 per cent, and ending with chemical precipitations, which have essentially no water content. There are many exceptions to the ranges indicated on this table, particularly if the porosity

TABLE 10

## Generalized resistivity ranges for rocks of different lithology and age

| Age | Marine sedimentary rocks | Terrestrial sedimentary rocks | Extrusive rocks (basalt, rhyolite) | Intrusive rocks (granite, gabbro) | Chemical precipitates (limestone, salt) |
|---|---|---|---|---|---|
| Quaternary and Tertiary age | 1–10 | 15–50 | 10–200 | 500–2000 | 50–5000 |
| Mesozoic | 5–20 | 25–100 | 20–500 | 500–2000 | 100–10,000 |
| Carboniferous Paleozoic | 10–40 | 50–300 | 50–1000 | 1000–5000 | 200–100,000 |
| Early Paleozoic | 40–200 | 100–500 | 100–2000 | 1000–5000 | 10,000–100,000 |
| Precambrian | 100–2000 | 300–5000 | 200–5000 | 5000–20,000 | 10,000–100,000 |

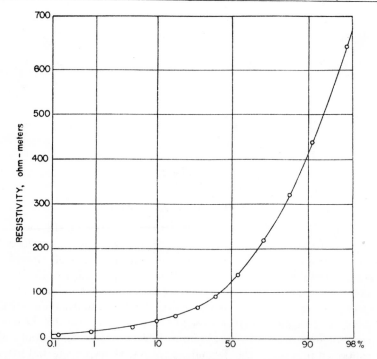

FIG. 23. Cumulative frequency curve for about 7000 resistivity values determined around radio stations. The horizontal scale is a linearized probability scale such that if the distribution of resistivity values were distributed normally (in the statistical sense), the cumulative frequency curve would plot as a straight line.

of normally porous rocks is decreased by metamorphism or if conductive minerals occur in high enough concentrations to lower the resistivity of otherwise resistant rocks.

The correlation between resistivity and age in sedimentary rocks is apparent from an inspection of reported values of earth resistivities measured in the United States. The most extensive tabulation of such measurements is a listing of earth resistivities determined about radio stations, published by the U.S. National Bureau of Standards (NBS Circular 546). These earth resistivity values are estimated from the rate of decay in field strength about a radio transmitter. Such surveys are required of all broadcast stations at frequent intervals by the Federal Communications Commission. Resistivities determined from

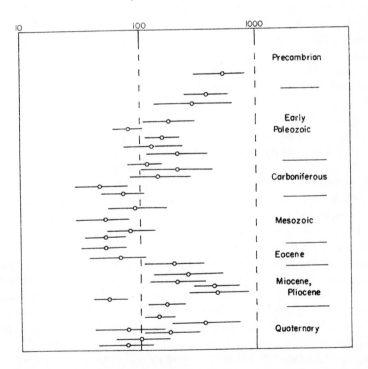

FIG. 24. Average resistivity for groups of sedimentary rocks with similar characteristics. Each circle indicates an average resistivity determined from 35 to 250 measurements about radio stations. The bar through each circle indicates the range within which 95 per cent of the values for that group of data fall.

these field-strength surveys represent the average values for rocks over an area of a few square miles about the broadcasting station to a depth of about a hundred feet.

A cumulative frequency curve for slightly over 7000 such resistivity values is shown in Fig. 23. The median resistivity for these 7000 values is 143 ohm-m. Half of the total number of rocks used for these measurements had resistivity values less than 143 ohm-m.

Selecting from these 7000 measurements those made over sedimentary rocks, the correlation between average resistivity and geologic age shown in Fig. 24 was obtained. Apparently, resistivity is not a simple function of age. Rather, maximum values are observed for rocks deposited during Precambrian and Cenozoic time, and minimum values are observed for rocks deposited during Mesozoic and Quaternary time. These variations probably reflect control by a combination of depositional environment and post-depositional changes in porosity and water content associated with lithification and diastrophism.

Some Quaternary alluvial deposits, particularly those in desert areas, have low resistivity, while alluvial deposits in humid areas have high resistivity (see listing of average resistivities in Table 11). The alluvial deposits have a large amount of pore space. In desert areas, the water in such rocks tends to be more saline than that in the rocks in more humid areas, because of the concentration of salts by evaporation and recharge of surface waters.

Miocene and Pliocene rocks in the United States are primarily fresh- and brackish-water deposits. Their resistivities are greater than those for similar Quaternary sediments because of the decrease in porosity on compaction.

The very low resistivities observed for rocks deposited during the last part of the Paleozoic and during the Mesozoic era reflect the fact that primarily marine rocks were formed during these eras, with deposition changing to mainly fresh-water facies in the early part of the Cenozoic era. Paleozoic and Mesozoic rocks are generally more compacted and have less pore space than the middle to late Cenozoic rocks, but the salinity of the connate water is such that the rocks are better conductors.

Resistivities become progressively larger as earlier Paleozoic and Precambrian times are considered. This is in part due to the greater intensity of distrophism since these early times and in part a result of a lithologic transition from predominantly fine-grain clastic deposits during the Carboniferous to predominantly chemical precipitate deposits during the Devonian and Silurian periods, with a corresponding decrease in porosity at the time of deposition.

Using the data listed in Table 11, a map showing the general distribution of high, low and moderate near-surface resistivities in the United States may be constructed (Fig. 25). The highest resistivities are found in the New England states, where intensely metamorphosed Precambrian rocks are exposed. The lowest resistivities are found in a region stretching through the mid-continent

FIG. 25. Areas of high, moderate and low near-surface resistivity in the United States, as indicated by resistivity measurements near radio stations.

from Texas to the high plains, where thick sections of Mesozoic shales form the sub-outcrop. The igneous core of the Rocky Mountain area has high resistivity, though not as high as that of the metamorphic rocks in New England.

## 8. DIELECTRIC CONSTANT AND ELECTRIC POLARIZATION

When alternating current methods are used to study earth resistivity, the data obtained in the field depend to a minor degree on the dielectric constant of the earth. In developing the theory for the behavior of alternating-current electric fields, it is necessary to know the dielectric constant. However, dielectric constant is only of secondary importance, and so, needs to be considered only briefly.

Electric polarization, the phenomenon involved in dielectric constant measurements, differs from electric current flow in that charges in a material travel only a short distance when an electric field is applied and then stop, rather than moving continuously as in conduction. The amount of polarization which takes place is measured in terms of the amount of charge which is separated and the distance through which it moves. Conduction takes place only in materials which have charge carriers which are free to move; polarization takes place in all materials.

TABLE 11

## Summary of resistivities measured about radio stations

| Area and age of rocks | Formation names | Lithology | Average resistivity (ohm·m) | 95% range in resistivity |
|---|---|---|---|---|
| Quaternary alluvium, Pacific coast | | Alluvium, lake deposits, beach sand, glacial drift | 75 | 43–130 |
| Quaternary alluvium, Rocky Mountains | Gila conglomerate | Alluvium, bolson deposits, lake beds, windblown sands | 99 | 58–170 |
| Quaternary alluvium, Mississippi Valley | | Alluvium | 182 | 105–308 |
| Quaternary sedimentary rocks, Gulf Coast states | Beaumont clay, Lissie formation | Sand and gravel | 78 | 40–155 |
| Quaternary sedimentary rocks, Atlantic coast states | | Sand and gravel | 340 | 182–645 |
| Quaternary limestones, Atlantic and Gulf coast states | Anastasia formation, Miami oolite, Key Largo limestone | Limestone | 143 | 106–193 |
| Tertiary sedimentary rocks, Pacific coast states (terrestrial) | | Terrestrial sediments, volcanics | 275 | 172–440 |
| Tertiary marine sedimentary rocks, Pacific coast | | Sandstone and shale | 167 | 114–247 |
| Tertiary volcanics, Rocky Mountains | Challis volcanics, Hinsdale formation, San Juan tuff, Silverton series, Potosi series | Latite lava, basalt, andesite, rhyolite, breccias | 167 | 114–247 |
| Miocene and Pliocene sedimentary rocks, Rocky Mountains | Santa Fe formation | Terrestrial sediments, lake beds | 80 | |

TABLE 11 (continued)

| Area and age of rocks | Formation names | Lithology | Average resistivity (ohm-m) | 95% range in resistivity |
|---|---|---|---|---|
| Miocene and Pliocene sedimentary rocks, Great Plains | Ogallala formation Arikaree formation | Channel deposits, sand, silt and gravel | 55 | 38–82 |
| Miocene sedimentary rocks, Gulf Coast states | Alum Bluff group Choctawhatchee fm. Hattiesburg clay Oakville formation | Fresh-water marls, sand, silt and gravel | 480 | 263–830 |
| Miocene sedimentary rocks, mid-Atlantic coast | Yorktown formation Duplin marl | Unconsolidated sand, clay, coquina rock, shell beds | 457 | 320–620 |
| Miocene sedimentary rocks, south Atlantic coast | Tampa limestone Catahoula sandstone | Limestone, sandstone | 209 | 119–357 |
| Miocene and Pliocene volcanics, Pacific northwest | Columbia River basalt Cascade andesite Yakima basalt Wenas basalt | Basalt, andesite, rhyolite flows and tuff | 257 | 132–500 |
| Oligocene sedimentary rocks, Great Plains | White River group Castle Rock conglomerate | Terrestrial clays, silts, and sandstones | 103 | |
| Eocene and Oligocene, sedimentary rocks, Gulf and Atlantic coast | Vicksburg group Jackson group Claiborne group Wilcox group Midway group | Marl, limestone, sand, and clay | 189 | 104–340 |
| Eocene sed.mentary rocks, Great Plains and Rocky Mountains | Green River formation Wasatch formation Denver formation Arapahoe formation Fort Union formation | Shale and limestone | 63 | 38–108 |

TABLE 11 (continued)

| Area and age of rocks | Formation names | Lithology | Average resistivity (ohm-m) | 95% range in resistivity |
|---|---|---|---|---|
| Upper Cretaceous sedimentary rocks, Great Plains | Montana group<br>Pierre shale<br>Foxhills sandstone<br>Laramie formation<br>Niobrara formation<br>Benton shale<br>Dakota sandstone | Shale, sandstone, lignite, chalk, calcareous shale | 49 | 33–71 |
| Cretaceous sedimentary rocks, Rocky Mountains | Mesa Verde formation<br>Colorado shale<br>Dakota sandstone<br>Morrison formation | Shale and sandstone | 80 | 53–143 |
| Cretaceous sedimentary rocks, Gulf and Atlantic coast | Ripley formation<br>Selma chalk<br>Eutah formation<br>Tuscaloosa formation | Marine sandstones, marls, clay, chalk | 410 | 130–1300 |
| Cretaceous sedimentary rocks, Texas and Oklahoma | Navarro formation<br>Taylor marl<br>Austin chalk<br>Eagle Ford formation<br>Woodbine formation | Marine sandstones, marls, chalk, clay | 48 | 28–82 |
| Lower Cretaceous sedimentary rocks, Texas and Oklahoma | Washita group<br>Fredericksburg group<br>Trinity group | Sandstone, anhydrite, limestone | 95 | 48–188 |
| Triassic rocks, New England states | Chicopee shale<br>Granby tuff<br>Longmeadow sandstone<br>Mount Toby conglomerate<br>Sugarloaf arkose | Intrusive diabase, basalt, tuff, shale, sandstone, conglomerate, arkose | 613 | 310–1220 |

TABLE 11 (continued)

| Area and age of rocks | Formation names | Lithology | Average resistivity (ohm-m) | 95% range in resistivity |
|---|---|---|---|---|
| Permian sedimentary rocks, mid-continent | Cloud Chief formation<br>Duncan formation<br>Woodward group<br>Enid formation<br>Wichita formation<br>Cottonwood limestone | Dolomite, limestone, gypsum, salt, shale, anhydrite, sandstone | 48 | 29–88 |
| Pennsylvanian sedimentary rocks, mid-continent | Pontotoc group<br>Nelagoney formation<br>Ochelata formation<br>Seminole conglomerate<br>Holdenville shale<br>Wetumka shale<br>Calvin sandstone | Sandstone and shale | 70 | 46–105 |
| Pennsylvanian sedimentary rocks, Great Lakes and northeast states | Monongahela formation<br>Conemaugh formation<br>Allegheny formation<br>Pottsville group | Sandstone, shale, coal, limestone, iron ore | 154 | 88–270 |
| Mississippian sedimentary rocks, midwestern states | Chester age<br>Meramec age<br>Osage age<br>Kinderhook age | Sandstone, shale, salt, coal, gypsum, dolomite, anhydrite, limestone | 230 | 109–490 |
| Mississippian sedimentary rocks, Ohio and Indiana | Meramec age<br>Osage age | Sandstone, shale, limestone | 116 | 83–135 |
| Carboniferous granite, Appalachians | | Granite intrusions | 420 | 313–555 |

TABLE 11 (continued)

| Area and age of rocks | Formation names | Lithology | Average resistivity (ohm-m) | 95% range in resistivity |
|---|---|---|---|---|
| Carboniferous and Devonian rocks, New England states | Cambridge slate<br>Roxbury conglomerate<br>Mattapan volcanics<br>Dighton conglomerate<br>Rhode Island formation<br>Wamsutta formation<br>Worcester phyllite | Volcanics slate, conglomerate, phyllite | 680 | 340–1380 |
| Upper Devonian sedimentary rocks, midwest and northeast states | Hamilton age<br>Marcellus age<br>Onondaga age<br>Helderberg age | Sandstone, shale, limestone | 135 | 75–244 |
| Upper Devonian sedimentary rocks, midwest and New England | Portage age<br>Catskill age | Black shale, sandstone | 223 | 127–395 |
| Silurian sedimentary rocks, New England states | Cayuga age<br>Lockport age<br>Clinton age | Limestone, dolomite, shale | 162 | 111–233 |
| Silurian sedimentary rocks, midwestern states | Cayuga age<br>Lockport age<br>Clinton age | Limestone, dolomite, shale, calcareous shale | 82 | 63–114 |
| Ordovician sedimentary rocks, northeastern states | Maysville age<br>Eden age<br>Trenton age<br>Chazy age<br>Beekmantown limestone | Shale, limestone, dolomite | 185 | 111–313 |
| Ordovician and Cambrian sedimentary rocks, Great Lakes area | St. Peter sandstone<br>Prairie du Chien dolo.<br>St. Croixan rocks | Sandstone, limestone, dolomite, conglomerate | 213 | 132–345 |

TABLE 11 (continued)

| Area and age of rocks | Formation names | Lithology | Average resistivity (ohm-m) | 95% range in resistivity |
|---|---|---|---|---|
| Ordovician and Cambrian rocks, western New England | Taconic sequences | limestone, shale, quartzite, sandstone | 303 | 147–525 |
| Cambrian rocks, Appalachian area | Conococheague ls. Honaker limestone Potsdam sandstone Waynesboro shale | | 385 | 340–1380 |
| Algonkian and Archean rocks, New England states | | Phyllite, schist, gneiss | 1200 | 250–590 |
| Keewenawan and Huronian rocks, Great Lakes area | Portage Lake series | Sandstone, conglomerate, basalt, rhyolite, slate, iron formation | 185 | 890–1670 |
| Algonkian and Archean rocks, Appalachian area | Wissahicken schist Cockeysville marble Setters formation | Schist, phyllonite, marble volcanics, limestone, granite | 500 | 119–389 |
| Algonkian and Archean rocks, southern Appalachians | | Granite, gabbro, gneiss, metabasalt, aporhyolite | 530 | 312–813 |
| Algonkian rocks, Montana | Wallace formation Helena limestone | Limestone, shale, quartzitic limestone | 320 | 323–870 |

Consider a hypothetical experiment in which a parallel-plate condenser is placed in a vacuum and connected to a battery as shown in Fig. 26. Charge

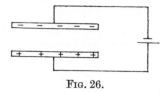

FIG. 26.

will flow from the battery and collect on the plates of the condenser as shown, the charge being held by coulomb forces operating across the insulating gap. In

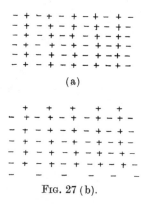

(a)

FIG. 27 (b).

such an experiment, if the charge collected on the condenser plates were measured, it would be found to be proportional to the area of the condenser plates and inversely proportional to the separation between them:

$$C = \varepsilon_0 \frac{A}{l} \tag{24}$$

where $C$ is the capacity (charge in coulombs stored per volt applied to the condenser), $A$ is the area of the condenser in meters and $l$ is the separation between the plates, also measured in meters. The constant of proportionality, $\varepsilon_0$, would be found to be $8 \cdot 85 \times 10^{-12}$ F/m. This constant is called the *permittivity of free space*.

If now in the hypothetical experiment described above a slab of some material substance is inserted between the plates of the condenser in Fig. 26, every charged particle in the material will be subjected to coulomb forces from the charges on the condenser plates. Electron orbits are shifted slightly towards the positive plate of the condenser and proton-bearing nuclei are shifted towards the negative plate. This separation of charge constitutes electric polarization of the material.

This polarization results in the collection of charge on the surface of the material but not on the interior (see Fig. 27). In Fig. 27 a, a neutral configuration of charges such as might exist in a non-polarized sample is shown. If an electrostatic field is applied so that the positive charges are shifted up one row and the negative charges down one row, the charge distribution shown in Fig. 27 b is obtained. The upper and lower surfaces accumulate an excess charge of one sign or the other, but the interior of the medium retains an electrically neutral charge configuration.

Displacement of an electron orbit takes place in a very short period of time, usually less than $10^{-9}$ sec, so that at any frequency used in electrical prospecting, it is safe to assume that electron polarization has time to take place.

The amount of polarization which takes place when an electric field is applied to a material is measured in terms of the surface charge density which accumulates per unit applied voltage gradient. In the hypothetical experiment described above (Fig. 26), if a slab of material is inserted between the condenser plates, a surface charge will collect on the surfaces of the material next to the condenser plates. This surface charge attracts an additional charge from the battery to the condenser plates, as shown in Fig. 28. The charge, $Q$, originally

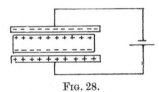

Fig. 28.

held on the condenser plates when the slab was not present is increased by an amount, $q$, equal to the surface charge on one surface of the slab. Again, it is found that the total charge held in the condenser is proportional to the area of the plates and inversely proportional to the separation, if the slab fills the space between the plates:

$$Q + q = \varepsilon \frac{A}{l} \tag{25}$$

where the constant of proportionality, $\varepsilon$, is found to be larger than the one measured with the condenser in a vacuum. This constant is measured in farads per meter and is called the *dielectric constant* for the material. Frequently, the ratio $\varepsilon/\varepsilon_0$ is used to indicate the relative degree of polarizability in a material. This ratio is called the *relative dielectric constant*, and is usually designated by the symbol, $K$.

The displacement of electron orbits is not the only mechanism by which polarization takes place. Ionic polarization takes place in materials with ionic bonds. In some materials, reorientation of polar molecules in an electric field is another mechanism by which polarization can take place, and in conducting materials, polarization may take place when mobile charges collect at boundaries where their mobility is reduced.

These mechanisms require greater operational times than does electron polarization. At very high frequencies, the polarizing charges do not have time to separate completely, and their motion merely constitutes an electric current which adds to the conductivity of the medium. As a result, both the conductivity and dielectric constant depend on frequency, as indicated by the curves in Fig. 29.

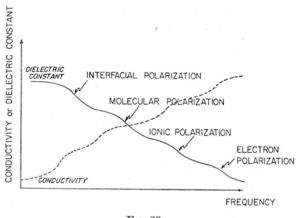

FIG. 29.

Ionic polarization takes place in materials with ionic bonds. When an electric field is applied across such a material, the cations are drawn towards the negative pole and the anions are drawn toward the positive pole. The ions do not move far, since a cation being displaced in one direction will meet an anion being displaced in the opposite direction. Most rock-forming minerals are ionic bonded and exhibit this type of polarization. Minerals formed from high-valency elements, such as iron and titanium, have the highest dielectric constants, since such ions are displaced further than ions with lower charges. However, values for dielectric constants in ionic bonded crystals do not cover a wide range, the ratio $\varepsilon/\varepsilon_0$ being between 4 and 10 for most minerals. Table 12 lists values for the dielectric constant of many common minerals measured at a sufficiently high frequency that the only polarization contributing to the dielectric constant is ionic or electronic in origin.

Some materials are made up of molecules in which the geometric center of the negative charges does not coincide with the geometric center of the positive charges. When an electric field is applied across such a material, these molecules tend to rotate about their centers of gravity until the negative side faces the

TABLE 12

## Dielectric constants of minerals (electron and ion polarization only)

| Mineral | Dielectric constant (pF/m)* |
|---|---|
| Galena, Pbs | 158 |
| Sphalerite, ZnS | 69·7 |
| | |
| Corundum, $Al_2O_3$ | 97·2 to 117 |
| Cassiterite, $SnO_2$ | 207 to 212 |
| Hematite, $Fe_2O_3$ | 221 |
| Rutile, $TiO_2$ | 274 to 1500 |
| Water, $H_2O$ | 721 |
| | |
| Halite, NaCl | 50·4 to 54·8 |
| Fluorite, $CaF_2$ | 55·4 to 60·0 |
| Sylvite, KCl | 38·8 to 54·8 |
| | |
| Aragonite, $CaCO_3$ | 57·1 to 86·0 |
| Calcite, $DaCO_3$ | 69·0 to 75·2 |
| Dolomite, $CaMg(CO_3)_2$ | 60·1 to 70·7 |
| | |
| Apatite, $Ca_5(F, Cl)(PO_4)_3$ | 65·5 to 92·8 |
| | |
| Anglesite, $PbSO_4$ | 644 to 4400 |
| Anhydrite, $CaSO_4$ | 50·5 to 55·7 |
| Barite, $BaSO_4$ | 69·5 to 109 |
| Celestite, $SrSO_4$ | 67·2 |
| Gypsum, $CaSO_4 . 2H_2O$ | 47·7 to 106 |
| | |
| Analcime, $NaAlSi_2O_6 . H_2O$ | 52·0 |
| Augite, $Ca(Mg, Fe, Al)(Al,Si)_2O_6$ | 61·0 to 76·0 |
| Beryl, $Be_3Al_2Si_6O_{18}$ | 50·2 to 58·3 |
| Biotite, $K(Mg, Fe)_3AlSi_3O_{10}(OH)_2$ | 54·8 to 82·3 |
| Epidote, $Ca(Al, Fe)_3(SiO_4)_3OH$ | 67·3 to 136 |
| Leucite, $KAlSi_2O_6$ | 63·0 |
| Muscovite, $KAl_3Si_3O_{10}(OH)_2$ | 54·8 to 70·7 |
| Orthoclase, $KAlSi_3O_8$ | 39·8 to 51·3 |
| Plagioclase | |
| var. albite | 48·2 to 49·1 |
| var. oligoclase | 53·3 to 53·6 |
| var. andesine | 54·8 to 57·2 |
| var. labradorite | 57·6 to 58·5 |
| var. anorthite | 62·3 to 64·0 |
| Quartz, $SiO_2$ | 36·4 to 37·8 |
| Sericite | 173 to 224 |
| Topaz, $Al_2SiO_4(F, OH)_2$ | 55·7 to 59·2 |
| Zircon, $ZrSiO_4$ | 76·0 to 106 |

* pF, picofarad or $10^{-12}$ farad.

positive pole of the field and the positive side faces the negative pole. The rotation of these molecules constitutes an electric current until all of the molecules have rotated into alignment with the polarizing field. The alignment results in a net surface charge, or polarization.

The length of time required for the dipole molecules to rotate into the polarized orientation depends on the state of the material. In a liquid, molecules may rotate rapidly; in a solid, rotation may be very slow.

The only common polar material in rocks is water. In the liquid state, water molecules become aligned with an electric field in about a thirtieth of a microsecond. The amount of polarization is large; water has a relative dielectric constant, $\varepsilon/\varepsilon_0$, of about 81·5 at low frequencies; at frequencies in excess of 30 Mc/s the constant is reduced to 4. In ice, water molecules take much longer to align with a polarizing field and as a result, molecular polarization is apparent only at relatively low frequencies. Figure 30 shows curves relating the dielectric constant of ice to frequency at various temperatures. At lower temperatures, the rigid structure of ice reduces the rate at which molecules can rotate.

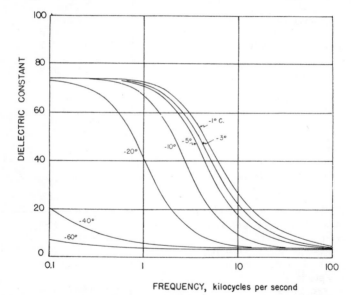

FIG. 30. The relative dielectric constant of ice as a function of temperature and frequency.

Other materials which exhibit molecular polarization include hydrocarbons such as are found in petroleum and coal. These materials occur infrequently in the earth.

A rock is a mixture of minerals, usually with some water, so the way in which dielectric constants combine in mixtures is important in determining the bulk dielectric constant for a rock. A logarithmic mixing rule is sometimes used to predict the dielectric constant for a mixture of materials:

$$\log \varepsilon = V_1 \log \varepsilon_1 + V_2 \log \varepsilon_2 \tag{26}$$

where $V_1$ and $V_2$ are the volume fractions of two materials forming a composite dielectric, $\varepsilon_1$ and $\varepsilon_2$ are the dielectric constants for each of the two components. This rule has a very limited applicability, unfortunately, since it is based on a system where little or no conduction takes place. When one of the components in a composite dielectric is conductive, large polarizations may be attributed to interfacial polarizations at the boundaries of the conductive phase. Since water is one of the component phases of most rock, and since water is a relatively good conductor, the mixing rule may not be applied except at such high frequencies that interfacial polarization cannot take place. Therefore, Eq. (26) applies with reasonable accuracy to water-bearing rocks at radio frequencies. The matter of dielectric constants in water-bearing rocks is discussed more thoroughly in a later section.

Since mixing rules are unreliable, the best approach to determining the dielectric constant of a rock is the experimental approach. It is usually found that even in dry rocks, interfacial polarization between minerals having slightly different conductivities leads to a dependence of dielectric constant on frequency in the frequency ranges ordinarily used in prospecting. Table 13 lists some dielectric constants measured on dry rocks as a function of frequency. The dielectric constant in the audio- and subaudio-frequency ranges may be several times larger than the dielectric constant measured at radio frequencies, where only ionic and electron polarization mechanisms are effective.

The dielectric constant in dry rocks also depends on temperature, and particularly, the portion of polarization due to interfacial polarization is enhanced at high temperatures. Figure 31 shows a series of curves relating dielectric constant to frequency for a variety of temperatures. At radio frequencies, where polarization is due to ionic and electronic mechanisms, there is virtually no dependence of dielectric constant on temperature. At low frequencies, there is a tenfold increase in dielectric constant as the temperature is varied from room temperature to 1000°C. The high temperatures cause some of the minerals in the rock to become relatively conductive, so that the amount of interfacial polarization is increased at high temperatures.

## 9. MAGNETIC PERMEABILITY

A third property of materials which affects to some extent the behavior of alternating electric fields is the magnetic permeability. The effect is slight except in rare cases, when rocks contain a high concentration of magnetic minerals. As with conductivity and dielectric constant, the magnetic permeability of a material can be defined in terms of a hypothetical experiment. If it

Table 13

### Dielectric constants of dry rocks as a function of frequency

| Rock | Dielectric constant in pF/m at a frequency of: | | | | | |
|------|--------|-------|--------|---------|-------|--------|
|      | 100 c/s | 1 kc/s | 10 kc/s | 100 kc/s | 1 Mc/s | 10 Mc/s |
| Dolomite (rock) | 105 | 85·6 | 77·0 | 71·8 | 70·0 | 68·3 |
| Kaolinite (rock) | 67·7 | 56·0 | 48·0 | 43·0 | 40·3 | 39·8 |
| Limestone | 91·8 | 82·9 | 79·8 | 78·5 | 77·0 | 75·8 |
| Sandstone (arkose) | 52·5 | 51·5 | 49·5 | 47·7 | 47·2 | 47·0 |
| Sandstone (greywacke) | 103 | 77·8 | 65·3 | 58·4 | 54·2 | 52·0 |
| Sandstone (quartzite) | 50·0 | 47·2 | 45·9 | 44·4 | 43·2 | 41·8 |
| Diabase | 119 | 99·5 | 88·1 | 80·7 | 73·6 | 68·8 |
| Diorite | 63·9 | 58·5 | 55·5 | 53·5 |  |  |
| Dunite | 88·7 | 75·0 | 69·3 | 67·3 | 65·3 | 63·5 |
| Gabbro | 133 | 102 | 90·8 | 85·5 | 80·8 | 77·8 |
| Granite | 75·0 | 66·9 | 63·4 | 62·1 | 60·8 | 59·1 |
| Argillite | 105 | 92·0 | 81·7 | 76·6 | 73·5 | 70·6 |
| Sillimanite schist | 86·0 | 80·8 | 77·7 | 75·2 | 72·9 | 71·5 |
| Hornblend schist | 91·5 | 87·8 | 84·3 | 81·9 | 80·6 | 78·7 |
| Talc schist | 279 | 201 | 139 | 96·0 | 75·0 | 67·0 |
| Serpentine (rock) | 89·8 | 69·0 | 61·2 | 58·5 | 56·7 | 55·3 |

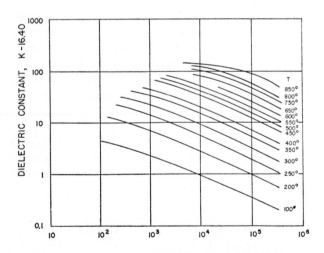

FREQUENCY, cycles per second

Fig. 31. The dielectric constant of a sample of peridotite measured as a function of frequency for the temperatures indicated. A constant, 16·40, has been subtracted from all values for relative dielectric constant so that the curves will not converge at the high frequencies.

were possible to study the behavior of isolated magnetic poles, it would be found that the force operating between two such poles was proportional to the product of pole strengths and inversely proportional to the distance between them squared:

$$F = \frac{1}{\mu_0} \frac{m_1 m_2}{r^2} \tag{27}$$

where $m_1$ and $m_2$ are the pole strengths in mks units and $r$ is the distance between the poles in meters. The constant of, proportionality, $\mu_0$, is expressed in henries per meter (H/m) in the mks system, and is observed to have a magnitude of $12 \cdot 56 \times 10^{-7}$ if the poles are isolated in free space.

If a material medium is considered for the hypothetical experiment, the observed value for permeability may be very slightly larger or smaller than the value for the free-space experiment, for most materials. The only case in which permeability departs significantly from the free-space value is that in which the material contains a large fraction of ferromagnetic material. The naturally

TABLE 14

**Magnetic permeability of minerals**

| Mineral | Ratio of magnetic permeability to that of free space | Magnetic permeability $\times 10^{-7}$ (H/m) |
|---|---|---|
| Magnetite | 5·0 | 17·6 |
| Pyrrhotite | 2·55 | 14·5 |
| Ilmenite | 1·55 | 13·3 |
| Hematite | 1·053 | |
| Pyrite | 1·0015 | |
| Rutile | 1·0000035 | |
| Anhydrite | 0·999987 | |
| Calcite | 0·999987 | |
| Fluorite | 0·999975 | |
| Quartz | 0·999985 | |
| Sylvite | 0·999986 | |
| Hornblend | 1·00015 | |
| Augite | 1·00167 | |

occurring ferromagnetic minerals are ilmenite, magnetite, ferrous ilmenite and pyrrhotite. Values for the magnetic permeability of some common minerals are listed in Table 14.

Of the minerals listed in Table 14, the only one which occurs commonly enough to alter the permeability of a large mass of rock significantly from the value for free space is magnetite. It has been observed that the permeability of a magnetite-bearing rock can be predicted approximately using the curve given in Fig. 32.

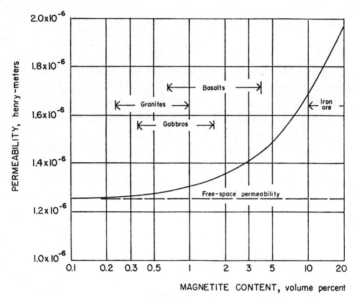

FIG. 32. Observed relationship between the amount of magnetite in a rock and magnetic permeability.

# REFERENCES

ARCHIE, G. E. (1942) The electrical resistivity log as an aid in determining some reservoir characteristics, AIME Tech. Paper 1422.

AUSTIN, I. G. and WOLFE, R. (1956) Electrical and optical properties of a semiconducting diamond, *Proc. Phys. Soc. (London)* B **69**, 329–338.

AYERS, M. L., DOBYNS, R. P. and BUSSELL, R. Q. (1952) Resistivities of water from subsurface formations, *Petrol. Engr.* **24** (13), B 36–B 48.

BACKSTOM, H. (1888) Elektrisches und thermisches Leitungsvermögen des Eisenglanzes *Öfvers. Vetensk-Akad. Förh., Stockh.* **8**, 533–551.

·BIRCH, F., SCHAIRER, J. F. and SPICER, H. C. (1942) *Handbook of Constants*, Geol. Soc. of Am. Spec. Paper No. 36.

BLOOM, H. (1959) Molten electrolytes, *Modern Aspects of Electrochem.*, No. 2, 161–261.

CARD, R. H. (1940) Correlation of earth resistivity with geological structure and age *Trans. AIME* **138**, 380–398.

CHAKRAVARTY, M. K. and KHASTIGIR, S. R. (1938) Direct determinations of the electrical constants of soil at ultra-high frequencies, *Phil. Mag.* **25**, 793–801.

CHEBOTAREV, I. I. (1955) Metamorphism of natural waters in the crust of weathering, *Geochimica et Cosmochim. Acta* **8**, 53–62, 137–170.

COSTER, H. P. (1948) The electrical conductivity of rocks at high temperatures, *Monthly Notices of the Roy. Astron. Soc. Geophys. Suppl.* **5**, 193–199.

COWNIE, A. and PALMER, L. S. (1952) The effect of moisture on the electrical properties of soil, *Proc. Phys. Soc. (London)* B **65**, 295–301.

CRAWFORD, J. G. (1951) Characteristics of oil-field waters of the Rocky Mountain region, *Subsurface Geologic Methods*, pp. 272–296, published by Colorado School of Mines.

CRONEMEYER, D. C. (1952) Electrical and optical properties of rutile single crystals, *Phys. Rev.* **87**, 876–888.

DE WITTE, L. (1955) A study of electric log interpretation methods in shaley formations, *Trans. AIME* **204**, 103–110.

DOMENICALI, C. A. (1950) Magnetic and electric properties of natural and synthetic single crystals of magnetite, *Phys. Rev.* **78**, 458–467.

DOSTOVALOV, B. N. (1937) Egmereniia dielektricheskoi postoyanoi i udelnovo soprotovleniia gorni porod. *Trav. inst. petrog. Loewinson–Lessing accad. URSS* **10**, 161–168.

DUNLAP, H. F. and HAWTHORNE, R. R. (1951) The calculation of water resistivities from chemical analyses. *J. Petrol. Technol.* **7**, 17.

DUTTA, A. K. (1953) Electrical conductivity of single crystals of graphite. *Phys. Rev.* **90**, 187–192.

FINKLEA, E. E. (1956) Formation evaluation of some limestone reservoirs with particular reference to well logging techniques. *Petrol. Technol.* **8**, 25–31.

FRENKEL, J. (1955) *Kinetic Theory of Liquids*, Dover Publications, New York, 488 p.

FRITSCH, Volker (1949) *Grundzüge der angewandten Geoelektrik*, Manzsche verlags- and Universitatsbuchhandlung, Vienna.

FUOSS, R. M. and KIRKWOOD, J. C. (1941) Electrical properties of solids. VIII. Dipole moments in polyvinyl chloride–diphenyl systems. *J. Am. Chem. Soc.* **63**, 385–394.

GUYOD, H. (1944) Electrical properties of oil bearing reservoirs, *Oil Weekly*, Nov. 13.

HARVEY, R. D. (1928) Electrical conductivity and polished mineral surfaces, *Econ. Geol.* **23**, 778–801.

HILL, H. J. and MILBURN, J. D. (1956) Effect of clay and water salinity on electrochemical behavior of reservoir rocks, *Trans. AIME* **207**, 65–72.

HOWELL, B. F., Jr. (1953) Dielectric constants of rocks and minerals, U.S. At. En. Com. Report NYO 3746, the Penn. State University, Mar. 20.

HOWELL, B. F., JR., KEITH, M. L. and LICASTRO, P. H. (1955) Dielectric constants of rocks and minerals, U.S. At. En. Com. Report NYO 3748, The Penn. State University, May 5.

HUGHES, H. (1955) The pressure effect on the electrical conductivity of peridote, *J. Geophys. Research* **60**, 187–191.

KELLER, G. V. (1953) Effect of wettability on the electrical resistivity of sand, *Oil Gas J.* **51** (34), 62–65.

KELLER, G. V. (1963) Electrical properties in the deep crust, IEEE Trans. on Ant. and Prop., vol AP–11, No. 3, pp. 344–357.

KIRBY, R. S., HARMAN, J. G., CAPPS, F. M. and JONES, R. N. (1954) Effective radio ground conductivity measurements in the United States, Nat. Bu. Stan. Circ. 546, U.S. Gov. Printing Office, Washington, D.C.

KONIGSBERGER, J. and REICHENHEIM, O. (1906) Uber die Elecktrizitatslitung einiger naturlich kristallisterten Oxyde und Sulfide und des Graphits, *Neues Jahrb. Mineral. Geol. u. Paläontal. II*, 20–49.

LIDIARD, A. B. (1957) Ionic conductivity, in *Handbuch der Physik*, Band 20, Electrische Leitungsphanomene II, 246–349.

MANNING, M. F. and BELL, M. E. (1940) Electrical conduction and related phenomena in solid dielectrics, *Rev. Mod. Phys.* **12**, 215–256.

MASON, B. (1958) *Principles of Geochemistry*, John Wiley, New York, 310 p.

McKELVEY, J. G., JR., SOUTHWICK, P. F., SPIEGLER, K. S. and WYLLIE, M. R. J. (1955) The application of a three element model to the S.P. and resistivity phenomena evinced by dirty sands, *Geophysics* 20, 913–931.

MILES, K. R. (1951) Origin and salinity distribution of artesian water in the Adelaide Plains, South Australia, *Econ. Geol.* 64, 193–207.

MOONEY, H. M. and RODNEY B. (1953) Magnetic susceptibility measurements in Minnesota, *Geophysics* 18, 383–394.

MOTT, N. F. and JONES, H. (1958) *The Theory of the Properties of Metals and Alloys*, Dover Publications, New York, 326 p.

NESTEROV, M. A. and NESTEROV, L. YA. (1947) Electrical resistivity of rocks at subzero temperatures *Materiali Vses. Nauk. Izled. Geol. Inst., Geof.* 11, 76–88.

NORITOMI, K. (1958) Migration of charged carrier in the case of electric conduction of rocks, *Sci. Repts. Tôhoku Univ., Fifth Ser.* 9 (9), 120–127.

NORITOMI, K. (1961) The electrical conductivity of rock and the determination of the electrical conductivity of the earth's interior, *J. Min. Coll, Akita Univ.* A 1 (1), 27–59.

NORITOMI, K. and AKIE, A. (1957) Studies on the electrical conductivity of a few samples of granite and andesite, *Sci. Repts. Tôhoku Univ., Fifth Ser.* 7 (3), 201–207.

PARASNIS, D. S. (1956) The electrical resistivity of some sulfide and oxide minerals and their ores, *Geophys. Prospecting* 4, 249–278.

PARASNIS, D. S. (1956) Elektrisk resistivitet hos nagra sulfidoch oxidmineral *Jernkontorets Ann.* 140, 494–511.

PAULING, L. (1960) *The Nature of the Chemical Bond*, Cornell University Press, Ithaca, New York, 644 pp.

PRIMAK, W. and FUCHS, L. H. (1954) Electrical conductivity of natural graphite crystals, *Phys. Rev.* 95, 22–30.

PUZIN, L. A. (1952) Connate water resistivity in Oklahoma, its application to electric log interpretation, *Petrol. Engr.* 24 (9), B67–B77.

TAKUBO, J. (1941) Versuche über die Dielektrizitätskonstanten einiger Mineralen und über das dielektrische Verhalten derselben bei Erhitzung, *Mem. Coll Sci. Kyoto Imp. Univ.* B 16, 95–154.

TARKHOV, A. G. (1947) Resistivity and dielectric constant of rocks in alternating current fields, *Materiali Vses Nauk.-izled. Geol. Inst., Geof.*, No. 12, pp. 1–42.

TOZER, D. C. (1959) The electrical properties of the earth's interior, in *Physics and Chemistry of the Earth*, vol. 3, Pergamon Press, New York, pp. 414–436.

WINSAUER, W. O., PERKINS, F. M., Jr., and BRANNON, H. R., Jr. (1954) Interrelation of resistivity and potential of shaley reservoir rock, *J. Petrol. Technol.* 6 (8), 28–34.

CHAPTER II

# ELECTRICAL WELL LOGGING

IN THE interpretation of geoelectric surveys, there is frequently a need for more exact knowledge of the electrical properties of the rock in an area under study than can be obtained by using available geologic information as a basis for prediction. Geophysical control in the form of knowledge of the electrical properties of the rocks is as essential to the interpretation of electrical survey data as drill-hole velocity surveys are in the interpretation of seismic data, density control in gravimetric surveys and magnetic susceptibility control in magnetic surveys. Geophysical surveys may be interpreted without such control, but a much more reliable interpretation may be made if control is available.

Electrical well logs are the best source of control for geoelectric surveys. In the United States, many hundred thousand oil wells and water wells have been electric logged in the past 30 years. Virtually all of these wells penetrated only sedimentary rocks, so that when electrical surveys are to be carried out in an area containing some sedimentary rocks, it is likely that electric logs are available. Information about the availability of logs can be obtained from the local office of the well logging service companies, the local offices of oil exploration companies or the local offices of state and federal geological surveys. In the oil-producing areas of the United States electric log libraries have been established, and copies of electric logs are available from these libraries in a very short time at a low price.

If a geoelectric survey is to be carried out in an area where electric logs are not available, every effort should be made to include electric logging as part of the exploration program. In mining districts, usually there are many exploration drill holes in which no electric logs have been run. Logs may be obtained in such holes using light weight portable equipment with relatively little effort. If no holes are available, consideration should be given to the drilling of holes especially for this purpose.

There are a variety of instrumental techniques for recording electrical resistivity logs which have been used over the past thirty years:

1. Single-electrode resistance logs;
2. Multi-electrode spacing logs;
3. Focussed current logs;

4. Micro-spacing and pad device logs;

5. Induction logs.

Not all these logs are equally suitable for determining the average electrical
properties of the rock penetrated by a well, but the choice of the type log
depends more on what is available than what is best suited to the problem.
Since it may be necessary to use any of these types of logs, it is wise to be
aware of the particular characteristics and limitations of each type.

## 10. SINGLE-ELECTRODE RESISTANCE LOGS

In resistance logging, a single electrode is lowered into a well on the end of an
insulated length of wire, as shown in **Fig. 33**. The electrical resistance between
this inhole electrode and a return electrode located at the surface is recorded
on a graphic recorder. The resistance is a function of the resistivity of the
material surrounding the electrode and of the shape and dimensions of the
electrode. It is necessary that the well bore be filled with water or drilling mud
so that there is electrical continuity between the inhole electrode and the rock
around the well.

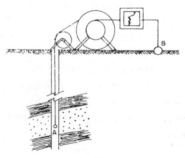

Fɪɢ. 33.

For the case of a spherical electrode, the grounding resistance may be
calculated fairly simply, particularly if the medium around the electrode is
homogeneous in electrical properties. Such an idealized situation may be
closely realized if the resistivity of a thick layer of rock is uniform and if the
well bore is filled with drilling mud with about the same resistivity as the rock.

In a completely uniform medium, current will spread out radially with a
uniform current density in all directions from an electrode, $A$, as shown in
Fig. 34. The grounding resistance may be calculated by dividing the earth
around the electrode into a series of thin, concentric spherical shells. The
total grounding resistance is found by summing the resistances through all such
shells extending from the surface of the electrode to infinity. The resistance

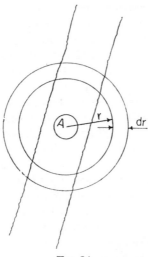

Fig. 34.

through a single shell is found using one of the defining equations for resistivity:

$$dR = \varrho\,\frac{l}{A} = \varrho\,\frac{dr}{4\pi r^2} \tag{28}$$

The total resistance is determined by integrating this expression for the resistance of a single thin shell over a range in $r$ extending from the surface of the electrode (radius $a$) to infinity:

$$R = \int_a^\infty \frac{\varrho}{4\pi r^2}\,dr = -\frac{\varrho}{4\pi r}\int_a^\infty = \frac{\varrho}{4\pi a} \tag{29}$$

A *geometric factor*, $K$, may be defined from this equation by combining the factors which depend on the geometry of the electrode:

$$K = 4\pi a \tag{30}$$

Every electrode or array of electrodes can be characterized by a particular geometric factor. It is a parameter which when multiplied by the measured resistance will convert the resistance to the resistivity for a uniform medium.

The measured resistance for a single-electrode logging system includes not only the grounding resistance of the inhole electrode, but also the resistance of the logging cable and the grounding resistance through the surface electrode. If the surface electrode has a large area, and if it is placed in a low resistivity medium such as the mud in the mud pit, its resistance will be small in comparison with the grounding resistance at the inhole electrode, and so, may be neglected. The cable resistance also is usually small and can be neglected. It is usually assumed that the measured resistance is entirely due to the grounding resistance at the inhole electrode. If this assumption cannot be made, the other resistances included in the measurement must be determined separately and subtracted from the total resistance. This can always be done, since the grounding resistance at the inhole electrode is the only resistance in the circuit which varies.

The grounding resistance for a single electrode is controlled primarily by the resistivity of the material within a short distance of the electrode, if the medium is not uniform. Usually, the drilling mud filling the bore hole has a lower resistivity than the rock around the bore hole. Under such conditions, a large portion of the current will flow up or down the mud column, rather than outward into the rock. The resistivity calculated from Eq. (29) will be larger than the resistivity of the mud, but less than the resistivity of the rock. In this case, we define an *apparent resisitivity* as being the product of the geometric factor calculated for a homogeneous medium and the resistance actually measured in a non-homogeneous environment.

The degree to which the apparent resistivity is affected by a relatively conductive mud in the drill hole can be determined using the resistance-shell approach indicated in Fig. 35. It is assumed that within the well, current flows entirely up or down the mud column, so that the equipotential surfaces are flat discs, and that outside the well the current flows radially, so that the equipotential surfaces are truncated spheres. It is necessary only to sum the resistances contributed by a series of these truncated spheres to determine the grounding resistance. We can calculate the resistance $R_1$ through the spherical

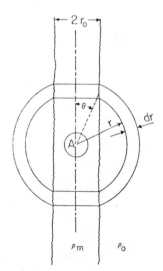

FIG. 35.

portion of each shell and the resistance $R_2$ through the truncated portion separately and then treat them as resistances in parallel:

$$dR_1 = \frac{4\pi \int_{\theta_1}^{\pi/2} r^2 \sin\theta \, d\theta}{\varrho_1 \, dr} \; ; \qquad \theta_1 = \arcsin \frac{r_0}{r} \tag{31}$$

$$dR_2 = \frac{2\pi r_0^2}{\varrho_m \, dr \cos\theta} \tag{32}$$

$$dR = \frac{dR_1 \cdot dR_2}{dR_1 + dR_2} = \frac{2\varrho_1 \varrho_m (r^2 - r_0^2)^{1/2} \, dr}{[4\pi r_0^2 \varrho_1 + 8\pi (r^2 - r_0^2) \varrho_m] \, r} \tag{33}$$

where $\varrho_m$ is the resistivity of the mud filling the well and $\varrho_1$ is the resistivity of the rock. After integrating Eq. (33), the grounding resistance turns out to be:

$$R = \frac{\varrho_m}{4\pi r_e} \left[ 1 - \frac{r_e}{r_0} + \frac{\varrho_1}{\varrho_m} \frac{r_e}{r_0} \int_{r_0}^{\infty} \frac{\left[\left(\frac{r}{r_0}\right)^2 - 1\right]^{1/2} d\left(\frac{r}{r_0}\right)}{\left[\left(\frac{r}{r_0}\right)^2 - \frac{1}{2}\frac{\varrho_1}{\varrho_m} - 1\right] \frac{r}{r_0}} \right] \tag{34}$$

The integral expression may be evaluated in a closed form by making a transformation in the variable of integration:

$$x = \left[\left(\frac{r}{r_0}\right)^2 - 1\right]^{1/2}$$

$$\frac{\varrho_a}{\varrho_0} = 1 + \frac{r_e}{r_0}\left( \frac{\frac{\varrho_1}{\varrho_m}}{\sqrt{\left(\frac{1}{2}\frac{\varrho_1}{\varrho_m} + 1\right)}} \cdot \frac{\pi}{2} - 1 \right) \tag{35}$$

This last equation expresses the ratio of apparent resistivity to the mud resistivity as a function of the well diameter and the true rock resistivity. Curves calculated from this equation for a variety of contrasts between mud resistivity and rock resistivity are shown in Fig. 36. The departure of apparent resistivity from true resistivity is extreme if the diameter of the well is more than twice the diameter of the electrode, or if the rock resistivity is more than ten times as large as the mud resistivity. These are serious limitations to the use of single-electrode resistance logs for quantitative measurements of resistivity in bore holes.

The accuracy of single-electrode resistance measurements can be improved if a long cylindrical electrode is used rather than a spherical electrode. The

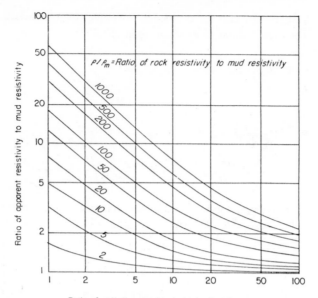

Fig. 36. The departure of apparent resistivity from true resistivity for resistance measurements made with a spherical electrode as a function of the well diameter and the contrast in resistivity between the rock and the mud filling the well bore.

grounding resistance of a cylinder can be determined in the same ways as that for a sphere if it is assumed that the equipotential surfaces about the electrode are ellipsoids of revolution. This approximation is reasonably good if the length of the electrode is several times greater than the diameter. Under these conditions, the geometric factor is:

$$K = 2{\cdot}73\,\frac{L}{\log 2L/d} \tag{36}$$

where $L$ is the length of the electrode and $d$ is the diameter.

The very great departure between apparent resistivity and true resistivity which may occur with single-electrode resistance logs means that such logs can be used only in a semi-quantitative manner in determining the electrical properties of rocks penetrated by a well. Single-electrode resistance logs are found frequently in areas where only water wells or seismograph shot holes have been logged. Usually other types of resistivity logs are run in oil wells where quantitative use is made of resistivity data. Resistance logs are useful primarily for correlations, or they may be used to determine the uniformity of the properties of the geoelectric section from hole to hole.

## 11. SPACING LOGS

The type of resistivity log most commonly used in oil wells until about 1960 used a four-electrode array, with either two or three electrodes fixed on a carrier (sonde) which was lowered into a well on the end of a multi-conductor cable. The electrodes not fixed on the carrier were placed at the surface. As in the single-electrode logging method, current is driven between an inhole electrode, $A$, and a surface current return, $B$, as shown in Fig. 37. The total resistance between these two electrodes is not measured, but rather, a measuring electrode is placed a short distance uphole from the current electrode, so that only the resistance contributed by the rock outside the equipotential surface passing through this measuring electrode is measured. If a single measuring electrode is carried on the inhole tool (Fig. 37a), the array is called a *normal* array. In European terminology, such an array is also called a *potential* array, since the potential at the electrode $M$ is measured.

In order to increase the resolution of a spacing log, two closely spaced electrodes $M$ and $N$ (Fig. 37b) are used rather than one. Such an array is called a *lateral* array, or in the European terminology, a *gradient* array, since the potential gradient is measured approximately. With a normal array, the distance between the $A$ and $M$ electrodes on the inhole tool is called the $AM$ *spacing*. For the lateral array, the distance from the current electrode, $A$, to the midpoint of the measuring pair, $M$ and $N$, is called the $AO$ *spacing*. For various reasons, the roles of the current electrodes $A$ and $B$ may be interchanged with the roles of the measuring electrodes, $M$ and $N$, in a lateral array. It can be shown this does not affect the value of measured resistivity,

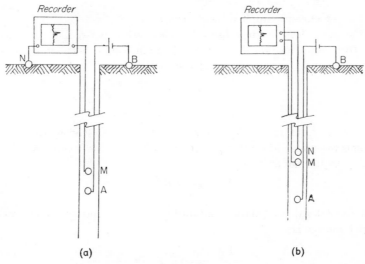

(a)                                        (b)

FIG. 37.

so it is assumed that the lateral array shown in Fig. 37 a is being used (electrode configuration $A$–$M$–$N$) when actually the reciprocal electrode array (configuration $M$–$A$–$B$) is being used.

The spacing factor determines the distance from the well to which resistivity is measured. A standard set of spacing logs for use in oil well logging consists of three curves; a short normal curve ($AM = 16$ in.), a long normal curve ($AM = 64$ in.) and a lateral curve ($AO = 18$ ft 8 in.). A variety of spacings other than these may be found on older logs, but in recent years these three spacings have been accepted as standards by the well logging service companies, and these are the only ones for which interpretation charts are available.

The relationship between measured resistance and resistivity is easy to determine if the electrodes are located in a completely uniform medium. Under such conditions, the current from the electrode $A$ will spread out radially in all directions. The equipotential surfaces are spheres centered on the electrode $A$. Using the same analysis as we did for a spherical electrode in a uniform medium, we find that the resistance contributed by the rock outside a sphere of radius, $r$, is:

$$R = \frac{\varrho}{4\pi r} \qquad (37)$$

Resistance is the ratio between potential and current (Ohm's law), so that the potential at the equipotential surface passing through a point at a distance, $r$, from the current electrode is:

$$U = \frac{\varrho I}{4\pi r} \qquad (38)$$

where $I$ is the current supplied by electrode $A$.

For the normal electrode array, the relationship between rock resistivity, current from the electrode $A$ and the voltage measured between the electrode $M$ in the hole and the electrode $N$ at the surface is:

$$\varrho_a = 4\pi \ \overline{AM} \ \frac{U}{I} \qquad (39)$$

For a non-uniform resistivity environment, the resistivity given by this equation is the *apparent resistivity measured with a normal array*. The geometric factor for the normal array is:

$$K = 4\pi \ \overline{AM} \qquad (40)$$

With the lateral array, the relationship between resistivity, current and measured voltage is:

$$\varrho = 4\pi \ \frac{\overline{AM} \ \overline{AN}}{\overline{AN} - \overline{AM}} \ \frac{U}{I} \cong 4\pi \ \frac{\overline{AO}^2}{MN} \ \frac{U}{I} \qquad (41)$$

The geometric factor for the lateral array is:

$$K = 4\pi \frac{\overline{AO}^2}{MN} \tag{42}$$

The apparent resistivity measured with a spacing log may depart appreciably from the true rock resistivity if the mud in the well bore is much more conductive than the rock. Figure 38 shows a series of curves relating apparent resistivity to true resistivity of a lateral spacing log used in a 12 in. well, and for normal spacing logs used in wells ranging from 6 to 12 in. in diameter. For relatively small contrasts in resistivity between the mud column and the rock around a well, the apparent resistivity is slightly larger than the true resistivity. For very large contrasts in resistivity between the mud and the rock, the apparent resistivity is less than the true rock resistivity. For the lateral spacing log, the apparent resistivity approaches a maximum value at very large contrasts between mud and rock resistivity, with the maximum value depending only on the mud resistivity and the diameter of the well, and not on the rock resistivity. The maximum resistivity measurable with a lateral spacing log is:

$$(\varrho)_{\max} = 2 \left( \frac{\overline{AO}}{d} \right)^2 \varrho_m \tag{43}$$

where $d$ is the well diameter and $\varrho_m$ is the mud resistivity. With an 18 ft 8 in. lateral spacing, the maximum apparent resistivity that can be measured in an 8 in. well is 6272 times as large as the mud resistivity. This limit is usually no problem in logging sedimentary rocks, where such large contrasts between mud resistivity and rock resistivity are not usually found.

There is no upper limit to the apparent resistivity that may be recorded with a normal spacing log. No matter how high the contrast in resistivity between the rock and the mud, an increase in rock resistivity will result in a increase in the measured apparent resistivity. However, even the apparent resistivity measured with a normal spacing will be lower than the true resistivity if the contrast is large.

The departure of apparent resistivity from true resistivity is less if large spacings are used. It would appear that if we use sufficiently large spacings, we can measure resistivity accurately, no matter how much greater the rock resistivity is than the mud resistivity. Unfortunately, there is a practical limitation to the size of the spacing which may be used. The curves shown in Fig. 38 apply only when the electrode array is situated in a bed whose thickness is many times greater than the electrode spacing. For beds with a thickness comparable to the electrode spacing, the apparent resistivity curve may be very complicated.

The apparent resistivity curve which would be recorded using a normal array of electrodes to log the resistivity of a relatively thin layer of rock with

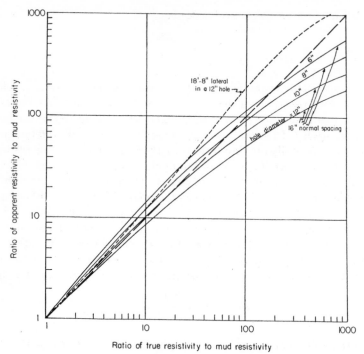

FIG. 38. Departure of apparent resistivity from the true resistivity of the rock around a well as a function of the contrast between rock resistivity and mud resistivity. The solid curves are calculated for a 16-in. normal spacing, while the dashed curve is calculated for a 224-in. lateral spacing. The dotted curve is the locus for no correction. For curves falling below this dotted line, apparent resistivity is less than true resistivity; for curves falling above the dotted line, apparent resistivity is greater than true resistivity.

high resistivity is shown in Fig. 39. For the case in which the resistant layer is several times thicker than the electrode spacing, the apparent resistivity increases gradually as the electrode array approaches the boundary of the layer. As the electrode array first penetrates the layer, the apparent resistivity assumes a constant value, and remains at this level until the second electrode of the array also enters the resistant zone. The apparent resistivity then rises sharply, coming close to the true resistivity of the resistant zone if the layer is thick enough. As the electrode array passes the midpoint of the layer, each of these steps is repeated in reverse, so that the recorded curve is symmetric about the midpoint of the layer.

The resistivity curves shown in Fig. 39 are calculated under the assumption that the effect of drilling mud in the well can be ignored. In practice, since electric logs are run in mud-filled holes, conduction through the drilling mud will tend to round the sharp corners shown by the curves in Fig. 39, and

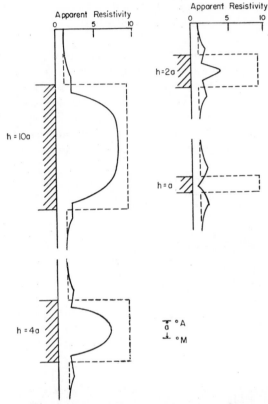

FIG. 39. Apparent resistivity logs which would be recorded with a normal array of electrodes passing through a thin layer with high resistivity. It is assumed that the layer has a resistivity of nine units; the surrounding medium, 1 unit. The presence of the well is neglected.

reduce the maximum amplitude of the deflections. The plateau values of resistivity which are recorded as the electrode array enters and leaves the resistant bed can rarely be identified on an actual log. Instead, the apparent thickness of a resistant zone is usually taken to be less than the actual thickness by an amount equal to the electrode spacing. If the thickness of the resistant zone is less than the electrode separation, no maximum will be observed opposite the bed. Thus, the normal array cannot detect the presence of resistant beds with a thickness less than the electrode spacing. Thin, resistant beds can contribute appreciably to the electrical anisotropy of a formation, so a normal spacing short enough to detect the presence of such beds must be used when the average electrical properties of a geoelectric unit are being calculated from electric logs. However, the use of very short spacings leads to errors in measured resistivity due to conduction through the mud column. As a compromise, usually several normal spacings are run simultaneously, one with a separation several times the well diameter and another with a separation ten to twenty times the well diameter.

The apparent resistivity curve recorded with a lateral array is not symmetrical about the midpoint of a resistant layer (see examples of lateral resistivity curves, Fig. 40). Not only are the curves recorded with a lateral array asymmetric, but the shape will be inverted if the measuring electrode pair is above the current electrode, rather than below it, as shown in Fig. 40. The apparent resistivity increases slightly as the lateral array approaches the resistant zone from underneath, and assumes a constant plateau value as the three electrodes

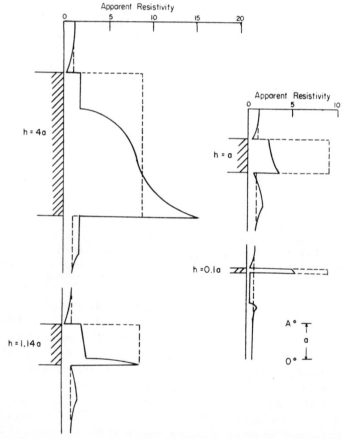

FIG. 40. Apparent resistivity logs which would be recorded with a lateral array of electrodes passing through a thin layer with high resistivity. It is assumed that the layer has a resistivity of nine units; the surrounding medium, 1 unit. The effect of the mud column has been neglected. A conductive mud column would cause the curves to be more rounded.

cross the lower boundary, providing the thickness of the layer is equal to several electrode spacings. The apparent resistivity rises abruptly to a value greater than the true resistivity of the layer when the measuring electrodes first enter the layer, and then approaches the true resistivity from the high side. As the array passes through the upper half of the layer, the apparent resistivity is less than the true resistivity, and once the array leaves the upper boundary, the measured resistivity drops to a value lower than the true resistivity outside the layer, and then recovers as the array travels further from the resistant zone.

If the spacing is larger than the thickness of the resistant zone, the zone is detected as an increase in apparent resistivity, but an anomalously low value of apparent resistivity is recorded for the interval during which the current electrode and measuring electrodes are on opposite sides of the resistant zone. This zone of low apparent resistivity constitutes a shadow zone, so that if there are several closely spaced thin resistant zones, only the uppermost one will be recorded on the electric log. The lower zones will be masked by the anomalous low values of apparent resistivity recorded beneath the uppermost layer.

The normal log has advantages in that the recorded curve is symmetric and relatively simple in shape, and the maximum resistivity which can be measured is unlimited. The lateral log has an advantage in that even thin resistant layers will appear on the log, unless their presence is masked by nearby layers. On the other hand, the normal log is blind to beds whose thickness is less than the electrode spacing, and the log recorded with a lateral array is complicated in shape. With the standard set of two normal logs and one lateral log supplied by most logging service companies, it is possible to determine the resistivity of a layer which is several feet or more in thickness with a reasonable degree of accuracy, using the various departure curves and correction curves which are available. In the evaluation of petroleum reservoir rocks, only a few tens of feet of log need to be corrected, so the amount of work involved is reasonable. In calculating the electrical properties of a complete section, the amount of work necessary to determine the correct resistivity of each bed shown on the electric logs is not within reason. Instead, a spacing log recorded with a short normal array should be used without corrections, and allowance should be made for the relatively low values of the coefficient of anisotropy and transverse resistance which will be obtained.

## 12. FOCUSSED CURRENT LOGS

The single-electrode resistance measurement discussed in Section 10 may be improved to the point where it can be used to determine resistivity quantitatively by adding Kelvin guard electrodes above and below the main current electrode, as shown in Fig. 41. All three of the inhole electrode are operated at the same potential, so that the current from the guard electrodes forces the current

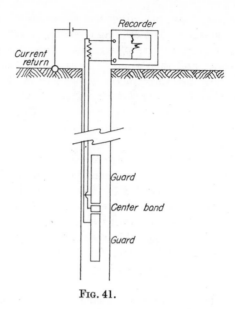

FIG. 41.

from the center electrode to be constrained in a disc-shaped area as it flows out into the rock. The guard current effectively focusses the center electrode current into a beam. Resistance measurements made with a guarded or focussed system give a more faithful representation of resistivity changes in the rock than does either the single electrode method or the spacing methods. Guarded electrode logs are offered commercially by the service companies under the trade names Laterolog 3 (Schlumberger), current-focussed logs (Lane–Wells), and Guard Log (Welex and Birdwell).

The two guard electrodes and the center band may be treated as a single electrode with the shape of an ellipsoid of revolution. The potential outside such an electrode is:

$$U = \frac{\varrho I}{4\pi L} \ln \frac{Z + L + [r^2 + (Z + L)^2]^{1/2}}{Z - L + [r^2 + (Z - L)^2]^{1/2}} \tag{44}$$

where $z$ and $r$ are cylindrical coordinates measured from the midpoint of the guard electrode array, and $L$ is the half-length of the assembly, measured from the midpoint to one end. As before, $\varrho$ is the resistivity of an assumed uniform medium, and $I$ is the current to the electrode assembly. Current flow lines are orthogonal to the equipotential surfaces in an isotropic medium, so the current flow lines follow surfaces which are hyperboloids of revolution about the electrodes. Only the resistance for the portion of the total current which flows from the center band is measured. This current is confined to the area between two hyperboloid surfaces which intersect the electrode system at the top and the bottom of the center band. This resistance may be computed by considering a series of concentric rings within this area, with the diameter of the rings varying incrementally from the diameter of the electrode to infinity. If this operation were carried out, it would be found that the resistance to current flowing from the center band is:

$$R = \frac{\varrho L/d}{2\pi l \left[\left(\dfrac{L}{d}\right)^2 - 1\right]^{1/2}} \ln\left\{\frac{L}{d} + \left[\left(\dfrac{L}{d}\right)^2 - 1\right]^{1/2}\right\} \qquad (45)$$

The geometric factor for the guard electrode array is:

$$K_g = \frac{2\pi l \left[\left(\dfrac{L}{d}\right)^2 - 1\right]^{1/2}}{\dfrac{L}{d} \ln\left(\dfrac{L}{d}\right) + \left[\left(\dfrac{L}{d}\right)^2 - 1\right]^{1/2}} \qquad (46)$$

where $L$ is the total length of the array, $l$ is the length of the center band and $d$ is the diameter of the electrode.

The guard electrode system is ideally suited to measuring the electrical properties of thin layers, even when the mud in the well is highly conductive. Beam widths of a few inches to a foot may be used in logging oil wells, with guard electrode lengths of the order of 5 ft. Longer guard electrodes insure more accurate resistivity measurements in resistant beds, but are a disadvantage in that a section of hole equal to a guard length cannot be logged in the bottom of the well.

Even with a guard electrode system, the apparent resistivity recorded opposite a thin resistant bed will be somewhat less than the true resistivity, though usually this departure can be ignored. When a guard array is located opposite a resistant zone, the current from the center band will tend to divert around the resistant zone, even though the guard current attempts to constrain the beam current. Correction factors for the apparent resistivity measured in thin resistant beds with the guard electrode system are given in Fig. 42. The correction becomes negligible when the bed thickness is 10 to 20 times the well diameter.

In the logging of oil reservoir rocks, it is considered that the guard electrode system is at a disadvantage since self potential logs cannot be run simultaneously with guard logs. Also, the large electrode surface used in the guard system requires a great deal of current, sometimes more than can be transmitted safely down the logging cable. The Laterolog-7 system was developed so that focussed current resistance measurements might be made with point electrodes, rather than extended electrodes. The Laterolog-7 system (and also the more recent Laterolog-8 system) uses seven small electrodes arrayed in the well as shown in Fig. 43. Guard electrodes $G_1$ and $G_2$ are located some distance up and

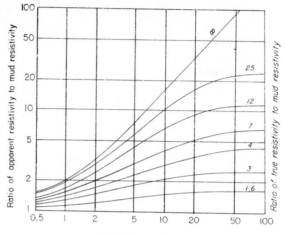

FIG. 42. Departure of apparent resistivity measured with a guard electrode array from the true resistivity for the case of a resistant layer with finite thickness.

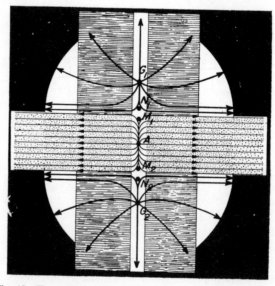

Fig. 43. (From Schlumberger Well Surveying Corporation.)

down the hole from the center electrode, A. Between each guard and the center electrode, a pair of electrodes, $M_1$–$N_1$ or $M_2$–$N_2$ is placed as shown. If no current flows up or down the hole from the center electrode, there will be no voltage detected by the electrode pairs $M_1N_1$ and $M_2N_2$. The voltage between each pair of these electrodes is monitored, and the current to the guard electrodes is varied to keep these voltages at zero. This variation in guard current keeps the beam of current from the center electrode focussed into the rock.

The geometric factor for the Laterolog-7 may be calculated by considering the array to be located in a completely homogeneous medium. If each of the seven electrodes is small enough to be treated as a point electrode, the potential at one of the $M$ monitoring electrodes will be:

$$U_M = \frac{\varrho}{4\pi} \left( \frac{I_A}{\overline{AM}} + \frac{I_G}{2\overline{G_1M_1}} + \frac{I_G}{2\overline{G_2M_1}} \right) \tag{47}$$

and at one of the $N$ monitoring electrodes:

$$U_N = \frac{\varrho}{4\pi} \left( \frac{I_A}{\overline{AN}} + \frac{I_G}{2\overline{G_1N_1}} + \frac{I_G}{2\overline{G_2N_1}} \right) \tag{48}$$

At balance, these two potentials are equal (no $MN$ voltage), so the relationship between guard current and center electrode current is:

$$\frac{2I_A}{I_G} = \frac{\overline{G_1M_1}\,\overline{G_2M_1}\,\overline{G_1N_1}\,\overline{G_2N_1}}{\overline{AM_1}\,\overline{AN_1}\,(\overline{G_2M_1}\,\overline{G_2N_1} - \overline{G_1M_1}\,\overline{G_1N_1})} = \eta \tag{49}$$

where $I_G$ is the total current to the two guard electrodes and $I_A$ is the current to the center electrode.

At balance, the potential is the same at electrode $M$ as at electrode $N$, and may be expressed in terms of the current from the center electrode only:

$$U_M = U_N = \frac{\varrho I}{4\pi} \left[ \frac{1}{\overline{AM}} + \eta \left( \frac{1}{\overline{G_1M_1}} + \frac{1}{\overline{G_2M_4}} \right) \right] \tag{50}$$

And as before, the factors containing only the dimensions of the electrode array may be grouped and called a geometric factor:

$$K = \frac{4\pi}{\dfrac{1}{\overline{AM}} + \eta \left( \dfrac{1}{\overline{G_1M_1}} + \dfrac{1}{\overline{G_2M_1}} \right)} \tag{51}$$

The Laterolog-7 system gives an accurate measurement of resistivity for beds with a thickness of twice the current beam thickness. Commonly, the beam thickness is 1 ft, so that for bed thicknesses of more than 2 ft, no corrections need be made for the departure of apparent resistivity from true resistivity.

The logs shown in Fig. 44 illustrate the superiority of the Laterolog-7 system over the conventional spacing arrays. The log obtained with a Laterolog-7 system is compared with a standard set of spacing logs for a relatively thin bed. The situation is complicated by the fact that a very saline mud with a resistivity of only 0·1 ohm-m filled the hole. Despite this, the Laterolog-7 curve comes within 75 per cent of the true resistivity of the resistant zone. Only very small deflections were recorded with the spacing logs.

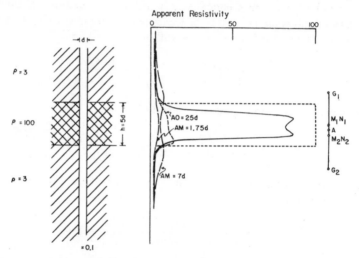

FIG. 44. Apparent resistivity curves recorded with Laterolog-7 and a standard set of spacing logs opposite a thin resistive bed with a highly conductive mud (from Pirson, 1963).

## 13. MICROLOGGING

Commonly in oil field logging, resistivity measurements are made with a very short electrode separation, with electrodes carried on a pad which is held against the wall of the well bore by a spring-loaded arm (see Fig. 45). A lateral or normal array may be used with spacings of 1 or 2 in., or the pad system may consist of a guard electrode array with a central button and having concentric ring electrodes as guards. In the oil industry, such micrologs are used to determine whether or not filtrate from the drilling mud has penetrated the rock. Permeable zones in the rock can be located by the fact that their resistivity is changed considerably by flushing.

Micro-spacing and micro-guard logs have a very high degree of resolution for thin resistant zones, far higher than even the Laterolog-7. However, the resistivity which is measured is that of a zone of rock about an inch thick lining the

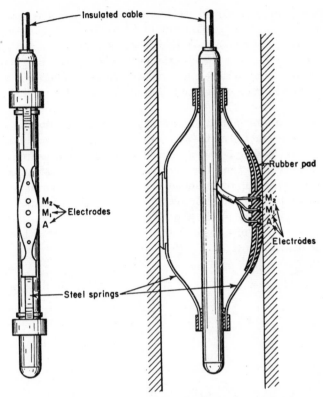

FIG. 45. (From Schlumberger Well Surveying Corporation.)

wall of the borehole. The resistivity of this rock is nearly always altered to some extent during drilling by breakage or by washing with the circulating drilling mud. A great number of fine layers with different resistivity may be logged with a micro-spacing, but this layering may not have been present before the hole was drilled. It is unlikely that micrologs can help in determining the geoelectric parameters for a section of rock.

## 14. INDUCTION LOGGING

In induction logging, current is excited in the ground around a borehole by a time-varying magnetic field generated with an induction coil. Induction logging differs from all the types of logging described so far in that galvanic contact between the logging system and the wall rock is not required. Induction logs may be run in dry holes or in holes filled with oil-base drilling mud, as well as in water-filled holes.

The primary magnetic field used in induction logging is generated by an alternating current with a frequency of 10 to 20 kc/s flowing in a small coil wrapped around the inhole tool. Eddy currents develop in the ground as the primary field varies, and these currents develop a secondary magnetic field which is proportional to the conductivity of the rock. The secondary magnetic field is measured with a second coil located from 1 to 4 ft from the transmitter coil.

If the problem is simplified to some extent, the relation between the voltage induced in the second coil and resistivity can be developed easily. If the problem is restricted to coupling between the coils at relatively lows frequencies (up to 20 kc/s), propagation time for the magnetic field and attenuation may be neglected. In setting up the problem, a system of polar coordinates, $z$, $R$ and $\varphi$, with the origin halfway between the two coils is used. The distance between the coils, $L$, is the spacing of the tool. The space around the induction tool is divided into a series of elementary toroids (rings), each with a square cross section, as shown in Fig. 46.

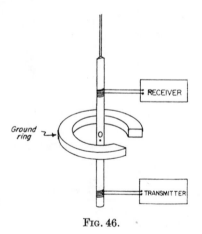

FIG. 46.

It is further assumed that the transmitting and receiving coils are small enough that the transmitter can be replaced with a small dipole magnet, and that the magnetic field is uniform throughout the area of the receiving coil. Since the coils are only about 3 in. in diameter, these approximations are realistic. The transmitter coil can be replaced with a dipole magnet having a magnetic moment:

$$M = i n_t A_t \tag{52}$$

where $i$ is the current flowing instantaneously in the coil ($i = I_0 \sin \omega t$), $\omega$ is the angular frequency of the current, $n_t$ is the number of turns in the coil, and $A_t$ is the area of the transmitter coil.

In Chapter III, Eq. (89), it will be shown that the field strength in a direction at an angle $\beta$ with the dipole axis is given by:

$$H_\beta = -\frac{2M}{r_t^3}\left[\cos\varphi_t\,\cos(\beta-\varphi_t)+\frac{1}{2}\sin\varphi_t\,\sin(\beta-\varphi_t)\right] \quad (53)$$

where the distance, $r_t$, and the angle, $\varphi_t$, are defined in Fig. 47. Picking for the moment a toroid which is located a distance $z$ above the midpoint of the pair of coils, we will consider the current induced by the time-varying magnetic

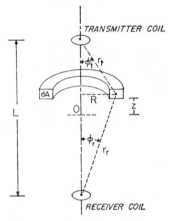

FIG. 47.

field generated by the transmitter coil. The voltage induced in a single-turn coil by a time-varying magnetic field is simply:

$$e = -\frac{d\theta}{dt} = -\frac{d}{dt}\int\mu H_z\,dA \quad (54)$$

where $\theta$ is the magnetic flux normal to the plane of the toroid, and which is found by integrating the vertical component of the magnetic field over the area of the toroid. Since we are interested only in the component of the field normal to the area of the toroid, the angle $\beta$ in Eq. (53) is zero, and the vertical component is given by:

$$H_z = \frac{2M}{r_t^3}\left(\cos^2\varphi_t - \frac{1}{2}\sin^2\varphi_t\right)$$

The integration indicated in Eq. (54) is straightforward, and produces the result:

$$e = -\frac{2\omega\mu\pi\,R^3 n_t A_t I_0\,\cos\omega t}{\left[R^2+\left(\dfrac{L}{2}-z\right)^2\right]^{3/2}} \quad (55)$$

This voltage causes an element of current to flow in the ring, according to Ohm's law:

$$\mathrm{d}I = \frac{e}{2\pi R}{\sigma\,\mathrm{d}A}$$

$$= -\frac{\sigma\omega\mu R^2 n_t A_t I_0 \cos\omega t\,\mathrm{d}A}{\left[R^2 + \left(\dfrac{L}{2} - z\right)^2\right]^{3/2}} \tag{56}$$

where $\sigma$ is the conductivity of the rock in the toroid and $\mathrm{d}A$ is the cross-section of the toroid. Each elementary toroid can be thought of as a one-turn coil generating a magnetic field at the receiver. Each of these coils has a large area, so the equation used to describe the field about the transmitter coil, which has a negligible area, cannot be used to describe the magnetic field due to current flow in the toroids. Using Ampere's law, it may be found that the magnetic field along the axis of a current-carrying coil is given by;

$$\mathrm{d}H_r = \frac{2\pi\,R\,\mathrm{d}I}{r_r^2}\sin\varphi_r$$

$$= -\frac{2\pi\,\sigma\omega\mu R n_t A_t I_0 \cos\omega t\,\mathrm{d}A}{\left[R^2 + \left(\dfrac{L}{2} - z\right)^2\right]^{3/2}\left[R^2 + \left(\dfrac{L}{2} + z\right)^2\right]^{3/2}} \tag{57}$$

where the distance $r_r$ and the angle $\varphi_r$ are defined in Fig. 47. Since this secondary magnetic field is also time-variant, it induces a voltage in the receiving coil:

$$\mathrm{d}e_r = -\mu n_r A_r \frac{\mathrm{d}H_r}{\mathrm{d}t}$$

$$= \frac{2\pi\,\sigma\omega^2\mu^2 R^3 n_t n_r A_t A_r I_0 \cos\omega t\,\mathrm{d}A}{\left[R^2 + \left(\dfrac{L}{2} - z\right)^2\right]^{3/2}\left[R^2 + \left(\dfrac{L}{2} + z\right)^2\right]^{3/2}} \tag{58}$$

where $n_r$ and $A_r$ are the number of turns and the cross-sectional area of the receiving coil, respectively.

The total voltage at the receiving coil is found by integrating the contributions of all elementary toroids. The parameters included in Eq. (58) may be grouped in two sets, one containing all those parameters which are determined by the construction of the logging tool, and the other including the geometry of the tool:

$$K_i = \frac{4\pi\omega^2 n_t n_r A_t A_r}{L} \tag{59}$$

$$B = \frac{L}{2} \frac{R^3}{\left[R^2 + \left(\dfrac{L}{2} - z\right)^2\right]^{3/2} \left[R^2 + \left(\dfrac{L}{2} + z\right)^2\right]^{3/2}} \tag{60}$$

Upon integration, Eq. (58) may be written as:

$$\sigma = K_i \frac{e}{i} \frac{1}{\int B \, dA} \tag{61}$$

The function $B$ is a weighting factor for the position of the various elementary toroids throughout space. If the medium around the induction tool is uniform in properties, the integral $\int B \, dA$ turns out to be unity (for an elementary proof, see Dakhnov, 1959). The function, $K_i$, is then merely a geometric factor for converting the measured ratio of received voltage to driving current into conductivity.

The voltage developed in the receiver of an induction tool is proportional to the rock conductivity, and to the square of the frequency used. Normally, induction logs are restricted to use in relatively conductive rocks (rocks with resistivities less than 100 ohm-m), since not enough response can be obtained in resistive rocks. Results might be obtained in resistive rocks bey using frequencies higher than those now used, but higher frequencies would lead to problems in that appreciable attenuation of the magnetic field might take place between the transmitter and receiver.

Induction logs provide accurate results even in relatively thin beds, since current is induced to flow in the bedding planes, rather than across the bedding planes, as in the case of spacing logs. Calculated logs for an induction system operating through thin conductive or resistive beds are shown in Fig. 48. Moderate corrections are required if the bed thickness is about the same as the coil spacing (the curves are calculated for a spacing equal to 2·5 times the well diameter, so the thinner beds shown have a thickness equal to 1·6 times the coil spacing). Better response may be obtained in thin beds by using a six-coil system (which the Schlumberger Well Surveying Corp. calls the 6FF40 tool). The additional coils are transmitter coils driven in such a way that the current induced near the receiver coil is increased, and currents induced further from the receiver coil cancel. Several of the curves shown in Fig. 48 are calculated for the 6FF40 system. There is essentially no departure of the measured resistivity from the true resistivity of the rock when the layer thickness is greater than the tool spacing.

The induction log may occasionally measure an erroneously low resistivity in permeable beds. Filtration of salt water from the drilling mud into permeable zones may reduce the resistivity of a toroidal zone about the borehole, and the induction log will measure the resistivity of this zone rather than that of the uncontaminated rock.

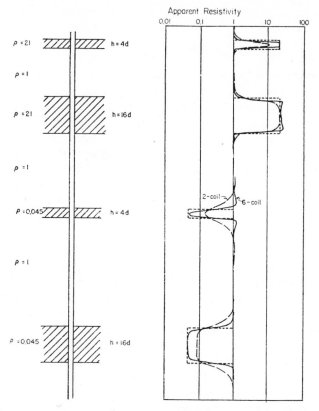

FIG. 48. Computed examples of the apparent resistivity curves which would be recorded with two-coil and six-coil induction systems for either thin conductive or thin resistive layers. The hole diameter is $d$; curves are computed for a coil separation equal to $2 \cdot 5\ d$.

## 15. CALCULATION OF GEOELECTRIC PARAMETERS FROM ELECTRIC LOGS

Electric logs are of interest in a geoelectric exploration program, since they may be used to provide control in the interpretation of surface surveys. Logs may be used to compute the five parameters for the geoelectric section defined in Section 6. From the preceding sections, we may arrive at the conclusion that accurate computations may be made from either laterologs or from induction logs, since both types of logs provide accurate resistivity data for beds whose thickness is several times the well diameter. It is not immediately obvious that spacing logs, which are generally the only type available from wells drilled prior to 1955, can be used in computing the geoelectric parameters. Such logs are subject to large errors if beds are not many feet thick, and even though corrections may be applied for the effect of bed thickness, such corrections are tedious, and probably not as accurate as might be desired.

Since most of the electric logs which are available were made with one of the spacing methods, they must be used much of the time in calculating the geoelectric parameters. Although errors are certainly involved when the section contains finely layered rocks, the practice has been to consider that the coefficient of anisotropy for rocks consists of two parts; a macroanisotropy, for the case in which the rocks comprising a geoelectric section are layered so coarsely that electric logs can be used to determine the true resistivity of each layer, and microanisotropy, or layering on too fine a scale for individual layers to be resolved with an electric log. The distinction between macroanisotropy and microanisotropy is largely artificial, depending on the resolution of the type of electric log used, and this depends on the size hole in which the logs were run and on the conductivity of the drilling mud. Even though the distinction is artificial, it is a useful device to emphasize that the values of geoelectric parameters calculated from logs are by no means absolute values. Generally, the value for longitudinal resistivity is more nearly correct than the other parameters; most errors are in the direction of underestimating the resistivity of resistant zones. The longitudinal resistivity is insensitive to errors made in picking the high values of resistivity, but the transverse resistivity is highly sensitive to such errors. As a result, both the transverse resistivity and coefficient of anisotropy are usually estimated on the low side.

In using electric logs to calculate the geoelectric parameters, the thickness and apparent resistivity for each layer recorded on the electric log may be compiled. However, if there are a great number of small units within the geoelectric section, it may be simpler to assume that the electric log represents a series of beds, each 1 or 2 ft thick, with each of these beds having a uniform resistivity equal to the value read from the log at the midpoint of the 1- or 2-ft interval (see Fig. 49 for an illustration of the meaning of this approximation). This assumption simplifies analysis when the logs can be converted to a digital form for machine computation on a high-speed computer.

A compilation of geoelectric parameters from well logs of wells in northeastern Colorado will serve to show the use of geoelectric parameters. Electric logs were selected for analysis from deep wells in the northeast quarter of Colorado to provide quantitative information about the sedimentary column covering the basement complex in the High Plains area. The sedimentary section in this area consists of Paleozoic to Tertiary beds at the west edge; only Paleozoic and lower Mesozoic beds at the east edge. There is a slight regional dip from the east into the Denver Basin, but essentially, beds are flat-lying. The section can be divided into 3 or 4 units on the basis of distinct resistivity contrasts, with resistivity generally increasing with depth. Exceptions to this sequence occur when a resistant surface layer is present, or high resistivity evaporites and carbonates are included in one of the units.

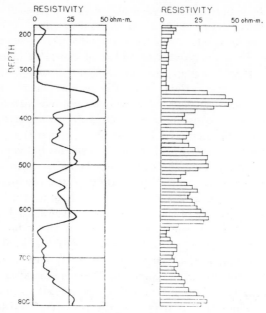

Fɪɢ. 49. Approximation used in calculating electrical parameters from resistivity logs.

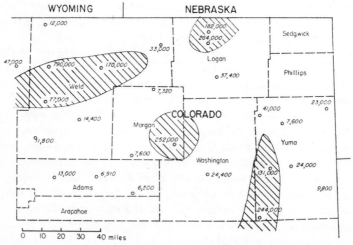

Fɪɢ. 50. Areal variation of the transverse resistance of all sedimentary rocks below the Pierre Shale in northeastern Colorado. Values of transverse resistance as computed from electric logs are indicated in ohms × m². Areas of unusually high transverse resistance are shaded.

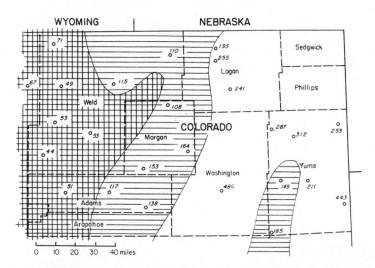

FIG. 51. Areal variations of the longitudinal conductance of all sedimentary rocks below the Pierre Shale in northeastern Colorado. Values of conductance as computed from electric logs are indicated in mhos. Areas of low and moderate conductance are indicated by shaded patterns.

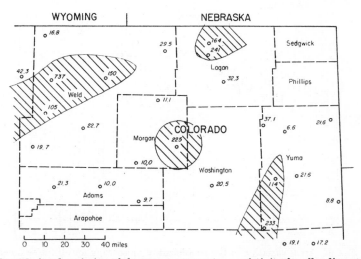

FIG. 52. Areal variation of the average transverse resistivity for all sedimentary rocks below the Pierre Shale in northeastern Colorado. Values of transverse resistivity as computed from electric logs are indicated in ohm-m. Areas of unusually high transverse resistivity are indicated by shading.

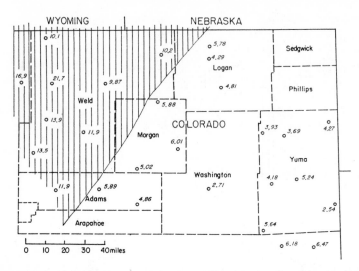

FIG. 53. Areal variation of the average longitudinal resistivity of all the sedimentary rocks below the Pierre Shale in north eastern Colorado. Values of longitudinal resistivity computed from electric logs are indicated in ohm-m. Areas of high longitudinal resistivity are indicated by shading.

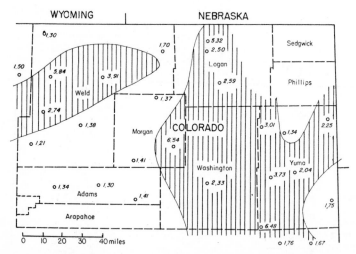

FIG. 54. Areal variation of the coefficient of anisotropy for all the sedimentary rocks below the Pierre Shale in northeastern Colorado. Values for the coefficient of anisotropy as computed from electric logs are indicated at each well location. Areas with an unusually high coefficient are shown with a shaded pattern.

Figures 50 to 54 are maps showing the areal variations in each of the geo-electric parameters for the portion of the section extending from the top of the Niobrara Formation (Upper Cretaceous) to the basement, but excluding the Pierre shale (Upper Cretaceous) and younger rocks. The resistivity curves used in this compilation were the short normal, induction and laterolog. Wherever possible, laterologs were used for the high resistivity carbonate sections. Induction logs were used instead of the short normal logs in conductive sections when they were available.

The values for each parameter are grouped into areas of high or low values, since contouring of such widely spaced data points would be misleading. Of particular interest are the areas of high anisotropy, transverse resistance and transverse resistivity. In most cases these values are high due to relatively thin zones, often not over 200 ft thick, collectively, of very highly resistive eva-porites and carbonates. These zones have resistivities approaching 5000 ohm-m, in contrast to the shale in the same section which has a resistivity of only 5 to 10 ohm-m. The high resistivity zones change electrical character abruptly. Areas of high transverse resistance lie adjacent to areas of low transverse resistance with no transition zones apparent. These high resistivity zones do not affect the conductance or longitudinal resistivity to any appreciable extent, and these parameters reflect primarily the increasing shale content of the section towards the cast.

# REFERENCES

DAKHNOV, V. N. (1959) Geophysical well logging, translated into English in *Quart. Colo. School Mines* **57**, no. 2, April, 1962.

DOLL, H. G. (1949) Introduction to induction logging and application to logging of wells drilled with oil base mud, *J. Petrol. Technol.*, June.

DOLL, H. G. (1951) The Laterolog: a new resistivity logging method with electrodes using an automatic focussing system, *Trans. AIME* **192**, 305.

DOLL, H. G. (1953) The Microlaterolog, *Trans. AIME* **198**, 17.

DUESTERHOEFT, W. C. (1961) Propagation effects in induction logging; *Geophysics* **26** (2), 192.

KELLER, G. V. (1949) An improved electrode system for use in electric logging, *Producers Monthly*, August.

LISHMAN, J. R. (1961) Salt bed identification from unfocused resistivity logs, *Geophysics* **26** (3), 320.

PIRSON, S. J. (1963) *Handbook of Well Log analysis*, Prentice-Hall, Englewood Cliffs. N. J., 326 pp.

SCHLUMBERGER, C. and M. and E. G. LEONARDON (1934) Electrical coring: a method of determining bottom hole data by electrical measurements, AIME, Geophysical Prospecting, p. 238.

*Schlumberger Documents* 3, 7 and 8 (1947, 1949, 1955), Schlumberger Well Surveying Corporation, Houston, Texas.

# CHAPTER 3

# GALVANIC RESISTIVITY METHODS

THE MOST commonly used methods for measuring earth resistivity are those in which current is driven through the ground using galvanic contacts. Generally, four-terminal electrode arrays are used, since the effect of material near the current contacts can be minimized. Current is driven through one pair of electrodes; the potential established in the earth by this current is measured with the second pair of electrodes. A great variety of electrode arrangements have been used to measure earth resistivities, but essentially they may be grouped in three classes:

    **1.** Arrangements in which the *potential difference* between two widely spaced measuring electrodes is recorded.

    **2.** Arrangements in which a *potential gradient or electric field intensity* is measured using a closely spaced pair of measuring electrodes.

    **3.** Arrangements in which the *curvature* of the potential function is measured using a closely spaced current electrode pair as well as a closely spaced measuring electrode pair.

Any one of these arrays may be used to study variations in resistivity with depth or lateral variations in resistivity. In studying the variation of resistivity with depth, as in the case of a layered medium, the spacings between the various electrodes are gradually increased. With larger spacings, the effect of material at depth on the measurements becomes more pronounced. In studying lateral variations such as might be associated with dike-like structures or faults, a fixed separation is maintained between the various electrodes and the array is moved as a whole along a traverse line. The first type of measurement is called a *vertical sounding*; the second, *horizontal profiling*. In some of the early literature, these two types of electrical surveys were called by the descriptive but rather optimistic names "electrical coring" and "electrical trenching", respectively.

## 16. INSTRUMENTATION

In resistivity measurements, instrumentation is usually simple. Current is generally provided from dry batteries. This current may take the form of a single long direct-current surge, or it may be a low frequency square wave with a commutator or system of relays being used to alternate the direction of current flow at time intervals ranging from a tenth of a second to tens of seconds. In some cases, silicon-controlled rectifiers may be used to turn the battery current on and off, rather than mechanical systems.

A low frequency sinusoidal current might also be used, but the necessity of constructing some sort of power oscillator to provide such a waveform makes it less practical than the simpler square wave generating devices. In any event, a sufficiently low frequency is used that induction effects and attenuation (to be discussed in later sections) can be neglected. The value of resistivity measured is that for direct current, even though a low frequency alternating current is actually used.

Current electrodes are generally steel or copper-clad steel stakes driven a few inches into the ground. In dry areas, the soil around the electrodes may have to be moistened to improve contact. Where bare rock is exposed at the surface it may not be possible to drive a stake into the ground, and in such a case, a current electrode may be formed by building a small mud puddle around a piece of copper screening.

Single stakes may have a contact resistance ranging from as low as 10 ohms in moist, clayey soil to as high as tens of thousands of ohms in dry, sandy soil or in soil which is composed primarily of humus or vegetable material. Contact resistance in ice or in frozen ground is even higher, ranging between 100,000 ohms and 100 megohms, depending on the temperature of the frozen material. If contact resistance must be reduced below that which can be obtained with a single stake, many stakes driven into the ground a few feet apart may be connected in parallel. The overall resistance is approximately equal to the average resistance for a single stake divided by the number of stakes connected in parallel.

The voltage between the measuring electrodes is usually measured with a potentiometer, though a voltmeter may be used. The pickup voltage is first converted to a d.c. voltage by synchronous rectification—a system using a commutator or relay to reverse the polarity of the voltage from the measuring electrodes each time the direction of current flow in the ground is reversed. A voltage as small as 1 mV may be measured with an accuracy of 1 per cent with the better quality potentiometers used in earth resistivity equipment.

In recent equipment, a transistorized amplifier which is tuned at the commutation frequency is used to amplify the pickup voltage before it is measured. Such a system discriminates against the low frequency electrical noise, either industrial or natural in origin, which limits the sensitivity that can be used in measuring equipment. With tuned receiving equipment, voltages as small as 10 $\mu$V may be measured with an accuracy of 2 per cent.

The requirements for electrodes in the measuring circuit are somewhat different than those for electrodes in the current circuit. Contact resistance is not nearly so important in the case of pickup electrodes as in the case of current electrodes. However, measuring electrodes must be stable electrically, particularly if slow commutation rates are used. When a copper or steel stake is driven into the ground, the potential difference between the metal in the electrode and the electrolytic solution in the soil pores may take minutes to reach equilibrium and may vary erratically during this time. When rapid commutation rates are used, the electrode potential variations may average out through many reversals, but such variations are serious if the reversal period for the current is measured in seconds.

A stable electrode may be obtained by using a non-polarizing electrode—an electrode consisting of a metal bar immersed in a solution of one of its salts carried in a fritted ceramic cup. Such electrodes are called "porous pots". The metal which is used may be copper and the solution copper sulphate, or silver metal in a silver nitrate solution may be used. If the solution carries an excess of salt in crystal form, it will always be saturated, and if current momentarily passes through the electrode, there is no change in the concentration of the electrolyte, and the electrode potential remains essentially constant. The ceramic cups used in porous pot electrodes must be permeable enough that water flows slowly through to maintain contact between the electrode and the soil moisture. This water must be replaced frequently.

Cables for connecting the current electrodes to the power source or the measuring electrodes to the measuring circuit present no special requirements. Lightweight cables are suitable except when very heavy currents are needed. It is important that a high quality insulation be used since leakage between the current circuit and the measuring circuit is one of the primary sources of error in resistivity measurements. Lightweight insulated magnet wire may be used for once-only measurements in areas where leaving such wire lying on the ground presents no hazard.

## 17. APPARENT RESISTIVITY

The simplest approach to the theoretical study of earth resistivity measurements is to consider first the case of a completely homogeneous isotropic earth. An equation giving the potential about a single point source of current can be developed from two basic considerations:

1. *Ohm's law*:

$$E = \varrho j \tag{62}$$

where $E$ is the potential gradient, $j$ is the current density and $\varrho$ is the resistivity of the medium.

2. *The divergence condition*:

$$\varDelta \cdot j = 0 \tag{63}$$

which states that all the current going into a chunk of material must leave the other side, unless there is a source or sink for current within the chunk. The divergence of the current density vector must be zero every place but at the current source.

These two equations may be combined to obtain Laplace's equation:

$$\Delta \cdot \boldsymbol{j} = \frac{1}{\varrho} \Delta \cdot \boldsymbol{E} = -\frac{1}{\varrho} \nabla^2 U = 0 \tag{64}$$

where $U$ is a scalar potential function defined such that $\boldsymbol{E}$ is its gradient. In polar coordinates, the Laplace equation is:

$$\frac{\partial}{\partial r}\left(r^2 \frac{\partial U}{\partial r}\right) + \frac{1}{r^2 \sin\theta} \frac{\partial}{\partial \theta}\left(\sin\theta \frac{\partial U}{\partial \theta}\right) + \frac{1}{r^2 \sin^2\theta} \frac{\partial^2 U}{\partial \psi^2} = 0 \tag{65}$$

If only a single source of current is considered, complete symmetry of current flow with respect to the $\theta$ and $\psi$ directions may be assumed, so that derivatives taken in these directions may be eliminated from Eq. (65):

$$\frac{\partial}{\partial r}\left(r^2 \frac{\partial U}{\partial r}\right) = 0 \tag{66}$$

This equation may be integrated directly:

$$r^2 \frac{\partial U}{\partial r} = C$$

$$U = -\frac{C}{r} + D \tag{67}$$

Defining the level of potential at a great distance from the current source as zero, the constant of integration, $D$, must also be zero. The other constant of integration, $C$, may be evaluated in terms of the total current, $I$, from the source. In view of the assumed symmetry of current flow, current density should be uniform through the surface of a small sphere with radius, $a$, drawn around the current source. The total current may be expressed as the integral of the current density over this surface:

$$I = \int_S \boldsymbol{j} \cdot \mathrm{d}s = \int_S \frac{E}{\varrho} \, \mathrm{d}s = \int_S \frac{C}{\varrho r^2} \, \mathrm{d}s = -\frac{2\pi C}{\varrho} \tag{68}$$

This equation may be solved for the constant of integration, $C$, and this value substituted in Eq. (67) for the potential function:

$$U = \frac{\varrho I}{2\pi r} \tag{69}$$

Potential functions are scalars, and so, may be added arithmetically. If there are several sources of current rather than the single source assumed so far, the total potential at an observation point may be calculated by adding the potential contributions from each source considered independently. Thus, for $n$ current sources distributed in a uniform medium, the potential at an observation point, M, will be:

$$U_M = \frac{\varrho}{2\pi}\left[\frac{I_1}{a_1} + \frac{I_2}{a_2} + \cdots + \frac{I_n}{a_n}\right] \tag{70}$$

where $I_n$ is the current from the $n$th in a series of current electrodes and $a_n$ is the distance from the $n$th source at which the potential is being observed.

Equation (70) is of practical importance in the determination of earth resistivities. The physical quantities measured in a field determination of resistivity are the current, $I$, flowing between two electrodes; the difference in potential, $\Delta U$, between two measuring points, M and N, and the distances between the various electrodes. Thus, the following equation applies for the ordinary four terminal arrays used in measuring earth resistivity:

$$\varrho = \left(\frac{U_M - U_N}{I}\right)\frac{2\pi}{\dfrac{1}{\overline{AM}} - \dfrac{1}{\overline{BM}} - \dfrac{1}{\overline{AN}} + \dfrac{1}{\overline{BN}}} = K\frac{\Delta U}{I} \qquad (71)$$

The minus signs for two of the four reciprocal distance terms arise since one of the current poles in a normal two-electrode current circuit must have a negative sense of current flow compared to the other.

Equation (71) can be used to compute the resistivity of the earth only if the earth is completely uniform. If the earth is not uniform, Eq. (71) may be used as a definition for the *apparent resistivity* of the earth. The apparent resistivity may bear no relation to any actual value of resistivity in a heterogeneous earth. It may be larger or smaller than the actual resistivity, or in rare cases, the apparent resistivity may be identical with one of the true resistivity values in a heterogeneous earth.

The quantities in Eq. (71) which represent only the effect of electrode separation distances may be segregated and expressed as a *geometric factor*:

$$K = \frac{2\pi}{\dfrac{1}{\overline{AM}} - \dfrac{1}{\overline{BM}} - \dfrac{1}{\overline{AN}} + \dfrac{1}{\overline{BN}}} \qquad (72)$$

The *Wenner* array, in which a potential difference is measured, is one of the most commonly used electrode arrays for determining resistivity. In the Wenner array, four electrodes are equally spaced along a straight line, as shown in Fig. 55. The distance between any two adjacent electrodes is called the array *spacing, a*. The geometric factor for the Wenner array is:

$$K = \frac{2\pi}{\dfrac{1}{a} - \dfrac{1}{2a} - \dfrac{1}{2a} + \dfrac{1}{a}} = 2\pi a \qquad (73)$$

The *Lee* modification of the Wenner array uses a third measuring electrode at the midpoint, 0, of the ordinary Wenner array. A potential difference is then measured between both M and N and the center electrode, 0, and apparent resistivities calculated for each half of the array. The advantage of such a system is said to be that horizontal changes in resistivity may be recognized by comparing the apparent resistivities measured with each half of the array. The geometric factor for one half of the Lee array is:

$$K = \frac{2\pi}{\dfrac{1}{\overline{AM}} - \dfrac{1}{\overline{BM}} - \dfrac{1}{\overline{AO}} + \dfrac{1}{\overline{BO}}} = \frac{2\pi}{\dfrac{1}{a} - \dfrac{1}{2a} - \dfrac{3}{2a} + \dfrac{1}{2a}} = 4\pi a \quad (74)$$

The *Schlumberger* array, which also is widely used in measuring earth resistivities, is designed to measure approximately the potential gradient. In the Schlumberger array, two closely spaced measuring electrodes are placed midway

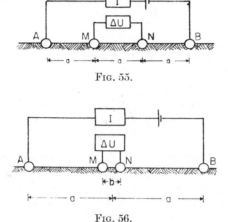

Fig. 55.

Fig. 56.

between two current electrodes, as shown in Fig. 56. The spacing for this array is taken as half the distance between the current electrodes. Using Eq. (72) to determine the geometric factor:

$$K = \frac{2\pi}{\dfrac{1}{a - \dfrac{b}{2}} - \dfrac{1}{a + \dfrac{b}{2}} - \dfrac{1}{a + \dfrac{b}{2}} + \dfrac{1}{a - \dfrac{b}{2}}} = \pi\left(\frac{a^2}{b} - \frac{b}{4}\right) \quad (75)$$

The potential gradient cannot be determined exactly, since it is defined as the limit of the ratio of voltage to spacing as two measuring electrodes are moved closer together. In the limiting case of zero separation, no voltage could be measured. In the Schlumberger array, the measuring electrodes are placed closely enough together that the ratio of measured voltage to separation approximately equals the voltage gradient at the midpoint of the current spread. The error introduced by having a measureable separation between the measuring electrodes can be evaluated by considering the geometric factor for the case in which the gradient, $E$, is actually measured.

According to Eq. (69), the potential gradient at a distance $a$ from a single current source in a homogeneous isotropic half-space is:

$$E = -\frac{\partial U}{\partial r} = \frac{\varrho I}{2\pi r^2} \tag{76}$$

Since $E$ is a vector quantity, the electric fields from several sources must be added vectorially. At the midpoint between two sources, the electric field due to each source has the same magnitude and direction, since the sources have opposite polarities. The total electric field at the midpoint is exactly twice the electric field contributed by a single source:

$$E = \frac{\varrho I}{\pi a^2} \tag{77}$$

Solving this expression for a geometric factor:

$$\varrho_a = \frac{\pi a^2}{I} \frac{\partial U}{\partial r} = \frac{\pi a^2}{I} \frac{\Delta U}{b}$$

$$K = \pi \frac{a^2}{b} \tag{78}$$

This expression is very similar to the expression for the geometric factor given in Eq. (75) for the case in which the electric field is only approximately determined. The degree of approximation can be evaluated by expressing the difference between the two expressions for the geometric factor as:

$$e = \frac{K_{\text{ex}} - K_{\text{ap}}}{K_{\text{ex}}} \times 100$$

where $e$ is the percentage difference between the two expressions, $K_{\text{ex}}$ is the geometric factor if the potential gradient is measured exactly, and $K_{\text{ap}}$ is the geometric factor in the case in which the potential gradient is measured only approximately with a pair of closely spaced electrodes. Substituting the appropriate expressions from Eqs. (75) and (78) in the above equation, it is possible to determine the maximum separation, $b$, which may be used in the Schlumberger array, so that the apparent resistivity will be within a given error limit, $e$, of the value that would be determined with a true measurement of electric field:

$$b^2 = \frac{4e}{1 + e} a^2 \tag{79}$$

Ordinarily, it is desirable that the geometric factor be within 5 per cent of the ideal value when the Schlumberger array is used. Placing this limit on the error, $e$, in Eq. (79), we find that the separation between the measuring electrodes should be less than 0·435 times the spacing factor, $a$.

Methods utilizing measurements of the curvature of a potential field are not as commonly used as either the Wenner or Schlumberger methods. Several varieties of methods using dipoles (closely spaced electrode pairs) for both the current and measuring circuits are used occasionally. In the *polar dipole* system, four electrodes are aligned, with both the current electrode pair and the measuring electrode pair closely spaced in comparison with the distance between the pairs (see Fig. 57). The exact expression for the geometric factor is:

$$K = \frac{2\pi}{\dfrac{1}{a + \dfrac{c}{2} - \dfrac{b}{2}} - \dfrac{1}{a - \dfrac{c}{2} - \dfrac{b}{2}} - \dfrac{1}{a + \dfrac{c}{2} + \dfrac{b}{2}} + \dfrac{1}{a - \dfrac{c}{2} + \dfrac{b}{2}}} \qquad (80)$$

If the separation between both pairs of electrodes is the same, $b$, this expression is simplified considerably:

$$K = \pi \left( \frac{a^3}{b^2} - a \right) \qquad (81)$$

In the limiting case in which the separation between the current electrode pair and the measuring electrode pair is reduced to zero simultaneously, it can be shown that the curvature of the potential field is being measured. First,

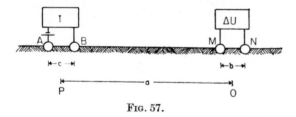

Fig. 57.

considering only the current from electrode A, the measuring electrodes would detect the potential gradient at the point O, the midpoint of the measuring electrode pair:

$$\frac{\Delta U_1}{\overline{MN}} = \frac{\partial U}{\partial r} \bigg|_{O,A}$$

Next, consider the potential difference between electrodes M and N due only to the current from electrode B:

$$\frac{\Delta U_2}{\overline{MN}} = \frac{\partial U}{\partial r} \bigg|_{O,B}$$

The total potential difference measured between electrodes M and N is:

$$\frac{\Delta U}{\overline{MN}} = \frac{\Delta U_1 - \Delta U_2}{\overline{MN}} \cong \frac{\partial U}{\partial r} \bigg|_{O,A} - \frac{\partial U}{\partial r} \bigg|_{O,B} \cong \overline{AB}\, \frac{\partial^2 U}{\partial r^2} \qquad (82)$$

The exact expression for the curvature of the potential field may be obtained also by taking the second derivative of the expression for potential in Eq. (69):

$$\frac{\partial^2 U}{\partial r^2} = \frac{\varrho I}{4\pi r^3} \tag{83}$$

Equating the left-hand expression in Eq. (82) with the right-hand expression in Eq. (83), we have:

$$\varrho_a = \frac{\pi r^3}{I} \frac{\partial^2 U}{\partial r^2} \cong \pi \frac{a^3}{bc} \frac{\Delta U}{I} \tag{84}$$

The geometric factor for an idealized polar dipole array is:

$$K = \pi \frac{a^3}{bc} \tag{85}$$

This expression is very similar to the one given in Eq. (20).

The measuring electrode pair need not be oriented in line with the current electrode pair. When the orientation of the current dipole is not the same as the orientation of the measuring dipole, as in the example shown in plan view

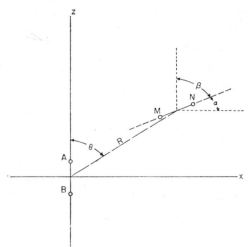

Fig. 58.

in Fig. 58, the geometric factor may be determined approximately as follows. The potential at point M due to the current sources at A and B is:

$$U_M = \frac{\varrho I}{2\pi} \left( \frac{1}{AM} - \frac{1}{BM} \right)$$

If the distance from the current dipole to the measuring point is large compared with the dipole length, we can use the distance, $R$, from the midpoint of the current dipole to the midpoint of the measuring dipole rather than the exact distances AM and BM:

$$AM = R - \tfrac{1}{2}b\cos\theta$$
$$BM = R + \tfrac{1}{2}b\cos\theta$$

And so, the potential at point M is approximately:

$$
\begin{aligned}
U_M &= \frac{\varrho I}{2\pi}\left[\frac{1}{R - \tfrac{1}{2}b\cos\theta} - \frac{1}{R + \tfrac{1}{2}b\cos\theta}\right]\\
&= \frac{\varrho I}{2\pi}\frac{b\cos\theta}{R^2 - \tfrac{1}{2}b^2\cos^2\theta}\\
&\cong \frac{\varrho I}{2\pi}\frac{b\cos\theta}{R^2} ;\quad R^2 \gg b^2
\end{aligned}
\tag{86}
$$

The dipole MN is used to determine the component of the gradient of this potential function along the line connecting points M and N. Since Eq. (68) is expressed in polar form, the components of gradient should be considered in radial and tangential directions. The radial component of the gradient at the midpoint of the measuring dipole is:

$$\frac{\partial U}{\partial r} = -\frac{\varrho I}{\pi}\frac{b\cos\theta}{R^3} \tag{87}$$

and the tangential component of the gradient is:

$$\frac{1}{R}\frac{\partial U}{\partial\theta} = \frac{\varrho I}{2\pi}\frac{b\sin\theta}{R^3} \tag{88}$$

These components must be projected onto the direction established by the axis of the measuring dipole, which is arbitrary. The component of gradient along the measuring dipole direction is:

$$\frac{\partial U}{\partial n} = \frac{\varrho I b}{\pi R^3}\left[\cos\theta\cos(\beta-\theta) + \frac{1}{2}\sin\theta\sin(\beta-\theta)\right] \tag{89}$$

The geometric factor for any ideal dipole array may be extracted from this expression:

$$K = \frac{\pi R^3}{bc[\cos\theta\cos(\beta-\theta) + \tfrac{1}{2}\sin\theta\sin(\beta-\theta)]} \tag{90}$$

This expression reduces to the one previously obtained for the polar dipole array if it is assumed that the angles $\theta$ and $\beta$ are both zero. Another special dipole array which is sometimes used is the *equatorially displaced dipole* array in which the current dipole and measuring dipole are parallel and their centers fall along a line perpendicular at their midpoints. In this case, the angles $\theta$ and $(\beta - \theta)$ have a value $\pi/2$, and Eq. (90) reduces to:

$$K = \frac{2\pi R^3}{bc} \tag{91}$$

It should be noted that the electric field at a given distance along the polar axis of a dipole source is twice the amplitude of the electric field measured at the same distance away along the dipole equator.

In each of the arrays described so far, all four electrodes are close enough to one another that each must be considered in calculating the geometric factor. Often, an array may be used in which one or two of the electrodes are placed so far from the rest of the array that the presence of these electrodes may be ignored. The current electrode, B, may be located far enough away from the rest of the electrodes that the terms 1/BM and 1/BN in the general expression for geometric factor are practically zero. In order for such an approximation to be valid, the electrode B must be at least 20 times further away from the rest of the electrodes than any of the other interelectrode distances. If there are horizontal resistivity contrasts, this distance must be increased by the ratio of the maximum to the minimum resistivity. When one current electrode is removed from the Wenner or Schlumberger array, the geometric factor is doubled.

A *single-pole* array is an array in which one current electrode and one measuring electrode may be placed far away from the other two (and from each other), so that the only term in the expression for geometric factor which has an appreciable size is the term 1/AM. The single-pole system measures the potential about a single-pole current source, to a good approximation.

Another variation in the standard arrays which may be made is the interchange of the measuring and potential electrodes. For example, the inside electrodes in the Schlumberger array may be used for current electrodes, and the outside electrodes for measuring electrodes. Such a change may be required for safety when very large electrode separations are used and it is not wise to have many kilometers of current cable laid out on the ground. According to the *theorem of reciprocity*, no change will be observed in the measured voltage if the roles of the measuring electrodes and of the current electrodes are interchanged. Reciprocity can be readily confirmed for an electrode array over a homogeneous earth. In the equation for geometric factor, the mutual distances are preserved, even though the roles of the electrodes are interchanged. Thus, the expression has the same numerical value for a reciprocal array as for the standard array.

## 18. HOMOGENEOUS ANISOTROPIC EARTH POTENTIAL

Even though the earth may be homogeneous on a large scale in some instances, this does not mean it is necessarily isotropic. The behavior of the single-pole potential function in a homogeneous, anisotropic medium is the next logical step in complexity beyond the completely homogeneous isotropic case. Consider a homogeneous but anisotropic half-space with the origin of a cartesian coordinate system at the point, A, on the surface of this half-infinite earth, A being the location of a current electrode. The resistivity of the medium is defined by two numerical values, $\varrho_l$ along the horizontal direction and $\varrho_t$ in the vertical direction. At each point in the medium, the current density vector has the components:

$$j_x = \frac{E_x}{\varrho_l} = -\frac{1}{\varrho_l}\frac{\partial U}{\partial x}$$

$$j_y = \frac{E_y}{\varrho_l} = -\frac{1}{\varrho_l}\frac{\partial U}{\partial y}$$

$$j_z = \frac{E_z}{\varrho_t} = -\frac{1}{\varrho_t}\frac{\varrho U}{\varrho z} \tag{92}$$

The usual requirement that the divergence of the current density vector is zero except at the electrode A must be satisfied:

$$\text{div } \mathbf{J} = \frac{\partial j_x}{\partial x} + \frac{\partial j_y}{\partial y} + \frac{\partial j_z}{\partial z} = \frac{1}{\varrho_l}\left(\frac{\partial^2 U}{\partial x^2} + \frac{\partial^2 U}{\partial y^2}\right) + \frac{1}{\varrho_t}\frac{\partial^2 U}{\partial z^2} = 0 \tag{93}$$

This equation may be transformed into Laplace's equation by a simple transformation in variables, which corresponds to stretching the anisotropic space into an isotropic space:

$$\xi = \varrho_l^{1/2}\,x$$

$$\eta = \varrho_l^{1/2}\,y$$

$$\zeta = \varrho_t^{1/2}\,z \tag{94}$$

When this transformation is made, Eq. (93) becomes Laplace's equation which can be solved by direct integration if it is expressed in polar coordinates with $R = (\xi^2 + \eta^2 + \zeta^2)^{1/2}$. Inasmuch as there is symmetry with respect to the two angular coordinates, Laplace's equation is:

$$\frac{\partial}{\partial R}\left(R^2\frac{\partial U}{\partial R}\right) = 0 \tag{95}$$

Integrating once with respect to $R$:

$$R^2\frac{\partial U}{\partial R} = C \tag{96}$$

where $C$ is a constant of integration. Dividing both sides of this equation by $R^2$, we have a form which can be integrated again, yielding an expression for the potential, $U$:

$$U = -\frac{C}{R} + D \tag{97}$$

The constant of integration, $D$, may be evaluated by requiring that the potential at a great distance from the source must be zero; therefore, $D$ must be zero. The constant, $C$, is evaluated by considering the total current flowing from the source electrode, A. The components of the current density vector in cartesian coordinates are:

$$j_x = -\frac{1}{\varrho_l}\frac{\partial U}{\partial x} = \frac{Cx}{\varrho_l^{3/2}(x^2 + y^2 + \lambda^2 z^2)^{3/2}}$$

$$j_y = -\frac{1}{\varrho_l}\frac{\partial U}{\partial y} = \frac{Cy}{\varrho_l^{3/2}(x^2 + y^2 + \lambda^2 z^2)^{3/2}}$$

$$j_z = -\frac{1}{\varrho_l}\frac{\partial U}{\partial z} = \frac{Cz}{\varrho_l^{3/2}(x^2 + y^2 + \lambda^2 z^2)^{3/2}} \tag{98}$$

The magnitude of the current density vector is:

$$|J| = (j_x^2 + j_y^2 + j_z^2)^{1/2} = \frac{C(x^2 + y^2 + z^2)^{1/2}}{\varrho_l^{3/2}(x^2 + y^2 + \lambda^2 z^2)^{3/2}} \tag{99}$$

The following relationships exist between the polar coordinates and the cartesian coordinates:

$$x^2 + y^2 = R^2 \sin^2\theta$$

$$z^2 = R^2 \cos^2\theta$$

In polar coordinates, the magnitude of the current density vector is:

$$|J| = \frac{C}{\varrho_l^{3/2} R^2 (\sin^2\theta + \lambda^2 \cos^2\theta)^{3/2}} \tag{100}$$

$$= \frac{C}{\varrho_l^{3/2} R^2 [1 + (\lambda^2 - 1) \cos^2\theta]^{3/2}}$$

Consider a hemisphere of radius $R_1$ drawn about the current electrode, A. The current is everywhere normal to the surface of this hemisphere, so the total current can be found by integrating the current density vector over the surface of the hemisphere:

$$I = \int_S |J|\, ds = \int_0^{\pi/2}\int_0^{2\pi} |J|\, R^2 \sin\theta\, d\theta\, d\psi$$

$$= \int_0^{\pi/2}\int_0^{2\pi} \frac{C \sin\theta}{\varrho_l^{3/2} [1 + (\lambda^2 - 1) \cos^2\theta]^{3/2}}\, d\theta\, d\psi \tag{101}$$

The integration with respect to $\psi$ is straightforward:

$$I = \frac{\pi}{2}\int_0^{2\pi} \frac{C \sin\theta}{\varrho_l^{3/2} [1 + (\lambda^2 - 1) \cos^2\theta]^{3/2}}\, d\theta \tag{102}$$

The integral in $\theta$ is a form contained in tables of integrals, if the transformation $x = \cos\theta$ is made:

$$I = \frac{\pi}{2}\int_0^{2\pi} \frac{C\, dx}{\varrho_l^{3/2} [1 + (\lambda^2 - 1) x^2]^{3/2}} = \frac{2\pi C}{\varrho_l^{3/2} \lambda} \tag{103}$$

The constant of integration, $C$, may now be evaluated and eliminated from the expression for the potential function:

$$U = \frac{I \lambda \varrho_l^{3/2}}{2\pi \varrho_l^{1/2} (x^2 + y^2 + \lambda^2 z^2)^{1/2}} \tag{104}$$

In an anisotropic medium, the potential $U$ is inversely proportional to the distance from the current source, but the constant of proportionality is a function of the angle $\theta$. The equation for an equipotential surface is:

$$x^2 + y^2 + \lambda^2 z^2 = \left(\frac{2\pi U}{I\lambda \varrho_l}\right)^2 = \text{a constant} \tag{105}$$

Equation (105) describes an ellipse revolved about the $z$ axis.

If, as in field measurements of earth resistivity, the potential is measured at a point at a distance, $a$, away from the current source along the surface (the $x$ or $y$ directions), it will be found to be:

$$U = \frac{I\lambda \varrho_l}{2\pi a} \tag{106}$$

The apparent resistivity for a single-pole array such as this is given by:

$$\varrho = 2\pi a \frac{U}{I} = \lambda \varrho_l \tag{107}$$

Thus, if resistivity is measured in the field with an array oriented parallel to the bedding planes (for instance, along the surface of flat lying beds), the measured resistivity is larger than the longitudinal resistivity by the ratio, $\lambda$. This, then, was the reason for defining the coefficient of anisotropy as we did in Chapter I.

Consider the case of vertically-dipping beds so that the potential could be measured at a distance, $a$, from the current electrode along the $z$ axis, and still with the measurement being made on the surface of the earth. The measured potential will be:

$$U = \frac{I\varrho_l}{2\pi a} \tag{108}$$

and the apparent resistivity is:

$$\varrho_a = 2\pi a \frac{U}{I} = \varrho_l \tag{109}$$

Thus, the longitudinal resistivity, and not the transverse resistivity is measured when an electrode array is oriented across the bedding planes. This is known as the *paradox of anisotropy*.

The effect of anisotropy must be considered if field measurements of resistivity are to be interpreted correctly. In particular:

1. If resistivity measurements are being used to determine the strike of steeply-dipping beds under alluvial cover, it must be remembered that the maximum apparent resistivity

will be measured along the strike of the beds and the minimum apparent resistivity will be measured in a direction at right angles to the strike direction.

2. In the resistivity sounding method, if it is applied over flat-lying beds or beds with shallow dip, the measured resistivity will be larger than the longitudinal resistivity by a ratio equal to the coefficient of anisotropy, and the interpreted thickness will be too great by the same ratio.

## 19. THE SINGLE OVERBURDEN PROBLEM

So far, we have considered the behavior of potential about a current source only for a completely uniform earth. This approach is useful in defining apparent resistivity and the geometric factors for various electrode arrays, but it is of limited use in the interpretation of practical field problems. The next logical step in building up a realistic model of the earth is the introduction of a single uniform layer, bounded by two horizontal planes, covering a semi-infinite uniform earth. This model has some practical utility, since there are many occasions when it is necessary to measure the thickness of a uniform overburden, or to determine the properties of a reasonably uniform earth covered by such an overburden. In fact, the things we learn from this fairly simple problem can be extended to more complicated layering situations, and the behavior of electrical potentials in the single overburden model is used as the basis for interpretation of measurements made over an earth consisting of any number of horizontal layers.

There are a number of mathematical approaches which have been used in calculating potential fields in a layered medium. The simplest approach is the use of geometric optics, based on the assumption that in many respects, electrical currents behave similarly to light rays. A second approach is to search for a special solution to Laplace's equation which satisfies an appropriate set of boundary conditions. The third approach is to integrate Laplace's equation directly, as has already been done for the uniform medium. These approaches will be considered in the order in which they are listed, since this is the order of mathematical sophistication involved.

An analogy may be drawn between the way in which current travels through the earth and the way light rays spread through space. For example, the density of current flows from a point source varies as the inverse square of the distance. Light rays behave similarly, with intensity decreasing as the square of the distance travelled. This analogy alone does not prove that the principles of geometric optics can be used to solve problems in the flow of electrical currents, but we might try to solve the optical problem and see if the solution satisfies the various requirements for an electrical potential function. If it does, we have a valid solution for the electrical problem. This does not prove that the optical analogy will necessarily be valid for every electrical potential problem. In fact, the optical analogy is known to apply only for a limited group of problems, of which, fortunately, the single overburden problem is one.

In setting up the optical analog, current sources are replaced with light sources and the planes separating regions with different resistivities are replaced with semi-transparent mirrors. Then, the correlation between reflection and transmission coefficients and the resistivities of the layers must be established. This can be done by considering a space divided into two semi-infinite media, each with its own resistivity, as shown in Fig. 59. Consider a current source at the point, A, located a distance, $h$, above a plane of separation between the two semi-infinite media. The resistivity on one side of the plane is $\varrho_1$ and on the other side, $\varrho_2$. The potential function must satisfy the following two conditions:

1. *The potential function, U, must be continuous across the boundary between the two media.* If it were not, there would be a finite difference in potential over a short distance across the boundary, which would correspond to an infinite potential gradient. According to Ohm's law, this would result in an infinitely large current density, which is not possible.

2. *The normal component of current flow through the boundary must be continuous.* This condition is a consequence of the requirement that current is conserved—all the current entering the boundary plane from one side must leave it from the other side. The component of current parallel to the boundary need not be continuous, and in general, it is discontinuous across the boundary when the resistivities on the two sides are different.

Consider now an optical analogy to the electrical problem described by Fig. 59. The source of current at A is replaced with a light source, and the boundary is replaced by a mirror with a reflection coefficient, $K$, and a transmission

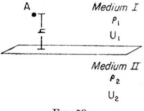

FIG. 59.

coefficient, $1 - K$. If the light is viewed from point $M_1$ on the top side of the mirror (see Fig. 60(a)), the original source is seen with its full intensity, and in addition, an image source is seen reflected from the mirror. This image source appears to be located behind the mirror at a distance, $h$, and has an intensity, $I'$.

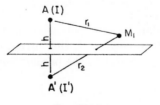

FIG. 60(a).

If the light is viewed from the dark side of the mirror (see Fig. 60(b)), only the primary source is seen, and this with a reduced intensity. This suggests we should seek expressions for two potential functions, one which is valid above the boundary plane, and the other which is valid beneath the plane. These

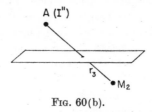

FIG. 60(b).

two functions will then be forced to match each other along the bounding plane, in order to satisfy the two conditions listed above.

The potential function for the region above the boundary plane consists of the sum of two terms, one representing the contribution from the primary source, and the other representing the contribution from the image source:

$$U_1 = \frac{\varrho_1 I}{4\pi r_1} + \frac{\varrho_1 I'}{4\pi r_2} \tag{110}$$

The potential function for the area below the bounding plane consists of a single term representing the contribution of the original source viewed darkly through the mirror:

$$U_2 = \frac{\varrho_2 I''}{4\pi r_3} \tag{111}$$

Along the boundary, the two potential functions must have the same numerical value, according to condition 1 listed above. At an observation point, $M_1$, along the boundary plane, the distances $r_1$, $r_2$ and $r_3$ will all be equal, so condition 1 can be expressed as:

$$\varrho_1(I + I') = \varrho_2 I'' \tag{112}$$

The second condition requires that the component of current density normal to the bounding plane, $j_n$, must be the same on both sides of the boundary·

$$j_{n,1} = j_{n,2}; \text{ at the boundary} \tag{113}$$

The current density is related to the potential functions $U_1$ and $U_2$ by using Ohm's law. At a point $M_1$ in medium 1, the current density vector is the sum of two vector components, one due to the primary source and the other due to the image source:

$$j_1 = \frac{I}{4\pi r_1^3} r_1 + \frac{I'}{4\pi r_2^3} r_2 \tag{114}$$

Let the angle between the $r_1$ direction and the normal to the surface be $\theta_1$; between the $r_2$ directions and the normal to the surface, $\theta_2$. Then, the component of current normal to the surface in medium 1 is:

$$j_{1,n} = \frac{I \cos\theta_1}{4\pi r_1^2} + \frac{I' \cos\theta_2}{4\pi r_2^2} \tag{115}$$

In the lower medium, there is only one component of current density, that due to the primary source with a decreased intensity:

$$j_2 = \frac{I''}{4\pi r_3^3} r_3 \tag{116}$$

The component normal to the boundary is:

$$j_{2,n} = \frac{I'' \cos\theta_3}{4\pi r_3^2} \tag{117}$$

where $\theta_3$ is defined as the angle between the $r_3$ direction and the normal to the surface. At the surface, $j_{1,n}$ must equal $j_{2,n}$. The three distances, $r_1$, $r_2$ and $r_3$ are equal along the boundary, so:

$$\cos\theta_1 = \cos\theta_2 = \cos\theta_3$$

and

$$I - I' = I'' \tag{118}$$

The unknown quantity $I''$ may readily be eliminated from Eqs. (112) and (118):

$$\varrho_1(I + I') = \varrho_2(I - I') \tag{119}$$

or

$$I' = \frac{\varrho_2 - \varrho_1}{\varrho_2 + \varrho_1} I = K_{1,2} I \tag{120}$$

where $K_{1,2}$ has the significance of a reflection coefficient for the boundary when viewed from medium 1. A similar solution for $I''$ in terms of $I$ gives:

$$I - KI = I''$$

or

$$I'' = (1 - K) I \tag{121}$$

The single-boundary problem serves only as a means for identifying the meaning of a reflection or transmission coefficient in terms of resistivity contrast. We may now proceed to the problem of two horizontal, planar boundaries, also known as the single-overburden problem. The geometry of the problem is shown in Fig. 61: a layer with thickness, $t$, and resistivity, $\varrho_1$, is bounded by two parallel planes separating it from a semi-infinite space above with resistivity, $\varrho_0$, and a semi-infinite space below with a resistivity $\varrho_2$. The upper half-space is the atmosphere in the practical case, but this need not be considered in formulating the problem. A current source, A, is located a distance, $h$, beneath the upper plane, and an observation point, M, is located the same distance

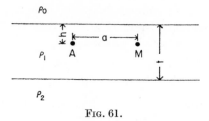

FIG. 61.

below the upper plane, but at a horizontal distance, $a$, from the current source.
In practice, the current source and the measuring point will be located at the
upper boundary to the layer, but it is easier to locate images when these points
are within the layer. The potential at point M due to the primary source at
A is:

$$U_0 = \frac{\varrho_1 I}{4\pi a} \tag{122}$$

If each of the boundary planes is replaced with a semi-transparent mirror,
with reflection coefficients as defined in Eq. (120), an observer at the point M
will see images of the primary source reflected from both the upper and lower
boundaries, as shown in Fig. 62. The apparent intensity of the upper image

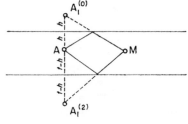

FIG. 62.

source is $IK_{1,0}$, where $K_{1,0}$ is the reflectivity of the upper plane when viewed
from underneath. The apparent intensity of the lower image source is $IK_{1,2}$,
where $K_{1,2}$ is the reflectivity of the lower plane viewed from above. The contributions to the potential function at the point M from these images are:

$$U_1^{(0)} = \frac{AIK_{1,0}}{4\pi(a^2 + 4h^2)^{1/2}}$$

$$U_1^{(2)} = \frac{\varrho_1 IK_{1,2}}{4\pi[a^2 + 4(t - h)^2]^{1/2}} \tag{123}$$

where the superscript refers to the medium in which the image source appears
to be and the subscript indicates that this is the first in a series of images. The
image series comes about, since in the optical case, there is no limit to the num-

ber of images formed by multiple reflection of light rays between the two
mirrors. Considering first only the paths in which the first bounce is off
the lower plane, we can examine the behavior of images for multiple reflections
(see Fig. 63). For a path with one reflection from each plane, the image source
appears to be in medium O, and has an intensity $IK_{1,2} K_{1,0}$. The contribution
to the total potential at point M from this image is:

$$U_2^{(0)} = \frac{\varrho_1 I K_{1,2} K_{1,0}}{4\pi (a^2 + 4t^2)^{1/2}} \tag{124}$$

The ray path with two reflections each from the upper and lower planes is
shown in Fig. 63(b). The image strength is reduced on each reflection by the
corresponding reflection coefficient, so the image strength is $IK_{1,2}^2 K_{1,0}^2$.
The image appears to be a distance $2t$ further above the top plane than the
preceding image. The contribution to the potential at M due to this image is:

$$U_3^{(0)} = \frac{\varrho_1 I K_{1,2}^2 K_{1,0}^2}{4\pi (a^2 + 16t^2)^{1/2}} \tag{125}$$

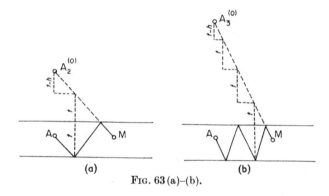

(a)                    (b)

FIG. 63 (a)–(b).

For each additional pair of reflections, one from the upper plane and one from
the lower plane, the corresponding image strength will be reduced by the factor
$K_{1,2} K_{1,0}$, and it will appear to be a distance $2t$ further behind the upper plane.
It is now possible to write an infinite series representing the potential due to
images corresponding to ray paths reflected an equal number of times from the
upper and lower planes, the first reflection being from the lower plane:

$$U_I = \sum_{n=1}^{\infty} \frac{\varrho_1 I K_{1,2}^n K_{1,0}^n}{4\pi [a^2 + (2nt)^2]^{1/2}} \tag{126}$$

Similarly, we may construct a series of images for paths with an odd number of reflections, the first being from the lower plane. The ray path and corresponding image for the first in such a series are shown in Fig. 63(c). The potential arising from this particular image is:

$$U_2^{(2)} = \frac{\varrho_1 I K_{1,2}^2 K_{1,0}}{4\pi[a^2 + \{2t + 2(t-h)\}^2]^{1/2}} \tag{127}$$

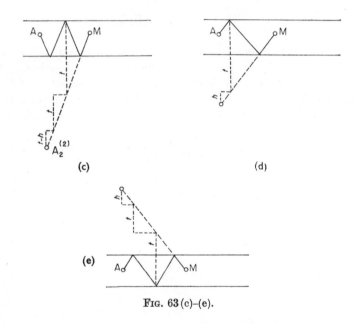

(c)                               (d)

(e)

FIG. 63(c)–(e).

Each additional pair of reflections will reduce the image strength by the factor $K_{1,0}K_{1,2}$, and the image will be located a distance $2h$ further below the lower plane. The potential due to an infinite series of such images is:

$$U_{\mathrm{II}} = \sum_{n=1}^{\infty} \frac{\varrho_1 I K_{1,2}^{(n+1)} K_{1,0}^n}{4\pi[a^2 + \{2nt + 2(t-h)\}^2]^{1/2}} \tag{128}$$

We have accounted only for half the possible multiple reflection paths within the layer. There may also be ray paths with an even or odd number of reflections with the first being from the upper surface rather than the lower surface. Figure 63(d) shows the first of a series of reflections from the upper plane. The series giving the potential due to images for paths with an even number of reflections is:

$$U_{\mathrm{III}} = \sum_{n=1}^{\infty} \frac{\varrho_1 I K_{1,0}^n K_{1,2}^n}{4\pi[a^2 + (2nt)^2]^{1/2}} \tag{129}$$

The first case with an odd number of reflections, the first one being from the upper plane, is shown by Fig. 63 (e). The potential due to an infinite sequence of such reflections is:

$$U_{IV} = \sum_{n=1}^{\infty} \frac{\varrho_1 I K_{1,0}^{(n+1)} K_{1,2}^n}{4\pi [a^2 + (2nt + 2h)^2]^{1/2}} \tag{130}$$

The total potential at the point M is found by adding all the terms included in Eqs. (122, 126, 128, 129 and 130):

$$U_M = \frac{\varrho_1 I}{4\pi a} \left\{ 1 + \frac{K_{1,0}}{[1 + (2h/a)^2]^{1/2}} + \frac{K_{1,2}}{[1 + \{2(t-h)/a\}^2]^{1/2}} \right.$$
$$+ \sum_{n=1}^{\infty} \frac{K_{1,2}^n K_{1,0}^n}{[1 + (2nt/a)^2]^{1/2}} + \sum_{n=1}^{\infty} \frac{K_{1,2}^{(n+1)} K_{1,0}^n}{[1 + \{2nt/a + 2(t-h)/a\}^2]^{1/2}}$$
$$\left. + \sum_{n=1}^{\infty} \frac{K_{1,0}^n K_{1,2}^n}{[1 + (2nt/a)^2]^{1/2}} + \sum_{n=1}^{\infty} \frac{K_{1,0}^{(n+1)} K_{1,2}^n}{[1 + \{2nt/a + 2h/a\}^2]^{1/2}} \right\} \tag{131}$$

This equation may be simplified considerably by stating that the points A and M are on the upper boundary $(h = 0)$, and that the upper half-space is air with an infinite resistivity. The reflection coefficient $K_{1,0}$ is then:

$$K_{1,0} = \frac{\varrho_0 - \varrho_1}{\varrho_0 + \varrho_1} = \frac{1 - \dfrac{\varrho_1}{\varrho_0}}{1 + \dfrac{\varrho_1}{\varrho_0}} = 1 \tag{132}$$

Under these conditions, Eq. (131) reduces to:

$$U_M = \frac{\varrho_1 I}{4\pi a} \left\{ 1 + 1 + \frac{K_{1,2}}{[1 + (2t/a)^2]^{1/2}} + \sum_{n=1}^{\infty} \frac{K_{1,2}^n}{[1 + (2nt/a)^2]^{1/2}} \right.$$
$$+ \sum_{n=1}^{\infty} \frac{K_{1,2}^{(n+1)}}{[1 + \{2(n+1)t/a\}^2]^{1/2}} + \sum_{n=1}^{\infty} \frac{K_{1,2}^n}{[1 + (2nt/a)^2]^{1/2}}$$
$$\left. + \sum_{n=1}^{\infty} \frac{K_{1,2}^n}{[1 + (2nt/a)^2]^{1/2}} \right\} \tag{133}$$

Three of the four series are identical with each other, and the fourth may be made identical by reducing the counter by one and including the lone term in the series:

$$\frac{K_{1,2}}{1 + (2t/a)^2} + \sum_{n=1}^{\infty} \frac{K_{1,2}^{(n+1)}}{[1 + \{2(n+1)t/a\}^2]^{1/2}} = \sum_{n=1}^{\infty} \frac{K_{1,2}^n}{[1 + (2nt/a)^2]^{1/2}} \tag{134}$$

The total potential function becomes simply:

$$U = \frac{\varrho_1 I}{2\pi a}\left[1 + 2\sum_{n=1}^{\infty}\frac{K_{1,2}^n}{[1 + (2nt/a)^2]^{1/2}}\right] \tag{135}$$

The potential consists of two parts: 1, that due to the first term in Eq. (135), which is the potential function for a homogeneous isotropic half-space and, so, is called the *normal potential*; and 2, that due to the infinite series, and which is called the *disturbing potential*. We shall find later that any galvanic potential can be expressed in this way—the sum of a normal potential and a disturbing potential.

We shall now examine the behavior of Eq. (135) under various extreme conditions. First of all, we shall examine the convergence of the series in Eq. (135). The Cauchy integral test for convergence is:

$$\int_{n}^{\infty} f(a)\,\mathrm{d}a < \infty$$

where $f(a)$ is the general expression for a term in the series which is being tested for convergence. Performing this test on the series in Eq. (135):

$$\int_{n}^{\infty}\frac{\mathrm{d}n}{[1 + (2nt/a)^2]^{1/2}} = a\log\left[2nt + \{(2nt)^2 + a^2\}^{1/2}\right]\Big|_{n}^{\infty} \to \infty$$

for the special case of $K_{1,2} = \pm 1$ (the lower half-space is an insulator or conductor). The series does not converge when the contrast in resistivity between the overburden and the lower half-space is very large. Convergence for special cases in which the contrast is not large can be shown readily.

The potential difference which would be measured with the Wenner array is:

$$\Delta U = U(a) - U(2a) = \frac{\varrho_1 I}{4\pi a}\left[1 + 4\sum_{n=1}^{\infty}\frac{K_{1,2}^n}{[1 + (2nt/a)^2]^{1/2}}\right.$$
$$\left.- 2\sum_{n=1}^{\infty}\frac{K_{1,2}^n}{[1 + (nt/a)^2]^{1/2}}\right] \tag{136}$$

when only one current electrode is considered. If both current electrodes are considered as contributing to the measured potential, the potential difference between the measuring electrodes is just twice that given by Eq. (136). Using this equation with the definition for apparent resistivity measured with the

Wenner array, it is possible to obtain an expression showing how the apparent resistivity varies with electrode spacing in the single overburden case:

$$\varrho_{a,w} = 2\pi a \frac{\Delta U}{I} = \varrho_1 \left[ 1 + 4 \sum_{n=1}^{\infty} \frac{K_{1,2}^n}{[1 + (2nt/a)^2]^{1/2}} - 2 \sum_{n=1}^{\infty} \frac{K_{1,2}^n}{[1 + (nt/a)^2]^{1/2}} \right]$$ (137)

The potential gradient, which is the parameter measured with the Schlumberger array, is found by taking an $a$ derivative of the function in Eq. (135). The derivative is evaluated for the spacing factor, $a$, and doubled to take into consideration that there are two current electrodes:

$$\frac{\partial U}{\partial a} = - \frac{\varrho_1 I}{2\pi a^2} \left[ 1 + 2 \sum_{n=1}^{\infty} \frac{K_{1,2}^n}{[1 + (2nt/a)^2]^{3/2}} \right]$$ (138)

The apparent resistivity measured with the Schlumberger array in the single overburden case is:

$$\varrho_{a,s} = - \frac{2\pi a^2}{I} \frac{\partial U}{\partial a} = \varrho_1 \left[ 1 + 2 \sum_{n=1}^{\infty} \frac{K_{1,2}^n}{[1 + (2nt/a)^2]^{3/2}} \right]$$ (139)

For the polar dipole array, it is necessary to find the curvature of the single-pole potential function given in Eq. (135). This is done by taking a second derivative with respect to $a$:

$$\frac{\partial^2 U}{\partial a^2} = \frac{\varrho_1 I}{\pi a^3} \left[ 1 - \sum_{n=1}^{\infty} \frac{K_{1,2}^n}{[1 + (2nt/a)^2]^{3/2}} + 3 \sum_{n=1}^{\infty} \frac{K_{1,2}^n}{[1 + (2nt/a)^2]^{5/2}} \right]$$ (140)

The apparent resistivity measured with a polar dipole array over a single overburden is:

$$\varrho_{a,d} = \varrho_1 \left[ 1 - \sum_{n=1}^{\infty} \frac{K_{1,2}^n}{[1 + (2nt/a)^2]^{3/2}} + 3 \sum_{n=1}^{\infty} \frac{K_{1,2}^n}{[1 + (2nt/a)^2]^{5/2}} \right]$$ (141)

Next, consider the behavior of these three Eqs. (137, 139 and 141), when the electrode spacing is very small compared to the overburden thickness. All the terms in the summations have a factor $t/a$ in the denominator which is large if $a$ is small. In the limit, each term in the series approaches zero, so the limit approached with all three equations as $a$ is made small is:

$$\varrho_{a,w} = \varrho_1$$
$$\varrho_{a,s} = \varrho_1$$
$$\varrho_{a,d} = \varrho_1 \quad \text{as} \quad a \to 0$$ (142)

This is reasonable and intuitively pleasing. If closely spaced electrode arrays are used, we expect that not much of the current will penetrate into the underlying half-space, and the measured resistivity ought to be nearly that of the overburden.

It is possible also to consider what happens when the spacing, $a$, is made very large in comparison with the layer thickness, $t$. If $a$ is sufficiently large, most of the terms $2nt/a$ in the various denominators will be small in comparison with 1, and so, may be ignored. This approximation will not be true for terms with large counter values, $n$, but if the reflection coefficient is less than $+1$, and more than $-1$, terms with large counter values will be small in the numerator, and so, may be ignored anyway. With these approximations, each infinite series reduces to the form $\sum K^n$, a series which has a closed form:

$$\sum K^n = (1 - K)^{-1} - 1$$

All three equations for apparent resistivity (137, 139 and 141) reduce to the same form when the spacing is large compared to the overburden thickness:

$$\varrho_a = \varrho_1 \left[ 1 + 2 \sum_{n=1}^{\infty} K_{1,2}^n \right]; \quad a \to \infty$$

$$= \varrho_1 \left[ 1 - 2 + \frac{2}{(1 - K)} \right]$$

$$= \varrho_1 \left[ 1 - 2 + \frac{2}{1 - \dfrac{\varrho_2 - \varrho_1}{\varrho_2 + \varrho_1}} \right]$$

$$= \varrho_2 \tag{143}$$

Since the series contained in Eq. (135) is not convergent when $K = \pm 1$, a different approach is required in determining the apparent resistivity under such conditions. Consider how current flows in the overburden if the lower

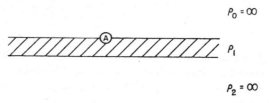

$$P_0 = \infty$$

$$P_1$$

$$P_2 = \infty$$

Fig. 64.

half-space is occupied by an insulator. At large distances from the current source, current must flow parallel to the bedding planes, since none of it can cross these boundaries. At some distance $a$ which is large compared to the overburden thickness, we can assume that the current density is uniform from top to bottom through the overburden (see Fig. 64). The electric field may be determined using Ohm's law:

$$E = \varrho_1 j$$

The electric field intensity may be related to the total current from the source electrode by integrating the current density over a complete equipotential surface, which at great distances from the electrode should be a ring with radius $a$ and height $t$:

$$I = \int_s j \cdot dS = 2\pi a t j$$

$$E = \frac{\varrho_1 I}{2\pi a t} \tag{144}$$

This value for the electric field should be doubled if the measurement is made midway between two widely spaced current electrodes, as is done with the Schlumberger array. Therefore, the apparent resistivity measured with a Schlumberger array in this limiting case is:

$$\varrho_{a,s} = \frac{\pi a^2}{I} \frac{\partial U}{\partial a} = \pi a^2 \frac{E}{I}$$

$$= \varrho_1 a/t \tag{145}$$

The relationship between apparent resistivity and electrode spacing is linear, with a slope of 1, and the relationship when extrapolated for small values of $a$ passes through the point $a/t = 1$, $\varrho_a/\varrho_1 = 1$. The ratio $\varrho_a/a$ is constant along the curve:

$$\boxed{\frac{a}{\varrho_a} = \frac{t}{\varrho_1} = S_1} \tag{146}$$

where $S_1$ is the longitudinal conductance of the overburden, as defined in Chapter I.

Equation (146) is probably the single most useful relationship that can be established for the interpretation of resistivity sounding data. In nearly every exploration problem, the lowermost layer reached by the current is an insulator, so that the portion of the sounding curve obtained with large electrode separations should approach the behavior predicted by Eq. (146). The total conductance of all the layers on top of the resistant basement rock can be determined directly by forming the ratio $\varrho_a/a$ for the portion of the data obtained with large electrode separations, if these data approach a line rising at 45° when apparent resistivity is plotted as a function of spacing on logarithmic coordinates. Usually, conductance is the quantity which can be determined with the best accuracy in resistivity sounding.

Relations similar to that given by Eq. (146) may be established for the other electrode arrays. The potential difference measured with a Wenner array is obtained by integrating the expression for electric field given in Eq. (144) bet-

ween the limits $a$ and $2a$. This quantity is doubled to find the potential difference due to two current electrodes:

$$\Delta U = 2 \int\limits_a^{2a} E \, \mathrm{d}a = 2 \int\limits_a^{2a} \frac{\varrho_1 I}{2\pi a t} \, \mathrm{d}a = \frac{\varrho_1 I}{\pi t} \ln 2 \qquad (147)$$

The apparent resistivity measured with a Wenner array when a conductive
overburden covers an insulating subsurface is:

$$\varrho_{a,w} = 2\pi a \frac{\Delta U}{I} = \frac{2a\varrho_1}{t} \ln 2 = 1{\cdot}38 \, a S^{-1} \qquad (148)$$

The apparent resistivity measured with large electrode separations with the
Wenner array is directly proportional to the spacing used, but in contrast
with the Schlumberger array, the constant of proportionality is 2 ln 2, rather
than unity. The conductance of the overburden is determined from the large
spacing portion of the sounding data using the equation:

$$\boxed{S = 1{\cdot}38 \, a/\varrho_{a,w}} \qquad (149)$$

The apparent resistivity which would be measured with a polar dipole array
over a conductive slab is found by taking a derivative of the expression for
electric field given in Eq. (144):

$$\frac{\partial E}{\partial a} = -\frac{\varrho_1 I}{2\pi t a^2} = -\frac{\partial^2 U}{\partial a^2}$$

and, according to the definition of apparent resistivity measured with the dipole
array:

$$\varrho_{a,d} = \frac{\pi a^3}{I} \frac{\partial^2 U}{\partial a^2} = \frac{a\varrho_1}{2t} = \frac{a}{2S} \qquad (150)$$

With each of the three basic electrode arrays, the conductance of all rocks
lying above an insulating basement rock can be read directly from the data,
if it is found that the sounding curve starts to rise at 45° when apparent resistivity is plotted as a function of spacing. Unfortunately, there is no such simple
interpretation for a resistant layer over a highly conductive half-space ($K_{1,2}$
$= -1$).

We should now consider more generally the apparent resistivity curves which
might be computed using Eqs. (137, 139 or 141). The sets of curves calculated
from each of these three equations are very similar, as might be expected since
the three equations behave similarly under various limiting conditions. A set
of curves relating the apparent resistivity that would be measured with the
Wenner array in the single overburden case to the electrode spacing
used for a variety of resistivity contrasts between the overburden and substra-

tum is shown in Fig. 65. Each curve varies smoothly and uniformly from an apparent resistivity equal to the overburden resistivity for small electrode separations to an apparent resistivity equal to the substratum resistivity for large electrode separations.

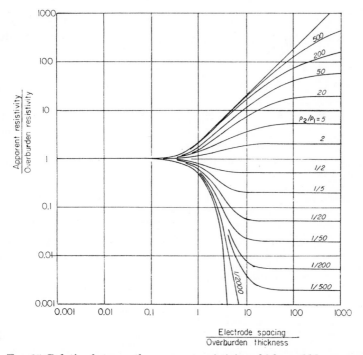

FIG. 65. Relation between the apparent resistivity which would be measured with a Wenner array over a single overburden and electrode separation for various contrasts in resistivity between the overburden and the substratum.

It should be noted that the curves for the cases in which the substratum is more conductive than the overburden are not symmetric to the curves for the cases in which the substratum is more resistant. With a conductive substratum, the apparent resistivity decreases much more rapidly with increasing electrode spacing than would be the case if these curves were the mirror image of the corresponding curves for a resistant substratum. Also, the apparent resistivity approaches the correct value for the second layer at smaller electrode separations when the second layer is more conductive rather than more resistive than the overburden.

The apparent resistivity curves for the various types of electrode arrays do differ from one another to a slight extent. These differences are most pronounced in the case of a high resistivity contrast between the overburden and the

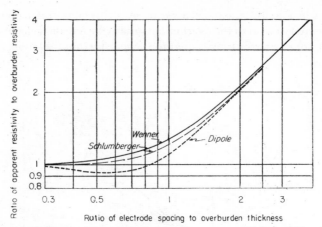

FIG. 66. Comparison of resistivity sounding curves which would be obtained at the surface of an overburden which rests upon an insulating substratum. The spacing factor for the Wenner array has been multiplied by 1·38 so the 45° asymptote will pass through the point 1·1. The spacing factor for the polar dipole array has been multiplied by $\frac{1}{2}$ so that its asymptote will pass through the same point.

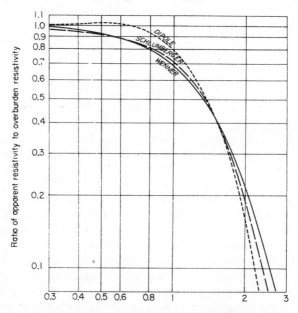

FIG. 67. Comparison of resistivity sounding curves which would be obtained at the surface of an overburden which rests upon a perfectly conducting substratum. The spacing factor for the polar dipole curve has been multiplied by the factor $\frac{1}{2}$.

substratum. A comparison of the three types of sounding curves (Wenner, Schlumberger and polar dipole) is shown in Figs. 66 and 67 for two cases; a conductive slab over an insulating substratum (Fig. 66), and a resistant slab over a conducting substratum (Fig. 67). The curves for the Wenner array and the polar dipole array in Fig. 66 have been shifted to the right and the left, respectively, until each has the same asymptote as the Schlumberger curve.

The apparent resistivity which would be obtained with a Schlumberger array for the case of a conductive overburden covering an insulating substratum (Fig. 66), departs from the actual resistivity of the overburden when the electrode spacing is $0.6t$, $t$ being the thickness of the overburden. (A significant departure may be considered to be 5 per cent, since this is a reasonable accuracy to be expected for field data.) The apparent resistivity approaches within 5 per cent of a rising asymptote for electrode spacings equal to five times the overburden thickness, or more. If the asymptotic behavior of apparent resistivity were to be used in interpretation, it would be necessary to use Schlumberger spacings up to twice the overburden thickness.

The apparent resistivity curve which would be obtained with a Wenner array under the same conditions shows a less abrupt transition from the resistivity of the overburden to resistivity values falling along an asymptotic curve rising at 45°. Spacings equal to three times the overburden thickness are needed to obtain the asymptotic portion of the curve.

The apparent resistivity curve which would be obtained with a dipole array differs from those obtained with the other two types of arrays in that for some spacings, the measured resistivity is lower than the overburden resistivity. A minimum resistivity is measured with a dipole spacing equal to half the overburden thickness, with the minimum apparent resistivity being 7 per cent less than the overburden resistivity. This minimum may or may not be apparent in actual field data, in view of the experimental errors usually contained in such measurements. The curve obtained with the dipole array shows the most abrupt transition from the resistivity for the overburden to the asymptotic behavior. The apparent resistivity curve approaches within 5 per cent of the asymptotic behavior for spacings about equal to the overburden thickness.

For the case of a resistive overburden resting on a highly conductive substratum, the apparent resistivity curves do not approach a linear asymptote, but rather, the slope increases continually as the spacing is increased. Of the three types of array considered in Fig. 67, the Wenner array produces the apparent resistivity curve which descends least rapidly as the spacing is increased (the curves for the Wenner array and the polar dipole array have not been shifted on Fig. 67 as on Fig. 66). The apparent resistivity curve obtained with a dipole array descends most rapidly. Both the curves for the Schlumberger and Wenner arrays depart uniformly from the value for the overburden resistivity, but the curve for the dipole array first rises to a maximum value of apparent resistivity which is 2 per cent higher than the overburden resistivity, and then decreases.

## 20. LOGARITHMIC CURVE MATCHING

All of the computed apparent resistivity curves which have been shown in the illustrations have been plotted on a logarithmic coordinate system. One of the advantages of logarithmic curve plotting is that it permits a wide range of values for the variables to be presented on a single graph. This is not the primary reason that apparent resistivity curves are plotted on a logarithmic coordinate system. Consider the equation which was developed in Section 19 for the apparent resistivity which would be measured over a single overburden:

$$\varrho_{a,s} = \varrho_1 \left[ 1 + 2 \sum_{n=1}^{\infty} \frac{(K_{23})^n}{\left[ 1 + \left( \dfrac{2nh}{a} \right)^2 \right]^{3/2}} \right]$$

This equation may be rewritten so that all the variables occur as dimensionless ratios: $K$, $a/h$ and $\varrho_a/\varrho_1$:

$$\frac{\varrho_{a,s}}{\varrho_1} = 1 + 2 \sum_{n=1}^{\infty} \frac{(K_{23})^n}{\left[ 1 + 4n^2 \left( \dfrac{h}{a} \right)^2 \right]^{3/2}} \tag{151}$$

The equations for the apparent resistivity measured with any of the electrode arrays can be expressed in terms of these same dimensionless ratios. This means that the theoretical curves computed with these equations can be plotted without regard to the system of units used, so long as they are consistent, and without regard to the absolute magnitudes of any of the resistivities, electrode spacings or bed dimensions. In practice, it is most convenient to measure the apparent resistivity in terms of the resistivity of the overburden, $\varrho_a/\varrho_1$, and the spacing in terms of the thickness of the overburden, $a/h$.

In normal field surveys, the proper values for $\varrho_1$ and $h$ are not known, usually being the object of the survey. Thus, dimensionless ratios expressed in terms of the overburden resistivity and thickness cannot be used in plotting field data. The coordinates for a point plotted on an apparent resistivity curve should be $(a/h, \varrho_a/\varrho_1)$. In a logarithmic coordinate system, the coordinates of the same point would be $(\log a - \log h, \log \varrho_a - \log \varrho_1)$. If a series of such points are to be plotted in logarithmic coordinates, the quantities $\log h$ and $\log \varrho_1$ are the same for all points. If values of 1 are arbitrarily assigned to the parameters $h$ and $\varrho_1$, each of the points along the apparent resistivity curve will be shifted a constant distance $(\log h)$ horizontally and a constant distance $(\log \varrho_1)$ vertically. The shape of a curve plotted in logarithmic coordinates is preserved, even when the ordinate and abscissa of each point along the curve are multiplied by arbitrary constants. *The preservation of curve shape in logarithmic coordinates is the basis for the curve-matching method of interpretation.*

A field curve, which is a plot of apparent resistivity as a function of electrode spacing obtained in a field survey, will have the same shape as a curve computed from theoretically derived expressions, provided both are plotted to the same logarithmic scales. *These field curves may be compared directly with a set of theoretical curves by superposition.*

In superposition, field data are plotted on a sheet of logarithmic graph paper which has exactly the same scales as the graph paper on which a set of theoretical curves have been plotted. The sheet with the field data (the *field curve*) is laid over the sheet with the theoretical curves and is moved around until the points on the field curve correspond, or match with one of the theoretical curves. The only restriction in moving the field curve around is that the coordinate axes of both sets of curves must be kept parallel.

The technique for curve matching is illustrated in Fig. 68, using the field data listed in Table 15:

TABLE 15

**Field data obtained from a Schlumberger sounding at Searles Lake, California**

| Spacing (ft) | Apparent resistivity (ohm-m) |
|---|---|
| 15 | 1·23 |
| 25 | 0·85 |
| 40 | 0·73 |
| 50 | 0·75 |
| 60 | 0·74 |
| 80 | 0·79 |
| 100 | 0·86 |
| 150 | 1·02 |
| 200 | 1·12 |
| 250 | 1·18 |
| 300 | 1·33 |
| 400 | 1·53 |
| 500 | 1·70 |
| 600 | 1·88 |
| 800 | 2·50 |
| 1000 | 2·85 |
| 1200 | 3·46 |
| 1400 | 3·95 |
| 1600 | 4·80 |
| 2000 | 5·80 |

These data have been plotted on logarithmic coordinates on Fig. 68. A set of two-layer theoretical curves have been superimposed on top of the field data, and moved around until a match was found between the field data and the

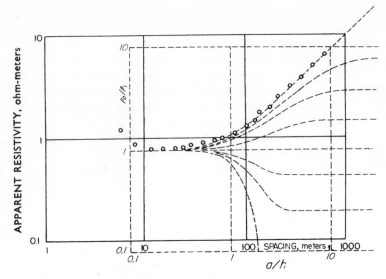

Fig. 68. Example of the interpretation of a field curve by superposition with
a set of two-layer (single-overburden) resistivity curves.

theoretical curves. The origin of coordinates for the theoretical curves ($a/h = 1$, $\varrho_a/\varrho_1 = 1$) is known as the *theoretical cross*. This point corresponds to a point on the field plot with corresponding values for $\varrho_a$ and $a$. In the example, these two points establish the conditions:

$a/h$    $=1$ on the theoretical curves when $a$ is 70 m on the field curve, and
$\varrho_a/\varrho_1 = 1$   on the theoretical curves when $\varrho_a$ is 0·72 ohm-m on the field curve.

These two conditions constitute a set of two equations with two unknowns, $h$ and $\varrho_1$. Solving these equations:

$$h = a = 70 \text{ m}$$
$$\varrho_1 = \varrho_a = 0·72 \text{ ohm-m}$$

## 21. INTERPRETATION OF TWO-LAYER CURVES BY ASYMPTOTES

Apparent resistivity curves may often be interpreted without having to go through the process of curve matching. If the lowermost layer apparent on a set of field data is an insulator, the extreme right-hand portion of the sounding curve will approach asymptotically a line rising at an angle of 45°. The ratio of spacing to apparent resistivity for any point along this line will be exactly $S$, the columnar conductance of all the rocks above the insulating layer, if the field data were obtained with either the Schlumberger of equatorial dipole arrays, or it will be exactly $2\,S\ln 2$ in case a Wenner array has used. The value of conductance can usually be determined from a set of field data with a higher precision than any of the other geoelectric parameters.

If the resistivity of the top layer is known, the thickness of the top layer may be found by multiplying the conductance by this resistivity. The resistivity of the top layer can be determined from the data obtained with short electrode spacings, provided the near-surface resistivity is reasonably uniform. In Fig. 69, the set of field data which had previously been interpreted using curve matching

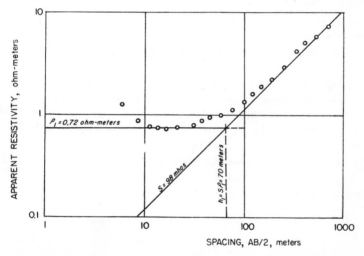

Fig. 69. Interpretation of the examples in Fig. 68 using the method of asymptotes.

procedures is now used as an example for asymptotic interpretation. The field data meet the requirement that the lowermost layer should appear to be an insulator, since the points obtained with the largest electrode separations appear to rise at about 45°. The line rising at 45° to which these data are tangent is shown. The ratio of spacing to apparent resistivity for any point along this line is constant, 98 mhos. The horizontal line indicates the value of resistivity chosen for the surface layer, 0·72 ohm-m. The two points obtained with the shortest electrode spacings have been ignored in making this estimate, since they reflect small scale resistivity variations within the first few meters of the surface. The intersection of the horizontal line with the inclined line yields graphically the thickness of the top layer, 70 m.

The asymptotic method of interpretation may be extended to any number of horizontal layers, provided the lowermost layer has a considerably larger value of resistivity than any of the overlying beds. This is the situation most commonly encountered in practice, since resistivity generally tends to increase with depth in the earth. This is true particularly if large enough electrode spacings are used that current penetrates to crystalline basement rock which almost always has a high resistivity. In interpreting a multi-layer field curve, a line rising at 45° is fitted to the right-hand portion of the field data, as in the two-layer case. The thickness of the individual layers cannot be determined in the same manner, since the conductance which is determined is that for all the layers above the basement:

$$S = \frac{h_1}{\varrho_1} + \frac{h_2}{\varrho_2} + \cdots + \frac{h_n}{\varrho_n} \tag{152}$$

It is necessary to determine the resistivity for each individual bed by some auxilliary means, before a complete interpretation can be made. Usually this is done with partial curve matching procedures, which are described in a later section.

The asymptote method may be applied even when field measurements have not been taken to large enough spacings to establish the asymptotic behavior. The minimum depth at which resistant rocks can occur may be established by fitting a line rising at 45° though the field point obtained with the largest electrode spacing used, thus establishing the least possible value for conductance of the rock above basement. Consider the set of field data in Fig. 70, obtained over a section including volcanic rocks at the surface, and underlain by conglomerate and granitic basement, in turn. For one reason or another,

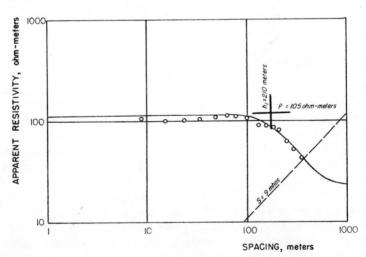

FIG. 70. Example of the use of the asymptote interpretation method for determining the least possible depth to resistant rock for a sounding in which the effect of basement rock is not apparent.

the electrode array was not expanded to the point where the high-resistivity granitic basement affected the sounding curve appreciably. In spite of this, a line rising at 45° may be drawn just to the right of the last point on the field curve, and the conductance corresponding to points along this line (9 mhos) is the least possible conductance which is compatible with the field data. There is no assurance that the conductance could not be considerably larger, but the least value for conductance is established.

In the example shown in Fig. 70, there appear to be two distinct layers above the basement. The conductance for these two layers is:

$$S = \frac{h_1}{\varrho_1} + \frac{h_2}{\varrho_2} \tag{153}$$

The value for $\varrho_1$ can be read directly from the field curve, since the resistivity measured with short spacings is uniformly 105 ohm-m. The thickness of the first layer may be determined by matching the data with the theoretical curves for a single overburden, as shown on Fig. 70. The resistivity of the second layer appears to be about 30 ohm-m. Using these parameters for the geoelectric properties of the upper two layers, Eq. (153) may be solved for $h_2$, the only quantity not determined:

$$h_2 = \varrho_2 \left( S - \frac{h_1}{\varrho_1} \right)$$
$$= 30 \left( 9 - \frac{210}{105} \right) = 210 \text{ m}$$

The minimum depth to basement is $h_1 + h_2$, or 400 m.

## 22. ACCURACY OF TWO-LAYER CURVE MATCHING

Inaccuracies in two-layer (single-overburden) curve matching may be of two types: those due to the use of two-layer curves when the data actually represent some other sort of resistivity distribution, and those due to the uncertainty involved in matching a theoretical curve with field data containing normal experimental errors, but truly representing a horizontal layering situation.

Errors in observation will nearly always cause some range in the way in which theoretical curves can be fitted to observational data, even when the field data were obtained under conditions closely approximating the ideal two-layer geometry assumed in the derivation of the theoretical expressions. One source of error is in the accuracy with which each of the physical measurements can be made; the accuracy with which electrode separations are measured or the accuracy with which current and voltage can be read from meters. Accuracies of ±5 per cent are readily obtainable with inexpensive instruments, and a considerably higher degree of accuracy in measuring the physical quantities is well within reason. Minor variations in resistivity which are too small in extent to signify by saying the mathematical model is incorrect lead to inconsistencies in the measured resistivity which are more important than the errors in measurement.

In the example in Fig. 68, an equally satisfactory match between the data and the theoretical curves could be obtained with slight shifts between the data and the theoretical curves. An interpretation of data of this sort cannot be objective unless the data points fall precisely along a single curve. Usually, an error in determining the vertical position of the matching curve, which determines the resistivity of the surface layer, is accompanied by a corresponding error in determining the horizontal position of the matching curve, which determines the thickness of the overburden. The errors in determining overburden thickness caused by errors in selecting the proper resistivity for the overburden can be estimated using the curves shown in Fig. 71. These curves relate the percentage error in determination of overburden thickness to the percentage error in assigning the overburden resistivity as a function of the contrast in resistivity

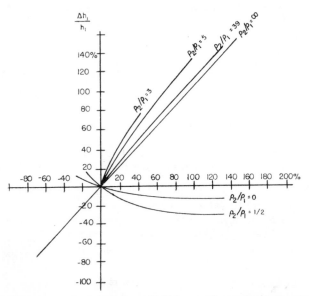

Fig. 71. Per cent error in interpreted thickness of a surface layer caused by errors in selecting a value of resistivity for the surface layer.

between the overburden and the bedrock. If the bedrock is much more resistive than the overburden, an error in assigning the overburden resistivity will cause an error in the interpreted thickness of the overburden which is in the same sense and equal in magnitude. For example, if the resistivity selected for the overburden is 20 per cent too high, the interpreted thickness for the overburden will also be 20 per cent too large.

If the resistivity of the bedrock is only several times greater than the resistivity of the overburden, the error in interpreted thickness will be proportionately greater than the error in the assigned resistivity. For example, for a resistivity contrast of 3 to 1, a 20 per cent error in assigning the resistivity of the overburden will cause nearly 40 per cent error in the interpreted thickness. On the other hand, a very large error may be made in assigning the overburden resistivity without causing a large error in the interpreted thickness, if the bedrock is more conductive than the overburden.

A second class of errors which may be included in the field data are those caused by horizontal variations in resistivity. One very simple example of errors generated in this way may be seen in the behavior of a resistivity sounding curve obtained with an electrode array which has been expanded parallel to a vertical contact between two regions with different resistivities (see

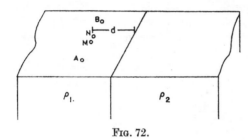

FIG. 72.

Fig. 72). The potential about a current electrode located a distance, $d$, from such a fault may be found using the method of images. If the vertical fault is replaced with a mirror having a reflection coefficient:

$$K = \frac{\varrho_2 - \varrho_1}{\varrho_2 + \varrho_1},$$

the potential on the side of the fault on which the current source is located is that due to the primary current source plus a disturbing potential due to an image source a distance $d$ on the far side of the fault plane:

$$U_1 = \frac{\varrho I}{2\pi} \left( \frac{1}{r} + \frac{1}{r'} \right) \tag{154}$$

The equations for the apparent resistivies measured with the various electrode arrays are:

1. For the Wenner array:

$$\varrho_{a,\,w} = 2\pi a \frac{\Delta U}{I} = \varrho_1 \left[ 1 + \frac{2K}{(1 + 4d^2/a^2)^{1/2}} - \frac{K}{(1 + d^2/a^2)^{1/2}} \right] \tag{155}$$

2. For the Schlumberger array:

$$\varrho_{a,\,s} = -2\pi a^2 \frac{E}{I} = \varrho_1 \left[ 1 + \frac{2K}{(1 + 4d^2/a^2)^{3/2}} - \frac{K}{(1 + d^2/a^2)^{3/2}} \right] \tag{156}$$

3. For the polar dipole array:

$$\varrho_{a,d} = \frac{\pi a^3}{I} \frac{\partial^2 U}{\partial a^2} = \varrho_1 \left[ 1 + \frac{K}{2(1 + d^2/a^2)^{3/2}} - \frac{K}{(1 + 4d^2/a^2)^{3/2}} \right. $$
$$\left. + \frac{3K}{(1 + 4d^2/a^2)^{5/2}} - \frac{3K}{2(1 + d^2/a^2)^{5/2}} \right] \qquad (157)$$

For the Wenner and Schlumberger arrays, the apparent resistivity will vary uniformly from the resistivity of the rock around the array for short spacings to the resistivity of the rock on the far side of the fault for large spacings, if the rock on the far side of the fault is less resistive than the rock on the near side. If the rock on the far side of the fault is more resistive than the rock on the near side, the apparent resistivity will increase as the spacing is increased, but it will not approach the resistivity for the rock on the far side. The highest apparent resistivity which can be measured is only twice the resistivity of the rock around the electrode array. This behavior is summarized by the curves sketched in Fig. 73.

The curves in Fig. 73 suggest that it might be easy to misinterpret a sounding curve run parallel to a vertical fault as being a sounding curve obtained over a

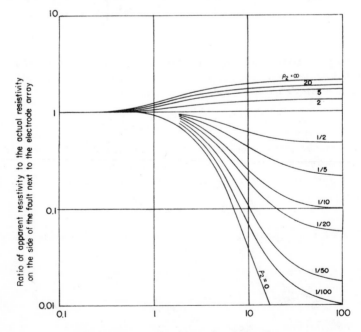

Ratio of spacing to offset distance from fault, $a/d$

FIG. 73. Resistivity sounding curves which would be measured using a Schlumberger array expanded parallel to a vertical fault.

single overburden. Vertical faults may be readily distinguished from horizontal layering, if sounding curves are obtained along a direction normal to the fault trace, as well as in a direction parallel to it. The equations for potential and apparent resistivity can be obtained in the same manner as those for an array expanded parallel to a fault, but the potential function will depend on whether the point at which potential is being observed is on the same side of the fault as the current source, or on the opposite side. If both the current source and the observation point are on the same side of the fault, the potential is given by Eq. (154). If both are not in the same medium, the potential is:

$$U = \frac{\varrho_1 I(1 - K)}{2\pi r}$$

Consider a Schlumberger array oriented normal to the fault trace, as shown in Fig. 74, with the center of the array at a fixed distance, $d$, from the trace. For

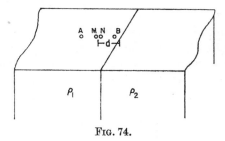

F_{IG}. 74.

small spacings with all the electrodes in the medium with resistivity, $\varrho_1$, the apparent resistivity is:

$$\varrho_a = \varrho_1 \left[ 1 - \frac{K}{2\pi} \left\{ \frac{1}{(2d/a - 1)^2} + \frac{1}{(2d/a + 1)^2} \right\} \right] \tag{158}$$

As the spacing of the array is expanded, the electrode B will cross the boundary between the two media when the electrode spacing, $a$, equals the offset distance, $d$. When electrode B is on the far side of the fault, the potential at the measuring electrodes appears to be derived from three sources: one source at the electrode A; an image source in medium 2 formed by the reflection of the electrode A at the boundary between the two media; and a source at electrode B reduced in strength by the factor $(1 - K)$. The voltage which would be measured with a Schlumberger array is:

$$U = \frac{\varrho_1 I}{2\pi} \left[ \frac{1}{\left(a - \dfrac{b}{2}\right)} - \frac{1}{\left(a + \dfrac{b}{2}\right)} \right] - \frac{\varrho_1 I(1 - K)}{2\pi} \left[ \frac{1}{\left(a + \dfrac{b}{2}\right)} - \frac{1}{\left(a - \dfrac{b}{2}\right)} \right]$$

$$+ \frac{\varrho_1 I K}{2\pi} \left[ \frac{1}{\left(a + 2d + \dfrac{b}{2}\right)} - \frac{1}{\left(a + 2d - \dfrac{b}{2}\right)} \right] \tag{159}$$

Or, taking the ratio $-U/b$ to be very nearly the electric field intensity, we have:

$$\frac{U}{b} = \frac{\varrho_1 I}{2\pi}\left(\frac{1}{a^2 - \dfrac{b^2}{4}}\right) + \frac{\varrho_1 I(1 - K)}{2\pi}\left(\frac{1}{a^2 - \dfrac{b^2}{4}}\right)$$
$$+ \frac{\varrho_1 I K}{2\pi}\left(\frac{1}{(a + 2d)^2 - \dfrac{b^2}{4}}\right) \tag{160}$$

Neglecting the $b^2/4$ term in each denominator and using the resulting expression for $U/b$ in the defining equation for resistivity measured with the Schlumberger array, we have:

$$\varrho_{a,s} = \frac{\pi a^2}{b}\frac{U}{I} = \varrho_1\left[1 - K\left\{1 - \frac{1}{\left(1 + \dfrac{2d}{a}\right)^2}\right\}\right] \tag{161}$$

Curves computed from Eqs. (158) and (161) for a variety of resistivity contrasts are shown in Fig. 75. The curves are greatly different from those which are obtained with an array expanded parallel to the fault. The apparent resistivity measured both with very short and very long spacings is nearly the same as the

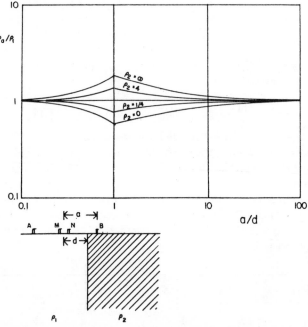

Fig. 75. Apparent resistivity curves which would be measured with a Schlumberger array expanded along a line normal to the trace of a vertical contact between regions with different resistivities.

resistivity of the medium in which the measuring electrodes are located. The maximum departure from this resistivity is observed when the spacing factor equals the offset distance between the center of the array and the fault plane. If the array were expanded along a line at an angle other than 90° with the fault, apparent resistivity curves similar to those shown in Fig. 75 would be obtained—the maximum departure would be observed as one current electrode crossed the fault trace.

It may be seen from a comparison of the curves in Figs. 73 and 75 that there would be little chance of confusing vertical faults with horizontal layering if resistivity soundings are made along several azimuths through a point. Comparison of resistivity soundings measured with the array expanded along three lines will show whether resistivity changes are caused by horizontal layering (all three curves will be essentially identical), or by horizontal resistivity changes (the three curves will differ in shape).

When the plane separating two regions with different resistivities is neither vertical nor horizontal, an exact mathematical solution for the potential function is difficult to obtain. The problem is particularly important in that we must know how much dip can be tolerated in the interpretation of apparent resistivity curves using the curves derived for horizontal layers. The image method has proved to be unsatisfactory for determining the potential function over an inclined plane. Combinations of images which will satisfy all the boundary conditions can be found only for a few special cases. These cases all require a reflection coefficient of $\pm 1$, and that the angle of dip be some submultiple of $\pi/2$ (Fig. 76).

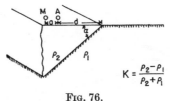

$$K = \frac{\rho_2 - \rho_1}{\rho_2 + \rho_1}$$

Fig. 76.

An exact mathematical expression for the potential about a point source of current over a single dipping contact has been developed by Skalskaya (1948) and by Maeda (1955). Maeda gives an expression for the potential function consisting of two integrals; one valid when the measuring point is in the same medium as the current source, and the other valid when the measuring point is on the far side of the trace of the dipping plane. The potential function in the first case is:

$$U = \frac{I\rho_1}{2\pi a}\left\{ 1 + \frac{4a}{\pi^2}\int_0^\infty\int_0^\infty \frac{K\sinh 2s(\pi - \alpha)\cosh s\theta}{\sinh \pi s + K\sinh s(2\alpha - \pi)}\,\mathscr{K}(tr_0)\,\mathscr{K}(tr)\cos tz\,dt\,ds \right\}$$

$$(162)$$

and in the second case:

$$U = \frac{I\varrho_1}{2\pi a} \left\{ 1 + \frac{4a}{\pi^2} \int\limits_0^\infty \int\limits_0^\infty \frac{K[\sinh s\pi - \sinh s(2\varkappa - \pi)] \cosh s(\pi - \theta)}{\sinh s\pi + K \sinh s(2\varkappa - \pi)} \right.$$

$$\left. \mathscr{K}(tr_0) \, \mathscr{K}(tr) \cos tz \, dt \, ds \right\} \tag{163}$$

where $t$ and $s$ are dummy variables of integration which have no physical significance, $\mathscr{K}(tr)$ indicates a modified Bessel function of the second type of argument $tr$, $a$ is the separation between the current electrode and the measuring point, and $d$ is the offset distance of the current electrode from the surface trace of the dipping plane. A cylindrical coordinate system is used such that the $z$ axis coincides with the surface trace of the dipping plane, the angle $\theta$ is

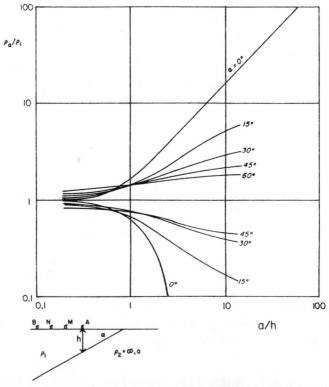

FIG. 77. Apparent resistivity curves which would be obtained with a Wenner array over a dipping bed with the array oriented normal to the strike. The position of electrode A is fixed relative to the trace of the dipping contact (taken from Maeda, 1955).

measured in the vertical plane, in a counter-clockwise direction from the horizontal, and $r$ is the radial distance of a point from the $z$ axis. The dip angle, $\alpha$, can vary from 0 to $\pi$.

Some curves computed by Maeda for a Wenner array are shown in Fig. 77. For an electrode array being expanded in the dip direction, the effect of dip is to make the resistivity contrast appear to be smaller than it actually is. The greater the dip, the greater is this reduction in the apparent resistivity contrast. However, for dips up to about 45°, little or no error will be made in the interpretation of the depth to the second layer if the interpretation is made with curves calculated for a single overburden with horizontal boundaries.

The curves shown in Fig. 77 are calculated for an expanding array in which the current electrode nearest the trace of the fault plane is fixed. In practice, an array is usually expanded about its midpoint, so that the electrode A will cross the fault trace when the spacing, $a$, is two-thirds the offset distance, and the electrode M will cross the trace when the spacing equals twice the offset. As each of these electrodes crosses the trace, the apparent resistivity curve will show a discontinuity like those apparent in Fig. 75.

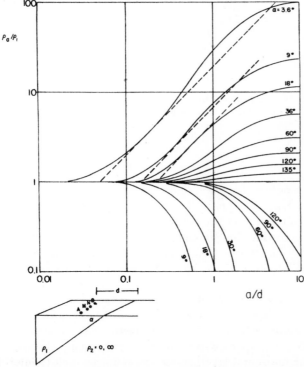

Fig. 78. Apparent resistivity curves which would be obtained with a Schlumberger array over a dipping bed with the array oriented parallel to the strike (taken from Kalenov, 1957).

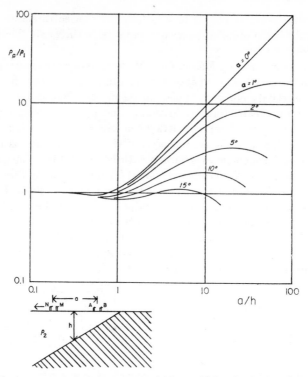

FIG. 79. Apparent resistivity curves which would be obtained with a dipole array over a dipping bed with the array oriented normal to the strike. The measuring dipole is moved away from the outcrop of the dipping bed, while the current dipole remains fixed (taken from Birdichevskiy and Zagarmistr, 1957).

In order to compute resistivities which would be measured with the Schlumberger or dipole arrays, the gradient and curvature of the potential functions given in Eqs. (162) and (163) would have to be determined. Kalenov (1957) has published curves for the Schlumberger array over a dipping plane, and Birdichevski and Zagarmistr (1958) have published similar curves for the dipole array. Selected curves from each of these sources are shown in Figs. 78 and 79.

Dip, apparently, has very little effect on the accuracy of depth determinations made with horizontal layer curves for data obtained with a Schlumberger array, though the resistivity contrast will be underestimated, as with the Wenner array. The dipole array appears to be more sensitive to dip than either of the other arrays. The horizontal layering curves cannot be used to interpret data obtained with a dipole array if the dip exceeds 10°. The sensitivity of the dipole array to dip is not necessarily an advantage. Wenner or Schlumberger soundings made over dipping beds may be interpreted using simple two-layer curves for dips up to about 45°. The dip angle may be determined by comparing depths interpreted from soundings made a short distance apart. With a dipole array, if the beds have a moderate dip, it is necessary to consider this in interpretation.

## 23. INTERPRETATION OF RESISTIVITY SOUNDINGS WHEN THERE ARE MORE THAN TWO HORIZONTAL BOUNDARIES

The single-overburden case is frequently inadequate to describe problems encountered in field surveys. Resistivity sounding data obtained in the field may be appear to represent two, three or even more layers covering resistive basement rock. There are four approaches which may be used in interpreting multiple-layer resistivity sounding data:

1. Complete curve matching, using curves computed for mathematical models with two, three or four layers covering an infinite, uniform substratum.

2. Partial curve matching, in which portions of the field data are matched with the curves computed for a single overburden.

3. Equivalent curve matching, in which all theoretical curves having similar shapes are grouped to form a single "equivalent" curve for comparison with field data.

4. Observation of the positions of the maxima and minima on the field data.

The first steps in interpreting multiple-layer curves rely on simple single-overburden theory. The thickness and resistivity of the top layer are estimated by matching the data measured with small electrode separations to the single-overburden theoretical curves. Also, the conductance of all the layers above the basement is estimated by drawing a line at 45° through the data point for the apparent resistivity measured with the largest electrode separation. If the last few points on the field-sounding curve do not indicate a resistant basement, the $S$-value so obtained will represent the least conductance the rock above basement may have.

Interpretation of the mid-portion of the field curve requires the use of one of the four methods listed above. For convenience in selecting the method of interpretation, sounding curves are classified according to their shape. A curve which has a minimum (see Fig. 80(a)) is called a type H curve, and indicates the presence of a three-layer sequence with the resistivity ratios varying as: $\varrho_1 > \varrho_2 < \varrho_3$. If the field curve shows a maximum ($\varrho_1 < \varrho_2 > \varrho_3$), it is classified as type K. For a curve which shows a uniform increase in resistivity ($\varrho_1 < \varrho_2 < \varrho_3$) or a uniform decrease in resistivity ($\varrho_1 > \varrho_2 > \varrho_3$), data are classified as type A or type Q, respectively.

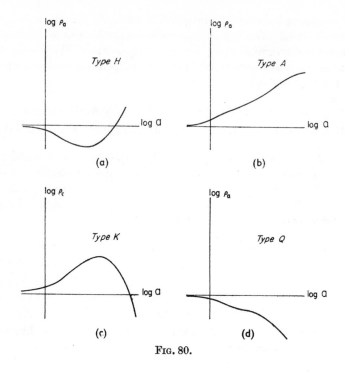

Fig. 80.

When there are more than three layers with different resistivities apparent on a field curve, several letters are used to classify the curve. For example, an HK curve indicates a sequence of resistivities ($\varrho_1 > \varrho_2 < \varrho_3 > \varrho_4$).

### 23 a. Complete curve matching

Precalculated catalogs of theoretical curves can rarely be used in complete curve matching in view of the large number of parameters needed to specify the contrasts in resistivities and ratios of thicknesses when several layers are present. With a single overburden, only one parameter is necessary, the ratio of resistivities for the overburden and the bedrock. With two layers resting on basement, three parameters are required to specify completely the combinations of resistivities and thicknesses: $\varrho_2/\varrho_1$, $\varrho_3/\varrho_2$ and $h_2/h_1$. With each additional layer, two more parameters are required. It is not possible to compile a catalog of curves for more than two layers and yet have the parametric values closely enough spaced to be of use.

Mooney and Wetzel (1957) have prepared a set of about 2300 curves for the Wenner array for two and three layers over an infinite substratum. La Compagnie Generale de Geophysique has published an album of 480 curves for the Schlumberger array for two layers resting on an infinite substratum (1957).

When a high-speed digital computer is available, exact curves may be computed to match any given set of field data. In using such a procedure, an approximate solution is obtained first by comparing the field data with the available two-and three-layer albums. Then, more exact curves are computed with trial and error adjustments being made in the assumed layering until an exact match with the field data is obtained.

The use of such a method requires the computation of the potential function over a many-layered earth, with perhaps as many as 25 to 50 layers being assumed when gradational changes in resistivity with depth are to be interpreted. The image method is far too cumbersome to apply to more than one or two layers. A more flexible solution of the boundary value problem is found by assuming a solution to Laplace's equation.

The geometrical model is shown in Fig. 81. The earth is divided into $n + 1$ horizontal layers by a series of parallel planes. Each layer is characterized by a resistivity, $\varrho_i$, and a thickness, $h_i$. A cylindrical coordinate system is used,

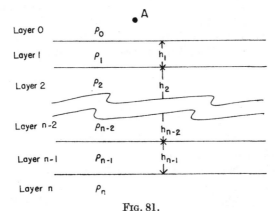

FIG. 81.

with the $z$ axis normal to the layering planes, and with the origin located at a current electrode, A, in the upper half-space, layer 0. In the practical case, the electrode A will be placed at the boundary between medium 0 and medium 1, with medium 0 being the air.

The conditions of continuity in potential and of the component of current density normal to the planes must apply at each boundary:

$$U_i\big|_{z=z_i} = U_{i+1}\big|_{z=z_i} \tag{164}$$

and

$$\frac{\partial U_i}{\partial z}\bigg|_{z=z_i} = \frac{\partial U_{i+1}}{\partial z}\bigg|_{z=z_i} \tag{165}$$

where $z_i$ is the $z$ coordinate giving the location of the plane separating layer $i$ from layer $1 + i$.

Fourier's method may be used in seeking a solution to Laplace's equation which will satisfy the boundary conditions, Eqs. (164) and (165). In this method, it is assumed that a solution can be found in the form of a product of two functions, each of which depends on only one variable:

$$U = f(r) \cdot \varphi(z) \tag{166}$$

No function in $\psi$ need be considered in view of the axial symmetry assumed for the problem.

To find if such a solution as that specified by Eq. (166) actually exists, it is substituted in Laplace's equation:

$$f''(r)\,\varphi(z) + \frac{1}{r}\,f'(r)\,\varphi(z) + f(r)\,\varphi''(z) = 0 \tag{167}$$

Dividing by $f(r)\,\varphi(z)$:

$$\frac{f''(r)}{f(r)} + \frac{1}{r}\frac{f'(r)}{f(r)} + \frac{\varphi''(z)}{\varphi(z)} = 0 \tag{168}$$

Since the functions $f$ and $\varphi$ were defined as being independent of one another, Eq. (168) will hold only if the ratio $\varphi''/\varphi$ is independent of the other two terms. If this ratio does not vary with the other two terms, it must be a constant, which may be called $m^2$. Then, Eq. (168) can be separated into two independent relationships:

$$\frac{\varphi''(z)}{\varphi(z)} - m^2 = 0 \tag{169}$$

$$\frac{f''(r)}{f(r)} + \frac{1}{r}\frac{f'(r)}{f(r)} + m^2 = 0 \tag{170}$$

Both of these differential equations are of forms commonly studied in physical problems. The solution to Eq. (169) may contain terms in $e^{-mz}$ and $e^{mz}$, while the solution to Eq. (170) may contain terms with Bessel's functions of zero order of both the first and second kinds, $J_0(mr)$ and $Y_0(mr)$.

The function $Y_0(mr)$ tends to become infinite when $r$ is small. Since this behavior is not acceptable for all points along the $z$ axis, all terms in the solution to Eq. (170) which contain $Y_0(mr)$ must have zero coefficients. The complete solution to Eq. (166) can contain only two types of terms; $J_0(mr) \cdot e^{-mz}$ and $J_0(mr) \cdot e^{mz}$. Any linear combination of terms of these two types is also a solution to the differential equation, so for most general solution, we must consider the sum of all such terms for all values of $m$, which was chosen arbitrarily. Since $m$ can vary continuously, this sum can best be expressed as an integral:

$$U = \int_0^\infty A(m) \cdot J_0(mr)\,e^{-mz}\,dm + \int_0^\infty B(m)\,J_0(mr)\,e^{mz}\,dm \tag{171}$$

where $A(m)$ and $B(m)$ are functions which must be evaluated by applying the boundary conditions.

There are $n + 1$ functions of the form given in Eq. (171), one for each layer in the assumed model of the earth. Each function is valid only within the layer for which it is determined, and each is characterized by a set of two arbitrary functions, $A_i(m)$ and $B_i(m)$. The boundary conditions (Eqs. (164) and (165)) provide a set of $2n$ equations for use in determining these $2(n + 1)$ parametric functions. The shortage of equations means that two additional conditions must be specified in order to obtain a unique solution for the $A$ and $B$ functions. These conditions are established by requiring that the potential function in medium 1 near the electrode A must approach the correct expression for the potential about a single-pole source in a uniform medium, and by requiring the potential function in the $n$th layer to approach zero for large values of $z$. All the A and B functions must be determined, even though we are interested only in the behavior of $U_0$, the potential observed at the surface of the earth. It turns out that the function $B_0$ which is necessary to express $U_0$ is dependent on all the A and B functions for the potential throughout the section.

Consider the potential function, $U_0$, observed in medium 0. It may be expressed as the sum of a normal potential, $U_0^*$, for a completely uniform medium, and a disturbing function, $W_0$, which represents the potential contributed by the layering:

$$U_0 = \frac{\varrho I}{4\pi} \cdot \frac{1}{(r^2 + z^2)^{1/2}} + W_0 \tag{172}$$

where the potential is being observed at a point in medium 0 with the coordinates $r$, $z$.

Equations (171) and (172) both presumably describe the potential function in the uppermost layer, so:

$$\frac{\varrho_0 I}{4\pi} \frac{1}{(r^2 + z^2)^{1/2}} + W_0 = \int_0^\infty A_0(m) \, J_0(mr) \, e^{-mz} \, dm + \int_0^\infty B_0(m) J_0(mr) \, e^{mz} dm \tag{173}$$

The Weber–Lipschitz identity for the Bessel's function is:

$$\int_0^\infty J_0(mr) \, e^{-mz} \, dm = \frac{1}{(r^2 + z^2)^{1/2}} \tag{174}$$

This identity could be applied to the first term to the right of the equality sign in Eq. (173) if $A_0$ were a constant instead of a function of $m$. A convenient sized constant turns out to be $A_0(m) = \varrho_0 I/4\pi$, so that Eq. (173) may be rewritten as:

$$W_0 = \int_0^\infty B_0(m) \, J_0(mr) \, e^{mz} \, dm \tag{175}$$

Since this equation in no way contradicts any of the conditions of the problem, the choice for a value for $A_0$ made above must be satisfactory. One of the arbitrary parametric functions has been determined.

In order to evaluate another of the parametric functions, consider the potential function in the lowermost layer. In this layer, large values of $z$ can be approached, since the lowermost layer is half-infinite in extent. The potential, $U_n$, must approach zero for large values of $z$, and in order for this to be true, the function $B_n(m)$ must be identically zero.

There remain only $2n$ functions $A_i$ and $B_i$ to be determined from a set of $2n$ boundary conditions. For convenience of notation, the quantities $A_i$ and $B_i$ would better be expressed as follows:

$$C_i(m) = \frac{4\pi}{\varrho_1 I} A_i(m)$$

$$D_i(m) = \frac{4\pi}{\varrho_i I} B_i(m) \tag{176}$$

The $n$ potential functions may now be written as:

$$U_0 = \frac{\varrho_0 I}{4\pi} \left[ \frac{1}{(r^2 + z^2)^{1/2}} + \int\limits_0^\infty D_0(m)\, J_0(mr)\, e^{mz}\, dm \right]$$

$$\vdots$$

$$U_i = \frac{\varrho_i I}{4\pi} \left[ \int\limits_0^\infty C_i(m)\, J_0(mr)\, e^{-mz}\, dm + \int\limits_0^\infty D_i(m)\, J_0(mr)\, e^{mz}\, dm \right]$$

$$U_n = \frac{\varrho_n I}{4\pi} \left[ \int\limits_0^\infty C_n(m)\, J_0(mr)\, e^{-mz}\, dm \right] \tag{177}$$

A set of $n$ equations is obtained when the requirement that potential be continuous across each boundary is met:

$$\varrho_0 \left[ \int\limits_0^\infty J_0(mr)\, e^{-mz_1}\, dm + \int\limits_0^\infty D_0(m)\, J_0(mr)\, e^{mz_1}\, dm \right]$$

$$= \varrho_1 \left[ \int\limits_0^\infty C_1(m)\, J_0(mr)\, e^{-mz_1}\, dm + \int\limits_0^\infty D_1(m)\, J_0(mr)\, e^{mz_1}\, dm \right]$$

$$\vdots$$

$$\varrho_i \left[ \int\limits_0^\infty C_i(m)\, J_0(mr)\, e^{-mz_i}\, dm + \int\limits_0^\infty D_i(m)\, J_0(mr)\, e^{mz_i}\, dm \right]$$

$$= \varrho_{i+1} \left[ \int\limits_0^\infty C_{i+1}(m)\, J_0(mr)\, e^{-mz_i}\, dm + \int\limits_0^\infty D_{i+1}(m)\, J_0(mr)\, e^{-mz_i}\, dm \right]$$

$$\vdots$$

$$\varrho_{n-1} \left[ \int\limits_0^\infty C_{n-1}(m)\, J_0(mr)\, e^{-mz_n}\, dm + \int\limits_0^\infty D_{n-1}(m)\, J_0(mr)\, e^{mz_n}\, dm \right]$$

$$= \varrho_n \left[ \int\limits_0^\infty C_n(m)\, J_0(mr)\, e^{-mz_n}\, dm \right] \tag{178}$$

A second set of $n$ equations is formed by equating the components of current density normal to the boundaries at each boundary:

$$-\int_0^\infty m J_0(mr)\, e^{-mz_1}\, dm + \int_0^\infty D_0(m)\, J_0(mr)\, e^{mz_1}\, m\, dm$$

$$= -\int_0^\infty C_1(m)\, J_0(mr)\, e^{-mz_1}\, m\, dm + \int_0^\infty D_1(m)\, J_0(mr)\, e^{mz_1}\, m\, dm$$

$$\vdots$$

$$-\int_0^\infty C_i(m)\, J_0(mr)\, e^{-mz_i}\, m\, dm + \int_0^\infty D_i(m)\, J_0(mr)\, e^{mz_i}\, m\, dm$$

$$= -\int_0^\infty C_{i+1}(m)\, J_0(mr)\, e^{-mz_i}\, m\, dm + \int_0^\infty D_{i+1}(m)\, J_0(mr)\, e^{mz_i}\, m\, dm$$

$$\vdots$$

$$-\int_0^\infty C_{n-1}(m)\, J_0(mr)\, e^{-mz_n}\, m\, dm + \int_0^\infty D_{n-1}(m)\, J_0(mr)\, e^{mz_n}\, m\, dm$$

$$= -\int_0^\infty C_n(m)\, J_0(mr)\, e^{-mz_n}\, m\, dm \quad (179)$$

All the integrations are carried out through the same limits, so all the terms within each equation may be grouped under a single integral sign:

$$\int_0^\infty \{[e^{-mz_1} + D_0(m)\, e^{mz_1}]\varrho_0 - [C_1(m)\, e^{-mz_1} + D_1(m)\, e^{mz_1}]\varrho_1\}\, J_0(mr)\, dm = 0$$

$$\vdots$$

$$\int_0^\infty \{[C_i(m)\, e^{mz_i} + D_i(m)\, e^{mz_i}]\varrho_i$$

$$- (C_{i+1}(m)\, e^{-mz_i} + D_{i+1}(m)\, e^{mz_i}]\varrho_{i+1}\}\, J_0(mr)\, dm = 0$$

$$\vdots$$

$$\int_0^\infty \{[C_{n-1}(m)\, e^{-mz_n} + D_{n-1}(m)\, e^{mz_n}]\varrho_{n-1} - (C_n(m)\, e^{-mz_n}]\varrho_n\}\, J_0(mr)\, dm = 0$$

and:

$$\int_0^\infty [-e^{-mz_1} D_0(m)\, e^{mz_1} + C_1(m)\, e^{-mz_1} - D_1(m)\, e^{mz_1}]\, J_0(mr)\, m\, dm = 0$$

$$\vdots$$

$$\int_0^\infty [-C_i(m)\, e^{-mz_i} + D_i(m)\, e^{mz_i} + C_{i+1}(m)\, e^{-mz_i} - D_{i+1}(m)\, e^{mz_i}]\, J_0(mr)\, m\, dm = 0$$

$$\vdots$$

$$\int_0^\infty [-C_{n-1}(m)\, e^{-mz_n} + D_{n-1}(m)\, e^{mz_n} + C_n(m)\, e^{-mz_n}]\, J_0(mr)\, m\, dm = 0 \quad (180)$$

These equations hold for all points along the various planes separating the layers. The only way Eq. (180) can be valid for an arbitrary value of $r$ is if the integrands are identically zero:

$$\varrho_0 e^{-mz_1} + \varrho_0 e^{mz_1} D_0 - \varrho_1 e^{-mz_1} C_1 - \varrho_1 D_1 e^{mz_1} = 0$$
$$-e^{-mz_1} + e^{mz_1} D_0 + e^{-mz_1} C_1 - D_1 e^{mz_1} = 0$$
$$\vdots$$
$$\varrho_i e^{-mz_i} C_i + \varrho_i e^{mz_i} D_i - \varrho_{i+1} C_{i+1} e^{-mz_i} - \varrho_{i+1} e^{mz_i} D_{i+1} = 0$$
$$-e^{-mz_i} C_i + e^{mz_i} D_i + C_{i+1} e^{-mz_i} - e^{mz_i} D_{i+1} = 0$$
$$\vdots$$
$$\varrho_{n-1} e^{-mz_n} C_{n-1} + \varrho_{n-1} D_{n-1} e^{mz_n} - \varrho_n e^{-mz_n} C_n = 0$$
$$-e^{-mz_n} C_{n-1} + D_{n-1} e^{mz_n} + e^{-mz_n} C_n = 0 \qquad (181)$$

This set of equations may be solved directly for all the $C$ and $D$ parameters by working backward from the last equation in set (181). Solving this equation for $C_n$:

$$C_n = C_{n-1} - D e^{2mz_n} \qquad (182)$$

The parameter $D_{n-1}$ may be found from the next to last equation in the set when this value for $C_n$ is substituted in that equation:

$$D_{n-1} = \frac{\varrho_n}{\varrho_{n-1}} \left( C_{n-1} - D e^{2mz_n} \right) e^{-2mz_n} - C_{n-1} e^{-2mz_n} \qquad (183)$$

This process is continued until a solution is found for $D_0$ (or $B_0$). The final expression for the potential at the earth's surface has been expressed by Sunde (1949) in a compact form which facilitates computation:

$$U_0(z = 0) = U_1(z = 0) = \frac{\varrho_1 I}{4\pi} \int_0^\infty [C_1(m) + D_1(m)] J_0(mr) \, dm \qquad (184)$$

The kernel function for the integration, $C_1(m) + D_1(m)$, may be found using a regression formula:

$$C_1(m) + D_1(m) = 2F_{n-1}(m) \qquad (185)$$

where:

$$F_1(m) = \frac{1 - G_1(m) \exp[-2m(z_n - z_{n-1})]}{1 + G_1(m) \exp[-2m(z_n - z_{n-1})]}$$

$$G_1(m) = \frac{\varrho_{n-1} - \varrho_n}{\varrho_{n-1} + \varrho_n}$$

$$\vdots$$

$$F_p(m) = \frac{1 - G_p(m) \exp[-2m(z_p - z_{p-1})]}{1 + G_p(m) \exp[-2m(z_p - z_{p-1})]}$$

$$G_p(m) = \frac{\varrho_{n-p} - \varrho_{n-p+1} F_{p-1}(m)}{\varrho_{n-p} + \varrho_{n-p+1} F_{p-1}(m)}$$

$$F_{n-1}(m) = \frac{1 - G_{n-1} \exp[-2m(z_2 - z_1)]}{1 + G_{n-1} \exp[-2m(z_2 - z_1)]}$$

$$G_{n-1}(m) = \frac{\varrho_1 - \varrho_2 F_{n-2}(m)}{\varrho_1 + \varrho_2 F_{n-2}(m)} \tag{186}$$

The behavior of the kernel function for the single overburden case is shown graphically in Fig. 82 for a variety of contrasts in resistivity between the overburden and the substratum. For more layers, the behavior of the kernel function will be more complicated, but the curve relating it to $mz_n$ will vary smoothly within the limits of the two extreme curves shown in Fig. 82.

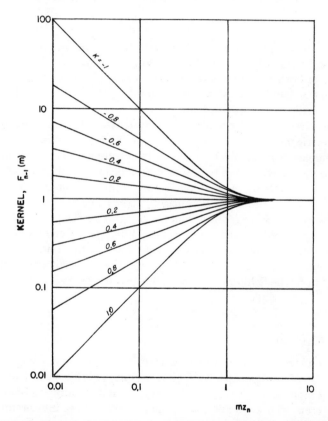

FIG. 82. Numerical values for the kernel function for the single overburden problem, plotted as a function of the dummy variable of integration, $m$. Parametric values for each curve are the values of reflection coefficient for the contrast between the overburden and the bedrock. The kernel function for a greater number of layers must lie between the two extreme curves shown here.

By using the Weber–Lipschitz identity for the Bessel's function see (Eq. (174)), Eq. (185) can be reduced to exactly the same form as that obtained with the image theory for the single overburden case (see Eq. (135)). Since Eq. (135) is non-convergent for reflection coefficients of $\pm 1$, Eq. (186) must also be non-convergent under the same conditions. In evaluating Eq. (186) in a practical problem, there is an advantage in calculating the electric field, $E$, rather than the potential, $U$, since the expression for the electric field converges even for infinite contrasts in resistivity. The electric field along the surface of the earth can be found from Eq. (186) by taking a derivative with respect to $r$:

$$E = -\frac{\varrho_1 I}{2\pi} \int_0^\infty F_{n-1}(m) \, J_1(mr) \, m \, \mathrm{d}m \qquad (187)$$

The calculation of apparent resistivity once the electric field has been computed is straightforward. For the Schlumberger array, the electric field is doubled to take both current-electrodes into account, and this value is then substituted in the defining equation for apparent resistivity measured with the Schlumberger array (Eq. 78).

The apparent resistivity for a Wenner array is found by numerical integration of the electric field between the limits $r = a$ and $r = 2a$, with the resulting potential difference being doubled to take both current electrodes into account.

The second derivative of the potential function, taken with respect to $r$, is not convergent. When the second derivative is taken, the Bessel function $J_1(mr)$ reverts to a zero order Bessel function, $J_0(mr)$, and the resulting integral does not converge any more rapidly than the integral for the potential function itself. The electric field expression in Eq. (187) may be differentiated numerically to obtain the apparent resistivity measured with a dipole array. For either the polar or equatorial displaced dipole arrays, the apparent resistivity may be computed from the Schlumberger apparent resistivity as follows:

$$\varrho_{a,d} = \varrho_{a,s} - \frac{a}{2} \frac{\partial \varrho_{a,s}}{\partial a} \qquad (188)$$

### 23 b. Partial Curve Matching

The trial and error computation of an exact matching curve using Eq. (187) requires the use of a large capacity digital computer. While such installations are becoming more and more available, they are not usually easily accessible to the field crew who are expected to make preliminary interpretations of field data. Preliminary interpretations must be made in the field so that soundings may be located in the best areas to obtain good results, and so that poor results may be recognized before a great deal of field effort has been expended. Partial curve matching is the procedure most commonly used for preliminary interpretation.

In partial curve matching, short segments of a resistivity sounding curve are selected for interpretation using the theoretical curves for the single overburden, usually starting with the shorter spacings and working towards the longer spacings. As each portion of the curve is interpreted the layers comprising the interpreted portion of the sounding curve are lumped together to form a fictitious uniform layer with a lumped resistivity, $\varrho_f$, and a lumped thickness, $h_f$. This fictitious layer is then used in place of the surface layers when the next portion of the curve is analyzed.

The partial curve matching procedure is best described using a specific example. The sounding data shown in Fig. 83 were obtained in the course of a ground-water survey near Phoenix, Arizona. The surface material was alluvium, covering basement rock, probably granite. The field curve has a minimum, so it is type H.

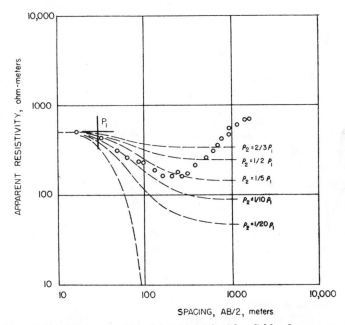

FIG. 83. Resistivity sounding data obtained with a Schlumberger array near Phoenix, Arizona. The first portion of the field data has been matched with the theoretical curves for a single overburden.

The first step in partial curve matching is the comparison of the portion of the curve obtained with short spacings with the theoretical curves for a single overburden. These theoretical curves were superimposed over the field data and moved around until the first portion of the field data matched with one of the curves, as shown in Fig. 83.

The location of the origin on the theoretical curves is shown as the point $P_1$ in Fig. 83. The locus of this point on the plot of field data indicates the resistivity of the first layer ($\varrho_1 = 550$ ohm-m) and the thickness ($h_1 = 46$ m). The resistivity contrast parameter for the particular theoretical curve which matches with the field data indicates the resistivity for the second layer. In this example, the best fit was obtained with a theoretical curve interpolated between $\varrho_2 = 1/5 \varrho_1$ and $\varrho_2 = 1/10 \varrho_1$, so that the second layer resistivity was interpreted to be 105 ohm-m.

We must now seek a way to combine the first two layers into a single fictitious layer so that the right-hand portion of the field data may be interpreted. The lowermost layer apparent in the field data appears to be highly resistant, so that for large electrode separations, it may be assumed that practically all the current is flowing in the upper two layers, parallel to the bedding planes. The current densities in the two layers should be inversely proportional to the resistivities in the two layers. The total conductance of the two layers is the sum of the individual conductances:

$$S_f = S_1 + S_2 \tag{189}$$

or, in terms of resistivities:

$$\frac{h_f}{\varrho_f} = \frac{h_1}{\varrho_1} + \frac{h_2}{\varrho_2} \tag{190}$$

This equation, which is to be used in defining the properties of the fictitious surface layer, contains three unknown quantities, $h_f$, $\varrho_f$ and $h_2$. We may also require that the thickness of the fictitious layer be equal to the sum of the thicknesses of the two real layers:

$$h_f = h_1 + h_2 \tag{191}$$

We now have two equations in three unknowns, still not enough to determine uniquely the parameters for the fictitious layer. However, we may solve these two equations for all possible pairs of values $\varrho_f$ and $h_f$ by assuming a range of values for $h_2$. For example:

| when $h_2 =$ | $h_f =$ | and | $\varrho_f =$ |
|---|---|---|---|
| 0 m, | 46 m | and | 550 ohm-m |
| 10 | 56 | | 313 |
| 20 | 66 | | 240 |
| 50 | 96 | | 171 |
| 100 | 146 | | 141 |
| 200 | 246 | | 123 |
| 500 | 546 | | 113 |

These points, when plotted on the field curve as shown in Fig. 84, define a curve which starts at $P_1$, the locus of the origin of the theoretical curves after they have been matched with the first portion of the field data, and which gradually approaches the resistivity found for the second layer. Such a curve is called an *auxiliary curve*.

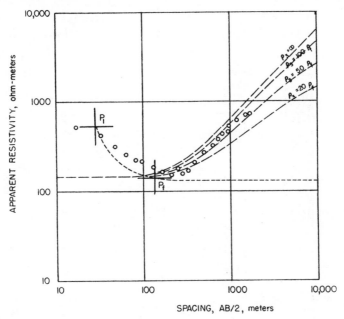

FIG. 84. Interpretation of the right-hand portion of the field data shown in Fig. 83. The auxiliary curve defines all possible combinations of $\varrho_f$ and $h_f$.

We are now prepared to interpret the terminal portion of the field data. Two conditions must be satisfied:

1. The right-hand portion of the field data must fall along one of the theoretical curves computed for a single overburden, and
2. The origin of the theoretical curves must fall along the auxiliary curve.

To accomplish this, the family of theoretical curves is superimposed over the field data and moved around until one of the curves matches the right-hand portion of the field data while the origin of the theoretical curves lies on the auxilliary curve. With the match selected as shown on Fig. 84, we have the interpretation:

$$\varrho_f = 140 \text{ ohm-m}$$
$$h_f = 130 \text{ m}$$
$$\varrho_3 = 74 \, \varrho_f = 9750 \text{ ohm-m}$$

The thickness of the second layer is 96 m.

It is not necessary that the auxiliary curve be computed each time a field curve is interpreted. When plotted on bi-logarithmic coordinates, the shape of an

148    ELECTRICAL METHODS IN GEOPHYSICAL PROSPECTING

auxiliary curve is a function only of the contrast in resistivity between the first and second layers. A family of auxiliary curves for the type H problem is shown in Fig. 85. Such curves must be plotted to the same logarithmic scales as the field data in order that they may be transferred directly to the field plot. In use, the origin of the auxiliary curve family is placed at the point

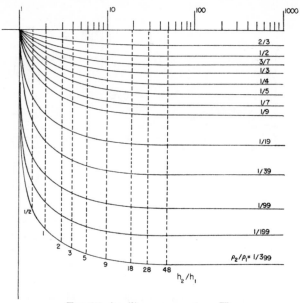

FIG. 85. Auxiliary curves, type H.

$P_1$, the locus of the origin of the theoretical curves for a single overburden when they are matched with the left-hand portion of the field data. The auxiliary curve corresponding to the resistivity ratio found from the initial match is traced onto the field data, and the interpretation is completed as above.

In establishing the conditions for the properties of a surface fictitious layer, it was assumed that the conductances for the two surface layers could be considered to be electrically in parallel. This assumption can be made only for type H curves, those in which the lowermost layer is resistant, so that the current is constrained to flow horizontally through the two upper layers. Different sets of conditions must be established for the other three classes o curves, A, K and Q, in order to use partial curve matching.

In the case of a type A curve ($\varrho_1 < \varrho_2 < \varrho_3$), we will require that the conductance and the transverse resistance of the fictitious layer will be the sums of the corresponding quantities for the real layers:

$$S_f = S_1 + S_2 \quad \text{or} \quad \frac{h_f}{\varrho_f} = \frac{h_1}{\varrho_1} + \frac{h_2}{\varrho_2} \tag{192}$$

$$T_f = T_1 + T_2 \quad \text{or} \quad h_f\varrho_f = h_1\varrho_1 + h_2\varrho_2 \tag{193}$$

Solving these two equations for the resistivity and thickness of the fictitious layer, we have:

$$h_f = [(S_1 + S_2)(T_1 + T_2)]^{1/2} \tag{194}$$

$$\varrho_f = [(T_1 + T_2)/(S_1 + S_2)]^{1/2} \tag{195}$$

The fictitious layer is thicker than the combination of the top two layers, and the factor by which it is thicker is exactly the coefficient of anisotropy for these two layers. (This is the reason this class of curves is designated by the letter A.)

$$h_f = \lambda(h_1 + h_2) \tag{196}$$

The auxiliary curves computed according to Eqs. (194) and (195) are shown in Fig. 86. In use, these curves are superimposed on a set of field data in such a way that the horizontal axis at the top of the graph lies along the value found for $\varrho_2$ by the initial curve match, and the left vertical axis of the graph passes

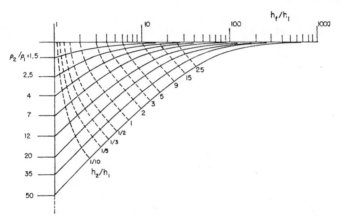

FIG. 86. Auxiliary curves, type A.

through the point, $P_1$, the origin for the single-overburden curves when the first portion of the field data is matched. The appropriate auxiliary curve (determined by the ratio in resistivities between the first two layers) is then traced onto the field plot.

Since the thickness of the fictitious layer is not equal to the sum of the thicknesses of the first two layers, it must be calculated using Eq. (196) after the right-hand portion of the field data has been interpreted. This computation may be carried out graphically, using the same family of auxiliary curves shown in Fig. 86. The dashed curves connect all points which have the same values for the ratio, $h_2/h_1$. After the right-hand portion of the data has been matched with a theoretical curve, the point $P_f$ indicating the proper values for the thickness and resistivity of the fictitious layer is marked on the field plot.

The auxiliary curves are superimposed a second time on the field plot, and the true thickness $h_2$ is found by noting the parameter for the dashed curve which passes through the point $P_f$.

This procedure is best demonstrated by example. Generally, type A curves are the most difficult to interpret, since they are not readily distinguished from single overburden curves unless the second layer is very thick. The diagnostic characteristic of a type A curve is that the apparent resistivity increases too slowly with increasing spacing to provide a good match with any of the single overburden curves. As an example, we shall consider a computed type A curve so that we know the correct interpretation. Such a curve is shown in Fig. 87 for a case in which the ratio of resistivities between the three layers is 1 : 3 : 10 and the ratio in thickness between the first two layers is 1 : 5. The theoretical curves for a single overburden are shown superimposed on the data. The location of the origin, $P_1$, is reasonably unique, but the selection of the theoretical curve for $\varrho_2/\varrho_1 = 3$ is somewhat arbitrary.

Using the auxiliary curves from Fig. 86, the curve corresponding to a resistivity contrast of 3 : 1 is traced on the field data, starting at point $P_1$ and rising asymptotically to a value $3\varrho_1$. Next, the theoretical curves for a single

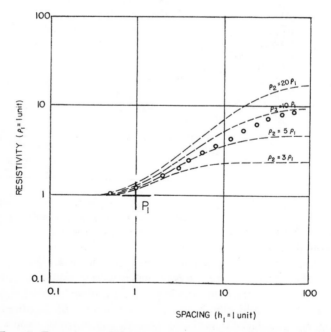

Fig. 87. Type A sounding curve computed for a sequence of resistivities 1 : 3 : 10 and a ratio of thicknesses between the first two layers of 1 : 5. The left-hand portion of the curve has been matched with the theoretical curves for a single overburden.

overburden are placed over the field data and moved around until a match with the right-hand portion of the data is obtained (the origin of the theoretical curves is restricted to lie along the auxiliary curve). The matching curve indicates a resistivity ratio, $\varrho_3/\varrho_f = 3\cdot7$. The locus of the point $P_f$ indicates the thickness and resistivity of the fictitious layer:

$$\varrho_f = 2\cdot7$$
$$h_f = 6\cdot5$$

The correct thickness for the second layer is found by placing the auxiliary curves in Fig. 86 over the field data a second time, and noting which of the dashed curves passes through the point $P$ . In this example, the dashed curve meeting this requirement has the parameter 5, so the thickness of the second layer is five times the thickness of the first layer (Fig. 88).

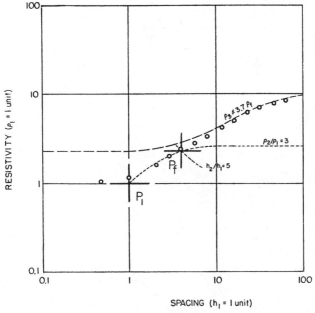

FIG. 88. Interpretation of the right-hand portion of the curve from Fig. 87.

For type K curves, the fictitious layer has been found empirically to have a greater thickness than the combined thickness of the top two layers, and the ratio by which the fictitious layer is thicker is greater than the anisotropy ratio. The anisotropy ratio must be increased by a factor, $\varepsilon$, which depends on the anisotropy ratio as shown in Fig. 89. With this requirement, the equations defining the resistivity and thickness of the fictitious layer are:

$$h_f = \varepsilon[(S_1 + S_2)(T_1 + T_2)]^{1/2} \qquad (197)$$

and

$$\varrho_f = [(T_1 + T_2)/(S_1 + S_2)]^{1/2} \tag{198}$$

A family of auxiliary curves computed according to Eqs. (197) and (198) is shown in Fig. 90.

An example of a K-type curve is shown in Fig. 91. These data were obtained over a section consisting of alluvium, welded tuff and bedded tuff. The left-hand portion of the data can be fitted with a theoretical curve for a single

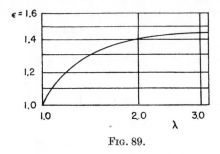

FIG. 89.

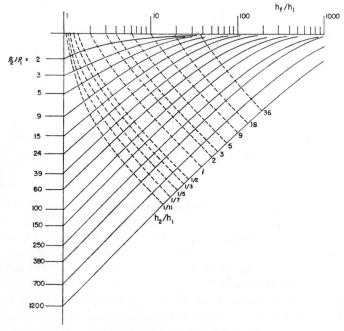

FIG. 90. Auxiliary curves, type K.

overburden, as shown in Fig. 91. The locus of the point $P_1$ provides the information:

$$h_1 = 42 \text{ m}$$

$$\varrho_1 = 235 \text{ ohm-m}$$

$$\varrho_2 = 50\,\varrho_1 = 11{,}700 \text{ ohm-m}$$

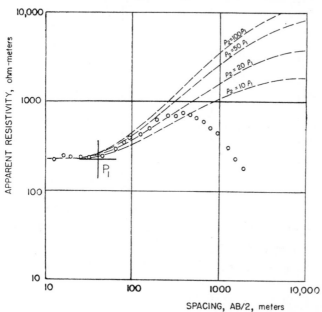

FIG. 91. Type K field data obtained over a section containing alluvium, welded tuff and bedded tuff. The left-hand portion of the data has been matched with the theoretical curves for a single overburden.

The right-hand portion of the data is interpreted using an auxiliary curve from the family in Fig. 90. The type K auxiliary curve for a resistivity ratio of 50 : 1 is shown plotted over the field data in Fig. 92. Next, the single-overburden curves were superimposed on the data to match the right-hand portion of the data. The locus of the point $P_f$ provides the information:

$$h_f = 525 \text{ m}$$

$$\varrho_f = 2500 \text{ ohm-m}$$

$$\varrho_3 = 1/50\,\varrho_f = 50 \text{ ohm-m}$$

The thickness of the second layer may be determined graphically by placing the auxiliary curves (Fig. 90) over the field data and noting which of the $h$-parametric curves passes through the point $P_f$. For the $h$-parametric curve traced on the example in Fig. 92, the second layer has a thickness 2·6 times

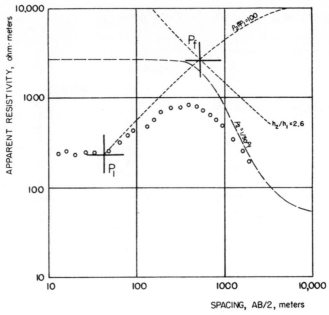

SPACING, AB/2, meters

FIG. 92. Interpretation of the right-hand portion of the field data shown in
Fig. 91.

greater than the thickness of the first layer. The complete solution provides the
information:

$$h_2 = 2 \cdot 6, \quad h_1 = 109 \text{ m}$$

In addition to the parameters interpreted above, it is possible to place a
minimum depth at which resistant rock can lie by drawing a line with a slope
of $+1$ through the last point of the field data. The locus of this line defines the
least value for the conductance of all the beds above the basement. In this
example, this least conductance is:

$$S_{\min} = \frac{a_{\text{last}}}{\varrho_{\text{last}}} = 10 \text{ mhos}$$

where the subscript "last" indicates that the quantities $\varrho$ and $a$ are taken from
the last point on the sounding curve. Using the results obtained from the
interpretation described above, the total conductance of the upper three
layers is:

$$S_t = \frac{h_1}{\varrho_1} + \frac{h_2}{\varrho_2} + \frac{h_3}{\varrho_3} = 0 \cdot 179 + 0 \cdot 0093 + \frac{h_3}{50}$$

The minimum value for the thickness of the third layer is found by substituting
the minimum value for $S_t$ (10 mhos) in this last equation. Doing so, we find
that the thickness of the third layer is at least 490 m.

The remaining type curve to consider is the Q-type curve, in which each of the beds has successively lower resistivities. It has been found empirically that the fictitious layer representing the first two real layers in such a sequence will have a thickness and resistivity which are smaller than the average values for the two real layers by the factor $1/\eta$:

$$h_f = \frac{1}{\eta} (h_1 + h_2) \qquad (199)$$

$$\varrho_f = \frac{1}{\eta} \frac{h_1 + h_2}{S_1 + S_2} \qquad (200)$$

The factor $\eta$ depends on both the resistivity contrast between the first two layers and on the ratio of their thicknesses (see Fig. 93). Auxiliary curves constructed according to Eqs. (199) and (200) are shown in Fig. 94.

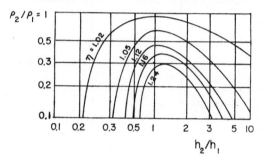

FIG. 93.

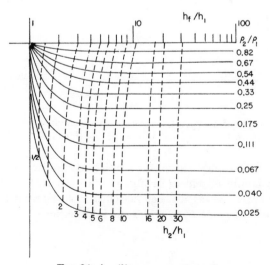

FIG. 94. Auxiliary curves, type Q.

As an example of the interpretation of a type Q curve, a set of data were computed for the case in which the second layer has a resistivity 1/3 that of the first layer, and is five times as thick as the first layer, and in which the third layer has a resistivity 1/10 that of the first layer. Type Q curves are frequently difficult to distinguish from a single overburden curve for a conductive substratum, so a computed example is more informative than an actual example. The match between the first portion of these computed data and the single-overburden curves is shown in Fig. 95. The location of point $P_1$ provides the information:

$$h_1 = 1\cdot0 \text{ (arbitrary units)}$$
$$\varrho_1 = 1\cdot0$$
$$\varrho_2 = 1/3\varrho_1 = 1/3$$

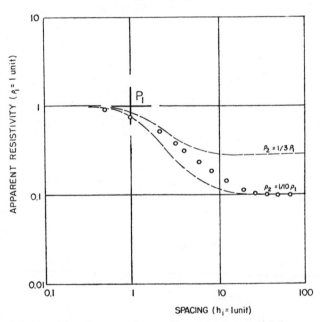

FIG. 95. Example of a type Q sounding curve computed for a resistivity sequence $1:1/3:1/10$ and for a ratio of thicknesses $1:5$.

The appropriate auxiliary curve (for a resistivity contrast of $1:1/3$) is then traced from the family of curves in Fig. 94 onto the data plot (see Fig. 96). The single-overburden curves are then matched with the right-hand portion of the data. The locus of point $P_f$ provides the information:

$$h_f = 3\cdot4$$
$$\varrho_f = 0\cdot33$$
$$\varrho_3 = 0\cdot32 \quad \varrho_f = 0\cdot105$$

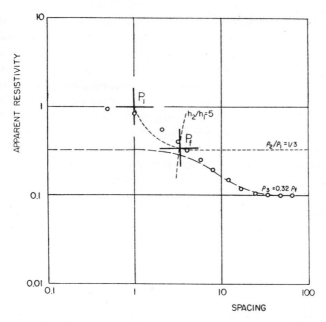

FIG. 96. Interpretation of the right-hand portion of the data from Fig. 95.

The final step in interpretation is to replace the auxiliary curves over the data plot and note which of the $h$-parametric curves passes through the point $P_f$. By so doing, we find that the second layer has a thickness five times that of the first layer.

The least depth at which a resistant layer may exist can be found from the value of $S$ corresponding to the last data point. In this example, the $S$ value for the last data point is 640 (arbitrary units). The total conductance of the first three layers is:

$$S_t = \frac{1}{1} + \frac{5}{0 \cdot 33} + \frac{h_3}{0 \cdot 105}$$

The minimum thickness for the third layer is found to be 65·6.

The partial curve matching technique works best for type H curves (those which show a minimum in apparent resistivity) and least well for type A curves (successively increasing resistivity with increasing spacing). Somewhat better results are usually obtained in interpreting type K curves than in interpreting type Q curves. Curve matching requires considerable subjective judgement on the part of the interpreter, and consistent results can be obtained only after considerable practice.

### 23 c. Equivalent curve matching

It is not always possible to obtain a unique solution to an interpretation problem using partial curve matching and fictitious layers. In many cases, the contrast in resistivity between the first and second layers may be great enough that the resistivity of the second layer cannot be established within reasonable limits by matching the first portion of the field data with single overburden curves. There has been no provision made so far for estimating the degree of uniqueness for an interpretation arrived at by partial curve matching, and this information is essential in evaluating fully an interpretation.

Usually, the precision with which earth resistivity measurements can be made is about $\pm 5$ per cent. Variations in near surface resistivity contribute at least this much scatter to measured resistivities, even under ideal conditions Unless a sounding curve varies by at least 5 per cent between two sets of layering parameters, it is not possible to distinguish between these two sets of conditions in interpreting the curve. When different sets of layering parameters provide the same sounding curve within 5 per cent, these sets of conditions are said to be *equivalent*.

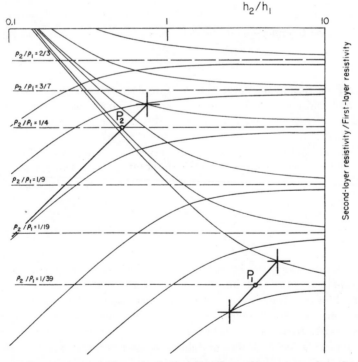

Fig. 97. Ranges in equivalence in $S_2$ for type H sounding curves when the third layer has the same resistivity as the first layer.

Graphs may be constructed to show the ranges in layering parameters which are equivalent, and such a graph is shown for H-type curves in Fig. 97. The horizontal lines indicate the resistivity contrasts for which theoretical three-layer curves have been computed (for the Schlumberger array). Each of these lines is bounded above and below by lines indicating the range through which the conductance of the second layer may vary while causing no more than a 5 per cent change in the sounding curve. The parameter $S$ is chosen to study equivalence in type H curves since we may expect that an increase in the thickness of the second layer has an effect which would be offset by a decrease in the resistivity of the second layer, over rather wide limits. Constant $S_2$ values define lines with a slope of $+1$ on Fig. 97.

For example, suppose the following interpretation had been obtained for a type H curve:

$$\varrho_2/\varrho_1 = 1/39$$
$$h_2/h_1 = 3 \cdot 7$$

These two values define the position of point $P_1$ on Fig. 97. The locus of all pairs of values, $\varrho_2/\varrho_1$ and $h_2/h_1$, which provide the same value for $S_2$ is shown by a line with slope $+1$ passing through the point $P_1$. The range in $S_2$ which will cause a 5 per cent change in the sounding curve is determined from the intercepts of this line with the two bounding curves. The range through which the resistivity and thickness of the second layer may vary (in this example) and yet provide essentially the same sounding curve is:

$0 \cdot 018 < \varrho_2/\varrho_1 < 0 \cdot 037$, for the resistivity contrast, and

$2 \cdot 5 < h_2/h_1 < 4 \cdot 8$, for the ratio in thicknesses.

The thickness or the resistivity of the second layer may be changed as much as 100 per cent, yet the measured resistivity will vary only by 5 per cent. This is by no means an extreme example; consider a second example in which the following interpretation had been made:

$$\varrho_2/\varrho_1 = 0 \cdot 25$$
$$h_2/h_1 = 0 \cdot 53$$

These values correspond to the point $P_2$ on Fig. 97. A line of constant $S_2$ passing through point $P_2$ never intersects the lower limit curve. The ranges through which the second layer parameters may vary without changing the interpretation are:

$$0 < \varrho_2/\varrho_1 < 0 \cdot 35$$
$$0 < h_2/h_1 < 7 \cdot 6$$

In this example, the only parameters which can be specified for the second layer are its conductance, $S_2$, and a maximum value for the resistivity. There exists a minimum second-layer conductance, depending on the contrast in resistivity between the first and second layers, for which a unique interpretation of a sounding curve in terms of second layer resistivity can be made. This minimum conductance for which unique interpretation is possible is given graphically in Fig. 98. As the contrast in resistivity between the first and second layers is increased, the thickness of the second layer required for complete equivalence increases at a faster rate.

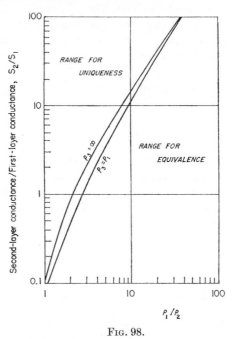

Fig. 98.

In the region of complete equivalence, interpretation can be made only in terms of the ratio of conductances between the first two layers. This suggests that it may be preferable to use three-layer curves for matching, with this ratio of conductances used as a curve parameter rather than the resistivity contrasts between the various layers. A family of H-type curves for layering cases which are equivalent in $S_2$ is shown in Fig. 99.

These curves may be used for curve matching in the same way the single-overburden curves are used: they are placed over the field data and moved around until one of the theoretical curves matches the data. If a match can be obtained, the resistivity and thickness are determined from the position of the origin of the theoretical curves on the data plot.

An example will serve to show the use of these equivalent curves in the interpretation of a type H curve. The sounding data plotted in Fig. 100 were obtained over a section consisting of a shale bed included between two limestone

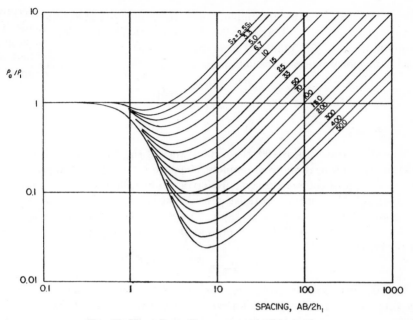

FIG. 99. Three-layer H curves, equivalent in $S_2$.

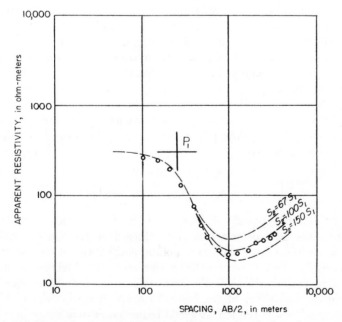

FIG. 100. Example of a sounding curve interpreted with three-layer curves equivalent in $S_2$.

layers. When these data are compared with the equivalent H curves in Fig. 99, a very good match is obtained with the curve for a ratio $S_2/S_1$ of 100. The position of the origin (point $P_1$) from the theoretical curves transferred to the field data plot provides information on the properties of the first layer:

$$h_1 = 240 \text{ m}$$
$$\varrho_1 = 380 \text{ ohm-m}$$
$$S_1 = h_1/\varrho_1 = 0\cdot63 \text{ mhos}$$

Since the ratio $S_2/S_1$ is 100, the conductance of the second layer is 63 mhos. The interpretation is not yet complete, since we must place an upper limit on the resistivity of the second layer. Referring to Fig. 98, we see that for a ratio $S_2/S_1$ of 100, the contrast in resistivities between the first two layers must be 40 or greater for equivalence to apply. The resistivity of the second layer (the shale) must be less than 10 ohm-m, and the maximum permissible thickness of the shale is:

$$h_2 < \varrho_{2,\,\text{max}} S_2 = 630 \text{ m}$$

Usually, it is possible to be more exact in interpretation than is permitted by equivalence, if anything is known about the properties of the equivalent layer. In this example, any combination of resistivity and thickness values less than the maximum values listed which provide the same conductance for the second layer would be a correct interpretation of the data, insofar as curve matching requirements are met. For instance, the second layer could be assigned a thickness of 1 cm and a resistivity of 0·00016 ohm-m. Such an interpretation would be difficult to accept if anything at all is known about the stratigraphy. In this example, it is known that the second layer is shale, and the minimum resistivity which a shale ordinarily can be expected to have is about 1 ohm-m. Therefore, in addition to the maximum limits placed on the resistivity and thickness of the layer by the conditions of equivalence, we may also place minimum values on these parameters using our knowledge of the electrical properties of rocks in general:

$$\varrho_{2,\text{min}} = 1 \text{ ohm-m}$$
$$h_{2,\text{min}} = \varrho_{2,\text{min}} S_2 = 63 \text{ m}$$

Each of the four types of sounding curves has its individual conditions for equivalence. The parameter which is considered in defining equivalence for type K curves (those with maximum points) is the transverse resistance of the second layer, $T_2$. We may expect that a decrease in the thickness of the middle layer in a K-section will have the same effect on apparent resistivity as a decrease in the resistivity of the layer. In order to maintain the shape of a sounding curve for a K-section, it would be necessary to increase the thickness of the middle layer as its resistivity was lowered, or vice versa.

Graphs showing the range in uncertainty in the interpretation of type K curves are given in Fig. 101. As in Fig. 97 (for H-type curves), the horizontal lines indicate the values of resistivity contrast between the first and second layers for which three-layer theoretical curves have been computed. Each of

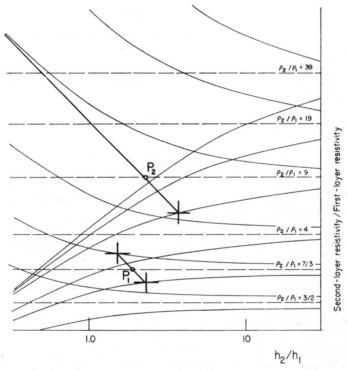

FIG. 101. Ranges in equivalence in $T_2$ for type K sounding curves for cases in which the third layer has the same resistivity as the first layer.

these lines is bounded by a pair of curves, one above and one below, which indicate the range within which the transverse resistance of the second layer may be varied while producing no more than a 5 per cent change in the shape of the sounding curve. Constant $T_2$ values define straight lines with a slope of $-1$.

Consider an example of the use of the curves in Fig. 101. Assume that three-layer theoretical curves have been used to interpret a set of data, with the result:

$$\varrho_2/\varrho_1 = 2 \cdot 33$$
$$h_2/h_1 = 1 \cdot 9$$

These two values determine the location of point $P_1$ on Fig. 101. All pairs of values for the ratios $\varrho_2/\varrho_1$ and $h_2/h_1$ which correspond to the same value for

$T_2$ lie along a line with slope $-1$ passing through point $P_1$ (such a line is shown on Fig. 101). The extreme range in values for the thickness and resistivity of the second layer which could have provided the same sounding curve are:

$$1\cdot5 < h_2/h_1 < 2\cdot3$$
$$2\cdot0 < \varrho_2/\varrho_1 < 3\cdot0$$

The values for thickness and resistivity of the second layer may have varied from the interpreted values by as much as 50 per cent with no change in the shape of the sounding curve.

Again, this is not an extreme example. Consider point $P_2$ on Fig. 101, which corresponds to the interpretation:

$$\varrho_2/\varrho_1 = 9$$
$$h_2/h_1 = 2\cdot2$$

A line representing a constant value for $T_2$ passing through point $P_2$ never intersects the upper limit curve, indicating that there is no value for $\varrho_2$ so large that it would change the shape of the sounding curve by 5 per cent, providing the transverse resistance remained the same.

For a given ratio of resistivities between the first and second layers, there is a minimum value for $T_2$ below which it is not possible to separate the effects of thickness and resistivity, and interpretation can be made only in terms of the transverse resistance of the second layer and a minimum limit for the resistivity. Numerical values for the minimum transverse resistance are given graphically in Fig. 102.

If the conditions for complete equivalence are met, a single three-layer family of curves based on $T_2$ as a parameter may be used for curve matching. Such a family of curves for the Schlumberger array is shown in Fig. 103. As with equivalent curves of type H, interpretation is made in terms of $T_2$ and a minimum value for $\varrho_2$ and maximum value for $h_2$ which are compatible with the requirements for equivalence.

An example of a type K sounding curve which may be interpreted using three-layer curves equivalent in $T_2$ is shown in Fig. 104. These sounding data were measured over a section consisting of alluvium, rhyolite flow rock and rhyolite tuff, in order downwards. The flow rock has a high resistivity in comparison with either the overlying alluvium or the underlying tuff. The data match well with a theoretical curve interpolated between the curves for $T_2 = 500\ T_1$ and $T_2 = 1000\ T_1$ on Fig. 103. The information provided by this interpretation is:

$$\varrho_1 = 200 \text{ ohm-m}$$
$$h_1 = 21 \text{ m}$$
$$T_1 = \varrho_1 h_1 = 4200 \text{ ohm-m}^2$$
$$T_2 = 700\ T_1 = 2{,}940{,}000 \text{ ohm-m}^2$$

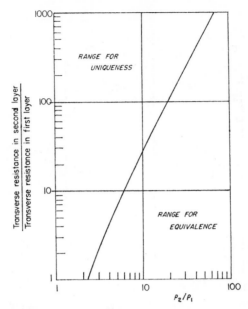

FIG. 102. Range of complete equivalence for type K curves.

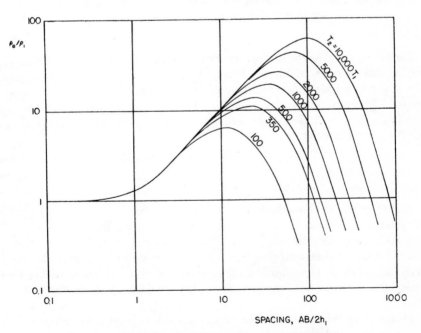

FIG. 103. Three-layer K curves, equivalent in $T_2$.

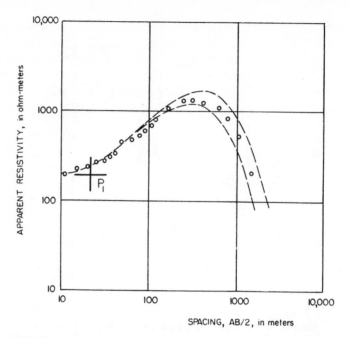

FIG. 104. Example of sounding data obtained over a type K section interpreted with three-layer curves equivalent in $T_2$.

Moreover, the range within which values for $\varrho_2$ and $h_2$ may fall is limited by the conditions for equivalence. When the ratio of transverse resistances for the upper two beds is 700 : 1, the ratio in resistivities between the first and second layers must be greater than 57 : 1 for equivalence to apply. Considering this limitation, we find that the minimum resistivity and maximum thickness for the second layer are:

$$\varrho_{2,\min} = 57 \; \varrho_1 = 11{,}400 \text{ ohm-m}$$

$$h_2 < T_2/\varrho_{2,\min} = 256 \text{ m}$$

As in the case of equivalent H curves, we must turn to some other source of information to obtain a more restricted interpretation. In this example, the flow rock comprising the second layer outcrops extensively on nearby mesas, where it has a thickness ranging from 100 to 150 m. Considering that the second layer apparent on the sounding curve should have a thickness within these limits, the corresponding range in resistivity would be from 20,000 to 30,000 ohm-m.

Determining the probable resistivity of a resistant layer causing equivalence in a type K sounding curve is not easily done. Generally, resistant beds vary in electrical properties to a greater extent than do conductive beds, and so, resistivity cannot be extrapolated from measurements made on nearby outcrops or in nearby wells with a high degree of reliability. Since the resistivity of the resistant bed cannot be defined within narrow limits, the thickness cannot be determined either. If determining the thickness of the resistant bed is essential, an electromagnetic sounding method should be used in conjunction with the galvanic sounding method (see Chapter VI).

In many cases, it may not be necessary to determine the thickness of the resistant layer in a type K section. In the example in Fig. 104, if the exploration problem had been the determination of the depth to resistant basement rock, the thickness of the second layer would be of relatively little importance. The sounding data were not taken at large enough electrode separations in this example to indicate the presence of basement, but a minimum depth to basement may be established. The least conductance the top three layers can have is found by drawing a line with slope +1 through the last point. This line corresponds to a minimum conductance of 10 mhos for the top three layers. The maximum conductance the second layer can have according to the inter-

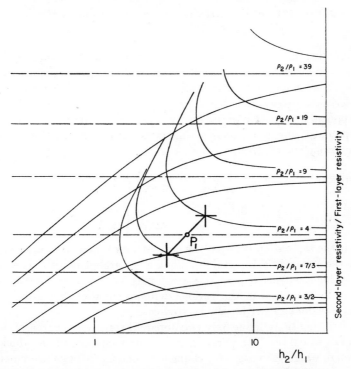

FIG. 105. Ranges in equivalence in $S_2$ for type A sounding curves. Lines of constant $S_2$ values have a slope +1 as shown by the line through point $P_1$.

pretation made above is 0·026 mhos. Thus, the presence of the resistant layer has almost no effect on the position of the asymptotic portion of the sounding data representing the effect of basement rock.

Types A and Q sounding curves also exhibit equivalence for wide ranges in properties for the second layer in such sections. Graphs showing the ranges in uncertainty in the interpretation of types A and Q curves are given in Figs. 105 and 106, respectively. For the A-curves, the conductance of the second layer $S_2$, is taken as the equivalence parameter. Constant values of $S_2$ define lines with slope $+1$, as indicated by the line passing through point $P_1$ on Fig. 105. The transverse resistance of the second layer, $T_2$, is taken as the equivalence parameter for the type Q curves. Constant values of $T_2$ define lines with slope $-1$, as indicated by the line passing through point $P_1$ on Fig. 106. The minimum values of $S_2$ for which A curves are equivalent, and the maximum values of $T_2$ for which Q curves are equivalent are shown graphically in Fig. 107.

Types A and Q curves are commonly measured over sections in which the change in resistivity with depth is gradational rather than occurring in steps. In many cases, such curves may be interpreted satisfactorily in terms of an exponential change in resistivity with depth. Consider that the resistivity of the earth may vary in the vertical direction according to the law:

$$\varrho_z = \varrho_1 e^{-az} \tag{201}$$

where $\varrho_1$ is the resistivity at the surface, and $\alpha$ is a parameter defining the rate of increase (or decrease) of resistivity with depth. If $\alpha$ is large and negative, resistivity increases rapidly with depth, while if $\alpha$ is large and positive, resistivity decreases rapidly with depth.

An exponential variation of resistivity with depth leads to a simpler solution for the potential function about a single-pole current source than does resistivity layering (Sunde, 1949):

For $\alpha$ positive:    $$U = \frac{\varrho_1 I}{2\pi a} \left[ e^{-\alpha a/2} - \frac{\alpha a}{2} E_i\left(\frac{\alpha a}{2}\right) \right] \tag{202}$$

For $\alpha$ negative:    $$U = \frac{\varrho_1 I}{2\pi a} \left[ e^{-\alpha a/2} - \frac{\alpha a}{2} E_i\left(\frac{\alpha a}{2}\right) + \alpha a \ln |\alpha a| \right] \tag{203}$$

where $a$ is the separation between the current source and the measuring point, and $E_i$ is the exponential integral function for the argument $\alpha a/2$. The apparent resistivity for a particular electrode array may be found by substituting this expression for potential in the various defining equations for apparent resistivity, as was done earlier for the layered case.

These two equations define only two apparent resistivity curves, rather than two families of curves. The parameter $\alpha$ is the reciprocal of the thickness of rock through which the resistivity changes by one logarithmic decrement. This is an increase of 2·7 times for negative $\alpha$ values, or a decrease to 37 per cent of the surface resistivity for positive $\alpha$ values.

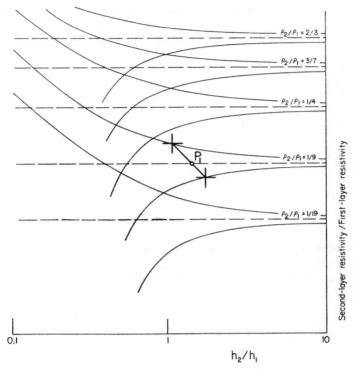

FIG. 106. Ranges in equivalence in $T_2$ for type Q sounding curves. Lines of constant $T_2$ values have a slope of $-1$, as shown by the line through point $P_1$.

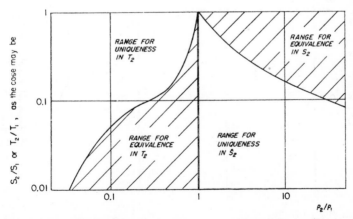

FIG. 107. Range of complete equivalence in types A and Q sounding curves. The right-hand half applies to type A curves; the left-hand half, to type Q curves.

Values for the two potential functions for several values of the ratio $\alpha a$ are:

| $\alpha_a$ | $\dfrac{2\pi a}{\varrho_1}\dfrac{U}{I}$ ; (negative $\alpha$) | $\dfrac{2\pi a}{\varrho_1}\dfrac{U}{I}$ ; (positive $\alpha$) |
|---|---|---|
| 0·1 | 0·939 | 0·828 |
| 0·2 | 0·950 | 0·722 |
| 0·3 | 0·973 | 0·642 |
| 0·4 | 1·017 | 0·575 |
| 0·6 | 1·135 | 0·469 |
| 1·0 | 1·422 | 0·326 |
| 1·4 |  | 0·235 |
| 2·0 | 2·209 | 0·149 |
| 3·0 | 2·83 | 0·073 |
| 4·0 |  | 0·049 |
| 10 |  | 0·00670 |
| 20 |  | 0·000045 |

These two curves are shown plotted on a bi-logarithmic coordinate system in Fig. 108. By matching data with similar curves for a particular electrode array, values for $\varrho_1$ and the reciprocal logarithmic decrement distance, $\alpha$, can be determined.

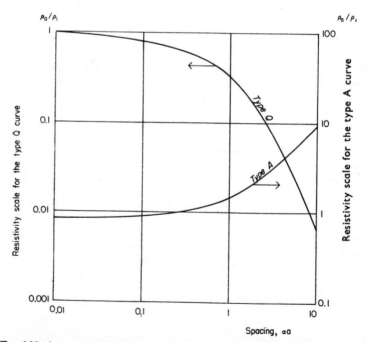

FIG. 108. Apparent resistivity curves for the single-pole array for an exponential increase or decrease in resistivity with depth.

The total transverse resistance may be computed for a section in which $\alpha$ is positive, as follows:

$$T = \int\limits_0^\infty \varrho z \, \mathrm{d}z = \varrho_1 \int\limits_0^\infty z e^{-\alpha z} \, \mathrm{d}z = \frac{\varrho_1 \Gamma(2)}{\alpha^2} = \frac{2\varrho_1}{\alpha^2} \qquad (204)$$

For a negative value of $\alpha$ (type A sounding curve), the total conductance of the section is:

$$S = \int\limits_0^\infty \sigma z \, \mathrm{d}z = \frac{1}{\varrho_1} \int\limits_0^\infty z e^{\alpha z} \, \mathrm{d}z = \frac{\Gamma(2)}{\varrho_1 \alpha^2} = \frac{2}{\varrho_1 \alpha^2} \qquad (205)$$

The use of these curves may be illustrated with a calculated example of a type Q sounding curve taken from Mooney and Wetzel (1957). Figure 109 shows a sounding curve for a sequence of four layers with the following parameters:

$$h_1 = 1 \cdot 0; \quad \varrho_1 = 1 \cdot 0$$
$$h_2 = 2 \cdot 0; \quad \varrho_2 = 0 \cdot 33$$
$$h_3 = 3 \cdot 0; \quad \varrho_3 = 0 \cdot 10$$
$$\varrho_4 = 0 \cdot 01$$
$$T_1 + T_2 + T_3 = 1 \cdot 43$$

This resistivity profile is shown graphically in Fig. 110. The sounding data can be fitted very well with the type Q exponential curve, as shown in Fig. 109, and the interpretation provides the results:

$$\varrho_1 = 1 \cdot 20$$
$$\alpha = 1 \cdot 27$$

This exponential variation of resistivity with depth is shown in Fig. 110, and is a good approximation to the actual stepwise resistivity function. The interpreted transverse resistance is:

$$T = \frac{2\varrho_1}{\alpha^2} = 1 \cdot 48$$

This is well within the 5 per cent equivalence range.

### 23 d. Interpretation by noting the position of maximum and minimum points

With either type H or type K curves (those curves which show a minimum or maximum apparent resistivity), an interpretation may be made by noting the electrode spacing for which the maximum or minimum apparent resistivity is measured. Such a procedure does not make full use of the sounding data, and so, is subject to more errors than the methods already outlined. The method

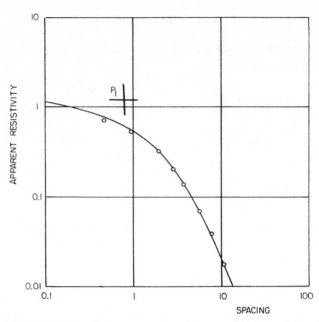

Fig. 109. Computed sounding curve, type Q, for a sequence of four layers with resistivities in the ratio 1 : 0·33 : 0·10 : 0·01, and with layer thicknesses in the ratio 1 : 2 : 3 : ∞. The computed points have been fitted with a theoretical curve for an exponential decrease in resistivity with depth.

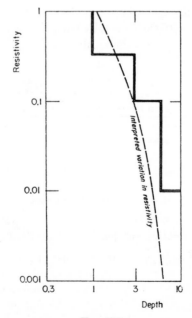

Fig. 110.

is useful for a rapid, preliminary interpretation, or for use with data in which there is a great deal of scatter to the measured resistivity values, so that it is not possible to obtain a satisfactory interpretation with one of the more precise interpretation methods.

Curves showing the relationships between the extremal points for a type K sounding curve as a function of the contrast in resistivity and the ratio of thicknesses for the first two layers are given in Fig. 111 and 112. In Fig. 111, the amplitude of the maximum apparent resistivity is plotted as a function of the resistivity of the second layer, for constant values of second-layer thickness. It is interesting to note that the maximum apparent resistivity is *not* proportional to the true resistivity of the second layer, when the second-layer thickness is held constant. Rather, the amplitude of the maximum apparent resistivity increases only in proportion to the square root of the resistivity contrast. Only for relatively small contrasts between the resistivities of the first two layers does the maximum apparent resistivity rise linearly with the second-layer resistivity.

Figure 112 shows the spacing for which the maximum apparent resistivity was measured plotted as a function of the ratio of thicknesses of the first two layers. Again, we see that the spacing for which the maximum resistivity is measured does not increase linearly with the thickness of the second layer, but only in proportion to the square root of the thickness. As the contrast in properties between the first and second layers is increased, the precision with which the parameters for the second layer can be measured decreases.

The use of these curves may be illustrated by reinterpreting the example of a K-type sounding curve originally considered in Fig. 91, and interpreted there using partial curve matching technique. The amplitude of the maximum, expressed as a ratio to the first-layer resistivity, and the spacing for which the maximum was observed, expressed as a ratio to the first-layer thickness, are:

$$\varrho_{max}/\varrho_1 = 3 \cdot 46$$
$$a_{max}/h_1 = 8 \cdot 83$$

First, a horizontal line is drawn across Fig. 111 at a value $\varrho_{max}/\varrho_1 = 3 \cdot 46$, and the intersection of this line with each of the curves in Fig. 111 provides a pair of values, $\varrho_2/\varrho_1$ and $h_2/h_1$. These values are plotted on a graph as shown in Fig. 113, defining a curve of trial solutions. Any point along this curve represents a pair of values for resistivity contrast and layer thickness ratio which would have caused the observed maximum value for apparent resistivity.

Next, a horizontal line is drawn across Fig. 112 at a value $a_{max}/h_1 = 8 \cdot 83$, and the intersections of this line with each of the curves in Fig. 112 defines pairs of values for $\varrho_2/\varrho_1$ and $h_2/h_1$ which are used to construct a second trial solution curve on Fig. 113. This curve contains all pairs of values for the second-layer resistivity and thickness which would have caused the maximum resistivity to be observed for the electrode spacing at which it was observed.

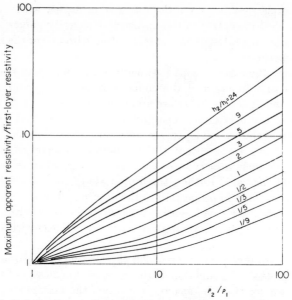

Fig. 111. Relation between the maximum apparent resistivity and the true resistivity in the second layer of a three-layer type K section (for the Schlumberger array).

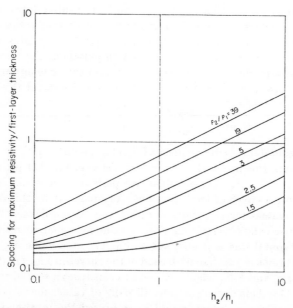

·Fig. 112. Relation between the spacing at which the maximum resistivity is measured and the thickness of the second layer from a type K sequence.

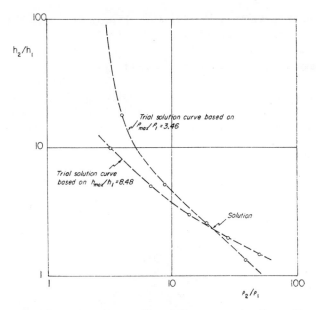

FIG. 113. Interpretation of a type K sounding curve using the position of the maximum point.

The intersection of these two curves represents the only combination of second layer thickness and resistivity that could have caused both the observed maximum apparent resistivity and the spacing at which it was observed. In this example, the two trial solution curves are nearly parallel to one another at the point of intersection, so there is some uncertainty in choosing the exact intersection. This uncertainty represents the effect of equivalence. If the parameters for the second layer had been in the range for complete equivalence, then, rather than intersecting, the two curves would have been tangent to one another, so that any combination of second-layer resistivity and thickness within the equivalence range would be an acceptable solution.

Each horizontal line drawn on Fig. 111 or 112 for an observed extremal defines only one trial solution curve. A series of trial solution curves may be constructed prior to interpretation, and drawn on a graph with a series of trial solution curves based on the maximum spacing curves. In interpretation, when a set of coordinates have been obtained from field data, the corresponding trial solution curves can be selected from such a graph and their intersection noted very quickly. There is no need to construct new trial solution curves for each interpretation problem, as was done in the example. Bogdanov (1952) has presented a series of monographs for use with various types of sounding curves, based on this principle.

Curves similar to those in Fig. 111 and 112 may be compiled also for H-curves, or for electrode arrays other than the Schlumberger.

Interpretation by noting the locus of an extremal point is subject to the same limitations imposed by the principle of equivalence as are any of the other methods of interpretation. In the range of complete equivalence, the position of an extremal point may be used to estimate the transverse resistance of the second layer (for type K curves) or the longitudinal conductance (in type H curves). The relationship between the location of an extremal and the

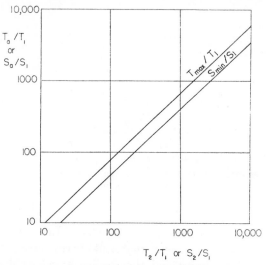

FIG. 114. Relation between maximum apparent conductance (or transverse resistance) and second-layer conductance (or transverse resistance) for completely equivalent type H (or type K) sounding curves.

transverse resistance (or conductance) of the second layer is shown by the curves in Fig. 114. On these two curves, the apparent transverse resistance, defined as:

$$T_a = \varrho_{max} a_{max} \tag{206}$$

and the apparent longitudinal conductance, defined as:

$$S_a = a_{min}/\varrho_{min} \tag{207}$$

are plotted as functions of the actual transverse resistance of longitudinal conductance of the second layer, whichever is appropriate. In both cases, the apparent values are very nearly proportional to the actual values. For resistance or conductance ratios between the first two layers of more than ten to one, the following relations hold:

$$T_2 = 1 \cdot 2\,T \tag{208}$$

$$S_2 = 2S_a \tag{209}$$

It is not necessary even to know the transverse resistance (or conductance) of the first layer to obtain an approximate value for $T_2$ (or $S_2$).

## 24. SUMMARY—INTERPRETATION OF SOUNDING CURVES

All the procedures for the interpretation of sounding curves described above are useful in interpreting field data, but each technique has its own place in the schedule of interpretation:

1. *Interpretation concurrent with field measurements.* It is desirable to make an approximate interpretation of the resistivity sounding data as the measurements are being made in the field. This interpretation must be done rapidly, so that the field manager knows when the electrodes have been carried to large enough separations that the target horizon appears on the sounding curves. Since this interpretation is made rapidly, and in the field, a minimum number of reference curves should be required. The interpretation may consist merely of noting such things as the surface resistivity ($\varrho_1$), the total conductance above the lowermost insulating layer, and the apparent transverse resistance or longitudinal conductance of the second layer, if type H or K curves are being recorded. This amount of interpretation requires the use of no master curves, which are readily lost in the field. If the field manager is aware of the general, electrical character of the horizons being mapped, this amount of preliminary interpretation will tell him when large enough spacings have been reached, and also, will indicate significant changes in the properties of these horizons so that additional sounding locations can be chosen accordingly.

2. *Preliminary interpretation.* A more exact preliminary interpretation using partial curve matching and matching with three-layer theoretical curves should be kept up to date at the field headquarters, so that interpretation is never more than a day or two behind the acquisition of data. Such information is necessary in efficient planning of the field operations, so that the density of sounding locations can be varied to meet the requirements of the interpretation program. If it is found that the parameters of the mapped horizons are varying but little, laterally, in one area, the density of soundings can be reduced. On the other hand, the density of soundings can be increased in an area where interpretation indicates variability in the character of the mapped horizons, while the field crew is still in the same area.

Another important factor to be evaluated as the field program is underway is the quality of the interpretation that can be made. It may be found that equivalence conditions are such that a satisfactory interpretation cannot be made without additional information. This may mean that exploration drill holes must be drilled to determine the electrical parameters for the equivalent horizons, or that an outcrop of the equivalent layer must be sought, or that the galvanic sounding method must be supplemented by some other method, such as electromagnetic sounding or seismic sounding.

3. *Final interpretation.* The final interpretation is best made in a central office, some time after the field survey has been completed. This permits re-evaluation of the sounding curves in light of all information available from the rest of the survey, from other geophysical surveys, and from geologic information. Interpretation at this time may be by curve matching, but preferably, if a computer is available, calculation of an exact theoretical curve by trial and error methods is the best approach.

## 25. RESISTIVITY PROFILING

The interpretation procedures discussed in the preceding three sections are based on an assumed lateral uniformity of layers—the absence of changes in thickness or resistivity of the rock layers being mapped. This assumption does not usually appear to hold true in an exploration problem since, usually, there is some lateral variation in the properties of one of the rock layers which is being sought. Gradual changes in the character of beds do not seriously interfere with the use of interpretation methods based on an assumption of horizontal uniformity. For steeply dipping contacts, or for studying lateral changes in the resistivity of a particular horizon, the resistivity profiling method is used. In resistivity profiling, the spacing between the electrodes is held constant, and the array is moved as a whole along a traverse line.

### 25a. Overburden surveys

One application of resistivity profiling which usually leads to quantitative results is its use in measuring the conductance of a conductive overburden. The problem consists of determining the depth of soil over bedrock, the sort of problem frequently encountered in engineering geology, and occasionally in mining geology. It can usually be assumed that a resistivity sounding curve would have the form shown in Fig. 115(a). The problem of determining the thickness of soil could be solved by making a series of soundings along a traverse line, but this represents a great deal more work than is necessary. Instead, a resistivity profile may be run using two fixed electrode separations. These separations are selected such that the shorter of the two, $a_1$, will measure essentially the soil resistivity even in areas where the soil cover is thinnest (see point $a_1$ in Fig. 115(a)). The longer of the two spacings is selected so that the apparent resistivity measured with it falls along the rising portion of the sounding curve, even in the areas where the soil cover is thickest (see point $a_2$ in Fig. 115(a)).

With this selection of spacings, the soil resistivity is measured with the shorter spacing:

$$\varrho_s = \varrho_1$$

and the soil conductance is measured with the longer spacing:

$$S_s = a_2/\varrho_2$$

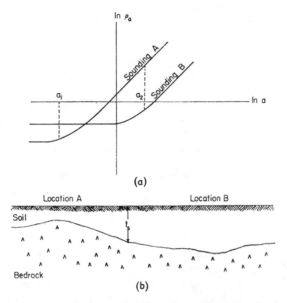

FIG. 115. The selection of electrode spacings for profiling in order to measure both the resistivity and the conductance of a variable soil layer.

These two measurements may be combined to determine the thickness of the soil layer:

$$t_s = \varrho_s S_s = \varrho_1 a_2 / \varrho_2 \qquad (210)$$

### 25 b. Horizontal profiling for vertical structures

Commonly, the resistivity profiling technique is used as a semi-reconnaissance tool in the search for vertical structures which are likely to be identified with large resistivity contrasts, such as fault contacts, dikes, shear zones, and veins. Generally, these problems arise in mineral exploration programs.

Any of the electrode arrays used in sounding may also be used in profiling, but the data obtained with some types of arrays are more readily interpretable than data obtained with other arrays. Generally, the fewer moving electrodes which are used, the better are the data. If one or two of the electrodes are placed far enough away from the rest so that they can be considered to be at infinity, the number of field personnel needed to move the array along the traverse is reduced, and more important, the complexity of the resistivity data is reduced.

In profiling, only half a Schlumberger, or Wenner, array may be used, as shown in Fig. 116. With the half-Schlumberger array, one of the current electrodes is placed a long distance off. With the half-Wenner array, one current

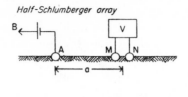

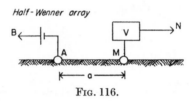

FIG. 116.

electrode and one measuring electrode are placed a long distance away from the traverse, and equally important, a long distance from each other.

The requirement that an electrode be "a long distance off" can be stated more explicitly. In the case of the half-Wenner array, the measured potential varies inversely as the distance from the current electrode. If the effect of the far electrodes is to be no more than 5 per cent of the measured potential, the far electrodes must be 20 times further away than the spacing between the two moving electrodes, if we are dealing with a homogeneous earth. For the half-Schlumberger array, the measured electric field varies inversely as the square of the distance from the current electrode, so that the distant current electrode need only be 4·5 times as far away as the near current electrode, in a homogeneous earth.

Since profiling is never done in homogeneous earth, we must consider the effect inhomogeneities may have on the distance requirements for the far electrodes. The measured quantity, whether it be potential or electric field, may vary from the value for a homogeneous earth by as much as the range in resistivities in an inhomogeneous earth. Since the range in resistivities normally expected may be from 100 to 1000 fold, the far electrode should be placed at a distance 100 to 1000 times greater than that required for a homogeneous earth.

Interpretation of resistivity profiles is usually qualitative in nature. The electrical structure being sought is located by the change in apparent resistivity associated with the structure. Rarely is any effort made to determine quantitatively the resistivity of the structure. However, the way apparent resistivity varies over a vertical boundary differs considerably between the various electrode arrays, and it is desirable to study the response of the various electrode arrays to various structures.

Since only one or two vertical boundaries are usually involved in a single interpretation problem, the image method described earlier is the easiest approach to determining the apparent resistivity recorded along a profile which crosses a vertical contact. Consider the problem shown in Fig. 117, that of a vertical dike of thickness, $t$, and of infinite extent in the direction normal to the page. A single current electrode, A, is moved along a profile normal to the dike and either the potential or electric field is measured at a distance $a$ from the current electrode. This represents the sort of measurement made with a half-

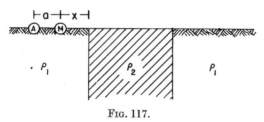

FIG. 117.

Wenner or half-Schlumberger array, respectively. If full arrays are used, the apparent resistivity profile can be obtained by adding the profiles obtained with two half arrays, with appropriate offsets.

The first case we will consider is that in which both the current and measuring electrodes are to the left of the dike. Two sets of image points which appear to be to the right of the dike are required. If the origin of coordinates is taken to be at the measuring electrode, and this is located a distance $x$ to the left of the fault, the observed potential at $M$ is:

$$U = \frac{I\varrho_1}{2\pi}\left[\frac{1}{a} + \frac{K_{12}}{2x+a} + \sum_{n=1}^{\infty}\frac{(1-K_{21}^2)K_{21}^{(2n-1)}}{(2x+2nt+a)} + \sum_{n=1}^{\infty}\frac{K_{12}(1-K_{21})\,K_{21}^{(2n-1)}}{(2nt-a)}\right]$$

(211)

where $K_{12}$ and $K_{21}$ are reflection coefficients defined as follows:

$$K_{12} = \frac{\varrho_2 - \varrho_1}{\varrho_2 + \varrho_1}$$

$$K_{21} = \frac{\varrho_1 - \varrho_2}{\varrho_1 + \varrho_2}$$

The electric field is found by taking a derivative of Eq. (211) with respect to $x$:

$$-E = \frac{I\varrho_1}{2\pi}\left[\frac{1}{a^2} + \frac{K_{12}}{(2x+a)^2} + \sum_{n=1}^{\infty}\frac{(1-K_{21}^2)\,K_{21}^{(2n-1)}}{(2x+2nt+a)^2}\right.$$
$$\left. + \sum_{n=1}^{\infty}\frac{(1-K_{21})\,K_{12}K_{21}^{(2n-1)}}{(2nt-a)^2}\right]$$

(212)

Apparent resistivities may be computed using these expressions for potential or electric field in the appropriate defining equation for the array under consideration.

When the current electrode is to the left of the dike and the measuring point is in the dike, the potential is:

$$U = \frac{\varrho_2 I(1 - K_{12})}{2\pi}\left[\frac{1}{a} + \sum_{n=1}^{\infty}\frac{K_{21}^{(2n-1)}}{(2nt - 2x + a)} - \sum_{n=1}^{\infty}\frac{K_{21}^{2n}}{(2nt + a)}\right] \quad (213)$$

and the electric field is:

$$-E = \frac{\varrho_2 I(1 - K_{12})}{2\pi}\left[\frac{1}{a^2} + \sum_{n=1}^{\infty}\frac{K_{21}^{(2n-1)}}{(2nt - 2x + a)^2} - \sum_{n=1}^{\infty}\frac{K_{21}^{2n}}{(2nt + a)^2}\right] \quad (214)$$

When both the current electrode and the measuring point are within the dike, the potential is:

$$U = \frac{\varrho_2 I}{2\pi}\left[\frac{1}{a} - \frac{K_{21}}{2x - a} + \sum_{n=1}^{\infty}\frac{K_{21}^{(2n-1)}}{(2nt - 2x + a)} - \sum_{n=1}^{\infty}\frac{K_{21}^{2n}}{(2nt + a)}\right.$$
$$\left. - \sum_{n=1}^{\infty}\frac{K_{21}^{(2n+1)}}{(2nt + 2x - a)} + \sum_{n=1}^{\infty}\frac{K_{21}^{2n}}{(2nt - a)}\right] \quad (215)$$

and the electric field is:

$$-E = \frac{\varrho_2 I}{2\pi}\left[\frac{1}{a^2} - \frac{K_{21}}{(2x - a)^2} + \sum_{n=1}^{\infty}\frac{K_{21}^{(2n-1)}}{(2nt - 2x + a)^2} - \sum_{n=1}^{\infty}\frac{K_{21}^{2n}}{(2nt + a)^2}\right.$$
$$\left. - \sum_{n=1}^{\infty}\frac{K_{21}^{(2n+1)}}{(2nt + 2x - a)^2} + \sum_{n=1}^{\infty}\frac{K_{21}^{2n}}{(2nt - a)^2}\right] \quad (216)$$

When the current electrode is within the dike and the measurement point is to the right of the dike, the potential is:

$$U = \frac{\varrho_1 I(1 - K_{21})}{2\pi}\left[\frac{1}{a} - \frac{K_{21}}{2x - a} - \sum_{n=1}^{\infty}\frac{K_{21}^{(2n+1)}}{(2x + 2nt - a)}\right] \quad (217)$$

and the electric field is:

$$-E = \frac{\varrho_1 I(1 - K_{21})}{2\pi}\left[\frac{1}{a^2} - \frac{K_{21}}{(2x - a)^2} - \sum_{n=1}^{\infty}\frac{K_{21}^{(2n+1)}}{(2x + 2nt - a)^2}\right] \quad (218)$$

When both the current electrode and measuring point are to the right of the dike, the potential is:

$$U = \frac{\varrho_1 I}{2\pi}\left[\frac{1}{a} - \frac{K_{12}}{(2x - 2t - a)} - \frac{(1 - K_{21})^2 K_{21}}{(2x - a)} - \sum_{n=1}^{\infty}\frac{(1 - K_{21})^2 K_{21}^{(2n+1)}}{(2x + 2nt - a)}\right.$$
$$\left. - \sum_{n=1}^{\infty}\frac{K_{12}K_{21}^{(2n-1)}(1 - K_{21})}{(4x - 2nt - a)}\right] \quad (219)$$

and the electric field is:

$$-E = \frac{\varrho_1 I}{2\pi}\left[\frac{1}{a^2} - \frac{K_{12}}{(2x - 2t - a)^2} - \frac{(1 - K_{21})^2 K_{21}}{(2x - a)^2} - \sum_{n=1}^{\infty}\frac{(1 - K_{21})^2 K_{21}^{(2n+1)}}{(2x + 2nt - a)^2}\right.$$
$$\left. - \sum_{n=1}^{\infty}\frac{K_{12}K_{21}^{(2n-1)}(1 - K_{21})}{(4x - 2nt - a)^2}\right] \quad (220)$$

An additional case must be considered if the width of the dike is less than the spacing, $a$, since the electrode array may straddle the dike.

The resistivity profiles which would be obtained with various electrode arrays crossing a single boundary (corresponding to large values of $t$ in the equations above) are shown in Fig. 118. The uppermost curve is the resistivity

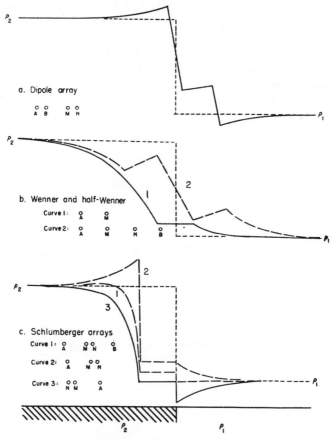

FIG. 118. Resistivity profiles recorded across a vertical contact between two zones with different resistivities.

profile which would be measured with a polar dipole array, with the measuring electrodes leading the current electrodes. The dipole profile is characterized by a small resistivity overshoot in the resistant bed, a small resistivity under-shoot in the conductive bed, and a plateau at the boundary equal in length to the electrode spacing.

The profiles in resistivity measured with a Wenner and half-Wenner array are shown by the middle set of curves. With the Wenner array, there is no overshoot or undershoot, merely a gradual change from one value of resistivity to the other, broken only by a plateau at the boundary equal in length to the spacing. With a full Wenner array, there is a discontinuity in the profile each time one of the four electrodes crosses the boundary.

The three profiles at the bottom of Fig. 118 show the apparent resistivity curves recorded with Schlumberger and half-Schlumberger arrays crossing a vertical contact. If the measuring electrodes lead the current electrode in a half-Schlumberger array, the profile will show a resistivity overshoot as the array approaches the boundary from the high resistivity side, followed by a plateau at the boundary equal in length to the spacing. If the measuring electrodes trail the current electrode, the resistivity profile will change smoothly until the plateau at the boundary is reached, following which there is a resistivity overshoot on the far side of the contact. With a full Schlumberger array, neither overshoot is observed.

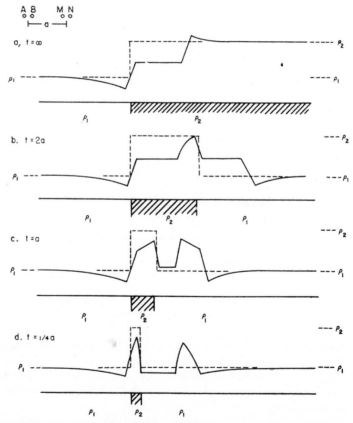

FIG. 119. Resistivity profiles which would be recorded with a dipole array crossing a resistant dike.

These curves illustrate to some extent desirable and undesirable characteristics for a resistivity profile. With some arrays (Wenner, half-Wenner and Schlumberger), a *symmetric* profile is obtained; that is, the shape of the profile does not depend on the direction from which the array approaches the contact. With the other arrays, non-symmetric profiles are obtained.

A second characteristic of these curves which is important is the number of discontinuities in the profile across a single boundary. Generally, there is a discontinuity for each moving electrode, except that in the dipole and Schlumberger arrays, dipole electrode pairs introduce only one discontinuity if the dipole spacing is small. Thus, with the full Schlumberger array, and the full Wenner array, there are three and four discontinuities per boundary,

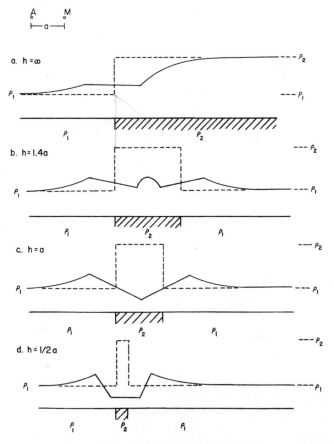

Fig. 120. Resistivity profiles which would be recorded with a half-Wenner array crossing a resistant dike.

respectively. With a dipole array, a half-Wenner array and the half-Schlumberger arrays, there are only two discontinuities at each boundary.

A third important characteristic for the resistivity profiles measured with an array is the behavior of the profile over a thin dike-like structure. Figures 119, 120 and 121 show the resistivity profiles which would be recorded with the dipole, half-Wenner and half-Schlumberger arrays, respectively, over dikes with a width comparable with the electrode spacing. The dipole curves are characterized by a double anomaly over a thin dike, with the distance between the two anomalies being equal to the dipole spacing. The curves for a half-Wenner array are characterized by a reversal in the sense of the anomaly for dike widths less than the electrode spacing. Thin resistant dikes appear as broader conductive zones. With the half-Schlumberger array, a single peak

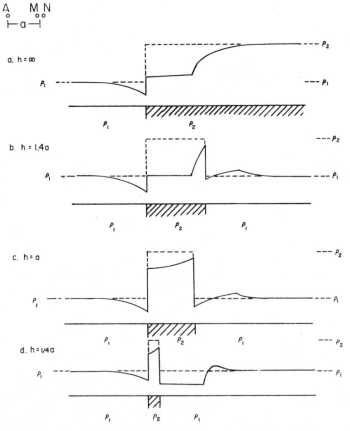

FIG. 121. Resistivity profiles which would be recorded with a half-Schlumberger array crossing a resistant dike.

is recorded in the proper sense, no matter how thin the dike is, though the main anomaly is accompanied by a broad anomaly in the opposite sense.

All the profiles presented in this section have been computed for electrode arrays which cross the resistivity structures normal to the strike direction. Unless a great deal is known about the structure before a survey is run, it is not usual that the profiles will happen to be measured normal to the strike direction. Figure 122 shows a series of profiles for half-Wenner and half-Schlumberger arrays which cross a vertical resistant dike at angles other than 90° from the strike direction. The profiles obtained by crossing a dike obliquely

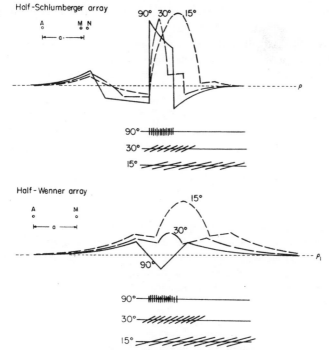

Fig. 122. Resistivity profiles which would be measured in crossing a dike-like structure at an oblique angle.

do not differ greatly from the profiles which would be obtained by crossing a wider dike in a normal direction. This suggests that better results may be obtained if survey lines are intentionally run obliquely to the regional strike when relatively thin dikes are being sought.

Profiles over structures other than vertical dikes have been computed. Blokh (1957) includes dipole profiling curves for dipping dike-like structures in his catalog of curves. Cook and Gray (1961) have published a catalog of profiles for the Wenner array crossing hemispherical structures, such as filled sinkholes.

## 26. APPARENT RESISTIVITY ABOUT STRUCTURES OF ARBITRARY SHAPE

The problem of calculating the apparent resistivity which will be measured about a body with arbitrary shape may be solved with little difficulty. A dot-chart, such as is used in gravity and magnetic interpretation may be constructed so that approximate values for the apparent resistivity about an arbitrarily shaped body may be computed quickly.

The physical statement of the problem to be solved is as follows: we have a semi-infinite medium of resistivity $\varrho_1$, within which there is a body of arbitrary shape with resistivity $\varrho_2$. A current electrode, A, and a measuring point, M, are located on the surface of the semi-infinite medium, separated by a distance, $a$. We wish to calculate the potential, $U$, at the measuring point, M.

The potential function can be determined using Ohm's law and the divergence condition, as was done earlier in this chapter. Ohm's law states that:

$$j = \frac{E}{\varrho} \tag{221}$$

The divergence of the current density vector must be zero everywhere except at the current source:

$$\nabla \cdot j = \frac{1}{\varrho} \nabla \cdot E + E \cdot \nabla \frac{1}{\varrho} \tag{222}$$

In the image analysis, it was assumed the second term in Eq. (222) was zero, and the distortion of the potential field caused by changes in resistivity within the medium was attributed to a sequence of image sources. In the present analysis, we will retain the second term, not assuming the resistivity is a constant.

The electric field, $E$, may be defined as the gradient of a scalar potential function, $U$:

$$E = -\nabla U \tag{223}$$

Substituting this expression for $E$ in Eq. (221), we have:

$$j = -\frac{1}{\varrho} \nabla \cdot \nabla U - \nabla U \cdot \nabla \frac{1}{\varrho} \tag{224}$$

or, rearranging terms:

$$\nabla^2 U = -\varrho \left[ \nabla \cdot j + \nabla U \cdot \nabla \frac{1}{\varrho} \right] \tag{225}$$

This is Poisson's equation; the two terms to the right of the equality sign may be identified with current sources or charge accumulations in the medium. The term $\nabla \cdot j$ represents the potential contributed by the current source at A, while the term $\nabla U \cdot \nabla(1/\varrho)$ represents the effect of current sources at places in the medium where the resistivity varies. The solution to Poisson's equation is given by Poisson's integral:

$$U = \frac{1}{2\pi} \int_{v} \frac{\text{(right-hand terms in Eq. 225)}}{|r|} \, dV \tag{226}$$

where $r$ is the distance from the point at which the terms in Eq. (225) are evaluated to the point where the potential is being observed. Substituting the terms from Eq. (225), Eq. (226) becomes:

$$U = \frac{1}{2\pi} \int_{v} \frac{-\varrho \left[ \nabla \cdot \boldsymbol{j} + \nabla U \cdot \nabla \dfrac{1}{\varrho} \right]}{|r|} \, dV$$

$$= -\frac{1}{2\pi} \int_{v} \frac{\varrho \nabla \cdot j}{|r|} \, dV - \frac{1}{2\pi} \int_{v} \frac{\varrho \, \nabla U \cdot \nabla \dfrac{1}{\varrho}}{|r|} \, dV \tag{227}$$

Equation (227) can be viewed as expressing the total potential in terms of a normal potential, $U_0$, given by the first integral, and a disturbing potential, $W$, given by the second integral. The first integral gives the potential due to a single-pole current source in a uniform half-space with resistivity $\varrho_1$. The second integral can be thought of as representing the effect of a charge distribution, $\nabla U \cdot \nabla(1/\varrho)$, which accumulates at discontinuities in resistivity since these are the only places that $\nabla(1/\varrho)$ is non-zero (continuous variations in resistivity are being eliminated from consideration, as a matter of convenience) The integral expression for the disturbing potential is:

$$W = \frac{1}{2\pi} \int_{v} \frac{\nabla U \cdot \nabla \dfrac{1}{\varrho}}{|r_2|} \, dV \tag{228}$$

where the integration is carried out over all space. It is interesting to note that the disturbing potential is a function only of the total potential, $U$, at the surfaces, of discontinuity in resistivity, and the distance $r_2$ from these surfaces to the measuring point, M (Fig. 123). Since the term $\nabla U \cdot \nabla(1/\varrho)$ is non-zero only on surfaces where resistivity changes, the volume integral may be replaced with a surface integral:

$$W = \frac{1}{2\pi} \int_{s} \frac{\nabla U \cdot \nabla \dfrac{1}{\varrho}}{|r_2|} \, dS \cdot \nabla t \tag{229}$$

where $S$ is the bounding surface between zones with different resistivities. $\nabla t$ is the effective thickness of the surface.

This integral may be evaluated by assuming the existence of an arbitrary source distribution, $\sigma(x, y, z)$ at the surface $S$. Designating the coordinates of the measuring point M as $(a, 0, 0)$, the integral for the disturbing potential becomes:

$$W = \frac{1}{2\pi} \int_s \frac{\sigma}{[(x - a)^2 + y^2 + z^2]^{1/2}} \, dS \qquad (230)$$

The discussion may be simplified by considering elements of surface oriented normal to the $x$ or $y$ axes. The surface integral then has the form:

$$W_x = \frac{1}{2\pi} \int \int \frac{\sigma(y, z)}{[(x_0 - a)^2 + y^2 + z^2]^{1/2}} \, dy \, dz \qquad (231)$$

for a surface element normal to the $x$ axis, or:

$$W_y = \frac{1}{2\pi} \int \int \frac{\sigma(x, z)}{[(x - a)^2 + y_0 + z^2]^{1/2}} \, dx \, dz \qquad (232)$$

for a surface element normal to the $y$ axis.

The charge density function, $\sigma(y, z)$ or $\sigma(x, z)$, may be evaluated by applying various boundary conditions. One of these conditions states that the normal

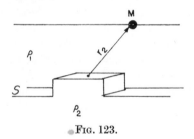

Fig. 123.

component of current density on either side of a surface element must be the same. For a surface element normal to the $x$ axis, this is:

$$\frac{1}{\varrho_1} \left( \frac{\partial U_1}{\partial x} \right)_{x \to x_0} = \frac{1}{\varrho_2} \left( \frac{\partial U_2}{\partial x} \right)_{x \to x_0} \qquad (233)$$

where $\varrho_1$ and $U_1$ are the resistivity and the potential on one side of the surface element, and $\varrho_2$ and $U_2$ are the corresponding quantities on the other side of the element. For an element oriented normal to the $y$ direction, the corresponding equation would be:

$$\frac{1}{\varrho_1} \left( \frac{\partial U_1}{\partial y} \right)_{y \to y_0} = \frac{1}{\varrho_2} \left( \frac{\partial U_2}{\partial y} \right)_{y \to y_0} \qquad (234)$$

Consider the evaluation of this boundary condition at a plane $S_1$ which separates regions with resistivities $\varrho_1$ and $\varrho_2$ as shown in Fig. 124. If the surface $S_1$ is oriented normal to the $y$ direction as shown, the current density normal to the surface is proportional to a derivative of the potential taken in that direction. The components of current density in the $y$ direction at points $P_1$ and $P_2$ in Fig. 124 are:

$$\frac{1}{\varrho_1} \frac{\partial U_1}{\partial y}\bigg|_{P_1} = \frac{1}{\varrho_1} \frac{\partial U_0}{\partial y}\bigg|_{P_1} + \frac{1}{2\pi\varrho_1} \int\int \sigma \frac{\partial(1/m_1)}{\partial y} dx\, dz\bigg|_{P_1}$$

and

$$\frac{1}{\varrho_2} \frac{\partial U_2}{\partial y}\bigg|_{P_2} = \frac{1}{\varrho_2} \frac{\partial U_0}{\partial y}\bigg|_{P_2} + \frac{1}{2\pi\varrho_2} \int\int \sigma \frac{\partial(1/m_2)}{\partial y} dx\, dz\bigg|_{P_2} \quad (235)$$

where $m_1$ and $m_2$ are the distances from the points $P_1$ and $P_2$ to the charged surfaces.

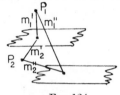

FIG. 124.

As the points $P_1$ and $P_2$ are brought into the surface $S_1$, these two expressions for current density become equal:

$$0 = \frac{1}{\varrho_1} \frac{\partial U_1}{\partial y}\bigg|_{P_1} - \frac{1}{\varrho_2} \frac{\partial U_2}{\partial y}\bigg|_{P_2}$$

$$= \frac{1}{\varrho_1} \frac{\partial U_0}{\partial y}\bigg|_{P_1} - \frac{1}{\varrho_2} \frac{\partial U_0}{\partial y}\bigg|_{P_2} + \frac{1}{2\pi} \int\int \sigma \left[\frac{1}{\varrho_1} \frac{\partial(1/m_1)}{\partial y} - \frac{1}{\varrho_2} \frac{\partial(1/m_2)}{\partial y}\right] dx\, dz$$

$$(236)$$

When the points $P_1$ and $P_2$ are very close to the surface, the term involving $U_0$ simplifies:

$$0 = \frac{\varrho_2 - \varrho_1}{\varrho_1\varrho_2} \frac{\partial U_0}{\partial y} + \frac{1}{2\pi} \int\int \sigma \left[\frac{1}{\varrho_1} \frac{\partial(1/m_1)}{\partial y} - \frac{1}{\varrho_2} \frac{\partial(1/m_2)}{\partial y}\right] dx\, dz \quad (237)$$

The remaining integral term may be evaluated by considering two groups of $m_1$ and $m_2$ distances; namely, the distances from points $P_1$ and $P_2$ to all points on the same surface at which the charge density is being evaluated will be designated as $m_1'$ and $m_2'$, while distances from $P_1$ and $P_2$ to points on other charged surfaces will be designated as $m_1''$ and $m_2''$. The remaining integral term is broken into two integrals to separate these two types of distances:

$$0 = \frac{\varrho_2 - \varrho_1}{\varrho_1\varrho_2} \frac{\partial U_0}{\partial y} + \frac{1}{2\pi} \int\int \sigma \left[\frac{1}{\varrho_1} \frac{\partial(1/m_1')}{\partial y} - \frac{1}{\varrho_2} \frac{\partial(1/m_2')}{\partial y}\right] dx\, dz$$

$$+ \frac{1}{2\pi} \int\int \sigma \left[\frac{1}{\varrho_1} \frac{\partial(1/m_1'')}{\partial y} - \frac{1}{\varrho_2} \frac{\partial(1/m_2'')}{\partial y}\right] dx\, dz \quad (238)$$

When the points $P_1$ and $P_2$ are close to the surface $S_1$, the surface appears to be an infinite sheet source, and the electric field outside such a sheet is:

$$E_1 = 2\pi\sigma = \frac{\partial U_1}{\partial y} = \frac{\partial}{\partial y}\left\{\frac{1}{2\pi}\int\int\frac{\sigma}{m_1'}\,\mathrm{d}x\,\mathrm{d}y\right\}$$

$$E_2 = -2\pi\sigma = -\frac{\partial}{\partial y}\left\{\frac{1}{2\pi}\int\int\frac{\sigma}{m_2'}\,\mathrm{d}x\,\mathrm{d}y\right\} \tag{239}$$

Beneath the sheet source, the direction of the electric field vector is opposite to that above the sheet, so Eq. (238) becomes:

$$0 = 2\pi\sigma\left(\frac{1}{\varrho_1}+\frac{1}{\varrho_2}\right) + \frac{1}{2\pi}\int\int\sigma\left[\frac{1}{\varrho_1}\frac{\partial(1/m_1'')}{\partial y} - \frac{1}{\varrho_2}\frac{\partial(1/m_2'')}{\partial y}\right]\mathrm{d}x\,\mathrm{d}z$$

$$+ \frac{\varrho_2-\varrho_1}{\varrho_1\varrho_2}\frac{\partial U_0}{\partial y} \tag{240}$$

Since the points $P_1$ and $P_2$ are very close together, the distances $m_1''$ and $m_2''$ are nearly the same:

$$\frac{\partial}{\partial y}\left(\frac{1}{m_1''}\right) = \frac{\partial}{\partial y}\left(\frac{1}{m_2''}\right) = \frac{\partial}{\partial y}\left(\frac{1}{m''}\right) \tag{241}$$

Equation (240) becomes:

$$0 = 2\pi\sigma\left(\frac{1}{\varrho_1}+\frac{1}{\varrho_2}\right) + \frac{1}{2\pi}\int\int\sigma\frac{\partial(1/m'')}{\partial y}\left[\frac{1}{\varrho_1}-\frac{1}{\varrho_2}\right]\mathrm{d}x\,\mathrm{d}z \tag{242}$$

Dividing this equation by $(\varrho_1+\varrho_2)$ and multiplying by $\varrho_1\varrho_2$, we have:

$$-2\pi\sigma = \frac{\varrho_2-\varrho_1}{\varrho_2+\varrho_1}\left[\frac{1}{2\pi}\int\int\sigma\frac{\partial(1/m'')}{\partial y}\,\mathrm{d}x\,\mathrm{d}z + \frac{\partial U_0}{\partial y}\right] \tag{243}$$

The quantity in front of the brackets is the reflection coefficient, $K$, for the resistivity contrast between $\varrho_1$ and $\varrho_2$. Similar expressions can readily be obtained for surface elements oriented normal to the $x$ and $z$ directions.

Equation (243) may be used to calculate the potential about a single-pole current source for any arbitrary arrangement of boundaries between areas with different resistivities. The charge density, $\sigma$, on the various surfaces is found by numerical solution of Eq. (243). Once the charges are evaluated, the potential at any point, and at the measuring point, M, in particular, is found from Poisson's equation:

$$U_M = U_0 + \frac{1}{2\pi}\int_r\frac{\sigma}{|r_2|}\,\mathrm{d}V \tag{244}$$

Equation (243) should be rewritten in finite difference form to illustrate the method for calculating the source density function, $\sigma$:

$$2\pi Q_i = K \left[ \frac{1}{2\pi} \int\int \left\{ \sum_{i=1}^{n} \sum_{j=1}^{n} \frac{\partial}{\partial x, y, z} \left( \frac{1}{m_{ij}''} \right) + \frac{\partial U_0}{\partial x, y, z} \right|_i \right] \quad (245)$$

The surface distribution of source has been replaced by a finite number, $n$, or point sources located at the points $(x_n, y_n, z_n)$. Each of these point sources is a function of all the other source sizes. If we select one of the sources, $Q_i$, to examine, its magnitude depends on the distances to each of the other sources:

$$m_{ij}'' = [(x_i - x_j) + (y_i - y_j)^2 + (z_i - z)^2]^{1/2} \quad (246)$$

By considering each of the $Q$ charges, we obtain a set of $n$ simultaneous algebraic equations:

$$2\pi Q_1 = K \left[ \frac{1}{2\pi} \left\{ Q_2 \frac{x_1 \Delta y \Delta z}{[(x_1 - x_2)^2 + (y_1 - y_2)^2 + (z_1 - z_2)^2]^{3/2}} \right. \right.$$

$$+ Q_3 \frac{x_1 \Delta y \Delta z}{[(x_1 - x_3)^2 + (y_1 - y_3)^2 + (z_1 - z_3)^2]^{3/2}} + \cdots + \frac{\partial U_0}{\partial x, y, z} \quad (247)$$

It is necessary to divide the surfaces into 50 to 100 elements in order to achieve a fairly high degree of accuracy. However, an approximate solution may be obtained by ignoring the integral in Eq. (240) which represents the interaction between source accumulations on various surfaces. With this simplification, the source density at a surface of separation is:

$$\sigma_y = \frac{K}{2\pi} \frac{\partial U_0}{\partial y} \qquad \text{(for $y$-oriented planes)}$$

$$\sigma_z = \frac{K}{2\pi} \frac{\partial U_0}{\partial z} \qquad \text{(for $z$-oriented planes)}$$

$$\sigma_x = \frac{K}{2\pi} \frac{\partial U_0}{\partial x} \qquad \text{(for $x$-oriented planes)} \quad (248)$$

Generally, only structures elongated in one direction provide a large enough anomaly in apparent resistivity to be of interest in interpretation. The direction of elongation can be specified arbitrarily as the $z$ direction. The disturbing potential caused by an infinitely long strip oriented normal to the $xy$ plane is found by inserting the source densities from Eq. (248) in Poisson's integral:

$$U = U_0 + \frac{1}{4\pi^2} \left[ \int_S \int_{-\infty}^{\infty} \frac{Kx \, dx \, dz}{(x^2 + y^2 + z^2)^{3/2} [(x - a)^2 + y^2 + z^2]^{1/2}} \right.$$

$$+ \int_S \int_{-\infty}^{\infty} \frac{Ky \, dy \, dz}{(x^2 + y^2 + z^2)^{3/2} [(x - a)^2 + y^2 + z^2]^{1/2}} \quad (249)$$

The integration in $z$ need be carried out only once, since the limits of integration are fixed. The integration in $y$ or $x$ must be carried out graphically for each problem, since the limits of integration for these two variables depend on the geometry of a particular problem. Contour charts showing the numerical value for the integrals in Eq. (249) per unit area, in the $xy$ plane, for $x$-oriented or $y$-oriented faces are given in Fig. 125. Integration in $x$ and $y$ can be done graphically by drawing the path of integration on these two contour maps, and performing the integration numerically.

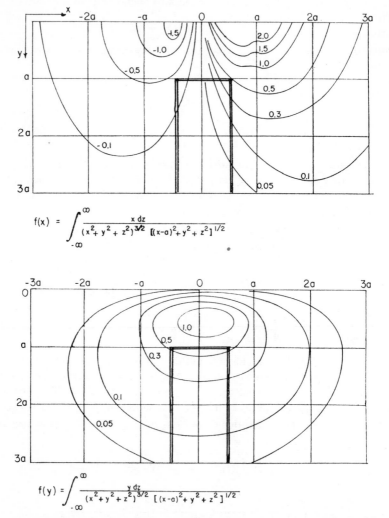

$$f(x) = \int_{-\infty}^{\infty} \frac{x\,dz}{(x^2 + y^2 + z^2)^{3/2}\,[(x-a)^2 + y^2 + z^2]^{1/2}}$$

$$f(y) = \int_{-\infty}^{\infty} \frac{y\,dz}{(x^2 + y^2 + z^2)^{3/2}\,[(x-a)^2 + y^2 + z^2]^{1/2}}$$

Fig. 125. Contour maps of the integral expressions used in computing the approximate single-pole potential function about a body of arbitrary shape.

As an example of the use of these graphs, consider the problem of a vertical dike, buried in a medium with resistivity $\varrho_1$, with the top of the dike at a depth $a$, equal to the spacing of a half-Wenner array, and with a width, $a$, also. This structure is drawn on the contour maps in Fig. 125, and the values for the $z$-integral function along the surfaces of the dike are plotted on a set of auxilliary graphs, as shown in Fig. 126. The area under these auxilliary curves is the value for the two integrals in Eq. (249), after being multiplied by the constant $K/4\pi^2$. The total potential is found by adding this quantity to the normal potential.

The value so found corresponds to a single location of the measuring point relative to the position of the dike. Other points on a resistivity profile are obtained by moving the position of the dike on the contour maps to the right or left, and repeating the integration. These moves correspond to moving the electrode array to the left or right over the dike. The resistivity profile for a half-Wenner array traversing this buried dike is shown in Fig. 127. It closely resembles the profile which would be obtained with a half-Wenner array over a dike of unit width which outcropped.

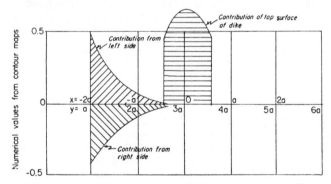

FIG. 126. Graphical integration for the dike superimposed on Fig. 125.

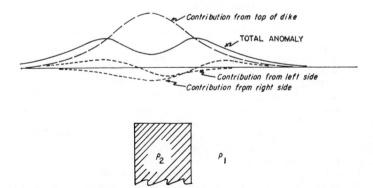

FIG. 127. Resistivity profile for a half-Wenner array crossing a buried dike.

## 27. SUMMARY—RESISTIVITY PROFILING

The resistivity profiling technique is not particularly effective when used alone. The method works well only in locating the contact between two extensive regions with different resistivities. Thin dike-like structures are difficult to locate, even when the resistivity contrast is high. The half-Schlumberger array provides the most diagnostic profile over a thin dike, in that the anomaly is always in the proper sense. Resistivity profiles recorded with the half-Wenner array may be misleading, since the anomaly measured over thin dikes is reversed in sense.

For thick dikes, the use of the Wenner or half-Wenner array is preferable to the use of the Schlumberger or dipole arrays, since the measured curves are symmetric and the maximum resistivity contrast which is measured is never greater than the actual resistivity contrast.

The structure being sought must usually be elongated in one direction, in order that the anomaly in apparent resistivity be large enough to be detected. For small bodies, thin dikes and other target structures which have a very high resistivity contrast with the host rock, the electromagnetic method provides a better chance of detection. Resistivity profiling is normally done as an adjunct to one of the more sensitive profiling methods, those utilizing either induced polarization or electromagnetic measurements.

## REFERENCES

ALFANO, L. (1959) Introduction to the interpretation of resistivity measurements for complicated structural conditions, *Geophys. Prospecting* **7**, 311–366.

BIRDICHEVSKIY, M. N. and ZAGARMISTR, A. M. (1958) Methods of interpreting dipole resistivity soundings, *Prikladnaya Geofizika* **19**, 57–107.

BLOKH, E. M. (1957) *Dipole Electrical Prospecting* Gosgeoltekizdat, Moscow, 238 pp.

BOGDANOV, A. E. (1948) *Graphical Methods for Constructing and Interpreting Three-Layer Electrical Sounding Curves*, Gostoptekizdat, Moscow.

COOK, K. L. and GRAY, R. L. (1961) Theoretical horizontal resistivity profiles over hemispherical sinks, *Geophysics*, **26** (June).

KALENOV, E. N. (1957) *Interpretation of Electrical Sounding Curves*, Gostoptekizdat, Moscow, 472 pp.

KRAEV, A. P. (1951) *Osnovi Geoelektriki*, Gosudarstveno Izdatelstvo Tekniko-Teoreticheskoi Literatur, Moscow.

MAEDA, K. (1955) Apparent resistivity for dipping beds, *Geophysics* **20**, 140–147.

MOONEY, H. M. and WETZEL, W. W. (1956) *The Potentials about a Point Electrode and Apparent Resistivity Curves for a Two-, Three-, and Four-layer Earth*, University of Minnesota Press, Minneapolis, 146 pp.

ROMAN, I. (1960) Apparent resistivity of a single uniform overburden, U.S. Geological Survey Professional Paper 365, GPO, Washington, D.C., 99 pp.

SKALSKAYA, I. P. (1948) The field of a point source of current situated on the earth's surface above an inclined plane, *J. Tech. Phys.* (U.S.S.R.) **18**, 1242–1254.

VOZOFF, K. (1960) Numerical resistivity interpretation; general inhomogeneity, *Geophysics* **25** (6), 1184–1194.

CHAPTER 4

# MAGNETO-TELLURIC RESISTIVITY METHOD

ELECTRICAL currents induced in rocks by fluctuations in the earth's magnetic field may be used to measure resistivity. If the time variations in the magnetic field can be treated as the magnetic component of a plane electromagnetic wave, a simple relationship can be shown to exist between the amplitude of the magnetic field changes, the voltage gradients induced in the earth and the earth resistivity. Also, since the depth to which an electromagnetic wave penetrates in a conductor depends both on frequency and on the resistivity of the conductor, resistivity may be computed as a function of depth within the earth if the amplitudes of the magnetic and electric field changes can be measured at several frequencies.

The magneto-telluric method for measuring resistivity has several important advantages over the galvanic methods. Since measurements are made with currents induced in the earth, no problem is encountered in determining the resistivity beneath a highly resistant bed, such as exists when the measurement is attempted with a galvanic method. A second advantage is that resistivities may be studied to great depths within the earth; measurements to similar depths using galvanic methods would require the use of tremendous power. One disadvantage of the magneto-telluric method is the instrumental difficulty met in trying to measure the amplitude of small, rapid changes in the magnetic field.

In discussing the magneto-telluric method, we shall first consider the nature of the magneto-telluric field, then discuss the theory for plane electromagnetic waves penetrating a conductive earth; then, methods for measuring the magnetic and electric field components, and finally, we shall consider methods for interpreting the resistivity values deduced from magneto-telluric data.

## 28. THE MAGNETO-TELLURIC FIELD

As is the case in the preceding chapters, we shall use an mks system of units in discussing the magneto-telluric field. However, virtually all the geophysical literature dealing with the magnetic field of the earth is written in the cgs system of units. Therefore, we need to consider the interrelations between the two systems of units.

The unit of measurement for the electric field is the *volt per meter*. In practice, the multiple unit, *millivolt per kilometer*, is commonly used, since the electric

197

component of the magneto-telluric field is measured with a pair of electrodes having a spacing of the order of a kilometer, and the voltages recorded over such a separation are in tens or hundreds of millivolts. The millivolt per kilometer is identical with one microvolt per meter, a unit which is commonly used in radio wave field intensity measurements.

Magnetic field *intensity* is defined as the *force* exerted on a magnetic *pole* of *unit strength* by a magnetic field. It is usually represented by the symbol $F$ when the earth's field is being considered. In the cgs system of units, which is used almost invariably by geophysicists, the unit of intensity is the *oersted*. The intensity of the earth's magnetic field, except for areas of anomalous local magnetization, ranges from a minimum value of about 0·25 oersted along the Atlantic coast of Brazil to a maximum value of 0·70 oersted along the north coast of Antarctica. For describing small changes in the earth's magnetic field, geophysicists have defined the *gamma*, which is $10^{-5}$ oersted. (The earth's total field intensity then ranges from 25,000 to 70,000 gammas.)

In the mks system of units, intensity is measured in *newtons per weber* (a newton is a force of about 0·2 lb; a weber is a unit of magnetic pole strength defined in such a way that a pole of one weber strength will repel a like pole, in a vacuum, with a force of $10^7/4\pi^2$ newtons when the two poles are separated a distance of one meter). Experiments by Biot and Savert (1820) lead to the realization that the magnetic field about a long, straight wire with current flowing through it is proportional to the amount of current and inversely proportional to the distance from the wire. In mks units, the relationship is:

$$|\mathbf{F}| = \frac{I}{2\pi a} \tag{250}$$

where $a$ is the distance from the wire at which the magnetic field is measured. This relationship between magnetic field strength and current permits the definition of another mks unit for intensity, the *ampere per meter*. One ampere of current flowing through a long, straight wire will generate a magnetic field intensity of one ampere per meter at a distance of one meter from the wire. The ampere per meter is more widely used than the newton per weber, but the quantity measured is numerically the same, whichever name is used for the unit.

The newton per weber (or ampere per meter) is a much smaller unit than the oersted:

1 newton/weber = 1 ampere/meter = $4\pi \times 10^{-3}$ oersted, or
1 oersted = 79·7 newtons/weber = 79·7 amperes/meter.

Magnetic *induction* is a measure of the *force* exerted on a *moving charge* by a magnetic field, whereas magnetic intensity is a measure of the force exerted on a magnetic pole by a magnetic field, whether the pole is moving or not.

Magnetic induction is related to magnetic intensity as:

$$B = \mu F \tag{251}$$

where $\mu$ is the magnetic permeability of the space in which the induction is being measured. Usually no distinction is made between magnetic induction and magnetic intensity when they are measured in cgs units because the permeability of most rocks is unity, or nearly so, when measured in cgs units. Therefore, the two quantities, intensity and induction, are numerically equal.

In the mks system of units, care must be taken to distinguish between the two quantities, since permeability, $\mu$, has the value $12.56 \times 10^{-7}$, or very nearly so, for most rocks. The quantity which is measured in practice depends on the type of magnetometer used. Some magnetometers measure the force on a magnetic pole by balancing this force against a gravitational force; such magnetometers measure the magnetic intensity (newtons per weber). Other magnetometers operate on the principle that time-variations in the magnetic induction induce voltage in a coil.

The mks unit of magnetic induction is defined in terms of Ampere's law:

$$F = \varrho \, v \times B \tag{252}$$

where $F$ is the force felt by a charge $v$ moving with a velocity $q$ in a field of magnetic induction strength, $B$. If charge is measured in coulombs, force in newtons and velocity in meters per second, the unit for magnetic induction is the *weber per square meter*. In the cgs system (dynes for force, statcoulombs for charge and centimeters per second for velocity) the unit for induction strength is the *gauss*. The weber per square meter is a much larger quantity than the gauss:

$$1 \text{ weber/m}^2 = 10,000 \text{ gausses}.$$

Ranges for each of these units which may be used in describing the earth's magnetic field are as follows:

| | | |
|---|---|---|
| cgs system: | magnetic induction in gauss | 0.25 to 0.70 |
| | magnetic intensity in oersteds | 0.25 to 0.70 |
| | magnetic intensity in gammas | 25,000 to 70,000 |
| mks system: | magnetic induction in webers/m² | 0.000025 to 0.000070 |
| | magnetic intensity in newtons/weber | 20 to 56 |

The magneto-telluric field can be defined generally as the time-varying portion of the earth's magnetic field which induces current flow in the earth. These fluctuations may occur over periods ranging from milliseconds to centuries. Variations with periods less than one millisecond are not usually used in resistivity determinations, since such rapid variations do not penetrate very far into the earth. Oscillations with periods longer than one day are usually excluded from the frequency range of interest, in view of the length of time required for a measurement. Within the frequency range which is used in magneto-telluric surveys, noise is contributed primarily by three types of sources; atmospheric electrical discharges, industrial noise from power distribution systems and micropulsations in the magnetic field.

A typical amplitude vs. frequency spectrum for magnetic variations is shown in Fig. 128. The variations have a minimum amplitude at frequencies near one cycle per second, increasing at both higher and lower frequencies.

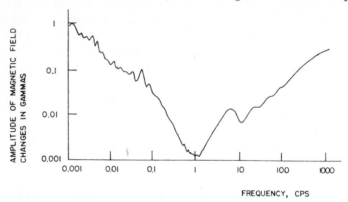

FIG. 128. Typical spectrum of amplitudes of electromagnetic noise in the extremely low frequency (ELF) range.

At frequencies above one cycle per second, the magneto-telluric field is generated almost entirely by lightning strokes. Locally, induction fields about power distribution systems may be an important noise source, but such magnetic noise is restricted to 60 c/s and harmonics.

During fair weather, the potential gradient outward from the surface of the earth is about 100 V/m, with a polarity such that the top of a vertical antenna will have a positive potential with respect to the ground. During thunderstorms, the field changes polarity, and at times, the gradient becomes large enough to break down the insulation resistance of the air, and a lightning stroke results.

The fair-weather voltage gradient causes a very slight electrical current to flow vertically through the atmosphere. Conduction in air does not follow Ohm's law; that is, the current density is not proportional to the potential gradient. Current depends on voltage gradient approximately as shown by the curve in Fig. 129. For very low current densities, current is proportional to voltage gradient, with charge carriers being supplied at random by ionization in the air. With increasing current densities, a saturation current is reached, so that an increase in gradient does not increase the current density. This occurs when each ion is driven so far in conducting current that it collides with neutral molecules in the air, with collisions inhibiting conduction. At very high voltage gradients, the charge carriers are accelerated to high velocities before

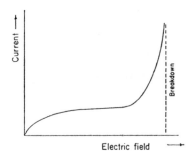

FIG. 129. Relation between current flow and voltage gradient in air.

they strike neutral molecules, attain sufficient energy to ionize the neutral molecules on collision, and start an ionization avalanche. When this occurs, there is a tremendous increase in the conductivity of the air.

The voltage gradient required to initiate this ionization in dry air at atmospheric pressure is of the order of three million volts per meter. In practice, breakdown probably takes place at much lower gradients in the atmosphere. The presence of water vapor or of radioactive dust will reduce the voltage gradient required for breakdown by increasing the amount of random ionization that takes place in the atmosphere.

The actual sequence of events in a lightning stroke consists of two steps; a leader stroke travelling from the cloud to ground, followed by a return stroke. Most strokes are from a thunder cloud to ground, though about 10 per cent of the lightning strokes in a storm may travel between two clouds. The leader stroke follows a path of maximum voltage gradient, proceeding in spurts 10 to 100 m long, and the time required for the stroke to reach the ground is of the order of one millisecond. The leader stroke contributes electric fields primarily with energy in the spectrum of frequencies above 20 kc/s.

The leader stroke serves to ionize a path from the cloud to the ground, through which a much larger current may follow in the return portion of the stroke.

The electric and magnetic field strength generated by a lightning stroke depends strongly on the distance from the storm location to the point of observation. In the case of a nearby storm, many individual strokes may be distinguished on a record of electric field strength, such as that shown in Fig. 130. These bursts contain high frequency components associated with the rapid rise in voltage at the leading edge of a signal, as well as relatively low frequency components contained in the slow decay, or tail, portion of the signal.

The shape of a sferic (signal from a lightning stroke) is changed markedly when it travels any distance. Some frequencies, particularly in the middle portion of the frequency spectrum, are lost and the individual sferics become reverbatory, as indicated in Fig. 130. Figure 131 shows the frequency spectrum

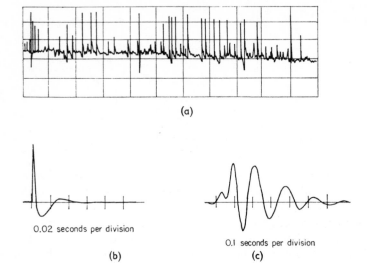

(a)

0.02 seconds per division

(b)

0.1 seconds per division

(c)

FIG. 130. Electric fields generated by atmospheric electric activity.

(a)    Record of voltage between electrodes at one kilometer separation while a thundershower was in progress at a distance of about 20 miles. The vertical scale is 10 mV signal per division; the horizontal scale is one minute in time per division.
(b)    Electric field at a distance of 10 miles from a single stroke of lightning.
(c)    Electric field at a distance of 100 miles from a single stroke of lightning.

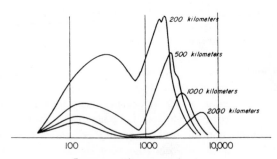

Frequency, cycles per second

FIG. 131. Amplitude spectra of electrical noise from atmospheric electric activity as a function of distance from the source. Noise levels are arbitrary from curve to curve, with all being referred to a common level at a frequency of 40 c/s (from Chapman and Macario).

(amplitude plotted as a function of frequency) for sferics measured at various distances from a storm, as found by Chapman and Macario.

Sferics from very distant thoundershowers provide a nearly uniform source of noise, inasmuch as there is nearly always a storm in progress someplace on earth. There are three major storm centers in the equatorial regions which have an average of 100 storm days per year, with small areas within these centers which average more than 200 stormy days per year (see Fig. 132). These centers are located in Brazil, central Africa and Malaya, distributed in such a way that during any hour of the day, there is probably a storm in progress in one of these centers.

Signals which have travelled several thousand kilometers from a storm center are modified considerably. The electromagnetic energy from distant lightning strokes travels in a guided-wave mode, with energy being reflected back and forth between the conductive surface of the earth and the ionized layers of air in the ionosphere. The height to the lowermost ionized layer (the "D" layer) is about 60 km during the day, but this layer disappears at night, and the height to the next layer (the "E" layer) is about 90 km. This leads to a diurnal variation in the nature of the signals observed from distant thunder showers. This energy excites a resonance in the earth–ionosphere cavity, providing noise level vs. frequency curves such as that shown in Fig. 133.

The level of noise which can be attributed to distant storm activity is nearly uniform, but not completely so. It is obvious from Fig. 132 that the distribution of storms about the world is not completely uniform, and in particular, there is somewhat less likelihood of storms when the Pacific half of the earth is sunlit than when the opposite half is sunlit. A maximum amount of storm-derived noise is noted at about 20:00 Universal Time (Greenwich Time), and a minimum is found at about 04:00 Universal Time.

A typical record of the 8 c/s noise derived from worldwide thundershower activity is shown in Fig. 134.

Sferics contribute very little to the naturally existing electromagnetic fields at frequencies less than 1 c/s; at these low frequencies, the natural electromagnetic field consists of periodic or nearly periodic variations in the earth's magnetic field. These variations cover the range in frequencies from one cycle per second to one cycle per eleven years, though variations with periods longer than a day are rarely used for determining earth resistivity. These variations represent a variety of phenomena, but all appear to arise fundamentally from some interaction between radiation or particle matter emitted by the sun and the earth's atmosphere and magnetosphere. A great deal has been learned about these interactions in recent years from satellite studies of the upper atmosphere, but even so, the phenomena are not completely understood at the present time.

When the long-period (periods longer than one second) variations in electric field or magnetic field are recorded, the most prominent component is a nearly

FIG. 132. Average annual number of days with thundershower activity.

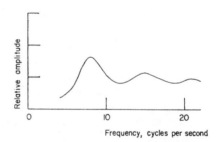

FIG. 133. Noise spectrum to be expected from distant worldwide thunder-shower activity (Raemer, 1961).

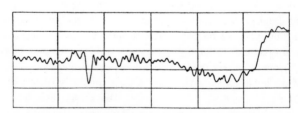

FIG. 134. Record of voltage between electrodes at a 1 km spacing showing activity at approximately 8 c/s which is probably generated by worldwide thundershower activity. Measurements were made in southern California. The vertical scale is 100 mV per division; the horizontal scale is one second of time per division.

periodic variation with periods in the range 10 to 40 sec. Usually, the period is close to 25 sec. These variations in the magneto-telluric field (simultaneous variations in electric and magnetic field strengths) take place continuously, though the amplitude varies with time both regularly and irregularly. This type of variation may be termed "25-sec continuous micropulsation activity", or for convenience, $P_{c,\,25}$. Some typical records showing these micropulsations are shown in Fig. 135.

The 25-sec $P_c$ activity has a characteristic diurnal pattern in amplitudes, with the amplitude being 10 to 20 times larger during the daylight hours than at night (see Fig. 136). Although very little is known in detail about the dependence of 25-sec $P_c$ activity on latitude, these micropulsations are known to be much larger in amplitude in the auroral latitudes than in the temperate latitudes. The peak amplitude in magnetic field variation is measured in tens of gammas in the auroral latitudes, but only in tenths of gammas in the southern United States. Despite this large variation in amplitude with both latitude and longitude (the diurnal amplitude pattern), individual events and packets of micropulsation activity appear to occur at almost exactly the same time over large portions of the earth.

The 25-sec micropulsation activity level has an amplitude pattern which repeats itself at approximately monthly intervals, related to the occurrence of magnetic storms. The amplitude of the $P_{c,\,25}$ activity increases abruptly at the beginning of a magnetic storm (see Fig. 137 for an example). The activity remains at a high level for several days, and then decreases gradually for several weeks to a quiet level preceding another storm.

In addition to the continuous micropulsations which occur in the 10- to 40-sec range of periods, transient series of pulsations with unique character also occur in this period range. These transient packets are designated as $P_{t,\,25}$ activity. They are also commonly called W-events, since a graphic record of the dominant character of a transient micropulsation looks like a W (see Fig. 138). The main W-event may be preceded or followed by a transient train of oscillations with as many as a dozen cycles. W-events are most readily seen in the period from 2 hr before until 2 hr after local midnight. They occur in sequences of three to five, separated by time intervals ranging from 10 to 20 min.

There are a number of period bands other than the 10–40 sec band which contain continuous micropulsation activity. In particular, there is continuous activity in a band of periods ranging from 80 to 300 sec. These periods are best seen on electric field records made at night, when the $P_{c,\,25}$ activity is low. Several examples of these longer period micropulsations are shown in Fig. 139. There is also the possibility that continuous micropulsation activity takes place in longer period bands, but such activity has not been studied.

Continuous micropulsations are sometimes observed in a period band ranging between the limits $\frac{1}{2}$–2 sec. This particular type of activity appears to correlate well with auroral displays. The pulsations have a modulated character

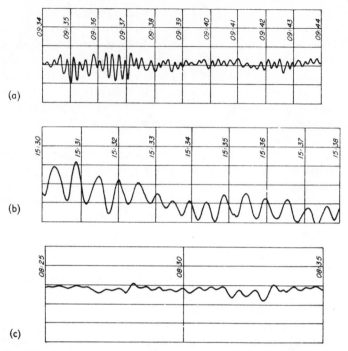

FIG. 135. Typical quiet-day telluric current activity in the 10- to 40-sec
period range.
(a) Hawaiian Volcano Observatory, September 3, 1962 (Hawaiian Standard
Time). The vertical scale is 0·1 mV per km per division.
(b) Bergen Park, Colorado, March 2, 1963 (Mountain Standard Time). The
vertical scale is 10 mV per km per division.
(c) China Lake, California, January 2, 1963 (Pacific Standard Time). The
vertical scale is 1 mV per km per division.

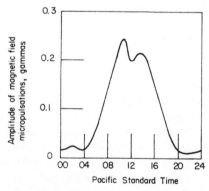

FIG. 136. Average diurnal amplitude behaviour of magnetic field micropulsations in California from March through September, 1958
(from Campbell, 1960).

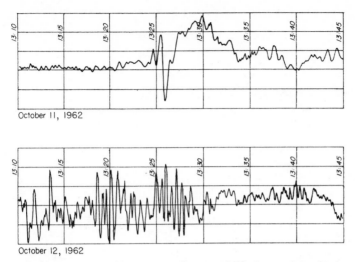

October 11, 1962

October 12, 1962

Fig. 137. Records of earth voltage gradient made during a magnetic storm which started about 13 : 25, October 11, 1962 (Mountain Standard Time). Records were made at Bergen Park, Colorado. The vertical scale is 20 mV per km per division.

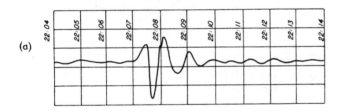

(a)

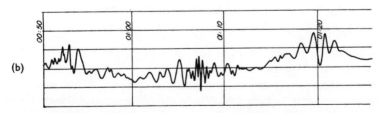

(b)

Fig. 138. Examples of transient micropulsation activity in the 10- to 40-sec period range (W events).
(a) Hawaiian Volcano Observatory, September 4, 1962 (Hawaiian Standard Time). The vertical scale is 0·1 mV per km per division.
(b) Bergen Park, Colorado, October 11–12, 1962 (Mountain Standard Time). The vertical scale is 10 mV per km per division.

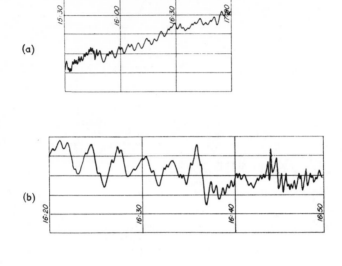

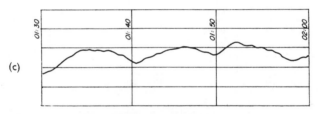

FIG. 139. Examples of continuous micropulsation activity in the period range
from 100 to 600 sec.
(a) Hawaiian Volcano Observatory, January 7, 1963 (Hawaiian Standard
Time). The vertical scale is 0·05 mV per km per division.
(b) Bergen Park, Colorado, October 12, 1962 (Mountain Standard Time).
The vertical scale is 10 mV per km per division.
(c) Bergen, Park, Colorado, October 10, 1962 (Mountain Standard Time).
The vertical scale is 10 mV per km per division.

(see Fig. 140), causing records to have the appearance of a string of pearls,
from whence the name "pearl activity" has arisen.

In addition to the nearly periodic micropulsation activity, there are erratic
variations in the magneto-telluric field which are particularly evident during
magnetic storms. Magnetic storms appear to be caused by material thrown out
by flares on the sun's surface. The material consists of a neutral plasma containing equal numbers of protons and electrons travelling outwards with velocities
of the order of 1500 km per sec and with densities of about $10^{10}$ ions per m³
when the flare ejecta reach the earth orbit. The oppositely charged particles

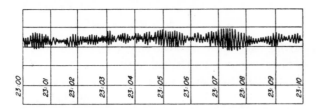

FIG. 140. Example of pearl-like micropulsation activity.
China Lake, California, January 10, 1963 (Pacific Standard Time).
The vertical scale is 1 mV per km per division.

constituting this plasma are deflected in opposite directions when the flare material enters the earth's magnetic field, so that a current is formed around the earth at altitudes ranging from 2 to 10 earth radii. The magnetic field associated with this ring of current around the earth enhances the horizontal component of the earth's magnetic field on the side facing the solar flare cloud (this also happens to be the sunlit side of the earth, since it takes about one day for the flare material to reach the earth).

It is observed that during the initial phase of a magnetic storm, which lasts from 2 to 4 hr, the horizontal component of the magnetic field is increased. This is followed by the main phase of the storm, during which the horizontal component decreases to a minimum value over a period of several hours. The field then returns to normal values over a period of several days.

Superimposed on this general storm character, there are usually a variety of erratic variations in magnetic field strength. These erratic variations are evident on the record of electric field strength shown in Fig. 137.

Storms occur about once a month, this being the period of rotation of the equatorial regions of the sun. Rotation brings centers of flare activity into view from the earth with this periodicity. The period may be somewhat longer if the center of flare activity is located away from the solar equator, since the polar regions do not rotate as rapidly as the equatorial regions.

The earth's magnetic field has a well-established diurnal variation with an amplitude of a few tens of gammas, ordinarily. The variation is related to local time, and has been attributed to a system of currents flowing in the ionosphere.

In summary, the magneto-telluric field of the earth contains continuous noise with a wide range in frequencies. At frequencies above one cycle per second, worldwide thundershower activity is the primary source of energy. Below 1 c/s, micropulsations in the earth's magnetic field account for the magneto-telluric field. All these events appear to be worldwide in character, so it is not unreasonable to treat the magneto-telluric field mathematically as a plane-wave field. This simplifies the mathematical analysis considerably.

## 29. BEHAVIOR OF PLANE-WAVE ELECTROMAGNETIC FIELDS IN THE EARTH

Since the earth's magnetic field is varying in time, currents may be induced in the earth which cannot be explained in terms of the direct current theory developed in Chapter III. We must make use of Maxwell's equations, which do take into consideration the existence of induction currents, in order to explain the behavior of the electric and magnetic field variations generated by the magneto-telluric field.

$$\text{curl } \boldsymbol{E} = -\frac{\partial \boldsymbol{B}}{\partial t} \tag{253}$$

$$\text{curl } \boldsymbol{H} = \boldsymbol{J} + \frac{\partial \boldsymbol{D}}{\partial t} \tag{254}$$

$$\text{div } \boldsymbol{B} = 0 \tag{255}$$

$$\text{div } \boldsymbol{D} = 0 \tag{256}$$

where $\boldsymbol{E}$ is the electric field vector, $\boldsymbol{H}$ is the magnetic field vector, $\boldsymbol{B}$ is the induction field vector and $\boldsymbol{D}$ is the displacement vector. The electrical properties of the medium are $\mu$, the magnetic permeability, $\varepsilon$, the dielectric constant and $\varrho$, the electrical resistivity. Curl indicates the cross product with the operator ($i\, \partial/\partial x + j\, \partial/\partial y + k\, \partial/\partial z$) and div indicates a dot product with the same operator. The proper units for each of the quantities in Eqs. (253) through (256) in mks units are:

$E$    v/m

$H$    A/m (or newtons per weber)

$D$    c/m³

$B$    webers per m²

$\varepsilon$    F/m, with the dielectric constant of free space being $8 \cdot 85 \times 10^{-12}$ F/m

$\mu$    H/m, with the permeability of free space being $12 \cdot 56 \times 10^{-7}$ H/m

These four equations, (253) through (256), may be reduced in complexity by expressing the displacement vector, $D$, and the current density vector, $J$, in terms of the electric field vector, $E$:

$$\boldsymbol{D} = \varepsilon \boldsymbol{E} \tag{257}$$

$$\boldsymbol{J} = \frac{\boldsymbol{E}}{\varrho} \tag{258}$$

Similarly, the magnetic induction may be expressed in terms of the magnetic field strength:

$$\boldsymbol{B} = \mu \boldsymbol{H} \tag{259}$$

With $B$ and $D$ eliminated in this way, Maxwell's equations become:

$$\text{curl } E = -\mu \frac{\partial H}{\partial t} \qquad (260)$$

$$\text{curl } H = \frac{E}{\varrho} + \varepsilon \frac{\partial E}{\partial t} \qquad (261)$$

$$\text{div } H = 0 \qquad (262)$$

$$\text{div } E = 0 \qquad (263)$$

An assumption that the physical properties of the medium depend neither on time nor position was required preceding this last step to avoid introducing derivatives of the properties of the medium with respect to either time or position. The variables $E$ and $H$ may be separated between Eqs. (260) and (261) by performing the curl operation on Eq. (260), and by differentiating Eq. (261) with respect to time and multiplying the various terms by $\mu$:

$$\text{curl curl } E = -\mu \text{ curl } \frac{\partial H}{\partial t} \qquad (264)$$

$$-\frac{\mu}{\varrho} \frac{\partial E}{\partial t} - \varepsilon\mu \frac{\partial^2 E}{\partial t^2} = -\mu \text{ curl } \frac{\partial H}{\partial t} \qquad (265)$$

Then, subtracting Eq. (265) from Eq. (264), we obtain an equation not containing $H$:

$$-\text{curl curl } E = \frac{\mu}{\varrho} \frac{\partial E}{\partial t} + \varepsilon\mu \frac{\partial^2 E}{\partial t^2} \qquad (266)$$

A vector identity for the operation curl curl is:

$$\text{curl curl } E = \text{grad div } E - \nabla^2 E$$

and since the divergence of $E$ is identically zero unless there are sources of current or changes in the electrical properties of the medium, Eq. (266) becomes:

$$\nabla^2 E = \frac{\mu}{\varrho} \frac{\partial E}{\partial t} + \varepsilon\mu \frac{\partial^2 E}{\partial t^2} \qquad (267)$$

The vector Laplacian represents three scalar Laplacian operations, one taken on each of the three orthogonal components of the Vector $E$. In a cartesian system of coordinates, these three Laplacian operations lead to the three equations:

$$\frac{\partial^2 E_x}{\partial x^2} + \frac{\partial^2 E_x}{\partial y^2} + \frac{\partial^2 E_x}{\partial z^2} = \frac{\mu}{\varrho} \frac{\partial E_x}{\partial t} + \varepsilon\mu \frac{\partial^2 E_x}{\partial t^2} \qquad (268)$$

$$\frac{\partial^2 E_y}{\partial x^2} + \frac{\partial^2 E_y}{\partial y^2} + \frac{\partial^2 E_y}{\partial z^2} = \frac{\mu}{\varrho} \frac{\partial E_y}{\partial t} + \varepsilon\mu \frac{\partial^2 E_y}{\partial t^2} \qquad (269)$$

$$\frac{\partial^2 E_z}{\partial x^2} + \frac{\partial^2 E_z}{\partial y^2} + \frac{\partial^2 E_z}{\partial z^2} = \frac{\mu}{\varrho} \frac{\partial E_z}{\partial t} + \varepsilon\mu \frac{\partial^2 E_z}{\partial t^2} \qquad (270)$$

In our particular problem, it is possible to select a cartesian coordinate system such that only one component of the electric field is non-zero. Unless there are lateral changes in resistivity, the current sheets induced in the earth by a varying magnetic field must be parallel to the surface of the earth, since no appreciable amount of current can be forced to flow across the air–earth boundary. This restriction permits only vertically travelling waves. We can then orient our coordinate system so that the $x$ direction is along the current flow direction. Then, there is no current flow or electric field in the $y$ or $z$ directions. With this choice of coordinates. Eqs. (268) through (270) reduce to a single equation:

$$\frac{\partial^2 E_x}{\partial z^2} = \frac{\mu}{\varrho}\frac{\partial E_x}{\partial t} + \varepsilon\mu\frac{\partial^2 E_x}{\partial t^2} \tag{271}$$

This is a common type of wave equation, for which there are a variety of special solutions. Experience has shown that the magneto-telluric field is quasi-periodic in the frequency range commonly used in electrical prospecting, so it seems logical to assume a solution which is periodic:

$$E_x \geqq A e^{+i\omega t} e^{\gamma z} \tag{272}$$

The exponent, $\gamma$, can be evaluated by substituting this assumed solution in the original Eq. (271). In doing this, the following derivatives are needed:

$$\frac{\partial^2 E_x}{\partial z^2} = \gamma^2 A e^{+i\omega t} e^{\gamma z} + \gamma^2 E_x$$

$$\frac{\partial E_x}{\partial t} = +i\omega A e^{+i\omega t} e^{\gamma z} = +i\omega E_x$$

$$\frac{\partial^2 E_x}{\partial t^2} = -\omega^2 A e^{+i\omega t} e^{\gamma z} = -\omega^2 E_x$$

Substituting these in Eq. (271), we find an expression for $\gamma$:

$$\gamma = \pm\left[\frac{i\omega\mu}{\varrho} - \varepsilon\mu\omega^2\right]^{1/2} \tag{273}$$

The parameter $\gamma$ is the reciprocal of a radian wave length, so we call it a wave number. At the very low frequencies dealt with in the magneto-telluric method, resistivity is more important than the dielectric constant or the magnetic permeability in determining variations in wave number. As an example, consider what values of wave number might pertain to a highly conductive sedimentary rock, or to a resistive igneous rock. A shale may have a resistivity of about 10 ohm-m, commonly, a magnetic permeability of $12 \cdot 56 \times 10^{-7}$ H/m and a dielectric constant of $10^{-6}$ F/m, if the effects of induced polarization, which will be discussed in Chapter VIII, are considered. Substituting these values in Eq. (273), we have:

$$\gamma^2 = i\omega\,\frac{12\cdot56 \times 10^{-7}}{10} - \omega^2 \times 12\cdot56 \times 10^{-7} \times 1 \times 10^{-6}$$

$$= i\omega \times 12\cdot56 \times 10^{-8} - \omega^2 \times 12\cdot56 \times 10^{-13} \tag{274}$$

Since the square of the wave number is a complex quantity, it is convenient to express $\gamma^2$ in terms of the polar form for a complex quantity, so that the absolute value of the wave number can be determined:

$$\gamma_1^2 = |\gamma^2|\,e^{i\varphi}$$

Taking the square root, we have:

$$\gamma = |\gamma|\,e^{i\varphi/2}$$

The magnitude of the wave number is:

$$|\gamma| = [(12\cdot56 \times 10^{-8}\omega)^2 + (12\cdot56 \times 10^{-13}\omega^2)^2]^{1/2} \tag{275}$$

and the phase angle is:

$$\varphi/2 = \arctan\frac{12\cdot56 \times 10^{-13}\omega^2}{12\cdot56 \times 10^{-8}\omega} \tag{276}$$

The frequencies used in the magneto-telluric method are nearly always less than 10 c/s, so the contribution to wavelength by the second term in Eq. (275), which represents the effects of displacement currents, is negligible, at least for rocks with the properties assumed for shale.

A granite would represent the opposite extreme in electrical properties. A reasonable set of properties for granite would be:

$$\varrho = 10^5 \text{ ohm-m}$$

$$\varepsilon = 1\cdot6 \times 10^{-10} \text{ F/m}$$

$$\mu = 12\cdot56 \times 10^{-7} \text{ H/m}$$

Substituting these values in Eq. (273), we have

$$\gamma^2 = i\omega\,\frac{12\cdot56 \times 10^{-7}}{10^5} - \omega^2 \times 12\cdot56 \times 10^{-7} \times 1\cdot6 \times 10^{-10}$$

$$\gamma^2 = i\omega \times 12\cdot56 \times 10^{-12} - \omega^2 \times 20\cdot1 \times 10^{-17} \tag{277}$$

Again, for any frequency which might be considered for the magneto-telluric method, the contribution of the second term, that due to displacement currents, would be negligible. Equation (273) may thus be simplified by ignoring the second term within the radical:

$$\gamma = \pm\left[\frac{i\omega\mu}{\varrho}\right]^{1/2} \tag{278}$$

Since the magnetic permeability of almost all rocks is very nearly that of free space, the wave number, $\gamma$, depends only on the resistivity of a medium and on the frequency. Curves showing the relationship between wave number, resistivity and frequency are given in Fig. 141.

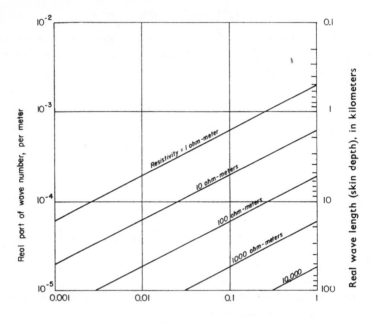

Frequency, cycles per second

FIG. 141. Wave numbers as a function of frequency and resistivity for cases in which displacement currents can be neglected. A scale for converting wave numbers to radian wave lengths or skin depths (the distance for 63·2 per cent reduction of the amplitude of an electromagnetic wave) is shown to the right.

The wave number consists of a real and imaginary part. Since it is present in the solution to the wave equation as an exponent, the imaginary part represents the oscillatory nature of the amplitude of the wave along the direction in which the wave is travelling, and the real part indicates an exponential decay in the amplitude along the travel path. The wave number may be separated into real and imaginary parts as follows:

$$\pm\gamma = \sqrt{\left(\frac{\omega\mu}{\varrho}\right)} \cdot \sqrt{+i} = \frac{1+i}{\sqrt{2}} \sqrt{\left(\frac{\omega\mu}{\varrho}\right)} = \left[\frac{\omega\mu}{2\varrho}\right]^{1/2} + i\left[\frac{\omega\mu}{2\varrho}\right]^{1/2}$$

The amplitude of a magnetotelluric wave is attenuated by the factor (1/e) or is reduced by 63·2 per cent in a travel distance given by 1·41 $\gamma_{\text{imag}}$. This same distance is a radian wavelength for the wave, or the distance through which the phase changes by one radian. The full wavelength is $2\pi$ times larger, but it is more convenient in magneto-telluric studies to speak in terms of radian wavelengths, $1/(\text{real part of } \gamma) = 1/(\text{imaginary part of } \gamma)$.

Since there are two roots to the radical expression for the complex wave number Eq. (278), there are two solutions to the original differential equation Eq. (271). Since any linear combination of special solutions will also be a solution to the differential equation, as a general solution, we should consider:

$$E_x = Ae^{+i\omega t + \gamma z} + Be^{+i\omega t - \gamma z} \tag{279}$$

We must now evaluate the arbitrary constants, $A$ and $B$.

The electric field given by Eq. (279) will cause currents to flow in a conductive medium. These currents will generate a magnetic field which will be time-variant in the same manner as is the current:

$$H_y = H_{y,0}e^{+i\omega t} \tag{280}$$

The time derivative of the magnetic field is:

$$\frac{\partial H_y}{\partial t} = +i\omega H_{y,0}e^{i\omega t} = +i\omega H_y \tag{281}$$

Using this expression for the derivative in Eq. (260) (Maxwell's first equation), we have:

$$\text{curl } \boldsymbol{E} = -i\mu\omega H_y$$
$$\frac{\partial E_x}{\partial z} = -i\mu\omega H_y \tag{282}$$

Using the expression for electric field strength found in Eq. (279), this last equation becomes:

$$H_y = \frac{-\gamma}{i\mu\omega}[Ae^{+i\omega t + \gamma z} - Be^{+i\omega t - \gamma z}] \tag{283}$$

The ratio of electric field strength to magnetic field strength can be defined as the *wave impedance* of a medium. It can be shown that for plane wave propagation, this ratio depends only on the electrical properties of the medium and on the frequency, and that dimensionally, the ratio is expressed in ohms. Combining Eqs. (279) and (283) to determine the wave impedance for this particular problem, we have:

$$Z = \frac{E_x}{H_y} = \frac{-i\mu\omega}{\gamma} \frac{Ae^{\gamma z} + Be^{-\gamma z}}{Ae^{\gamma z} - Be^{-\gamma z}} \tag{284}$$

It will prove convenient to convert the fraction in this last equation to a hyperbolic function, which it already resembles in form. The defining equations for hyperbolic sine and cosine are;

$$\cosh x = \tfrac{1}{2}(e^x + e^{-x})$$
$$\sinh x = \tfrac{1}{2}(e^x - e^{-x})$$

The expression in Eq. (284) is not quite that for a hyperbolic function, but can be manipulated so that it is. The first step is to divide both the numerator and denominator in the fraction by the term $\sqrt{(AB)}$:

$$Z = \frac{i\mu\omega}{\gamma} \frac{\sqrt{\left(\dfrac{A}{B}\right)} e^{\gamma z} + \sqrt{\left(\dfrac{B}{A}\right)} e^{-\gamma z}}{\sqrt{\left(\dfrac{A}{B}\right)} e^{\gamma z} - \sqrt{\left(\dfrac{B}{A}\right)} e^{-\gamma z}} \qquad (285)$$

Then using the identity $\sqrt{(A/B)} = e^{\ln \sqrt{(AB)}}$, we have the correct form for the hyperbolic cotangent:

$$Z = \frac{i\mu\omega}{\gamma} \frac{\exp\left[\ln \sqrt{\left(\dfrac{A}{B}\right)} + \gamma z\right] + \exp\left[-\ln \sqrt{\dfrac{A}{B}} - \gamma z\right]}{\exp\left[\ln \sqrt{\left(\dfrac{A}{B}\right)} + \gamma z\right] - \exp\left[-\ln \sqrt{\left(\dfrac{A}{B}\right)} - \gamma z\right]}$$

$$= \frac{i\mu\omega}{\gamma} \coth\left[\gamma z + \ln \sqrt{\left(\dfrac{A}{B}\right)}\right] \qquad (286)$$

The constants $A$ and $B$ may be eliminated by considering initial or boundary conditions, or they may be eliminated by considering only ratios of wave impedances at two different positions. We shall use the second approach, rather than limit the generality of the solution by imposing additional conditions. First, we evaluate Eq. (286) to find the wave impedance at a specific location, $z_1$, and solve this expression for $\ln \sqrt{(A/B)}$

$$\ln \sqrt{\left(\frac{A}{B}\right)} = \coth^{-1}\left(\frac{\gamma Z_1}{i\mu\omega}\right) - \gamma z_1 \qquad (287)$$

Next, we evaluate Eq. (286) to find the wave impedance at another location, $z_2$, and replace the term $\ln \sqrt{(A/B)}$ with the expression found in Eq. (287):

$$Z_2 = \frac{i\mu\omega}{\gamma} \coth\left[\gamma z_2 + \ln \sqrt{\left(\frac{A}{B}\right)}\right]$$

$$= \frac{i\mu\omega}{\gamma} \coth\left[\gamma(z_2 - z_1) + \coth^{-1}\left(\frac{\gamma Z_1}{i\mu\omega}\right)\right] \qquad (288)$$

This equation is valid only if the two points, $z_1$ and $z_2$ are in the same medium. In a layered medium, they would have to be in the same layer. The value of Eq. (288) is that we may compare the wave impedance at the top of a layer with the impedance at the bottom of the same layer. We shall now specify that the origin of our cartesian coordinate system is at $z_2 = 0$, or on the surface of a layered earth, and that $z_1$ was chosen equal to $h_1$, the thickness of the top layer. Then, the wave impedance at the surface, where it is usually measured, is related to the wave impedance at the bottom of the first layer as follows:

$$Z_2(z = 0) = -\frac{i\mu\omega}{\gamma}\coth\left[\gamma h_1 + \coth^{-1}\left(\frac{\gamma Z_1(z = h_1)}{i\mu\omega}\right)\right] \qquad (289)$$

The simplest case to be considered is that of a completely homogeneous earth. In this case, the thickness of the first layer becomes infinite, as does the argument for the hyperbolic cotangent in Eq. (289). When this happens, the value for the hyperbolic cotangent approaches unity, and Eq. (289) becomes simply:

$$Z_{(0)} = \frac{i\omega\mu}{\gamma} = [+i\omega\mu\varrho]^{1/2} \qquad (290)$$

As in the galvanic resistivity methods, we may define an *apparent resistivity* based on the value of wave impedance measured over a uniform earth. Solving Eq. (290) for resistivity, we obtain the defining equation for apparent resistivity measured with the magnéto-telluric method:

$$\varrho_a = -i\frac{Z^{z'}}{\omega\mu} = -\frac{i}{\omega\mu}\left(\frac{E}{H}\right)^2 \qquad (291)$$

The defining equation gives the apparent resistivity as a pure imaginary number. This merely indicates that there is a 45° phase difference between the oscillations in the magnetic and electric field intensities, and when the ratio of these two quantities is formed and squared, the term $i$ must be introduced. The phase difference between the $E$ and $H$ components of the magneto-telluric field need not be considered if a homogeneous earth is being considered, since the phase difference is constant. However, we shall see that in more general cases, the phase relation between the two field components varies, and may be used to determine resistivity distributions, just as variations in the magnitude of the wave impedance may be used.

In an $n$–layered earth, continuity conditions which must hold at each boundary permit us to express the wave impedance observed at the surface in terms of the wave impedances in each of the lower layers. Both the electric field, $E_x$, and the magnetic field, $H_y$, must be continuous across the boundaries between layers, since there are neither charge accumulations nor free poles allowed at these boundaries. The wave impedance at the bottom of the first layer must equal the wave impedance at the top of the second layer, and so on. As an example, consider a medium consisting of only two layers having resistivities $\varrho_1$ and $\varrho_2$ and thicknesses $h_1$ and $\infty$. The wave impedance at the surface, expressed in terms of the impedance at the base of the first layer is:

$$Z_{(0)} = \frac{i\mu\omega}{\gamma_1}\coth\left[\gamma_1 h_1 + \coth^{-1}\frac{\gamma_1 Z(h_1)}{i\mu\omega}\right] \qquad (292)$$

The wave impedance at the top of the second layer ($z = h_1$) expressed in terms of the impedance at the bottom of the second layer ($z = \infty$) is:

$$Z(h_1) = \frac{i\mu\omega}{\gamma_2} \coth\left[\gamma_2(\infty - h_1) \coth^{-1} \frac{\gamma_2 Z(\infty)}{i\mu\omega}\right] = \frac{i\mu\omega}{\gamma_2} \qquad (293)$$

Combining these last two equations, we have an expression for the wave impedance observed at the surface of a two-layer earth:

$$Z_{(0)} = -\frac{i\omega\mu}{\gamma_1} \coth\left(\gamma_1 h_1 + \coth^{-1} \frac{\gamma_1}{\gamma_2}\right)$$

$$\varrho_a = -\frac{i\omega\mu}{\gamma_1^2} \coth^2\left(\gamma_1 h_1 + \coth^{-1} \frac{\gamma_1}{\gamma_2}\right)$$

$$= \varrho_1 \coth^2\left[\sqrt{\left(\frac{-\omega\mu}{\varrho_1}\right)} h_1 + \coth^{-1} \sqrt{\left(\frac{\varrho_2}{\varrho_1}\right)}\right] \qquad (294)$$

This equation is far simpler to evaluate than the corresponding equation for the galvanic apparent resistivity over a two-layer earth. Curves showing the relationship between apparent resistivity and frequency are given in Fig. 142 for the two layer case. These curves have the following characteristics which should be recognized:

1. At high frequencies (short wavelengths), the computed resistivity is approximately that of the surface layer. High frequencies in the magneto-telluric method correspond to short electrode separations in the galvanic method.

2. At low frequencies (long wavelengths), the computed resistivity is approximately that for the substratum.

3. At intermediate frequencies, there are a number of maxima and minima for computed resistivity which may be either greater than or less than the corresponding maximum or minimum values of resistivity actually existing in the earth. These small maxima and minima represent constructive or destructive interference between waves reflected from the second layer and the primary magneto-telluric waves.

4. The calculated impedances deviate significantly from the value for the first layer only for radian wavelengths greater than 0·83 times the thickness of the surface layer.

5. The curves for apparent resistivity for the case of a more resistant substratum are mirror images of the curves for the case of a more conductive substratum for the corresponding contrast in resistivities. (This was not true for galvanic resistivity curves).

6. If the contrast in resistivity between the top layer and the substratum is large, the computed resistivity is proportional to the square of the wavelength (for resistant substratum) or inversely proportional to the square of the wavelength (for a conductive substratum).

In Eq. (294), the apparent resistivity is computed by considering the magnitude of the impedance. In practice, the ratio of peak amplitudes of oscillations in the electric and magnetic field components is taken, without regard for the phase difference between the field components. The phase shift between magnetic and electric field components may be found by taking the ratio of the real and imaginary components of the impedance given in the first line of Eq. (294).

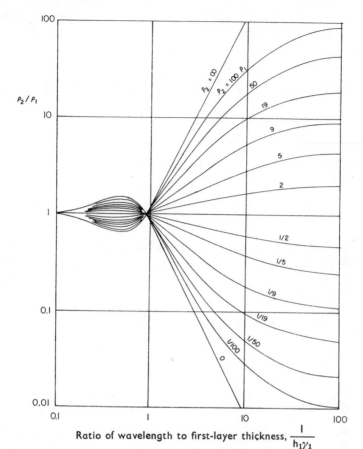

FIG. 142. Apparent resistivity calculated from magneto-telluric measurements over a two-layer earth.

Consider the identity:

$$\coth (x \pm iy) = \frac{\coth x \cot y \pm i}{\cot y \pm i \coth x} \tag{295}$$

In order to separate real and imaginary parts, this fraction must be rationalized:

$$\coth (x + iy) = \frac{\coth x (\cot^2 y + 1)}{\cot^2 y + \coth^2 x} + i \frac{\coth y (1 - \coth^2 x)}{\cot^2 y + \coth^2 x} \tag{296}$$

Using this identity, the first line in Eq. (294) may be separated into real and imaginary parts:

$$\text{REAL } \{Z_{(0)}\} = -\sqrt{\left(\frac{\omega\mu\varrho}{2}\right)} \left[ \frac{\coth x (\cot^2 y + 1)}{\cot^2 y + \coth^2 x} + \frac{\cot y (1 - \coth^2 x)}{\cot^2 y + \coth^2 x} \right] \tag{297}$$

$$\text{IMAG}\ \{Z_{(0)}\} = \sqrt{\left(\frac{\omega\mu\varrho}{2}\right)}\left[\frac{\coth x\,(\cot^2 y + 1)}{\cot^2 y + \coth^2 x} - \frac{\cot y\,(1 - \coth^2 x)}{\cot^2 y + \coth^2 x}\right] \quad (298)$$

where:

$$x = h_1 \cdot \text{REAL}\ \{\gamma_1\} + \coth^{-1}\sqrt{\left(\frac{\varrho_2}{\varrho_1}\right)} = h_1\sqrt{\left(\frac{\omega\mu}{2\varrho}\right)} + \coth^{-1}\sqrt{\left(\frac{\varrho_2}{\varrho_1}\right)} \quad (299)$$

$$\gamma = h_1 \cdot \text{IMAG}\ \{\gamma_1\} = h_1\sqrt{\frac{\omega_y}{2\varrho}} \quad (300)$$

The phase angle between electric and magnetic field components is found by taking the ratio of Eqs. (297) and (298):

$$\tan\varphi = -\frac{\coth x\,(\cot^2 y + 1) + \cot y\,(1 - \coth^2 x)}{\coth x\,(\cot^2 y + 1) - \cot y\,(1 - \coth^2 x)} \quad (301)$$

Curves showing the relationship between this phase angle and frequency are given in Fig. 143 for a variety of resistivity contrasts between a single overburden layer and a uniform substratum.

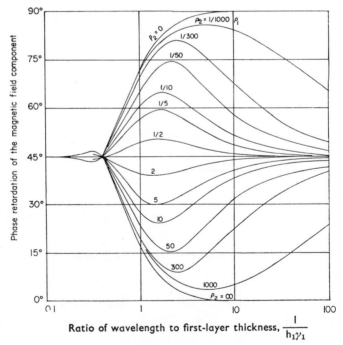

FIG. 143. Phase retardation of the magnetic component of a magnetotelluric field observed over a layered earth consisting of a single overburden layer and a uniform substratum.

The magnetic field may lag the electric field by phase angles ranging from 0° (for a conductive layer over an insulating bedrock) to 90° (for a resistant layer over a conductive bedrock). The following characteristics of the phase curves should be noted:

1. With short wavelengths (high frequencies), the phase retardation of the magnetic field is 45°, and does not depend on frequency.

2. With very long wavelengths (low frequencies), the phase retardation is also 45°. Deviations from 45° retardation occur only for intermediate wave lengths, ranging from 2 to 50 times the first-layer thickness.

3. Phase retardation is decreased from 45° if resistivity increases with depth and is increased if resistivity decreases with depth.

As with the galvanic resistivity method, it is of interest to examine the characteristics of the apparent resistivity curves for various extreme values of resistivity and wavelength to see if simple interpretation rules exist. The asymptotic behavior of the right-hand portions of the curves shown in Fig. 142 for high resistivity contrasts may be studied by rearranging the expression for wave impedance over a two-layer earth as follows. In Eq. (294) (first line), the right-hand side is of the form coth $(x + y)$ which may be expanded using the following identity:

$$\coth (x + y) = \frac{\coth x \coth y - 1}{\coth x + \coth y} \tag{302}$$

Using this identity, Eq. (294) may be rewritten as:

$$Z_{(0)} = \frac{-i\omega\mu}{\gamma_1} \frac{\dfrac{\gamma_1}{\gamma_2} \coth \gamma_1 h_1 - 1}{\dfrac{\gamma_1}{\gamma_2} + \coth \gamma_1 h_1} \tag{303}$$

Since we are considering only the right-hand portion of the apparent resistivity curves, the wavelengths being considered are considerably greater than the thickness of the first layer:

$$\gamma_1 h_1 \ll 1$$

so that in place of the hyperbolic cotangent, we may use the approximation:

$$\coth \gamma_1 h_1 \approx \frac{1}{\gamma_1 h_1}$$

Making this approximation, the equation for wave impedance becomes:

$$Z_{(0)} = \frac{-i\omega\mu}{\gamma_1} \frac{1 - \gamma_2 h_1}{\gamma_1 h_1 + \dfrac{\gamma_2}{\gamma_1}} \tag{304}$$

We may now consider two special cases: the case in which the lower medium is much more resistant than the overburden, and the case in which the lower medium is much more conductive than the overburden. Let us consider first the case in which the bedrock resistivity is much greater than the resistivity of the surface layer. This means that the wave number, $\gamma_2$, for the lower medium is much less than the wave number for the surface layer, and so, the second term in the numerator of Eq. (304) may be neglected. Expressing the wave numbers in terms of their definitions in electrical properties, Eq. (304) becomes:

$$Z_{(0)} \approx \frac{-i\omega\mu}{\gamma_1^2 h_1 + \gamma_2} = \frac{-i\omega\mu}{\dfrac{-i\omega\mu h_1}{\varrho_1} + \sqrt{\left(\dfrac{i\omega\mu}{\varrho_2}\right)}} \tag{305}$$

When the resistivity of the second layer is large in comparison with that of the first layer, the second term in the denominator of Eq. (305) may be neglected in comparison with the first term:

$$Z_{(0)} \approx \frac{\varrho_1}{h_1} \tag{306}$$

The apparent resistivity, as defined in Eq. (291), is:

$$\varrho_a = \frac{Z_{(0)}^2}{\omega\mu} = \frac{1}{\omega\mu S_1^2} \tag{307}$$

or, solving for the first-layer conductance:

$$\boxed{S_1 = \frac{1}{[\omega\mu\varrho_a]^{1/2}} = \frac{1}{1\cdot12 \times 10^{-3} \sqrt{(\omega\varrho_a)}}} \tag{308}$$

In using the equation, it should be remembered that the frequency is measured in radians per second.

In the case of a conducting layer covering an insulating substratum, the computed apparent resistivity varies inversely as the square of the longitudinal conductance of the surface layer. This property of a telluric sounding curve is used in interpretation in the same manner as the 45° asymptote is used in interpreting galvanic resistivity curves.

In the second case, that in which the lower half-space is very much more conductive than the first layer, the first term in the denominator of Eq. (304) may be neglected in comparison with the second term. The equation for wave impedance under these conditions reduces to:

$$Z_{(0)} = \frac{-i\omega\mu}{\gamma_1} \frac{\gamma_1 - \gamma_2\gamma_1 h_1}{\gamma_2} \tag{309}$$

$$\approx + i\omega\mu h_1$$

Using Eq. (291) to compute apparent resistivity, we have:

$$\varrho_a = \omega\mu h_1^2 \tag{310}$$

Solving for the thickness of the surface layer, we have:

$$h_1 = (\varrho_a/\omega\mu)^{1/2} = 800(\varrho_a/\omega)^{1/2} \qquad (311)$$

The apparent resistivity computed from wave impedance measurements made over an insulating bed covering a conductive semi-infinite half-space is directly proportional to the square of the thickness of the layer. It is interesting to note in comparison that there is no simple asymptotic behavior for the galvanic resistivity curve measured over an insulating slab.

The equation for wave impedance at the surface for a three-layer geometry is only moderately more complicated than the equation for two layers (Wait, 1953a). From Eq. (288), we may write the equation for the three-layer case:

$$Z_{(0)} = \frac{-i\mu\omega}{\gamma_1} \coth\left[\gamma_1 h_1 + \coth^{-1}\left\{\frac{\gamma_1}{\gamma_2}\coth\left(\gamma_2 h_2 + \coth^{-1}\frac{\gamma_2}{\gamma_3}\right)\right\}\right] \qquad (312)$$

The equation may be extended to an arbitrary number of layers merely by matching wave impedances at each boundary, starting from the bottom of the

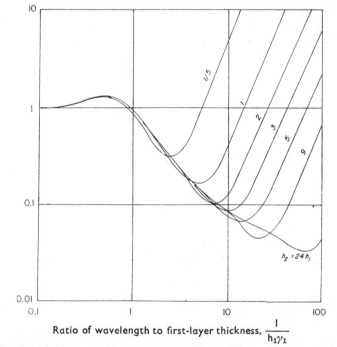

Ratio of wavelength to first-layer thickness, $\dfrac{1}{h_1\gamma_1}$

Fig. 144. Apparent resistivity curves computed from impedance measurements made at the surface of a three-layer earth of type H. The curves are computed for the case in which the middle layer has a resistivity 19 times less than that of the surface layer, and in which the lowermost layer is a perfect insulator.

section and moving towards the surface. The general equation for an $n$-layered geometry is:

$$Z_{(0)} = \frac{-i\mu\omega}{\gamma_1} \coth\left[\gamma_1 h_1 + \coth^{-1}\left\{\frac{\gamma_1}{\gamma_2}\coth\left(\gamma_2 h_2 + \coth^{-1}\left[\frac{\gamma_2}{\gamma_3}\cdots\right.\right.\right.\right.$$

$$\left.\left.\left.\left.\cdots \coth^{-1}\left\{\frac{\gamma_{n-2}}{\gamma_{n-1}}\coth\left(\gamma_{n-1}h_{n-1} + \coth^{-1}\frac{\gamma_{n-1}}{\gamma_n}\right)\right\}\cdots\right]\right)\right\}\right]$$    (313)

The three-layer geometries may be divided into four classes, as in the case of galvanic resistivity curves, on the basis of the sequence of resistivity contrasts between the three layers:

$$\text{Type H:}\quad \varrho_1 > \varrho_2 < \varrho_3$$
$$\text{Type A:}\quad \varrho_1 > \varrho_2 > \varrho_3$$
$$\text{Type K:}\quad \varrho_1 < \varrho_2 > \varrho_3$$
$$\text{Type Q:}\quad \varrho_1 < \varrho_2 < \varrho_3$$

Apparent resistivity curves are given for one set of contrasts for each of these cases in Fig. 144 through 147. Curves for type H resistivity contrasts, those in

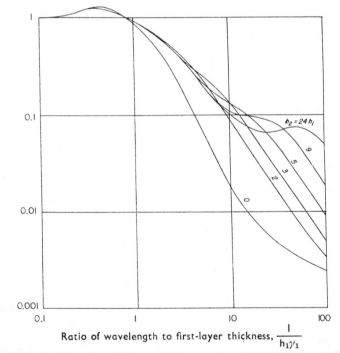

FIG. 145. Apparent resistivity curves computed for impedance measurements made at the surface of a three-layer earth of type A. The curves are computed for the case in which the middle layer has a resistivity 1/19th that of the surface layer, and in which the lowermost layer has a resistivity 1/500th that of the surface layer.

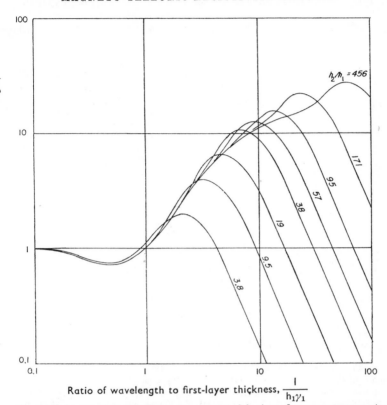

FIG. 146. Apparent resistivity curves computed for impedance measurements made at the surface of a three-layer earth of type K. The curves are computed for the case in which the middle layer has a resistivity 19 times greater than that of the surface layer, and in which the lowermost layer is perfectly conducting.

which the middle layer has the lowest resistivity, exhibit a minmium apparent resistivity at intermediate wavelengths, while curves for type K resistivity contrasts, those for which the middle layer has the highest resistivity, exhibit a maximum apparent resistivity at intermediate wavelengths. Type A curves show a monotonous decrease in apparent resistivity, except for a minor maximum value which develops for constructive interference between the primary magneto-telluric field and waves reflected from the second layer. Likewise, type Q curves show a monotonous increase in apparent resistivity, except for the effects of destructive interference at certain wavelengths.

Asymptotic expressions for the wave impedance measured over a three-layer earth may be obtained using the same approximations used for the two-layer case. Equation (312) may be rewritten in the form:

$$Z_{(0)} \approx \frac{h_1 + Z(2)}{1 + S_1 Z(2)} \tag{314}$$

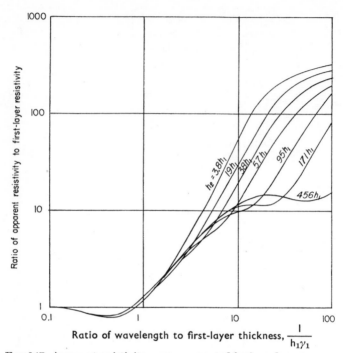

FIG. 147. Apparent resistivity curves computed for impedance measurements made at the surface of a three-layer earth of type Q. The curves are computed for the case in which the middle layer has a resistivity 19 times larger than that of the surface layer, and in which the lowermost layer has a resistivity 200 times that of the surface layer.

when only the right-hand portion of the data are being considered (the wave length is very long compared to the combined thickness of the two surface layers). The parameter $Z(2)$ is the wave impedance which would be measured at the surface of the second layer in the three-layer case. If the bottom layer is a perfect insulator, Eq. (314) reduces to:

$$Z_{(0)} \approx \frac{1}{S_1 + S_2} \tag{315}$$

The apparent resistivity curve is asymptotic to a line rising with a slope of 2. This is the same limit that is approached in the two-layer case, if the surface layer were assumed to have a conductance equal to the sum of the two surface-layer conductances in the three-layer case.

If the bottom layer is very conductive, Eq. (314) reduces to:

$$Z_{(0)} \approx h_1 + h_2 \tag{316}$$

The apparent resistivity curve is asymptotic to a line descending with a slope of $-2$. This is the same limit that is approached in the two-layer case, if the surface layer were assumed to have a total thickness equal to the sum of the two surface-layer thicknesses in the three-layer case.

When the resistivity contrasts in three-layer cases of type $H$ or $K$ are large, the problem of equivalence arises, as it does in galvanic resistivity sounding curves. For example, in type $H$ curves such as those shown in Fig. 144, the minimum apparent resistivity computed from impedance measurements depends only on the ratio of conductances between the first two layers, and not on the resistivity contrast, so long as the resistivity contrast is large. The ratio of conductances between the first two layers is proportional to the square of the ratio of minimum apparent resistivity to first-layer resistivity:

$$\frac{S_1}{S_2} \approx 5 \left( \frac{\varrho_{min}}{\varrho_1} \right)^2 ; \quad S_1/S_2 > 4, \quad \varrho_2 \ll \varrho_{min} \tag{317}$$

Thus, the magneto-telluric method for determining resistivity is restricted in much the same way as the galvanic resistivity method when one of the layers in a layered section of rock is very highly conductive.

When the resistivity of the second layer is very great, the amplitude of the maximum apparent resistivity depends only on the thickness of the resistant layer and not on its resistivity:

$$\frac{h_2}{h_1} \approx 1/5 \left( \frac{\varrho_{max}}{\varrho_1} \right)^2 ; \quad h_2/h_1 > 4, \quad \varrho_2 \gg \varrho_{max} \tag{318}$$

In the case of galvanic soundings through a highly resistant layer, only the transverse resistance of the second layer could be determined. The non-uniqueness of the magneto-telluric method in the same circumstances is of a different sort. A combination of magneto-telluric data and galvanic sounding data is theoretically capable of providing a complete interpretation in the case of a highly resistant bed. The product $h_2\varrho_2$ can be found from the galvanic data, while the sum $h_1 + h_2$ can be found from magneto-telluric data. This is enough information to provide a unique solution for type $K$ layering.

## 30. INSTRUMENTATION USED IN MEASURING THE MAGNETO-TELLURIC FIELD

Determination of earth resistivity using the magneto-telluric field as a source of power requires that wave impedances be measured over a frequency spectrum several decades in width. In designing field equipment, it is necessary to know what frequency intervals must be observed in order to obtain data for the construction of a sounding curve similar to the ones shown in Fig. 142. The highest frequency which would be of use in studying a two-layer case would be that at which the wave interference effects is seen; a frequency which can be specified in terms of the ratio of wavelength to first-layer thickness as follows:

$$\left(\frac{\lambda_1}{h_1}\right)_m = 5\cdot 3 \tag{319}$$

The wave length can be expressed in terms of frequency and the resistivity of the first layer, so that the frequency corresponding to the interference maximum or minimum can be determined as follows:

$$\lambda_{1,m}^2 = (5\cdot 3h_1)^2$$

$$\frac{2\varrho_1}{\mu\omega_m} = (5\cdot 3h_1)^2$$

$$\omega_m = \frac{2\varrho_1}{\mu(5\cdot 3h_1)^2} \; ; \quad \omega = 2\pi f$$

$$f_m = 9\cdot 05 \times 10^3 \times \frac{1}{S_1 h_1} \tag{320}$$

As an example, consider the shortest periods which would be of use in determining the thickness and resistivity of a sedimentary basin which is assumed to contain about 1000 m thickness of rocks with an average resistivity of 10 ohm-m. The maximum frequency which would be of use, according to Eq. (320), would be 0·09 c/s (micropulsation periodicities of 11 sec and longer would be required). Wavelengths ten times longer than those for the maximum frequency would be required to establish the rising or falling portion of the sounding curve, while wavelengths 100 times longer than those for the maximum frequency would be required to establish the resistivity of the second layer. Since wavelength is proportional to the square root of period, this means that periods ranging from 11 sec to 1100 sec would be required to determine the thickness of sediments, and periods ranging from 11 to 110,000 sec would be required to determine the resistivity of the basement. (There are only 86,400 sec in a day).

An alternate problem which might be studied with the magneto-telluric method is that of determining the thickness of a resistant layer, such as volcanic rock or an igneous thrust sheet, over more conductive rock, such as sediments. Consider the problem of determining the thickness of a resistant overburden when it has a thickness of 1000 m and a resistivity of 1000 ohm-m. The maximum usable frequency computed from Eq. (320) is 9·05 c/s. Micropulsations with periods ranging from 0·1 to 10 sec would be required to establish the thickness of the resistant layer, and periods ranging up to 1000 sec would be required to establish the resistivity of the second layer.

With these considerations, we see that the instrumentation required in measuring the magneto-telluric field should be capable of detecting and measuring oscillations in the electric and magnetic fields with periods ranging from about 0·1 sec to one day. Longer period oscillations may be of interest in studying the properties deep in the crust or within the mantle; audio frequencies may be of interest in studying resistivity variations at depths of a few hundred feet, but the magneto-telluric method is generally considered most useful in the range of frequencies above one cycle per day and below audio frequencies.

To be generally useful, instrumentation should be capable of measuring the amplitude of the magneto-telluric field during quiet periods, as well as during active periods. Sensitivity requirements then are for a magnetometer which can detect variations in the magnetic field of the order of tens of gammas with a diurnal periodicity, and variations of the order of tens of milligammas with periods of a few seconds. The sensitivity of equipment for detecting variations in the electric field is less exacting, since electrode separation may be increased to compensate for lack of sensitivity. However, in temperate latitudes, it is necessary to be able to detect electric field oscillation amplitudes ranging from tenths of millivolts to tens of millivolts per kilometer.

It is a fairly simple matter to measure the electric field in the earth with the necessary precision. The commonest technique uses three electrodes, or ground contacts, in the form of an L, with lengths of the arms ranging from several tenths of a kilometer to several kilometers. If measurements are being made over resistant rock, such as igneous or metamorphic rocks, the electric field strengths will be large and the shorter electrode separations will be adequate. If measurements are being made over very conductive rock, electrode separations of several kilometers may be required to obtain a large enough voltage to measure accurately.

With the L-spread, the corner electrode is used as a common ground for two recording channels (see Fig. 148). It is essential that no current flows through the recording system, since if it does, current from one arm of the L spread may return to the ground through the other arm of the L spread. When this happens, a voltage appearing along one arm of the spread will be recorded also, but with reduced amplitude, on the recording channel connected to the other arm of the spread. This *cross-sensitivity* leads to errors in relating the proper component of electric field variation with the component of magnetic field variation measured at the same time. If the recording system in use has a relatively low resistance, a four-terminal cross array of electrodes, such as that shown in Fig. 148 (b), should be used. With a four-terminal system, neither recording circuit has a common ground.

Almost any type of electrode may be used to make contact with the ground if variations with periods shorter than hundreds of seconds only are to be measured. A lead plate buried in a shallow trench, and moistened with salt water is a good simple electrode. However, if variations with periods ranging up to a day are to be measured, some precautions should be taken to avoid variations in electrode potential due to temperature changes. Electrode potential, which is the potential drop between an electrode and the electrolyte in contact with it, depends on temperature. If both electrodes used in measuring an earth

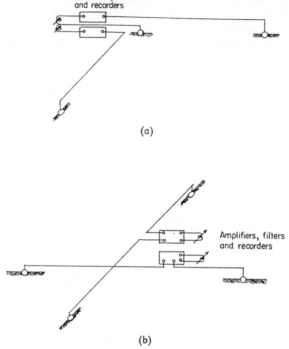

FIG. 148. Electrode configurations which may be used in measuring two orthogonal components of an electric field oscillation.

(a) Three-terminal L array with the corner electrode used as a common ground point for both recording channels.

(b) Four-terminal cross array with the two recording channels having no common ground point.

voltage are identical in behavior, and if both electrodes vary in temperature in the same way, the changes in electrode potential at each electrode will cancel out. In practice, it is unreasonable to expect that both electrodes will be subjected to the same changes in temperature, since they are widely spaced, unless some provision is made to assure that this is so. This may be done by burying electrodes several feet in the ground, so that diurnal temperature changes, as well as the changes in temperature caused by variation in cloud cover, wind velocity, and similar factors, will not penetrate to the electrode.

If measurements are to be made as soon as electrodes are emplaced, there is some advantage to using non-polarizing electrodes; electrodes which consist of a metal immersed in a saturated solution of one of its salts. Combinations commonly used are copper electrodes in solutions of copper sulfate and zinc electrodes in solutions of zinc sulfate.

In view of the small changes in voltage which are to be measured, some form of amplification is required. This may be accomplished with a direct-current amplifier, in view of the low frequencies being measured. It is important that the short term drift of the amplifier be considerably smaller than the voltage variations to be measured in order that the drift not be confused with true voltage variations. Long-term drift is not a problem unless long period variations are to be recorded. Voltage amplification ratios up to 3000 may be obtained readily using chopper-stabilized amplifiers to avoid drift problems. In such amplifiers, the low-frequency voltages appearing at the input are converted to higher-frequency voltages using either a mechanical or electronic alternator (a device which alternates the polarity of the input at rates ranging from 60 to 400 times per second). Amplification is then accomplished in the same manner as in a.c. amplifiers and the output is converted back to d.c. (or low frequency) using an alternator operating synchronously with the input alternator. Such amplifiers may be used for frequencies up to several cycles per second. At higher frequencies, conventional a.c. amplifiers may be used, since the short-term drift at these higher frequencies is not usually a problem.

The voltages from the electrode arrays may be filtered before amplification in some cases. The extent of filtering which is done before the data are recorded depends on the type of data analysis which is planned. If data are recorded on magnetic tape, generally as little filtering as possible is done since it may be performed more effectively when the tapes are replayed at an analysis center. If data are recorded on paper, it is usually necessary to record only a narrow band of frequencies on a single record. If broad-band signals are recorded, usually only a single frequency, or perhaps several, are readily apparent since they are far larger than signals in other frequency bands. In order to see variations in the magneto-telluric field in frequency bands outside the dominant frequencies, it is necessary to reject these dominant frequencies with filters.

In addition to filtering, it is usually necessary to remove a static self-potential level which ordinarily exists between a widely spaced pair of electrodes. The static potential difference between a pair of electrodes spaced at one kilometer is ordinarily of the order of some hundreds of millivolts. This level must be removed before the voltage between the electrodes can be amplified, if the static level would exceed the dynamic range of the amplifier and recording system. The static potential may be cancelled with a potentiometric circuit, or rejected with a capacitive input to the recorder system. If a capacitive input system is used, the time constant must be much longer than the longest period to be recorded. A typical potentiometric circuit for rejecting the static potential difference is shown in Fig. 149.

The measurement of magnetic field micropulsations is considerably more difficult than the measurement of electric field oscillations, and only in recent years has equipment become available which makes it possible to detect such micropulsations of normal amplitude. Magnetometers have been constructed using a wide variety of physical phenomena for detecting magnetic field changes, but only four have been used extensively up to the present time in studying the magneto-telluric field. These are the magnetic balance, the flux-gate magnetometer, the induction coil and the optical pumping magnetometer. The last is the most sensitive for long-term magnetic field variations.

Of the four methods for measuring magnetic field mentioned above, the magnetic balance is the oldest method and the simplest. A bar magnet is suspended in such a way that torque due to the magnetic field is balanced against gravity. Figure 150 shows an arrangement for measuring the horizontal component of the magnetic field with a magnetic balance. A bar magnet is

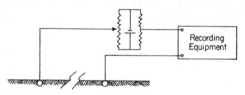

Fig. 149. Potentiometric circuit used in cancelling the static potential difference between a pair of electrodes.

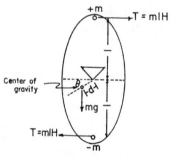

Fig. 150. Principle of a magnetic balance for measuring changes in the horizontal component of the ambient magnetic field. The torque on a magnet due to the ambient field, $2\ mlH$, is balanced against a gravitational torque, $wgd \sin\theta$, where $w$ is the mass of the magnet. The center of gravity is located below and to one side of the point of suspension.

balanced vertically on a knife edge, as shown. The magnet is free to rotate, and the torque due to the horizontal component of the ambient magnetic field would tend to rotate the magnet clockwise. The magnet is balanced in such a way that the center of gravity is below the point of suspension and to the left. The torque due to the shift of the center of gravity from the point of suspension is designed to exactly balance the torque due to the magnetic field acting on the bar magnet. If the horizontal field increases slightly, the magnet will rotate in a clockwise direction. This rotation increases the component of weight contributing to the gravitational torque, and when the magnet has rotated far enough, the two torques come into balance. The amount of rotation is very small, being less than a degree. The rotation is detected using a light beam reflected from a mirror mounted on the magnet, and the motion of the light beam is recorded on a strip of moving film.

Magnetic balances are commonly sensitive enough to provide record deflections of one millimeter for magnetic field changes of 2–50 gammas. The chief advantage of a magnetic balance is its simplicity. Disadvantages include sensitivity to temperature changes, seismic accelerations, and lack of sensitivity to short-period variations in magnetic field. The sensitivity to temperature arises from changes in the magnetic moment and the position of the center of gravity, though these can be compensated to a large degree. The sensitivity to seismic accelerations comes about since the value of $g$ is apparently changed when seismic accelerations are added to the normal gravitational acceleration. The long period is a result of the small forces being balanced, as well as the large mass of the magnet system. The natural period of sensitive magnetic balances is of the order of several hundred seconds, so magnetic variations with periods longer than this, only, can be measured.

In the flux-gate magnetometer, the fact that the magnetic permeability of a ferromagnetic material depends on the magnetic field strength is used in measuring the magnetic field strength. A typical curve relating permeability to field strength is shown in Fig. 151. The rate of change of permeability with increasing field strength is very slow for some ferromagnetic materials, very rapid for others.

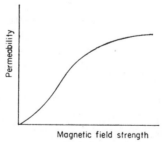

FIG. 151. Typical curve relating permeability to ambient field strength for a ferromagnetic material.

In practice a material having a much sharper transition or "knee" between the linear and the saturation region than that shown in Fig. 151 should be used. Permalloys or $\mu$ metal having a high initial permeability and a fairly narrow hysterisis loop are suitable. The length-to-diameter ratio for the core should be fairly high to avoid a great reduction in effective permeability because of demagnetization.

A number of types of flux-gate magnetometers have been developed. The following discussion is limited to the so-called "second harmonic" type. The magnetic core is wound with one or more coils and is excited by a sinusoidal alternating current of sufficient amplitude to drive the core into the saturation region of its characteristic curve. If there is an ambient field through the coil, it adds to the alternating field during half of the cycle and opposes it during the other half. As a result, the intervals during which the core is in saturation are different for the positive half-cycles than for the negative half-cycles. A Fourier analysis of the field through the coil and the voltage across the coil shows that this difference between intervals of saturation creates voltages which are even harmonics of the driving current (odd harmonics are present, even in the absence of an ambient field, provided the core is driven into its non-linear

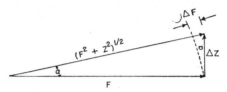

Fig. 152. Vector diagram showing the change in total field strength caused by adding a small vertical field change to a horizontally directed total field vector.

region). The second harmonic voltage appearing across the flux-gate is selected by means of a filter, amplified, rectified, and fed to a recorder. This second harmonic voltage is proportional to the ambient field, provided that the amplitude of the driving field is much larger than the field to be measured.

In order to measure small variations in the earth's field with the required accuracy, it is necessary to cancel most of the ambient field. This can be done conveniently by passing direct current through the coil in such a direction as to oppose the earth's field. Instability of this direct current is a potential source of noise which, however, usually is not serious unless very long periods are to be measured. The upper limit on frequency response usually is a few cycles per second depending on the driving frequency, the filters, the demodulation circuits, and the recorder. The maximum sensitivity which may be obtained with a flux-gate magnetometer depends on noise generated within the core material and in the associated electronic circuits. Noise can be reduced to about 0·02 gammas or less with careful design. Advantages of flux-gate magneto-meters for magneto-telluric measurements are that they measure only that component of the field parallel with the core and that they are readily available and reliable. Disadvantages are the high noise level, as compared with some other types, and long term drifts in the d.c. bias current and in the electronic circuitry.

Induction magnetometers differ from the other three types used in measuring magneto-telluric field variations in that they measure the rate of change of field strength, rather than the field itself. The e.m.f. induced in a coil of wire is simply:

$$\text{e.m.f.} = -n\frac{d\varphi}{dt} = -n\mu_0 A\frac{dH}{dt}\cos\theta \qquad (321)$$

if the coil can be assumed to have negligible resistance, inductance and capacitance ($a$ is the number of turns of wire, $A$ is the area of the coil, $\varphi$ is the total amount of magnetic flux through the coil and $\theta$ is the angle between the magnetic field, $H$, and the axis of the coil). So long as these assumptions may be made, the voltage output of an induction coil is proportional to the oscillation frequency of the magnetic field. For the low field strengths and low frequencies of interest in magneto-telluric measurements, a very large number of turns must be used to provide a measureable voltage output from a coil. From Eq. (321), we see that a coil with 50,000 turns and having a cross sectional area of one square meter will produce an output voltage of $3 \cdot 14\,\mu\text{V}$ for a magnetic field change of one gamma at one cycle per second (one gamma field change in space corresponds to an induction field change of $10^{-9}$ webers per m²). In as much as nearly 40 km of wire is required to construct such a coil, and since light-weight wire must be used to keep the coil weight down, the resistance of the wire in such a coil is not small, amounting to thousands or tens of thousands of ohms.

In addition to a large resistance, such a coil will also have an appreciable inductance and capacitance between windings. The equivalent circuit for such a coil is shown in Fig. 153. The capacity between windings will serve to reduce the output voltage from the coil at frequencies above a critical frequency

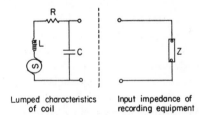

Lumped characteristics        Input impedance of
of coil                       recording equipment

FIG. 153. Equivalent circuit for an induction coil used for measuring the rate of change of a magnetic field.

determined by the time constant calculated from the resistance and capacity of the coil. For the coil described in the preceding paragraph, this time constant may be as low as 1/20 sec.

The sensitivity of an induction coil may be increased or its resistance decreased by winding the coil on a core of high permeability material. The sensitivity of such a coil is given by:

$$e = -K_c n A\frac{dB}{dt} \qquad (322)$$

where $K_e$ is the ratio by which the magnetic flux linkage through the coil is increased when the core is present. The parameter $K_e$ is usually less than the relative permeability of e material, which is the ratio of core permeability to the permeability of free space:

$$K_e \leqq K_m = \mu/\mu_0 \qquad (323)$$

This ratio may be as large as $10^5$ for special high permeability alloys, such as superpermalloy. However, because of demagnetization the effective permeability, $K_e$, for a specific core may be much less than the relative permeability given by Eq. (323) unless the form of the core is that of a long thin rod.

Figure 154 shows curves relating the effective permeability of long cylindrical cores to the ratio of length to diameter for several high permeability materials (Kalmakov and Zelentsov, 1962). Consider a short rod, one in which the ratio of length to diameter is two. These curves indicate that the effective permeability, and thus the gain in sensitivity for a coil wrapped on such a core is about 6·5, no matter whether the core material is soft iron with a relative permeability of 120 or the alloy superpermalloy with a relative permeability of 100,000. However, for a rod whose length is 1000 times as great as its diameter, the effective permeability for a soft iron core will be nearly 120 and that for a superpermalloy core will be nearly 100,000. Therefore, to take full advantage of the flux-concentrating characteristics of a high susceptibility core, a coil should be wrapped in the form of a solenoid on a long thin rod.

A solenoid consisting of 50,000 turns of wire wrapped on a permalloy rod with a diameter of about 0·75 cm (cross sectional area of about 0·5 cm$^2$) and a length of 7·5 m will have the same voltage output as the air-core coil with a cross sectional area of 1 m$^2$ described earlier. The resistance of the wire in such a coil will be much less, inasmuch as the length of wire required has been reduced by about 130 times. The coil now has a time constant about 130 times less than that for the air-core coil, and so, has a significant voltage output for magnetic field variations at higher frequencies.

It is apparent that full advantage cannot be taken of the increase in effective permeability for high permeability cores, since there is a practical limit to the length of core which may be used. For field use, a solenoid longer than a few meters becomes very difficult to use.

The maximum voltage output will be observed at a frequency:

$$\omega_{\max} = \left( \frac{1 + R/R_m}{LC} \right)^{1/2} \qquad (324)$$

where $R$, $L$ and $C$ are the resistance, capacity and inductance of the coil, and $R_m$ is the input resistance of the equipment used to measured the voltage from the coil. At this resonant frequency, the voltage output can be larger than that given by Eq. (321) depending on the load and the parameters of the coil. For frequencies below about $1/10$ the resonant frequency, the voltage output will be very nearly that computed from Eq. (321). At high frequencies, the

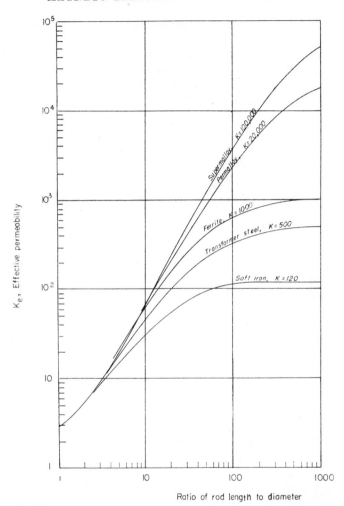

Fig. 154. The effective permeability (ratio of increase in flux density) of high permeability rods as a function of the ratio of rod length to diameter.

output will decrease in inverse proportion to increasing frequency. Thus, an induction coil is useful primarily for detecting magnetic field variations over a narrow band of frequencies, usually within half a decade of frequency change on either side of the resonant frequency for a coil.

The characteristics of a coil may be changed by changing the number of turns, the dimensions or the permeability of the core material. Changing the number of turns will not change the maximum output from a coil, as shown by the curves in Fig. 155. The output at low frequencies will be increased but the output at high frequencies will be decreased. The net result of increasing the number of turns on a coil will be to shift the output curve towards lower frequencies.

An increase in the product of effective permeability and cross sectional area, $K_eA$, will provide higher output at all frequencies, but will shift the frequency for maximum output slightly downwards. At low frequencies, the increase in output voltage will be the same as the increase in $K_eA$, but at high frequencies, the output will be increased only by half the increase in $K_eA$ (see Fig. 156).

The narrow frequency range through which an induction coil provides output voltage may be either an advantage or a disadvantage. It is a useful characteristic if one wishes to measure magnetic field variations over a narrow frequency range without interference from field variations outside this frequency range. The principle advantage of the induction coil is its high output at high frequencies. The principle disadvantages are the low output at low frequencies, the sensitivity to coil movement (seismic disturbances), and the difficulty of obtaining a precise calibration.

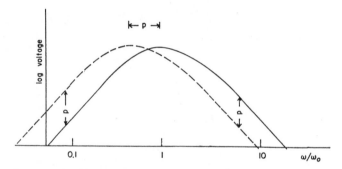

FIG. 155. Effect on output of a coil of increasing the number of turns by the ratio, $p$.

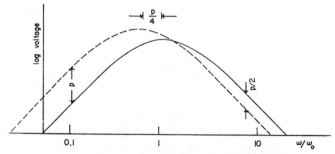

FIG. 156. Effect on the output of a coil of increasing the product $K_eA$ by the ratio, $p$.

Optical pumping magnetometers are the newest type of sensitive magneto-meter to become available, and provide the best means for measuring long period variations in magnetic field strength. Optical pumping magnetometers take advantage of a rather complicated internal energy transfer phenomenon in atoms which depends on the ambient magnetic field strength. The principle of optical pumping magnetometers has been discussed by Bloom (1960). In essence, an optical pumping magnetometer takes advantage of the energy differences between three levels of energy within an atom (see Fig. 157). The energy contained in an electron may vary in large steps, as an electron moves from a smaller orbit to a larger one, or vice versa. The energy for a transition

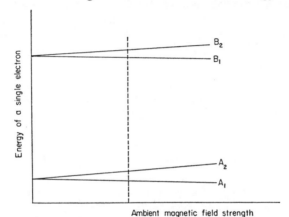

FIG. 157. Energy levels which may be contained in an electron. The energy level $A$ represents the energy most probably contained by an electron at room temperature. It may vary with magnetic field strength, depending on whether the magnetic moment of the electron is aligned with or against the ambient field. The energy level $B$ is the level occupied by an electron after it has adsorbed energy from a light beam, and before it radiates energy to return to state $A$.

from a low to a high energy level may be taken from a light beam, while on transition from a high level to a low level, light is emitted. In addition, within any one orbit, there are small differences in energy level, depending on the orientation of an electron's axis with respect to the direction of the ambient magnetic field. The stronger the ambient magnetic field, the greater will be the difference between these energy states representing the effects of orientation (see Fig. 156, showing the "splitting" of energy levels in the presence of a magnetic field).

The energy contained in an electromagnetic wave, such as light, is given by:

$$\varepsilon = hf \tag{325}$$

where $\varepsilon$ is the energy, $f$ is the frequency of oscillation and $h$ is Planck's constant $6 \cdot 624 \times 10^{-27}$ erg sec. If a tube of gas is irradiated with a monoenergetic light beam whose frequency corresponds exactly to the energy difference between one of the lower energy states and one of the higher energy states (for example, the difference between states $A_1$ and $B_1$ in Fig. 157), light will be absorbed by the gas as electrons are raised from state $A_1$ to state $B_1$. This absorption of energy may reduce the amount of light passing through the absorption cell by as much as 20 per cent. Electrons in the higher energy state will drop back to a lower state spontaneously, emitting light as they do. However, they may drop back to either state $A_1$ or $A_2$. If the electrons return to state $A_1$, they will be raised back to state $B_1$ by absorbing energy from the monoenergetic light beam. On the other hand, if the electrons go to state $B_1$, they will tend to remain there since the monoenergetic light beam contains the wrong energy to change their energy level to a permissible value. Thus, ultimately, a high proportion of electron energies will be at level $A_2$, a small proportion at level $A_1$. Thus, the monoenergetic light beam has been used to "pump" electrons from energy level $A_1$ to energy level $A_2$.

Once most of the electrons have been placed at energy level $A_2$, there is little absorption of energy from the light beam through the absorption cell. In order to measure magnetic field strength, the electrons have to be moved from level $A_2$ back to level $A_1$, with the energy required to do this being measured. The amount of energy needed to move an electron from level $A_2$ to level $A_1$ is small, and corresponds to radio-frequency oscillations rather than to optical frequencies. In order to determine the difference in energy level between states $A_1$ and $A_2$, a coil is wrapped around the absorption cell, and energized with current from a variable frequency oscillator. The frequency is varied until maximum absorption takes place in the absorption cell, and this represents the maximum transfer of electrons from state $A_2$ to $A_1$, as the monoenergetic light beam is raising them to state $B_1$. The energy between levels $A_1$ and $A_2$ is directly proportional to the ambient magnetic field strength, so the frequency for maximum absorption is also directly proportional to the magnetic field strength.

An optical pumping magnetometer consists of a monoenergetic light source, an absorption cell, a radio-frequency coil wrapped around the absorption cell and a means of measuring the radio frequency (see Fig. 158). The magneto-meters which are commercially available use either a rubidium or helium gas in the absorption cell. The light source is a tube of heated gas of the same material, so that the particular wavelength light desired is among the wave-length radiations emitted by the source. The one energy level required (that to transfer electrons from level $A_1$ to $B_1$) is separated from the other, similar wavelengths emitted by such a source ($A_1$ to $B_2$ or $A_2$ to $B_1$) with a high resolution filter. Some electronic means is provided so that the radio frequency is varied to maintain a maximum absorption through the absorption cell. The frequency is then measured with a high-accuracy frequency meter. It is not unreasonable to measure radio frequencies to within about 0·1 c/s.

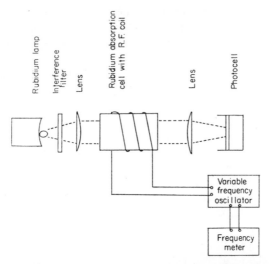

Fig. 158. Generalized block diagram of an optical pumping magnetometer. Light from a rubidium vapor lamp is filtered and passed through an absorption cell containing rubidium gas. A coil wrapped around the absorption cell is energized with a radio frequency current, with frequency continuously being varied to maintain minimum transmission of light through the absorption cell. The frequency is monitored.

The difference in energy between the states $A_1$ and $A_2$ for a rubidium gas is about 350,000 c/s in a field of 1/2 gauss intensity (the approximate strength of the earth's magnetic field), so that a resolution of 0·1 c/s in measuring frequency corresponds to a resolution of about 1/70 gamma in measuring magnetic field strength. The splitting of energy levels in helium gas is even greater, with the energy difference corresponding to a frequency of 2·8 Mc/s in a field of one gauss. A resolution of 0·1 c/s in frequency measurement corresponds to a precision of measurement for field strength of about 1/280 gamma.

Optical pumping magnetometers have a great advantage over the other types of magnetometers in the sense that the measurement of field strength is absolute, not subject to the many uncertainties of maintaining calibration that other types of magnetometers are subject to. Optical pumping magnetometers may be used to measure long-term field changes without concern for drift problems. One major disadvantage is the complexity and cost of the equipment required to obtain an accurate measurement of frequency.

In computing resistivity from magneto-telluric field data, it is necessary to have the amplitude of a magnetic micropulsation in the horizontal plane at right angles to the direction in which the electric field is measured. If changes in total field are measured, as with the optical pumping magnetometer, the vector projection of a horizontally directed micropulsation will mean that a much smaller amplitude will be measured, particularly when the dip angle of the total field is high. If the dip angle is 60°, the amplitude of a micropulsation projected onto the total field direction will be reduced to $\frac{1}{2}$. Also, variations in the vertical component of the field, which cannot be related to horizontally directed electric fields may be confused with horizontal micropulsations. Therefore, it is essential that only the horizontal component of the magnetic field be measured.

A total field magnetometer, such as an optical pumping magnetometer, may be converted to a horizontal component magnetometer by cancelling approximately the vertical component of the ambient field using either an auxiliary magnet or an induction coil. It can be shown that the vertical component need be cancelled only approximately in order to make a total field magnetometer insensitive to field variations in the vertical direction. Consider the vector diagram shown in Fig. 152. The vertical component has been cancelled completely, so that the total field vector is in the horizontal direction and has a magnitude $F$. If a vertical directed variation field, $\Delta Z$, is added to this horizontal field as shown, the total field will be tilted slightly and increased slightly. The increase in magnitude of the total field is:

$$\Delta F = \Delta Z \sin \alpha \tag{326}$$

where $\alpha$ is the angle that the vector has been tilted from the horizontal. The fractional change in total field is:

$$\frac{\Delta F}{F} = \frac{\Delta Z}{F} \sin \alpha$$

$$= \frac{\dfrac{\Delta Z^2}{F}}{(F^2 + \Delta Z^2)^{1/2}}$$

$$= \frac{\dfrac{\Delta Z^2}{F^2}}{\left(1 + \dfrac{\Delta Z^2}{F^2}\right)^{1/2}} \approx \left(\frac{\Delta Z}{F}\right)^2 \tag{327}$$

Consider the case in which the horizontal field vector has a magnitude of 30,000 gammas. A change in the vertical direction of 30 gammas (one part per thousand) will change the total field vector by only 0·03 gammas (one part per million). Therefore, the vertical field need be nulled only with a precision of one part per thousand to provide adequate insensitivity to variations in the vertical direction. It is a simple matter to determine whether the vertical field has been nulled, since a minimum total field is observed when cancellation is exact.

If only the vertical component of the ambient field is nulled, a total field magnetometer becomes a magnetometer sensitive only to variations directed in a magnetic northerly direction. The magnetometer has no sensitivity to variations in a magnetic easterly direction for the same reason it is insensitive to variations in the vertical direction. If it is necessary to measure variations in the horizontal plane which are not directed towards magnetic north, the north component of the ambient field may be cancelled with auxiliary magnets or induction coils, but a primary field must also be added which is directed in the direction in which it is wished to observe variations.

## 31. INTERPRETATION OF MAGNETO-TELLURIC DATA

Techniques for the interpretation of magneto-telluric data are not well developed, inasmuch as the method has had only limited use in recent years. The problems encountered in interpretation are of two sorts; first, the amplitudes of the magnetic and electric field variations must be determined at some specific frequency, then, the computed apparent resistivities must be compared with theoretically derived curves.

In practice, a great deal of difficulty may be encountered in trying to find accurate amplitude values for the field components at a given frequency. One approach, if the primary measurements have been recorded on magnetic tape, is to replay the information, sharply filtering the various frequencies which appear in the data. An alternate approach which is useful for broad-band recording of data on graphic records is the computation of power spectra, using techniques described by Blackman and Tukey (1958).

Once reliable values have been obtained for the amplitudes of the magnetic and electric field variations, apparent resistivity values may be computed from the equation given earlier, or from the equation:

$$\varrho_a = 0{\cdot}2T\left(\frac{E}{H}\right)^2 \tag{328}$$

where the various parameters are expressed in working units rather than mks units (period, $T$, in sec; electric field, $E$, in mV/km magnetic field, $H$, in gammas). Peak values of $E$ and $H$ are used, ignoring the phase difference between them. Phase information is rarely used, up to the present time, in view of the uncertainties introduced in phase measurements by phase shift in the measuring and filtering circuits used in recording the data.

If apparent resistivities are computed for frequencies (or periods) extending over several decades, it may be possible to use the logarithmic curve matching procedures described under the chapter on galvanic resistivity measurements. Briefly, the procedure in interpreting a set of magneto-telluric data using only the two-layer theoretical curves (Fig. 142) would be as follows:

1. The computed values of apparent resistivity are plotted on a bi-logarithmic graph, apparent resistivity being plotted as a function of the square root of period.

2. The initial part of the field curve, that recorded for high frequencies, is fitted with the theoretical curves. The origin from the theoretical curves (the point $\varrho_a/\varrho_1 = 1$, $1/h_1\gamma_1 = 1$) is noted on the field curve. From the location of this point on the field curve, the following relationships are established:

$$\varrho_a/\varrho_1 = 1 \quad \text{when} \quad \varrho_a = D$$

therefore:

$$\varrho_1 = D$$

and

$$1/h_1\gamma_1 = 1 \quad \text{when} \quad T_a = E$$

Expressing the wave number in terms of resistivity, we have:

$$h_1\gamma_1 = h_1 \left( \frac{2\pi\mu}{2\varrho_1 T} \right)^{1/2}$$

so,

$$T = \frac{h_1^2 \pi\mu}{\varrho_1} \quad \text{when} \quad T = E$$

or:

$$h_1^2 = \frac{E\varrho_1}{\pi\mu} \tag{329}$$

Thus, the resistivity of the surface layer must be determined before the thickness can be calculated.

3. The first two layers are then replaced with a single fictitious layer whose properties are determined from Hummel's equations:

$$\frac{H_f}{\varrho_f} = \frac{h_1}{\varrho_1} + \frac{h_2}{\varrho_2}$$

$$H_f = h_1 + h_2$$

Hummel's equations, which assume the conductances of the two surface layers to be electrically in parallel, are valid both for the case of the surface layers being more conductive than the lower layers or the case of the surface layer being less conductive than the lower layers. This is true, since the currents induced by the magneto-telluric field flow only in horizontal planes, and if only long wavelengths are considered, skin-depth attenuation can be neglected.

In order to determine the thickness and resistivity of the fictitious layer, it is necessary to know both the resistivity and thickness of each of the two real layers. However, we obtain values for only three of these four parameters from the initial curve match, with the thickness of the second layer not being determined. The right-hand portion of the field data might be interpreted by trial and error, assuming various values for $h_2$, calculating the parameters for the fictitious layer, and seeing if the right-hand portion of the field data can be fitted satisfactorily with the two-layer theoretical curves. The trial and error procedure can be simplified using the auxiliary curves described in Chapter III. For a given ratio of second-layer resistivity to first-layer resistivity, the locii of points for all possible values of $h_2$ forms a curve which can be plotted on bi-logarithmic coordinates for direct transfer to the field data plot. Such curves for the case in which the second layer is more conductive than the first and third layers have already been presented in Chapter III (Fig. 35). The appropriate auxiliary curve (the one with the correct ratio of resistivities for the first two layers) is traced on the field data plot, with the 1·1 point of the locus curves superimposed on the field data plot at the point where the 1·1 point from the theoretical curves fell for the initial match. Curves for Hummel's equation in the case of a resistant second layer were not given in Chapter III, since they are not used in interpreting galvanic resistivity data. The Hummel

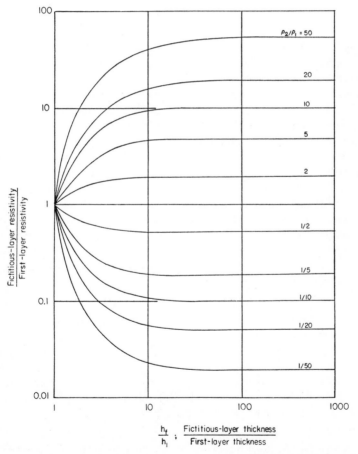

FIG. 159. Hummel's auxiliary curves for use with magneto-telluric data.

auxiliary curves for both this case and the one already covered in Fig. 85 are plotted on Fig. 159.

4. After the proper auxiliary curve has been transferred to the field data plot, this graph is again superimposed over the two-layer theoretical curves. The field data plot is moved around over the theoretical curves, keeping the 1·1 point along the auxiliary curve, until the right-hand portion of the field data matches with one of the theoretical curves. The parameter for the particular curve with which the match is made is noted, as well as the fictitious layer resistivity and thickness. Then, the third-layer resistivity and the second-layer thickness can be calculated using Hummel's equations:

$$\varrho_3 = \left(\frac{\varrho_3}{\varrho_f}\right)\varrho_f$$

$$h_2 = H_f - h_1$$

Interpretation by this curve matching procedure can be used only if wave impedances have been measured over several decades of frequency. Commonly, wave impedances can be measured only over a very limited range in frequency, either because of some inadequacy in measuring equipment, or because a wide spectrum of frequencies is not available in the natural magneto-telluric field. In each cases, it may be possible to make an interpretation using Eq. (308) or (311). This may be done if enough is known about the structure of the area where measurements have been made to ensure that the resistivity contrast between the surface layers and the basement is large, and that the wave impedances are measured for the range in periods for which the theoretical curves (Fig. 142) are rising or falling with a slope of 2. Such is commonly the case if measurements are made of the natural magneto-telluric field in the 10- to 40-sec period range over a thick sequence of conductive sedimentary rocks.

Birdichevskiy (1960) gives an example of the interpretation of magneto-telluric data under conditions where such an assumption can be made. The records of electric and magnetic field from which wave impedances were cal-

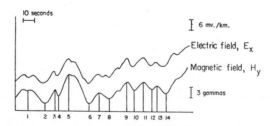

FIG. 160. Record of magneto-telluric field variations used for computing apparent resistivity (from Birdichevskiy). The numbers indicate consecutive half-periods of the oscillations picked from the record.

culated are shown in Fig. 160. The amplitudes were taken as the range recorded between one peak and the succeeding trough, while the period was taken as twice the time interval between these points. These values, along with the apparent resistivities computed from them, are listed in the following table:

| Number of event | $H_y$ gammas | $E_x$ mV/km | $T$ sec | $\varrho_a$ ohm-m | $\sqrt{T}$ |
|---|---|---|---|---|---|
| 1 | 4·55 | 6·85 | 42 | 19·1 | 6·48 |
| 2 | 3·52 | 8·55 | 30 | 35·2 | 5·48 |
| 3 | 1·97 | 3·88 | 10 | 7·8 | 3·16 |
| 4 | 7·21 | 13·7 | 23 | 16·6 | 4·80 |
| 5 | 9·62 | 19·9 | 50 | 42·8 | 7·08 |
| 6 | 3·84 | 8·05 | 17 | 15·0 | 4·12 |
| 7 | 5·76 | 13·8 | 38 | 43·8 | 6·16 |
| 8 | 3·26 | 6·39 | 16 | 12·3 | 4·00 |
| 9 | 3·12 | 6·85 | 28 | 27·1 | 5·30 |
| 10 | 2·76 | 5·41 | 16 | 12·3 | 4·00 |
| 11 | 1·49 | 3·52 | 13 | 14·5 | 3·61 |
| 12 | 2·76 | 5·02 | 12 | 8·0 | 3·46 |
| Average | | | | 21·2 | 4·39 |

The measurements had been made over a sedimentary basin in western Siberia. There is some suggestion that the value of apparent resistivity increases with period, though the data are scattered. On the assumption that the data fall along a magneto-telluric curve which is rising with a slope of 2, the conductance of the surface layers may be computed using Eq. (308):

$$S = \frac{1}{1\cdot12 \times 10^{-3}(\omega\varrho_0)^{1/2}}$$

$$= \frac{1}{1\cdot12 \times 10^{-3} \times 0\cdot571 \times 4\cdot6} = 341 \text{ mhos} \qquad (330)$$

A galvanic sounding made in this same area gave a value of 350 mhos for the conductance of the sediments.

As with the galvanic method, the magneto-telluric method is subject to errors in interpretation if a lateral change in resistivity is confused with a layered geometry. Some calculations of the effect of vertical structures such as dikes or faults, on the values of impedance computed from the magneto-telluric field strength have been given by Rankin (1962) and by d'Erceville and Kunetz (1962). Some typical curves showing the variation of apparent resistivity at various distances from a vertical fault-like boundary separating regions with different resistivity are given in Fig. 161. It is interesting to note that if the magneto-telluric field is studied on the high resistivity side of the fault, the presence of the fault has almost no effect. If the field is measured on the low resistivity side, close to the fault, the computed resistivities may be much less than the true resistivity. At the fault, the ratio of apparent resistivity to the resistivity on the resistant side equals the square of the resistivity contrast.

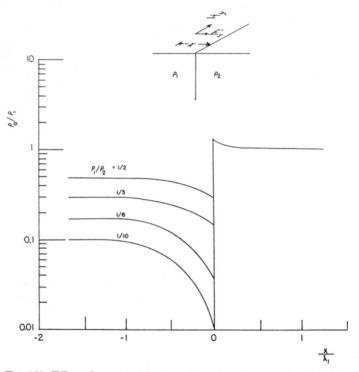

FIG. 161. Effect of a vertical fault on the value of apparent resistivity computed from magneto-telluric data (from d'Erceville and Kunetz).

# REFERENCES

ALFVÉN, H. and FALTHAMMA, C.-G. (1963) *Cosmical Electrodynamics*, Clarendon Press, Oxford, 228 p.

BENIOFF, H. (1960) Observations of geomagnetic fluctuations in the period range, 0·3 to 120 sec, *J. Geophys. Research* **65** (5), 1413–1422.

BERDICHEVSKIY, M. N. (1960) (Electricheshagen razvedka metodom telluricheskikh tokov), Electrical exploration with telluric currents (in English translation), *Quart. Colo. School Mines*, Jan., 1965.

BLACKMAN, R. B. and TUKEY, J. W. (1958) *The Measurement of Power Spectra*, Dover Publications, New York, 190 p.

BLOOM, A. L. (1960) Optical pumping, *Sci. American* **203** (4), 72–97.

GAGNIARD, L., (1953) Basic theory of the magneto-telluric method of geophysical prospecting, *Geophysics* 18 (3), 605–635. (See also, discussion by J. R. Wait of this article.)

CAMPBELL, W. H. (1959) Studies of magnetic field micropulsations with periods of 5 to 30 sec. *J. Geophys. Research* 64 (11), 1819–1826.

CAMPBELL, W. H. (1960) Natural electromagnetic noise below the ELF range, *J. Research Natl. Bur. Standards* D 64 (4), 409–412.

CANTWELL, T. and MADDEN, T. R. (1960) Preliminary report on crustal magneto-telluric measurements. *J. Geophys. Research* 65 (12), 4202–4205.

CHAPMAN, S. and BARTELS, J. (1940) *Geomagnetism*, vol. 1, Clarendon Press, Oxford, 542 p.

D'ERCEVILLE, I. and KUNETZ, G. (1962) The effect of a fault on the earth's natural electromagnetic field, *Geophysics* 27 (5), 651–665.

DUFFUS, H. J. SHAND, J. A., WRIGHT, C. S., NASMYTH, P. W. and JACOBS, J. A. (1959) Geographical variations in geomagnetic micropulsations, *J. Geophys. Research* 64 (5), 581–584.

FLEMING, J. A. (1949) *Terrestrial Magnetism and Electricity*, Dover Publications, New York, 794 p.

GREEN, A. W., JR., LIST, B. H. and ZENGEL, J. F. P. (1962) The theory, measurement and applications of very-low-frequency magneto-telluric variations, *Proc. I.R.E.* 50 (11), 2347–2363.

HAZARD, D. L. (1947) *Directions for Magnetic Measurements.* U.S. Dept. of Commerce, Coast and Geodetic Survey, Serial 166, 129 pp.

HORTON, C. W. and HOFFMAN, A. A. J. (1962) Power spectrum analysis of the telluric field at Tbilisi, U.S.S.R., *J. Geophys. Research* 67 (9), 3369–3372.

JACKSON, C. M., WAIT, J. R. and WALTERS, L. C. (1962) Numerical results for the surface impedance for a stratified conductor, National Bureau of Standards, Technical Note 143, Dept. of Commerce, Office of Technical Services, Washington, D.C.

JACOBS, J. A., LOKKEN, J. E. and WRIGHT, C. S. (1963) Notation and classification of geomagnetic micropulsations, *J. Geophys. Research* 68 (14), 4373–4374.

KOLMAKOV, M. V. and ZALENTSOV, I. A. (1962) The construction of induction pick-ups for magneto-telluric investigations, *Bull. Acad. Sci. U.S.S.R., Geophys. Ser.*, no. 10, pp. 860–867.

MATSUSHITA, S. (1963) On the notations for geomagnetic micropulsations, *J. Geophys. Research* 68 (14), 4369–4372.

NAWROCKI, P. J. and PAPA, R. J. (1963) *Atmospheric Processes*, Prentice-Hall, Englewood Cliffs, N.J., 674 p.

NEVES, A. S. (1957) The generalized magneto-telluric method, Ph.D. Thesis, Mass. Inst. of Tech., 106 pp.

NIBLETT, E. R. and SAYN-WITTGENSTEIN, C. (1960) Variation of electrical conductivity with depth by the magneto-telluric method, *Geophysics* 25 (5), 998–1008.

PRICE, A. T. (1962) The theory of magneto-telluric methods when the source field is considered, *J. Geophys. Research* 67 (5), 1907–1918.

RAEMER, H. R. (1961) On the spectrum of terrestrial radio noise at extremely low frequency. *J. Research Natl. Bur. Standards* D 65, 541–543.

RANKIN, D. (1962) The magnetotelluric effect of a dike, *Geophysics* 27 (5), 666–676.

SANTIROCCO, R. A. and PARKER, D. G. (1963) The polarization and power spectrums of $P_c$ micropulsations in Bermuda, *J. Geophys. Research* 68 (19), 5545–5558.

SMITH, H. W., PROVAZEK, L. D. and BOSTICK, F. X., JR. (1961) Directional properties and phase relations of the magneto-telluric fields at Austin, *J. Geophys. Research* 66 (3), 879–888.

SRIVASTAVA, S. P., DOUGLASS, J. L. and WARD, S. H. (1963) The application of the magneto-telluric and telluric methods in central Alberta, *Geophysics* 28 (3), 426–446.

VOZOFF, K., HASEGAWA, H. and ELLIS, R. M. (1963) Results and limitations of magneto-telluric surveys in simple geologic situations, *Geophysics*, 28 (5), 778–792.

WAIT, J. R. (1953b) Receiving properties of a wire loop with a spheroidal core, *Can. J. Tech.* **31**, 9–14.

WAIT, J. R. (1961) Theory of magneto-telluric fields, *J. Research Natl. Bur. Standards* D **66**, 509–541.

WARD, S. H. (1959), AFMAG—airborne and ground, *Geophysics* 24 (4), 761–789.

WESCOTT, E. M. and HESSLER, V. P., (1962) The effect of topography and geology on telluric currents, *J. Geophys. Research* 67 (12), 4813–4823.

YUNGUL, S. H. (1961) Magneto-telluric three-layer sounding curves, *Geophysics* **26** (4), 465–473.

# CHAPTER 5

# TELLURIC CURRENT METHODS

THE magneto-telluric method for measuring earth resistivity requires a very difficult measurement—the measurement of magnetic field variations with amplitudes as small as a hundredth of a gamma. The telluric method for studying changes in earth resistivity utilizes the same natural magneto-telluric field as a source of power, but requires only that the electric field component of the field be measured simultaneously at several locations. In this way, the need to measure small changes in magnetic field strength is avoided, but in simplifying the measurement technique, the possibility of determining earth resistivities in absolute terms is also surrendered. The telluric method may be used to study ratios of resistivities between locations, but cannot be used to determine the absolute value of resistivity, anywhere, without auxiliary information.

The telluric method can logically be thought of as a special application of the more general magneto-telluric method, but in fact, the telluric method was developed prior to the development of the theory for the magneto-telluric method. The telluric method was first applied in exploration geophysics during the 1930's by Conrad and Marcel Schlumberger, and has been used extensively in Europe and Africa, particularly since 1946. Between the years 1941 and 1955, 565 crew-months of effort were devoted to telluric current surveys by the Compagnie Generale de Geophysique of Paris, France (d'Erceville and Kunetz, 1962). The telluric current method was first used in routine exploration in the U.S.S.R. in 1954, and by 1959, the extent of use had grown to the point where 24 exploration teams were in the field, covering an area of 120,000 km$^2$ per year (Birdichevskiy, 1960). Use of the method in the United States has been very limited—23 crew-months of the work reported by Compagnie Generale de Geophysique was done in the United States. The relatively small usage of the method in the United States reflects the fact that the telluric method is primarily a reconnaisance method, used to best effect in areas where geology is poorly known.

## 32. THEORY FOR THE TELLURIC CURRENT METHOD

The development of the telluric current method has been based on the assumption that the electric field measured at the surface of the earth with a pair of electrodes is developed by a sheet of direct current flowing in a conductive

251

layer of sedimentary rocks covering an insulating basement. The total current flowing in the sediments must therefore be conserved, and the current density should be inversely proportional to the thickness of the sedimentary cover. The electric field measured at the surface will be proportional to the resistivity of the sediments (Ohm's law) and inversely proportional to the thickness of conductive rocks. Remembering that the ratio of thickness to resistivity has been defined as the longitudinal conductance, $S$, of a sequence of rocks, this means that the measured electric field is inversely proportional to the conductance of the sedimentary column.

In practice, it is impossible to separate the voltage due to direct current flow in the ground from voltage due to differences in electrode potential between the electrodes used. Instead, earth voltages caused by micropulsations in the frequency range from 0·1 to 0·01 c/s are measured under the assumption that the behavior of the current at these frequencies will not differ significantly from the behavior of a direct current. The limitations imposed by this assumption can be checked using the theory developed for the magneto-telluric method.

Ordinarily, the resistivity contrast between a column of sedimentary rocks and the basement complex is large, with the basement resistivity being 10 to 1000 times greater than the resistivity of the sediments. If the contrast is sufficiently high, it may be assumed that the apparent resistivity computed from magneto-telluric measurements would fall along a line rising with a slope of 2 on the theoretical curves in Fig. 142. If the contrast between sediment and basement resistivity is of the order of 1000 to 1, the magneto-telluric resistivity will differ but little from this line (see Fig. 162) over a frequency range of about two decades. If the contrast is only 100 to 1, the departure of the apparent resistivity from the asymptotic curve is significant, while for a contrast of only 10 to 1, no portion of the magneto-telluric resistivity curve approaches a slope of 2.

The locus of the asymptotic curve which rises at a slope of 2 is given by Eq. (308):

$$S_1 = \frac{1}{[\omega\mu\varrho_a]^{1/2}}$$

In terms of electric and magnetic field strengths, this equation is:

$$-i\left(\frac{E}{H}\right)^2 = \left(\frac{1}{S_1}\right)^2 \tag{331}$$

The ratio $E/H$ is inversely proportional to the conductance of the surface layer if the contrast in resistivities is high enough, and the ratio $E/H$ does not

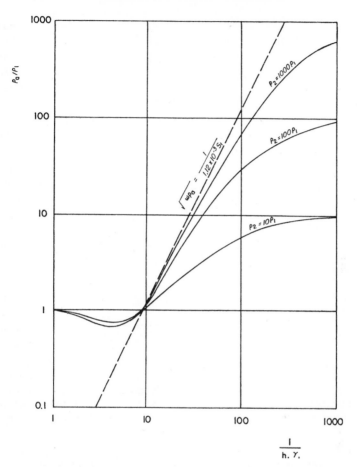

$$\frac{1}{h. \, \gamma.}$$

FIG. 162. Magneto-telluric resistivity curves for the high contrasts in resistivity required for the application of the telluric current method.

depend on frequency, if the frequency is such as to cause the parameter $1/h_1\gamma_1$ to fall between the limits 10 and 100. Since in the direct current approach, it is argued that the electric field alone, as well as the ratio $E/H$, is inversely proportional to the conductance of the surface layer, the magnetic field component of the magneto-telluric field must be independent of variations in the conductance of the surface layer. This can be shown readily, applying Gauss' law. Consider the case shown in Fig. 163; a surface layer of variable thickness, and therefore, variable conductance, is underlain by an infinitely resistant basement complex. According to Stokes' law, a line integral of the magnetic field normal to the plane in which the line integration is performed is proportional

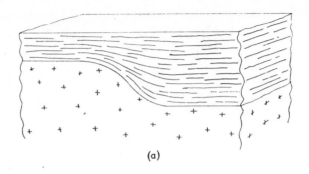

(a)

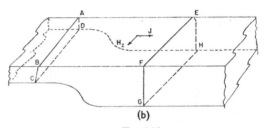

(b)

FIG. 163.

(a) Earth model assumed in developing the theory for telluric current methods.
A section of sedimentary rocks with variable thickness and conductance
covers basement rock which is infinitely resistant.

(b) Integration paths for the application of Gauss' law to show that the horizontal component of the magnetic field does not depend on the thickness or
conductance of the sedimentary cover.

to the total current flowing across the area outlined by the path of integration:

$$\oint_{ABCD} j \cdot \mathrm{d}A = \frac{1}{4\pi} \int_{ABCD} H_x \, \mathrm{d}l = \frac{1}{4\pi} \left[ \int_A^B H_x \, \mathrm{d}x + \int_C^D H_x \mathrm{d}x \right] \qquad (332)$$

where $j \cdot \mathrm{d}A$ is the total current flowing through the area outlined by $ABCD$
in Fig. 163(b). The component of magnetic field along the path of integration
is $H_x$ for the two path segments $AB$ and $CD$, while along the path segments
$BC$ and $DA$, there is no component of $H_x$ in the direction of the integration
path. The total contribution to the integral is from the path segments $AB$
and $CD$.

Consider that the total magnetic field consists of two portions; a static
portion which corresponds to the earth's primary field, and an anomalous
portion caused by current flowing in the conductive surface layer. For the
purposes of this discussion, it may be assumed that the static field is everywhere

the same, and may be ignored by subtracting a constant value from all measurements. In practice, this is done in field measurements since only the amplitude of long period micropulsations is measured, and the total field is not.

The physical problem shown in Fig. 163 may be simplified to the extent that there is uniform current density in the sedimentary layer at regions $ABCD$ and $EFGH$, so that the value for the horizontal component of the magnetic field does not vary along the path segments $AB$, $CD$, $EF$ or $GH$. Moreover, in view of the symmetry, the magnetic field along path $CD$ should be equal in magnitude but opposite in direction to the magnetic field along path $AB$. Therefore, Eq. (332) becomes:

$$\int_{ABCD} j \cdot \mathrm{d}A = \frac{1}{2\pi} \overline{AB} \, H_{x,1} \qquad (333)$$

and by a similar argument, integration along path $EFGH$ yields:

$$\int_{EFGH} j \cdot \mathrm{d}A = \frac{1}{2\pi} \overline{EF} \, H_{x,2} \qquad (334)$$

If the two paths of integration are taken far enough from the change in thickness of the surface layer, it may be assumed that the horizontal component of current is the same at both locations, so that:

$$H_{x,1} = H_{x,2} \qquad (335)$$

Thus, it has been proved that with the proper restrictions, the magnetic field due to current flow in the sedimentary column is constant, not depending on the thickness of the column.

If, instead of having the thickness of the column change, we consider a model in which only the resistivity of the sedimentary rocks changes in the $y$ direction (in the direction of current flow), the current flow is always parallel to the surface of the earth, and the equality of magnetic fields measured at different locations holds exactly.

The model considered so far is highly restrictive; no variations in thickness of the conductive layer are permitted near the point of measurement, and the structure must be two dimensional. If the current spreads in the $x$ direction (normal to the direction of current flow assumed in Fig. 163), then the path of integration $EF$ need not necessarily be the same as the path of integration $AB$, and the magnetic field components at these two locations will not be equal. However, this simplified model does show that in some cases, the magnetic field need not be measured, and changes in conductance of a sedimentary column may be equated approximately to the inverse ratio of measured electric fields.

In practice, the telluric current method requires that variations in natural electric field be measured simultaneously at two points, one of which is usually chosen to be a base station, and which is maintained at one location while measurements are made at a series of temporary field stations throughout the area where a survey is being conducted. Since neither the direction of the telluric current vector nor the trend of regional structure is known ahead of time in most cases, two components of electric field are measured at both the base station and the field stations, so that the direction of the total electric field vector can be computed. Preferably, the directions in which electric field are measured should be orthogonal, so that the resultant electric field vector can be computed with a minimum of difficulty.

The restriction that there can be no changes in thickness of conductive layers is unnecessarily restrictive, and electric field variations may be readily computed for constant current sheets flowing in conductive layers whose thickness varies in a variety of ways. These problems are usually solved by the method of conformal transformation. In this approach, a two-dimensional structure is mapped in a complex plane, $z = x + iy$. The electric field vector in the complex plane is then:

$$E_z = E_x + iE_y \tag{336}$$

The complex potential function is defined as:

$$U_z = V + iU \tag{337}$$

where $V$ is a functional description of current density and $U$ is a functional description of the electric potential. These two functions must satisfy the Cauchy–Rieman criterion:

$$\frac{\partial U}{\partial x} = \frac{\partial V}{\partial y} ; \quad \frac{\partial U}{\partial y} = -\frac{\partial V}{\partial x} \tag{338}$$

in order that the potential and current contours will be orthogonal to each other, as is required by our physical knowledge of the problem. The electric field vector (336) may be expressed in terms of the complex potential function as follows:

$$
\begin{aligned}
E_z &= -\frac{\partial U}{\partial x} + i\frac{\partial U}{\partial y} \\
&= -\frac{\partial U}{\partial x} - i\frac{\partial V}{\partial x} \\
&= -i\left[\frac{\partial V}{\partial x} - i\frac{\partial U}{\partial x}\right] \\
&= -i\frac{\partial U_x}{\partial x} \tag{339}
\end{aligned}
$$

where $U_z$ is thus defined as a modified complex potential.

The significance of a conformal transformation is shown by Fig. 164. If current flows in a conductive layer of varying thickness, the potential function

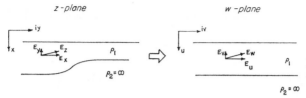

FIG. 164. Significance of conformal transformation in calculating potential functions.

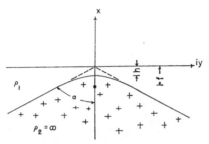

FIG. 165. Representation of an anticlinal basement structure with a hyperbolic surface.

may be found by making an appropriate transformation of variables from the $xy$-plane into a $uv$-plane which distorts the layer until it has a uniform thickness. Commonly, a boundary value problem can be solved by transforming from the $z = x + iy$ plane to the $w = u + iv$ plane. In simple cases, it is possible to determine the complex potential function $U_z$ without transformation to the $w$-plane, but in general, the voltage gradient, $E_w$, must be determined in the $w$-plane and then transformed back to the $z$-plane. This inverse transformation is simply:

$$E_z = -i \frac{\partial \bar{U}_x}{\partial x} = -i \frac{\partial U_z}{\partial w} \frac{dw}{dz} = E_w \frac{dw}{dz} \tag{340}$$

One problem which may be solved with conformal transformation techniques is that of current flow in a layer over an anticlinal uplift in the surface of the basement complex (see Fig. 165). Symmetrical anticlines can be approximated with boundaries having the equation for a hyperbola:

$$\frac{x^2}{h^2} - \frac{y^2}{f^2 \sin^2 \frac{\alpha}{2}} = 1 \tag{341}$$

where $f$ is the the depth to the focal point of the hyperbolic basement surface, $h$ is the depth to the crest of the anticline and $\alpha$ is the angle between the flanks of the anticline and the vertical. The angle $\alpha$ is given by:

$$\alpha = 2 \arctan \sqrt{\left[\left(\frac{f}{h}\right)^2 - 1\right]} \tag{342}$$

The complex potential function in the $z$-plane is known to be:

$$\bar{U}_z = \frac{2I\varrho}{\pi - \alpha} \arcsin \frac{z}{f} \tag{343}$$

where $I$ is the total current flowing in a slice of unit width normal to the structure.

The electric field, computed from the first relationship in Eq. (339) is:

$$E_z = -i \frac{\mathrm{d}\bar{U}_x}{\mathrm{d}x} = -\frac{\cos t - i \sin t}{[4x^2y^2 + (f^2 - x^2 + y^2)^2]^{1/4}} \frac{2iI\varrho}{\pi - \alpha} \tag{344}$$

where:

$$t = \frac{1}{2} \arctan \frac{2xy}{f^2 + y^2 - x^2}$$

The horizontal component of $E_y$, which is the only component which can exist at the earth's surface is the imaginary part of this expression (from Eq. 336):

$$E_y = -\frac{2I\varrho}{\pi - \alpha} \frac{\cos t}{[4x^2y^2 + (f^2 - x^2 + y^2)^2]^{1/4}} \tag{345}$$

At the earth's surface, $x = 0$, so:

$$E_y = \pm \frac{2I\varrho}{\pi - \alpha} \frac{1}{(f^2 + y^2)^{1/2}} \tag{346}$$

Several electric field profiles calculated for various dip angles for the anticlinal flanks are shown in Fig. 166. These have been normalized to a base station located directly over the crest of the structure. As might be expected from the symmetry of the problem, the maximum ratio of electric fields is found when one of the two observation stations is located directly over the crest of the anticline. It is interesting to note Eq. (346) may be extended to the case of a uniformly dipping structure by considering that the offset distance, $y$, at which electric fields are being measured is large compared to the focal depth, $f$. Under these circumstances, Eq. (346) becomes:

$$E_y = \frac{2}{\pi - \alpha} \cdot \frac{I\varrho}{y} \frac{1}{\left[\left(\frac{f}{y}\right)^2 + 1\right]^{1/2}} \approx \left(\frac{2}{\pi - \alpha}\right) \frac{I\varrho}{y} \tag{347}$$

Considering the ratio of two electric field strength measurements, at a distance $\Delta y$ apart, we have:

$$\frac{E_{y,2}}{E_{y,1}} = \frac{y_1}{y_1 + \Delta y} = \frac{1}{1 + \frac{\Delta y}{y}} \tag{348}$$

The ratio of electric fields is independent of dip angle, and depends only on the distance from the outcrop trace of the rising basement surface, $y$.

Another problem which may be solved by the method of conformal transformation is that of a vertical step-like displacement in the surface of the basement complex, such as a fault might cause. We will assume that the rock above this basement complex has a uniform resistivity, $\varrho$, and that a sheet of current is flowing normal to the step with a current density, $I$, per meter width. The structure is shown plotted in the complex $z$-plane in Fig. 167.

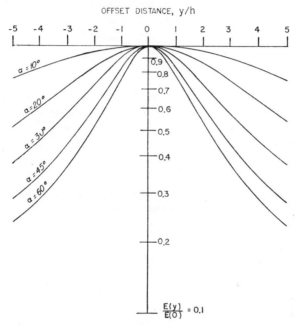

FIG. 166. Electric field variations caused by a sheet of current crossing a hyperbolic anticlinal structure of the basement surface.

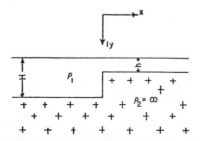

FIG. 167. Representation of a step offset of the basement surface.

The conductive area in Fig. 167 may be transformed into the upper half of a $uv$-plane using Swartz's transformation. This leads to a solution for the electric field at the earth's surface of the form:

$$E_x = \frac{I\varrho}{\pi c} \left[ \frac{u + \left(\dfrac{H}{h}\right)^2}{u + 1} \right]^{1/2} \tag{349}$$

where $c$ is a constant introduced in Swartz's transformation. This constant may be evaluated by considering the electric field at a great distance from the step. In so doing, the constant is found to be:

$$c = \frac{H}{\pi}$$

In order to be of use, Eq. (349) must be expressed in terms of the $xy$-plane instead of the $uv$-plane. The relation between $x$ and $u$ is:

$$x = H \left[ \frac{u + 1}{u + \left(\dfrac{H}{h}\right)^2} \right]^{1/2} \tag{350}$$

Curves computed from Eqs. (349) and (350) for various step heights are shown in Fig. 168. It is interesting to note that the presence of the step has virtually no effect if the electric field is measured on the upthrown side, but that the

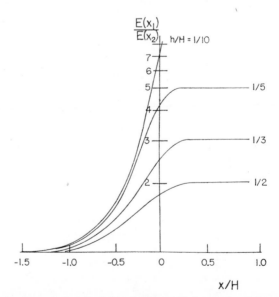

Fig. 168. Profiles of electric field which would be observed for a sheet of current flowing across a step rise in the basement surface.

transition in electric field is gradual if measurements are made on the down-thrown side of the step.

Curves have been computed for a variety of basement relief forms which can be represented by simple formulas, but the two examples shown in the previous paragraphs are adequate to show the behavior of telluric currents in a sedimentary column of varying thickness.

So far, we have considered only the effect of gross changes in the character of the sedimentary column. We should also consider what effect small, near-surface contrasts in resistivity may have on the measurement of telluric electric fields. One problem which can be handled with the method of conformal trans-

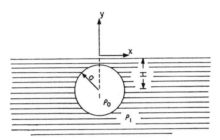

FIG. 169. The parameters defining a right-circular cylinder in a transverse telluric current field.

formations, thus yielding an exact result, is that of a horizontal circular cylinder with current flowing normal to its axis. The geometry is shown in Fig. 169. The solution to this problem has been given by Tikonov (1942) and is of the form:

$$E_x = j\varrho_1 \left[ 1 + 8 \sum_{m=1}^{\infty} \beta^m \frac{\gamma^{2m}}{(1 + \gamma^{2m})^2} \frac{\left[1 + \left(\dfrac{a}{H}\right)^2\right]\left[1 - \left(\dfrac{a}{H}\right)^2 - \dfrac{(1 - \gamma^{2m})^2 x^2}{(1 + \gamma^{2m})^2 H^2}\right]}{\left[1 - \left(\dfrac{a}{H}\right)^2 + \dfrac{(1 - \gamma^{2m})^2}{(1 + \gamma^{2m})^2}\left(\dfrac{x}{H}\right)^2\right]^2} \right]$$

(351)

where:

$$\beta = \frac{\varrho_0 - \varrho_1}{\varrho_0 + \varrho_1}$$

$$\gamma = \frac{H}{a} - \left[\left(\frac{H}{a}\right)^2 - 1\right]^{1/2}$$

The parameter $\beta$ is merely the reflection coefficient defined earlier under galvanic resistivity methods. The current density used in Eq. (351), $j$, is the normal current density which would be present if the cylinder were not disturbing the pattern of current flow. The expression for electric field depends only on three dimensionless parameters, if ratios between two observation points are considered. These parameters are the contrast in resistivity between the cylinder and the surrounding rock, the ratio of cylinder radius to the depth of burial, $a/H$, and the ratio of the offset distance at which measurements are made to the depth of burial, $x/H$. Profiles for electric field ratios across cylinders buried at various depths are shown in Fig. 170. The variation in electric field may be quite large if the cylinder differs considerably in resistivity from the surround-

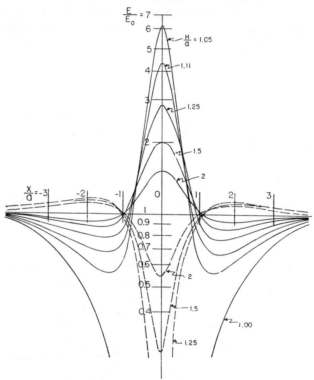

Fig. 170. Curves showing the variation of electric field intensity for current flow normal to a cylinder. Solid curves apply for a resistant cylinder; dashed curves apply for a conductive cylinder.

ing rock, and if it is not buried deeply. However, it should be noted that in practice, electrode separations ranging from 300 m to 1 km are used in measuring electric fields. With such electrode separations, the probability that one electrode will be directly over a cylinder decreases with decreasing size of the cylinder when the cylinder is much smaller in dimensions than the electrode separation.

The variations in electric field caused by bodies of other shapes, such as spheres and ellipsoids, may be computed, but the example of the cylinder is sufficient to show the degree to which near-surface features may effect the telluric electric field.

In all the examples considered so far, it has been assumed that both the direction of current flow and the direction in which electric field measurements are made are normal to the strike of a two dimensional structure. This assumption in no way limits the generality of the curves given in Fig. 166, 168 and 170. In general, the direction in which the induced current generated by magnetic field microspulsations will flow is arbitrary.

Consider the geometry shown in Fig. 171, a fault-like step in the surface of an infinitely resistant basement complex. If the point 0 is located far to the right of the step, a varying magnetic field will generate the same current density in any direction in the conductive layer, and the measured electric field will be independent of the direction of current flow. If telluric voltages are observed over a long period of time, and if a magnetometer is provided to measure the amplitude of magnetic field fluctuations, it would be possible to plot the electric field generated per gamma of magnetic field as a function of the direction of the magnetic field vector. Since the induced current is independent of the direction of the inducing field, such a plot is a circle, as shown in Fig. 172(a).

If the point of observation is only a short distance to the right of the step, the electric field will depend on the direction in which the magnetic field induces current flow. For the component of current flow induced in the direction normal to the step, the curves shown in Fig. 168 will apply; the electric field will be larger than that observed far to the right of the step. On the other hand, the electric field due to current induced parallel to the step will be the same a short distance to the right of the step as at a large distance to the right of the step.

For the location a short distance to the right of the step, we may consider an inducing magnetic field variation by separating it into components directed along the direction normal to the step and the direction parallel to the step. If the direction of the inducing field vector makes an angle $\alpha$ with the direction normal to the step, the two components of electric field induced in the ground will be:

$$E_x = aH_0 \cos\alpha$$
$$E_y = bH_0 \sin\alpha \tag{352}$$

where $a$ and $b$ are the constants of proportionality relating induced electric field intensity to the strength of the inducing magnetic field along the two principle directions. The total electric field is:

$$E_{\text{total}}^2 = a^2 H_0^2 \cos^2\alpha + b^2 H_0^2 \sin^2\alpha$$
$$\left(\frac{E_1}{H_0}\right)^2 = a^2 \cos^2\alpha + b^2 \sin^2\alpha \tag{353}$$

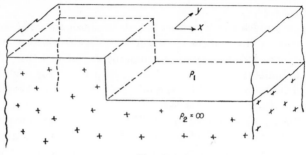

Fɪɢ. 171.

(a)

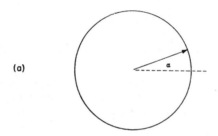

(b)

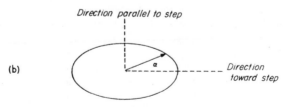

Fɪɢ. 172.

(a) Amplitude of the electric field induced per gamma amplitude magnetic micropulsation as a function of the direction of the inducing field for a measurement location far to the right of the step in Fig. 171.

(b) Amplitude of the electric field induced per gamma amplitude magnetic micropulsation as a function of the direction of the inducing field for a measurement location a short distance to the right of the step in Fig. 171.

This last equation is the equation for an ellipse with semi-axes of lengths $a$ and $b$. In general, then, we shall expect that an inducing magnetic field of constant strength but arbitrary direction would generate an elliptical pattern of electric field strengths, such as that shown in Fig. 172(b).

## 33. ANALYSIS OF TELLURIC CURRENT MEASUREMENTS

The object in analysis of telluric current measurements is the determination of patterns such as those shown in Fig. 172. The observed data consist of pairs of records made simultaneously at two locations, one a base station and the other a field station. These stations may be located only a few kilometers apart, or they may be tens of kilometers apart. If the data from only one station were to be considered, there would be no information on the strength of the inducing magnetic field variations, and it would not be possible to separate the effect of changing field strength from the effect of geologic control on the measured electric field. If the position of the tip of the electric field vector were to be plotted as a function of time, it would describe a very complicated path, such as that shown in Fig. 173.

The use of a fixed base-station serves to remove the uncertainty about source field strength. It is assumed that the variation in the inducing magnetic field will be of the same intensity at the same instant at both the base station and at a field station, and any difference in amplitude of the observed electric fields will be due to differences in the wave impedance of the earth between the base station and the field station.

At any given instant, the electric field vector at the base station and at the field stations may be represented by a system of linear homogeneous equations:

$$E_{\text{BASE}} = aE_x\mathbf{i} + bE_y\mathbf{j}$$
$$E_{\text{FIELD}} = cE_u\mathbf{i} + dE_v\mathbf{j} \tag{354}$$

where $aE_x$ and $bE_y$ are the amplitudes of the changes in electric field strength measured at that instant at the base station and $cE_u$ and $dE_v$ are the amplitudes of the same orthogonal components of the electric field measured at the field station at the same instant. The constants $a$, $b$, $c$ and $d$ are determined both by the strength and polarization of the primary inducing field and by the electrical properties at the base and field stations.

If the measured electric fields are indeed induced by a constant primary electromagnetic field variation, then there are certain characteristic patterns of behavior followed by the two electric field vectors described in Eq. (354):

1. The ratio of amplitudes of a single component at both the base station and the field station (for example, the north component measured at each station) will be a constant, no matter what the orientation, of the complete electric field vector, and no matter what the amplitudes.

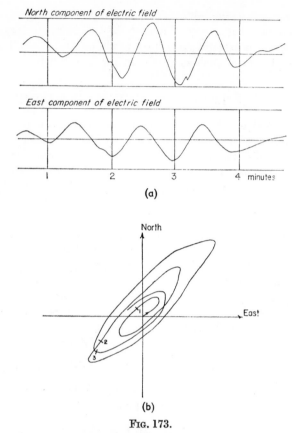

FIG. 173.

(a) Recording of two components of earth voltages induced by magnetic micropulsations.

(b) Plot of the curve traced out by the tip of the electric field vector recorded in Fig. 173(a).

2. The total electric field vectors at the two stations need not be in constant ratio, but in general, the ratio between the two electric field strengths will depend on the direction of polarization of the primary field.

3. The direction of the total electric field vector need not, and generally will not be the same at both the field station and the base station.

4. For a sequence of electric field vectors oriented along different azimuths at the base station, the same sequence of electric field vectors recorded at a field station will lie in the same angular order (if the third of a set of electric field vectors falls between the second and fourth of the set at the base station, it will also fall between the second and fourth at the field station, even though the exact direction of each vector differs from base station to field station).

In interpreting electric field measurements, it would be desirable to determine electric field patterns such as those shown in Fig. 171. If the electric field intensity per gamma primary field variation could be determined as a function of azimuth, this ratio would be proportional to the square root of apparent resistivity for each corresponding direction. However, the advantage of the telluric method lies in the fact that the magnetic field intensity need not be measured. Rather, the ratio between the amplitudes of the electric field intensity measured simultaneously at the base and field stations is formed.

A typical set of records obtained with the telluric current method is shown in Fig. 174. These records show the usual micropulsation activity in the frequency range from 0·02 to 0·1 c/s which is used in telluric current surveys. Records such as these must be obtained over a long enough time interval that

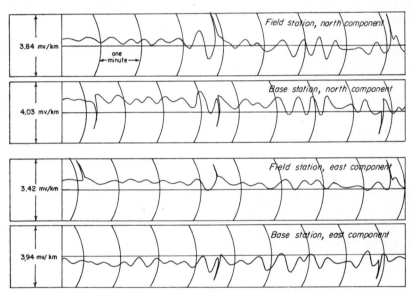

FIG. 174. Set of electric field measurements made during a telluric current survey.

correlatable events are noted on both the base and field station records, and preferably long enough that several such correlatable events representing several different primary field directions are noted. The minimum time interval required varies with the activity of the magnetotelluric field; during active intervals, excellent records may be obtained within a few tens of minutes; during quiet periods, records may have to be made over an interval of several hours. Even with the longer recording period, it may not be possible to obtain a variety of primary field directions. Frequently, the primary magneto-telluric field may have a persistent polarization over very long periods. Under such conditions, it may not be possible to obtain information for a variety of azimuths within a reasonable length of time.

In comparing records between a base station and a field station, it is extremely important that a precise time correlation can be made. The precision of time comparison between the two records should be within one second, or somewhat better. A variety of time correlation techniques are available, with the choice between techniques depending on operating conditions:

1. Direct radio communication between the base station and each field station may be used. With this method, the operator at the base station and the operator at the field station may place fiducial marks on the records at the same instant, choosing the instant to be close to the arrival of telluric current events which appear to be useful in analysis. However, radio communication over the longer distances which may be used in telluric current surveys may not be sufficiently reliable, so this method is used primarily when the separation between base and field stations does not exceed 10 to 20 miles.

2. Time marks may be made on the records using radio time signals from a station such as WWV. This requires that the survey be conducted in areas where reliable reception of such radio time signals can be obtained during the daylight hours, when the maximum amplitude telluric-current events occur.

3. Precise chronometers may be used at both the base station and the field stations, with fiducial marks being made on the records at prearranged times. Excellent chronometers are available which have a precision far better than one second per day. However, to avoid loss of data from catastrophic chronometer failures, such chronometers should be checked against primary time standards at frequent intervals. Such primary time standards usually may be provided by radio time-stations, even in areas where reception is unreliable, since only intermittent reception is necessary to provide chronometer checks.

4. If only approximate timing can be marked on the records, such as might be provided by a good watch, more exact time comparison between records may usually be obtained by correlating the arrival of high-frequency sferics. However, such a correlation technique requires that there be a few sferics, but not too many, for exact correlation. In practice, time correlation of sferics should be used only as a check on the validity of a more reliable timing system.

It is not possible to separate the long-term electric field caused by steady state current flow from the voltage due to differences in electrochemical potential at the measuring electrodes, so only differences in electric field may be read from records such as those shown in Fig. 174. In so doing, it is necessary that the base level from which these differences are measured is the same on the record from the field station as on the record from the base station. In addition, it is essential that there be no change in the long-term or electrochemical potential differences over the time interval during which electric fields are compared.

In selecting amplitude values from a set of telluric current records, a set of micropulsations which have very similar character at both the base station and the field station is selected. Then, a base line is drawn through each record, with care being taken that the base line is drawn at the same level on each record, relative to the amplitudes of the electric field variations. Finally, sets of amplitudes are read from each of the four traces, at time intervals of several seconds to several tens of seconds. This process is repeated for other micropulsation events, until a set of from 20 to 50 groups of four values each are compiled for each field station–base station pair.

If pairs of values from the base station and field station are used to construct two families of vectors, the results might be as shown in Fig. 175. Vectors vary in length by a ratio of two or three to one, with most vectors grouping in some principle direction (east–west in the example).

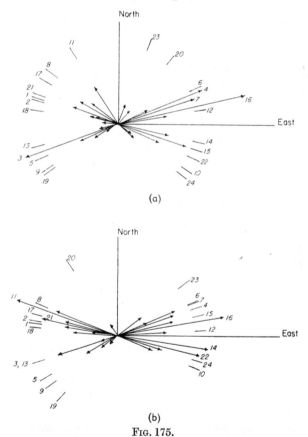

(a)

(b)

FIG. 175.

(a) Set of 24 field vectors recorded at a base station.
(b) Set of electric field vectors recorded at corresponding times at a field station. The vectors do not fall in exactly the same angular sequence as required by theory, due to errors in selecting record amplitudes and assigning base line values to the electric field records.

Since the magnetic field strength inducing the electric field for each vector is not known, an assumption is necessary before an elliptical pattern such as that shown in Fig. 172( b) can be constructed for the data. The assumption usually made is that the electric field induced at the base station does not depend on the azimuth of the inducing field, and so, the pattern there should correspond to the circular pattern shown in Fig. 172( a). If this assumption can be made, then the magnetic field amplitude at the base station is proportional to the electric field at the base station, with the constant of proportionality being independent of azimuth. Similarly, the magnetic field variation at the base station is assumed to be the same as the magnetic field variation at the field station, so the electric field amplitude at the field station can be related to the magnetic field variation at the base station.

Each component of each electric field vector at the field station is normalized for variations in amplitude of the primary magneto-telluric field by dividing it by the corresponding amplitude of the corresponding component of the electric field recorded at the base station. This normally leads to a pattern of data through which an ellipse may be drawn as shown in Fig. 176.

Interpretation of elliptical patterns such as that shown by the data in Fig. 176( b) may be based on the various characteristic dimensions, as follows:

1. The linear dimension of an ellipse (any radius) will be proportional to the ratio of inverse conductances between the base station and the field station:

$$r = S_{\text{base}}/S_{\text{field}} \tag{355}$$

where $r$ is a radial of the ellipse, referred in length to a base circle of unit radius.

2. The area of the ellipse at a field station will be proportional to the ratio of the products of the maximum and minimum resistivities at the base and field stations, providing the thickness of the conductive rocks is the same at both locations:

$$A_{\text{field}} = \frac{h_{\text{base}}^2}{(\varrho_{\min} \varrho_{\max})_{\text{base}}} \bigg/ \frac{h_{\text{field}}^2}{(\varrho_{\min} \varrho_{\max})_{\text{field}}} \tag{356}$$

3. The major axis of a field ellipse will be directed up dip, or towards the direction in which the conductance of the conductive layer decreases, providing the base station is further from such a horizontal change than the field station. The opposite is true if the base station is closer to a lateral discontinuity in conductance than is the field station.

4. The degree of ellipticity is greater the closer a field station is to a lateral discontinuity in conductance, providing the base station is further from the lateral change than is the field station, and provided the field station is on the conductive side of the discontinuity. Ellipticity dissappears for stations located on the low conductance side of a discontinuity.

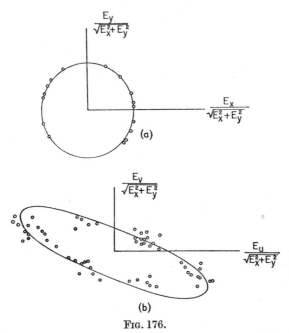

FIG. 176.

(a) Circular pattern assumed for electric field vectors recorded at a base station.

(b) Typical elliptical pattern described by electric field vectors recorded at a field station when they are normalized to the unit electric field vector circle assumed for the base station.

These interpretation rules are based on several assumptions, one of which is that the assumption of a circular vector pattern at the base station is reasonably valid. If this assumption is not warranted, it may be difficult to interpret properly the elliptical pattern computed for field station data. The effect of assuming a circular vector pattern for the electric field at a base station when actually the vector should be assigned an elliptical behavior pattern is shown in Fig. 177. In this illustration, it was assumed that an electromagnetic field with random polarization, but having an amplitude independent of azimuth (Fig. 177(a)) induces electric fields in an earth in which the apparent resistivity depends strongly on azimuth. If the electric field induced per gamma were measured, an elliptical pattern such as that shown in Fig. 177(b) would be obtained. If the apparent resistivity at a field station does not depend as strongly on azimuth as does the apparent resistivity at the base station, then the electric field induced per gamma change in the magnetic field might be as shown in Fig. 177(c). However, the electric field at the field station is not expressed in terms of the amplitude of the magnetic field variation, but rather in terms of the amplitude of the electric field recorded at the same time at the base station.

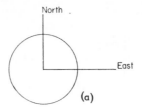

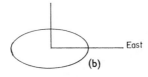

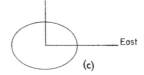

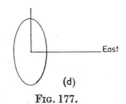

Fig. 177.

(a) Primary electromagnetic field assumed to have equal intensity components along all directions.

(b) Electric field pattern induced by the electromagnetic field if the apparent resistivity depends strongly on azimuth. Here, the apparent resistivity is considerably greater in the east-west direction than in the north-south direction.

(c) Electric field pattern induced by the electromagnetic field at a field station where the apparent resistivity does not depend as strongly on azimuth as it does for the base station considered above.

(d) Ratio of field station pattern to base station pattern.

In so doing, the electric field pattern computed for the field station would be as shown in Fig. 177(d); the major axis would be at right angles to the proper direction. Thus, ellipticity can be used in interpretation only in areas where the base station does not have a large ellipticity of its own.

This problem may be avoided by locating the base station in an area far removed from local changes in conductance of the surface layers, if enough is known about the structure to do so. In most cases, it may be necessary to refer the field station patterns to a more appropriate base location after the survey is completed.

A hypothetical geological structure which might be investigated with the telluric current method is shown in Fig. 178. It consists of a circular symmetri-

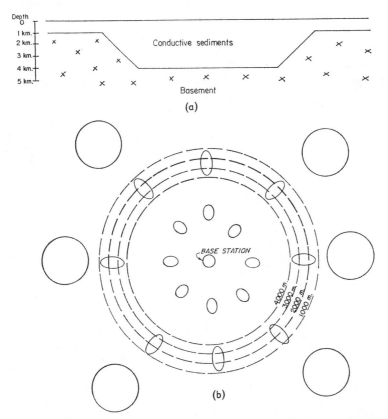

FIG. 178.
(a) Simplified basin structure used to demonstrate the interpretation of telluric current surveys.
(b) Plan map showing structure contours on the basement surface and the telluric electric field patterns which would be observed if they were referred to a base station in the center of the basin.

cal basin, with the thickness of sediments in the basin thinning by a ratio of
4 : 1 over the flanks. For simplicity, it may be assumed that the resistivity of
the sediments is constant throughout the basin. The electric field patterns
which would be recorded if all the field stations were referred to a base station
in the middle of the basin are shown. The ellipticity of the patterns increases
as the locations of the field stations are brought close to the boundaries of the
deep portion of the basin, and then disappears for station locations on the flanks.
In all cases, the major axis of the electric field ellipse points up dip. The area
of the ellipses on the flanks of the structure is 16 times the area of the base-
station circle.

Figure 179 shows the patterns which would be obtained over the same basin
if the base station had been selected in an area of high ellipticity, rather than
at the center of the basin. The interpretation of the pattern is obviously much
less simple.

If no auxiliary data are available, only changes in the conductance of the
surface rocks can be observed with the telluric current method. For quantitative
interpretation, it is necessary to have some other source of information about
the thickness of conductive rocks. This auxiliary information may be obtained

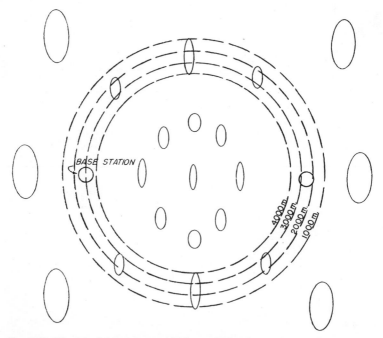

Fig. 179. Electric field patterns which would be observed over the structure
shown in Fig. 178 if measurements were referred to a base station in an area
of high ellipticity.

from a wildcat well, from seismic depth determinations, from a resistivity or magneto-telluric sounding, or under favorable conditions, from a gravity survey.

It is not always possible to obtain electric field measurements from enough azimuths to construct the elliptical patterns which have been shown in the examples given so far. In many cases, the electric field may be polarized so that all measurements are made with electric fields directed in almost the same direction. With a linearly polarized primary magneto-telluric field, it is possible only to form the ratio of amplitudes for this one azimuth between base stations and field stations. This ratio of amplitudes is proportional to the inverse ratio of surface-layer conductances between stations, and a contour map of the amplitude ratio provides information on the changes in conductance. However, the additional information about the direction in which conductance is changing which can be determined from ellipticity when elliptical patterns can be drawn, cannot be extracted from amplitude data alone.

The graphical procedure outlined above for determining the elliptical pattern for a field station may not provide the desired accuracy for estimating the ratio of conductances. The "triangle" method of interpretation may be used along with the construction of ellipses to provide a more quantitative estimate of the ratio of areas than can be obtained merely by drawing an ellipse through a scattering of data points, such as was shown in Fig. 176(b).

In the triangle method, each field station vector is normalized to the base station by dividing the appropriate component at the base station by the corresponding component recorded at the base station. Then, triangles are formed by taking two vectors at a time at both the field station and base station to form two triangles (see Fig. 180). The angle between a pair of vectors selected to form a triangle should lie between 45° and 135°, if possible. Each vector should be used only once in constructing a triangle, to avoid giving undue weight to a measurement which may be in error.

The area of each triangle is:

$$A_{\text{BASE}} = [\tfrac{1}{2}\sin\varDelta\theta]$$

$$A_{\text{FIELD}} = \left[\frac{1}{4}\left(\frac{E_{u,1}^2 + E_{v,1}^2}{E_{x,1}^2 + E_{y,1}^2}\right)\cdot\left(\frac{E_{u,2}^2 + E_{v,2}^2}{E_{x,2}^2 + E_{y,2}^2}\right)\right]^{1/2}\sin\varDelta\varphi \qquad (357)$$

where $\varDelta\theta$ and $\varDelta\varphi$ are the angles between corresponding pairs of vectors used in constructing a triangle:

$$\varDelta\theta = \arctan\frac{E_{y,1}}{E_{x,1}} - \arctan\frac{E_{y,2}}{E_{x,2}}$$

$$\varDelta\varphi = \arctan\frac{E_{v,1}}{E_{u,1}} - \arctan\frac{E_{v,2}}{E_{u,2}}$$

The ratio of areas between triangles formed at the field station and the base station should be formed for 20 to 50 pairs of vectors. The standard deviation

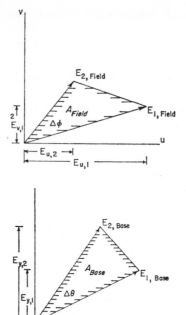

Fɪɢ. 180. Definition of quantities used in the triangle method for reducing telluric current data.

of the area ratios so formed should be computed, and area ratios which fall outside a reasonable error limit should be dropped before the average of the rest of the values is taken. The average ratio:

$$K = \frac{A_{\text{FIELD}}}{A_{\text{BASE}}} \tag{358}$$

is then inversely proportional to the square of the ratio of conductances between the two stations, the base station and the field station.

The fact that there is scatter to the electric field data indicates not only that there are errors involved in measuring electric field intensities, but also that the assumptions on which the telluric current method is based may not be fully met. Errors arise principally in two ways; the resistivity contrast between the sediments and the basement rock may not be large enough to prevent some appreciable amount of current from flowing in the basement, or the depth of the conductive rocks may be so great that some attenuation of the magneto-telluric field takes place before it penetrates to the basement. In either case, the electric field may depend to some extent on frequency, and magneto-telluric events with different frequency contents may give different results.

# REFERENCES

BIRDICHEVSKIY, M. N. (1960) Elektricheskaya razvedka metodom telluricheskikh tolkov; Gostoptekizdat, Moscow, 237 pp. (English translation in *Quart. Colo. School-Mines*, Jan., 1965).

BOISSONAS, E. and LEONARDON, E. G. (1948) Geophysical exploration by telluric currents with special reference to a survey of the Haynesville salt dome, Wood County, Texas, *Geophysics*, vol. 13, no. 3.

BONDARENKO, A. P. (1953) O svyazi zemnik tokov i geomagnetnik variatzii, *Doklady Akad. Nauk, SSSR*, vol. 89, no. 3.

D'ERCEVILLE, I. and KUNETZ, G. (1962) The effect of a fault on the earth's natural electromagnetic field, *Geophysics* **27** (5), 651–665.

FRIEDL, K. and KANTAS, K. (1959) Geophysical results in Austria with special regard to telluric measurements, *Geofis. pura e appl.*, no. 2.

KUNETZ, G. (1957) Anwendung statistischer Eigenschaften der Erdströme in der praktischen Geophysik; *Freiberger Forsch.*, January.

KUNETZ, G. and CHASTENET DE GERY, J. (1956) Conformal representation and various potential distribution problems in media of different permeabilities, *Rev. inst. franç. pétrole et Ann. combustibles liquides* **11**, 1179–1192.

LIY-SHU (1963) Calculating the parameters used in telluric current prospecting, *Geophysics* **28** (3), 482–485.

MAINGUY, M. and GREPIN, A. (1952) Some practical examples of interpretation of telluric maps in Languedoc, *Eur. Assoc. of Exp. Geophys.*, May.

NEWMAN, L. R. and KALANTAROV, P. L. (1948) *Teoreticheskiye osnovi elektrotekniki*, vol. 3, Gosenergoizdat.

POPOV, U. N. (1958) Ob interpretatsii nabliudenii telluricheskikh tokov, from *Razvedochnaya i Promislovaya Geofizika*, vol. 24, Gostoptekizdat.

SHEINMAN, S. M. (1958) O vosmoshnosti isplozovedki polei telluricheskikh tokov i dalnikh radiostatzii dlya geologicheskogo kartirovania, from *Tr. Vses. in-ta Metodiki i Tekniki razvedki*, Gostoptekizdat.

TIKONOV, A. N. (1942) K voprosi o vliyanii neodonorodnosti zemnoi kori na pole telluricheskikh tokov, *Doklady Akad. Nauk, SSR, ser. Geofiz.*, no. 5.

TIKONOV, A. N. (1950) Ob opredelenii elektrichskikh karaktiristikh glubinnikh sloev zemnoi kori, *Doklady Akad. Nauk, SSSR*, vol. 73, no. 2.

TIKONOV, A. N. and LIPSKAYA, N. V. (1952) O variatsiakh zemnogo elektricheskogo polya, *Doklady Akad. Nauk, SSSR*, vol. 87.

TIKONOV, A. N. and SHAKSUVAROV, D. N. (1956) Vozmoshnosti ispolsovaniya impedansa yestyestvennogo elektromagnetnogo polya Zemli dlya izsledovania yeye verknikh sloev, *Izvest. Akad. Nauk, SSSR, ser. Goefiz.*, no. 4.

TUMAN, V. S. (1951) The telluric method of prospecting and its limitations under certain geologic conditions, *Geophysics*, vol. 16, no. 1.

UTZMANN, R. and FAVRE, B. (1954) Influence de la non-cylindricite des structures sur le champ tellurique, *Rev. inst. franç. pétrole et Ann combustibles liquides*, vol. 12, no. 2.

# CHAPTER 6

# INDUCTION METHODS

## 34. INTRODUCTION

There are many geoelectrical prospecting techniques in which a controlled time-varying magnetic field is used as a power source. Such methods are usually referred to as *electromagnetic methods*, though they should more properly be termed induction methods to distinguish them from other geoelectrical techniques which also employ measurements of electromagnetic field strength (the magneto-telluric method or the radio-wave method, for example). The frequencies used with induction methods are generally higher than those used in the magneto-telluric method but lower than the frequencies used in radio-wave studies; usually, induction methods employ frequencies in the range from 100 to 5000 c/s. Maxwell's equations must be used rather than Laplace's equation to describe the behavior of currents and electric fields within a conductive medium.

A time varying magnetic field may be generated by driving an alternating current through a loop of wire, or through a wire grounded at both ends. If any conductive material is present within the magnetic field so generated, induced or eddy currents will flow in closed loops in paths normal to the direction of the magnetic field. These eddy currents, in turn, generate their own magnetic fields so that at any point in space the total magnetic field may be thought of as consisting of two parts: a *primary* or normal field due to the source current and a *secondary* or disturbing field due to eddy currents induced in conductors.

The resultant magnetic field is usually measured in terms of the voltage induced in a loop of wire used as a receiver. The functioning of an induction system may be compared with that of a transformer having three windings, one of the windings being shorted to simulate the effect of a conductive body (see Fig. 181). The voltage induced along the length of the shorted turn is proportional to the rate of change of flux linking the turn:

$$(\text{e.m.f.})_s \propto - i\mu_0\omega I_p e^{i\omega t} \qquad (359)$$

where $I_p e^{i\omega t}$ is an oscillatory current with frequency $f = \omega/2\pi$ flowing in the primary winding. The secondary current (that flowing in the shorted turn) and

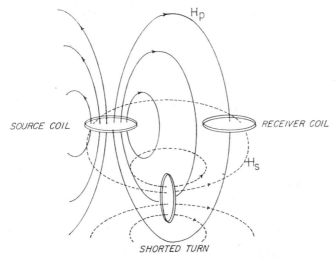

F_IG. 181. Transformer analogy of an induction prospecting system. A primary magnetic field $H_p$ is generated by an oscillatory current flowing in the source-coil. This field generates current in a shorted turn, representing a conductive earth. This current generates a secondary magnetic field, $H_s$. Both fields are measured by the voltage induced in a receiver coil.

the secondary magnetic field depend on this induced voltage, as well as on the resistance, $R$, and the inductance, $L$, of the shorted turn:

$$H_s \propto I_s \propto - \frac{i\mu_0 \omega I_p e^{i\omega t}}{R + i\omega L} \tag{360}$$

The voltage induced in the receiver loop is proportional to the negative rate of change of the flux coupling it which, in turn, is proportional to the vector sum of the primary and secondary magnetic fields at the point of observation:

$$(\text{e.m.f.})_{\text{receiver}} \propto - i\omega B \propto - i\omega\mu H_t \propto - i\omega\mu(H_p + H_s) \tag{361}$$

The secondary magnetic field need not in general be in phase (or 180° out of phase) with the primary field, nor does it need to have the same direction as $H_p$ at the point of observation, so $H_t$, the total field vector is elliptically polarized in the general case. For a complete description of the total field, $H_t$, it would be necessary to measure the amplitude of each of the three spatial components and the phase angle between each of these components and the primary current. Alternately, the portion of each spatial component which is in phase and the portion which is out of phase with the primary current may be measured. These quantities may be referred to using a variety of names; inphase and out-of-phase components or quadrature components, or real and imaginary components, with the understanding that the components are taken relative to phase and not to vector direction.

At sufficiently low frequencies, the resistive term in the denominator of Eq. (360) is more important then the inductive term, so that the secondary current and its magnetic field are about 90° out of phase with the primary field. As the frequency is increased, the inductance term becomes increasingly important until at very high frequencies it dominates the expression and the secondary field becomes almost 180° out of phase with respect to the primary field at the shorted turn. The secondary field will almost cancel the primary field within the shorted turn under this condition.

A solid conductive body behaves in a similar way to the shorted turn the main difference being that the distribution of eddy currents within the body depends not only on frequency but also on the size, shape and conductivity of the body. At low frequencies, eddy currents flow deep within the conductor; at high frequencies, the skin effect restricts current flow mainly to a region near the surface of the body. At very high frequencies, the primary field is almost cancelled by the secondary field due to surface currents in the conductive body so that the field within the conductor is nearly zero. This condition is sometimes referred to as eddy-current saturation; the anomalous secondary field at the receiver loop cannot be increased by further increasing the frequency.

Electromagnetic fields depend on the magnetic permeability of earth materials, as well as on the conductivity. Ordinarily, the range of effects which can be attributed to variations in permeability is small compared to the range of effects caused by variations in conductivity. However, distortion of the magnetic field caused by anomalously large values of magnetic permeability in the earth is sometimes observed over highly magnetic rocks. Consider a mass of rock having a high magnetic permeability but no conductivity (see Fig. 182). The body develops an induced magnetization in accordance with its permeability, physical dimensions and the strength and direction of the primary field. The voltage induced in the receiver coil is a function of the vector sum of the primary field and the secondary field about the magnetized body. The induced magnetization is oriented in such a direction as to add to the primary field within the magnetized body. The secondary field due to a magnetized body is always 180° out of phase with respect to the phase of the secondary field about a conductive body at high frequencies.

In Fig. 181, the relationship between the directions of the primary and secondary fields is such that the secondary field tends to cancel the primary field at the receiver loop. In Fig. 182, the relationship is such that the secondary field due to magnetization adds to the primary field. There are combinations of positions for the loops in relation to the position of the disturbing mass such that the secondary field may either aid or oppose the primary field at the receiver loop. Thus, the real (or inphase) component of the secondary field about a body may be either positive or negative depending on whether it is in phase or

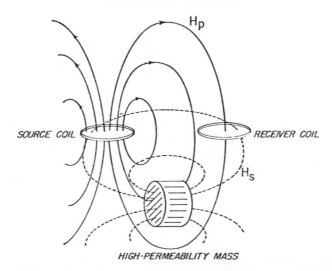

FIG. 182. Generation of a secondary magnetic field by a mass with high magnetic permeability but no conductivity. A primary magnetic field, $H_p$, is generated by an oscillatory current flowing in the source coil. A polarization field is developed by this primary field in the high permeability mass. The vector sum of the primary field and the polarization field is measured by the voltage induced in a receiver coil.

180° out-of-phase with respect to the primary field at the point of observation. Similarly, the out-of-phase component of the secondary field may be positive or negative depending on whether it is 90° or −90° out-of-phase with respect to the primary field.

## 35. INSTRUMENTATION

Instrumentation for making electromagnetic field measurements is much more sophisticated than that required for making galvanic resistivity measurements. A variety of commercial equipment is available, but the application of induction methods to new or unique problems is often limited because of a lack of suitable instrumentation. Development of electromagnetic measuring equipment for specialized applications such as airborne surveying is often expensive and time consuming.

Basically, electromagnetic instrumentation consists of providing a source for an alternating magnetic field, one or more receiving loops and a means for measuring the voltages induced in the receiving loops. One commonly used power source is a long grounded-wire through which an alternating current from a rotary generator or oscillator–amplifier source is driven. The wire used and the electrodes are similar to those used in galvanic resistivity studies. The length of wire used may range from 1 km to as much as 5 km. An alternator driven by a gasoline engine is the most commonly used power source for the alternating current. An electronic oscillator and power amplifier may be used to provide a more stable current than can be obtained with a rotary generator, and to provide a means for changing the frequency readily. Currents ranging from one-half to several amperes are required.

In other methods, the source for the magnetic field consists of a large loop of insulated wire. Single-turn horizontal loops (loop axis directed vertically) may have dimensions of a half mile or more. Such loops are formed from flexible wire so that they may be stored on portable reels. Large vertical loops (loop axis directed horizontally) may have 50 turns of wire supported by a rectangular or triangular folding frame, with dimensions of several meters each side. The power required to operate a large loop is about the same as that required for a long grounded-wire. The inductance of multi-turn loops is moderately large, and they are usually tuned to their operating frequencies with a series capacitor to provide an impedance match with the power source.

Small hand-held loops may be used as a power source if the distance from the source to the receiver loops is to be no more than a few hundred feet. Air-core coils consist usually of about 100 turns of wire wrapped on a circular form with a diameter of about 1 m. Single-turn loops made of heavy tubing with a matching transformer are sometimes used. Coils wrapped in the form of solenoids about a ferrite or other high-permeability core material are used in place of air-core loops in some of the more modern equipment (see Fig. 155 in Chapter IV). Hand-carried transmitter loops are excited by small battery-powered oscillator-amplifiers. The power used is about 0·5 to 5 W.

Receiver loops are usually small hand-held coils similar in appearance to the small transmitter coils but consisting of many more turns of light-weight wire. Receiver loops, and sometimes the transmitter loops as well, are shielded electrostatically by wrapping them with strips of tin foil, or by painting the coil form with conductive paint. This shielding is grounded at some appropriate point in the circuit in order to eliminate capacitive coupling between the loops and the operator or between the loops and the earth. Such capacitive coupling causes errors in measurement.

If it is necessary to determine the direction of the field, the coils may be provided with some means for measuring inclination of the coil axis as well as the azimuth. The direction of the field may be determined by rotating such a receiver coil until no voltage is induced in it. Usually, an amplifier with headphones or a meter is all that is required to determine the null position of such a coil.

When the magnitude and phase relations of one or several components of the electromagnetic field are to be measured, a ratiometer (a.c. potentiometer) is usually employed. A simplified circuit diagram of a typical ratiometer is shown in Fig. 183. The reference voltage used in the ratiometer is obtained from a small coil which is either attached to the frame of the transmitter coil or

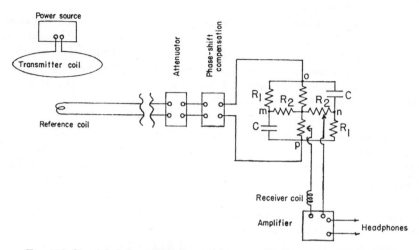

FIG. 183. Simplified circuit diagram of an a.c. potentiometer or ratiometer
used in measurements of electromagnetic fields.

placed near it. This reference voltage may be carried to the receiving equipment
over a two-conductor cable. A phase shift circuit is provided to correct for
phase shifts within the equipment. An attenuator is provided so that the
level of the reference voltage can be varied for different separations between the
transmitter and receiver. The sizes of the resistor, $R$, and the capacitor, $C$, are
such that the voltage observed between points $m$ and $n$ in the circuit is
exactly 90° out of phase with respect to the voltage observed between points
$o$ and $p$. This phase relationship holds if;

$$\frac{1}{\omega R_1 C} = \left[ \frac{R_1}{R_1 + R_2} \right]^{1/2}$$

(362)

Allowance must be made for the fact that the voltage between circuit points
$m$ and $n$ is less than the voltage between points $o$ and $p$:

$$\frac{V_{MN}}{V_{OP}} = \frac{i}{2} \frac{\left[ \left( \frac{1}{\omega C} \right)^2 + R_1^2 \right]}{\frac{R_1}{\omega C} \left( \frac{R_1}{2R_2} + 1 \right)}$$

(363)

In making a measurement, the sliders on the two resistors are adjusted until
the sum of the voltages across the two resistances is equal to the voltage
induced in the receiver coil, a condition which is recognized by observing a null
tone in the earphones. The inphase and out-of-phase ratios for the electromagnetic field intensity at the receiving coil relative to the primary field at the
reference coil can then be determined from the positions of the dials for the
two potentiometric resistances.

Other types of ratiometer circuits are also used. Some make use of inductances and resistors rather than capacitors and resistors in the phase shifting network, while other ratiometers are designed to measure amplitude ratios and phase angle rather than the inphase and out-of-phase components. Since the values of resistance and capacitance required for balance depend on the frequency being used, some means for switching circuit elements must be provided if a ratiometer is to be used at several operating frequencies.

The resolution of a typical ratiometer is in the range 0·1 to 1 per cent, with the accuracy of measurement being about the same. Some slow shift in null values for the circuit elements may occur if the resistance or capacitance of any of the elements is temperature dependent.

The receiving coil may be parallel tuned at the operating frequency, so that the voltage induced in it is amplified by the Q-factor of the circuit. This also provides discrimination against noise at frequencies other than the operating frequency. However, if a tuned receiving coil is used, the magnitude and phase of the voltage output from the coil will depend on changes in inductance of the coil or capacitance of the tuning capacitor. Also, if a tuned receiving coil is used, the voltage output will be sensitive to small changes in the frequency of the transmitter. These disadvantages can be avoided by placing the tuning capacitor directly across the input to the amplifier.

The design and construction of amplifiers to provide detectable signals for a null detector present no serious problems. Such amplifiers are tuned to the operating frequency to minimize signals from atmospheric sources and from power distribution lines. It is not particularly difficult to design amplifiers in which the noise level is less than the thermal noise developed in the receiving coil and external noise seen by the coil (Slichter, 1955). Since external noise is usually the main noise component in electromagnetic equipment, the only means available for increasing the ratio of signal level to noise level is to increase the strength of the electromagnetic source field or to decrease the bandwidth of the receiving equipment. Linearity and gain stability in the amplifier are not important factors in a null-measuring system. When high operating frequencies are used, the signal frequency is mixed with a local oscillator frequency to provide a heterodyne frequency to the headphones which may be recognized more readily than the operating frequency.

When the distance between the source and receiver is large, it may not be convenient to use a cable to transmit the reference signal to the receiver and undesirable phase shifts may take place along long cables. The reference signal may be transmitted to the receiver using a radio link. The ratiometer is replaced by a phase-sensitive voltmeter accurate to perhaps one-half per cent. Phase reference information may be transmitted readily by a radio link, but it is more difficult to transmit amplitude reference information. Therefore, when a radio link is used, the current into the transmitter loop or wire must be stable, so that there will be no variations in the field strength at the receiver-location caused by variations in source current.

Further details of instrumentation will be discussed in later sections, particularly under airborne methods.

## 36. PRIMARY FIELDS

It is usually necessary to calculate the primary electromagnetic field strength, or at least the ratio of primary field components to the current flowing in the source wires, when electromagnetic surveys are interpreted. With any method, it is important to understand the geometry of the primary field. The primary field (that present in the absence of conductors) about any arbitrary configuration of current-carrying wires may be computed with little difficulty by applying Ampere's law or the Biot–Savart law for static magnetic fields, providing the distance between the sources and the point of observation is much less than a wave length in free space. This condition is usually satisfied in the case of induction prospecting systems; for example, the free-space wave length at 5000 c/s is 60 km.

Ampere's law may be written as:

$$dH = \frac{I \, dl \sin\varphi}{4\pi r^2} \tag{364}$$

where a current-carrying wire has been divided into a series of short lengths, $dl$, so that the portion of the static field, $dH$, contributed only by the current flowing in this element of wire can be considered. The magnetic field is given at a point at a distance, $r$, from the wire. The current flowing through the wire is $I$ and the angle between the direction of the element of wire and the line drawn from the observation point to the element of wire is $\varphi$ (see Fig. 184).

In the case of a long straight wire, integration of this expression to obtain the total magnetic field at the point P is most easily carried out by considering the angle, $\beta$, rather than the angle $\varphi$. In Fig. 184, the angle $\beta$ is formed between a perpendicular line of length, $x$, dropped from the point P to the current-carrying wire, and the line, $r$, connecting the observation point to the line

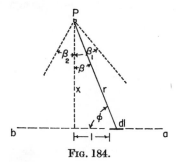

Fig. 184.

element, $dl$. The distance of the element, $dl$, from the point where the perpendicular intersects the current-carrying wire is $l$. Trigonometric relationships for a right triangle permit us to say:

$$\sin\varphi = \cos\beta$$

$$r = \frac{x}{\cos\beta}$$

$$l = x \tan\beta$$

$$dl = \frac{x}{\cos^2\beta}\,d\beta$$

The total magnetic field at P is obtained by integrating Eq. (364) over the length of the wire, from the angle $\beta_1$ to the angle $\beta_2$:

$$H = \int_{\beta_1}^{\beta_2} \frac{Ix \cos^3\beta\,d\beta}{4\pi x^2 \cos^2\beta}$$

$$= \frac{I}{4\pi x}(\sin\beta_2 - \sin\beta_1) \tag{365}$$

The magnetic field so calculated is expressed in amperes per meter (or newtons per weber) if current is measured in amperes and the distances in meters. If the length of the wire is very great compared to the offset distance $x$, both $\beta_1$ and $\beta_2$ will be approximately 90°, so Eq. (365) simplifies to:

$$H = \frac{I}{2\pi x} \tag{366}$$

In practice, it is not possible to obtain a single length of wire carrying current. Some provision for the return of the current to the source must be made. This may be done by forming the wire in the shape of a large square loop, with the distance, $x$, being small at one side only. In this case, the contribution to the magnetic field by current flowing in the other three sides of the loop must be calculated by the method used above and added to the field strength given by Eq. (365).

The primary field about a small square loop may be computed in the same way. Consider the loop shown in Fig. 185, with the magnetic field being observed

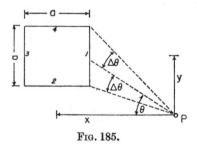

FIG. 185.

at a point, P, whose distance from the center of the loop, $r$, is large compared to the width of the loop, $a$. Applying Eq. (365), we may calculate the contribution of one side of the loop to the total magnetic field at the point P:

$$H_1 = \frac{I}{4\pi\left(x - \dfrac{a}{2}\right)}\left[\sin\left(\theta - \Delta\theta\right) - \sin\left(\theta + \Delta\theta\right)\right] \qquad (367)$$

Using appropriate trigonometric identities for the sums and differences of angles, this equation may be rewritten as:

$$H_1 = \frac{Ia}{4\pi}\frac{\left(x - \dfrac{a}{2}\right)}{\left[y^2 + \left(x - \dfrac{a}{2}\right)^2\right]^{3/2}}$$

where:

$$\sin\Delta\theta \cong \Delta\theta = \frac{a\cos\theta}{\left[y^2 + \left(x - \dfrac{a}{2}\right)^2\right]^{1/2}}$$

$$\cos\theta = \frac{\left(x - \dfrac{a}{2}\right)}{\left[y^2 + \left(x - \dfrac{a}{2}\right)^2\right]^{1/2}} \qquad (368)$$

The contributions to the total magnetic field by current flowing in the other three sides of the coil are:

$$H_3 = -\frac{Ia}{4\pi}\frac{\left(x + \dfrac{a}{2}\right)}{\left[y^2 + \left(x + \dfrac{a}{2}\right)^2\right]^{3/2}}$$

$$H_2 = \frac{Ia}{4\pi}\frac{\left(y - \dfrac{a}{2}\right)}{\left[x^2 + \left(y - \dfrac{a}{2}\right)^2\right]^{3/2}}$$

$$H_4 = -\frac{Ia}{4\pi}\frac{\left(y + \dfrac{a}{2}\right)}{\left[x^2 + \left(y - \dfrac{a}{2}\right)^2\right]^{3/2}} \qquad (369)$$

The total field at the observation point is:

$$
H_t = \frac{Ia}{4\pi} \left\{ \frac{\left(x - \dfrac{a}{2}\right)}{\left[y^2 + \left(x - \dfrac{a}{2}\right)^2\right]^{3/2}} - \frac{\left(x + \dfrac{a}{2}\right)}{\left[y^2 + \left(x - \dfrac{a}{2}\right)^2\right]^{3/2}} \right.
$$

$$
\left. + \frac{\left(y - \dfrac{a}{2}\right)}{\left[x^2 + \left(y - \dfrac{a}{2}\right)^2\right]^{3/2}} - \frac{\left(y + \dfrac{a}{2}\right)}{\left[x^2 + \left(y + \dfrac{a}{2}\right)^2\right]^{3/2}} \right\} \quad (370)
$$

However, if $r^2$ is substituted for $x^2 + y^2$ and the relations $\cos \theta = x/r$ and $\sin \theta = y/r$ are used, this equation may be rewritten as:

$$
H_t = \frac{Ia}{4\pi r^3} \left\{ \frac{\left(x - \dfrac{a}{2}\right)}{\left[1 - \dfrac{a}{r}\cos\theta + \dfrac{a^2}{4r^2}\right]^{3/2}} - \frac{\left(x + \dfrac{a}{2}\right)}{\left[1 + \dfrac{a}{r}\cos\theta + \dfrac{a^2}{4r^2}\right]^{3/2}} \right.
$$

$$
\left. + \frac{\left(y - \dfrac{a}{2}\right)}{\left[1 - \dfrac{a}{r}\sin\theta + \dfrac{a^2}{4r^2}\right]^{3/2}} - \frac{\left(y + \dfrac{a}{2}\right)}{\left[1 + \dfrac{a}{r}\sin\theta + \dfrac{a^2}{4r^2}\right]^{3/2}} \right\} \quad (371)
$$

The ratio $a/r$ can be taken to be much smaller than unity, so the various denominators in Eq. (371) are all approximately unity. The total magnetic field due to current in the square coil, providing the observation point is at a large distance compared to the radius of the coil, is simply:

$$
H_t = \frac{Ia^2}{2\pi r^3} \quad (372)
$$

We should note at this point that the magnetic field from a long wire source or close to the side of a square loop varies as the inverse first power of the distance from the wire. The field about a small loop decreases as the cube of the distance.

The current return in long-wire methods is usually provided through the ground by earthing the two extreme ends of the current cable. It can be shown that the magnetic field at the surface caused by direct current entering the ground at an electrode is the same as the field which would be produced by the same current flowing in a wire extending vertically downward from the earth's surface to infinity, with no easthing point. This magnetic field can be calculated from Ampere's law Eq. (364), with the result:

$$H_x = \frac{Iy}{4\pi(x^2 + y^2)}$$

$$H_y = \frac{Ix}{4\pi(x^2 + y^2)}$$

$$H_z = 0 \tag{373}$$

In the case of alternating currents, this expression is valid if the current is changing slowly enough that at any instant the field about the wire is the same as in the static case. If the rate of variation is sufficiently fast that the field varies in phase with the current flowing in the wire, the correct expression is much more complicated, but the contribution to $H_z$ from current flowing in the ground is still negligible.

Although these expressions for the primary field are nearly always adequate for computing the primary fields about long wires and small loops, more general expressions for the primary field are useful in calculating the secondary fields generated by eddy current flow in a conductive earth. Also, the relationship between induction and radio-wave methods is more readily apparent if we now develop the general expressions for the primary field.

If the distance between the source and the observation point approaches a wavelength, the primary field cannot be described simply with Ampere's law. To find the field under such conditions, it is convenient to make use of the *retarded magnetic vector potential* or the *wave potential*. In the electrostatic problems discussed in the preceding sections, it was convenient to introduce a potential function from which the electric field could be obtained by differentiation, by taking the gradient. With magnetic induction, a similar procedure can be followed, with the difference that the potential is a vector and the induction is found by taking the curl. The two potential functions are not the same.

By virtue of the way in which it is defined, the vector potential function, $A$, has certain analytic properties:

1. The vector potential is a continuous function of position.
2. The vector potential satisfies the partial differential equation

$$\nabla^2 A = -j \tag{374}$$

where $j$ is the current density vector, and providing only steady currents are being considered.

3. For the special case of steady currents, the vector potential satisfies the equation

$$\text{div } A = 0 \tag{375}$$

4. Also, for steady currents, the equation

$$\text{curl } \boldsymbol{B} = \mu_0 \boldsymbol{j} \tag{376}$$

holds.

5. The vector potential at any point in space is unique, with the possible exception of an additive constant, as in the case of scalar potential functions. In vector notation, the integral form of Ampere's law for static fields is:

$$\boldsymbol{H} = \int \frac{\boldsymbol{I} \cdot \mathrm{d}\boldsymbol{l} \times \mathrm{d}\boldsymbol{r}}{4\pi \, |r'^2|} \tag{377}$$

with the parameters being defined as before in Fig. 184. If this magnetic field strength were to be calculated from a vector potential function, $\boldsymbol{A}$, two steps would be involved:

$$\boldsymbol{H} = \nabla \times \boldsymbol{A} \tag{378}$$

where:

$$\boldsymbol{A} = \int_V \frac{\boldsymbol{I} \, \mathrm{d}v}{4\pi r'} \tag{379}$$

The differential equation for $\boldsymbol{H}$, considering only static fields, is:

$$\nabla \times \boldsymbol{H} = \boldsymbol{j} \tag{380}$$

so that:

$$\nabla \times \nabla \times \boldsymbol{A} = \boldsymbol{j} = -\nabla^2 \boldsymbol{A} + \nabla(\nabla \cdot \boldsymbol{A}) \tag{381}$$

Since we have assumed only steady currents, Eq. (375) holds, so that:

$$\nabla^2 \boldsymbol{A} = -\boldsymbol{j} \tag{382}$$

This is the vector potential equivalent of Poisson's equation. If only regions with no current flow are being considered, the vector potential, $\boldsymbol{A}$, need only be a solution to Laplace's equation, $\nabla^2 A = 0$.

Consider a small current carrying element (or electric dipole) with a moment $I \, dl$ oriented in the $z$ direction. From Eq. (379), we have for steady currents:

$$A_z = \frac{I \, \mathrm{d}l}{4\pi r'} \tag{383}$$

If the electromagnetic field is varying sinusoidally with respect to time, Eq. (380) becomes:

$$\nabla \times \boldsymbol{H} = i\omega\varepsilon\boldsymbol{E} + \boldsymbol{j} \tag{384}$$

Again, let the magnetic field be the curl of a vector potential function, $\boldsymbol{A}$, so that we have:

$$\boldsymbol{H} = \nabla \times \boldsymbol{A}$$

$$\nabla^2 A + \gamma^2 A = -\boldsymbol{j}$$

$$\boldsymbol{E} = -i\omega\mu\boldsymbol{A} + \frac{1}{j\omega\varepsilon} \nabla(\nabla \cdot \boldsymbol{A}) \tag{385}$$

where $\gamma = \omega \sqrt{(\varepsilon\mu)}$. It should be noted that we no longer require the divergence of $A$ to be zero. The vector potential function, $A$, must satisfy the equation

$$\nabla^2 A_z + \gamma^2 A_z = 0 \tag{386}$$

except at the origin, where the current dipole is located. Considering the nature of the source for $A_z$, it should be a spherically symmetric function, and so there may be some advantage in using a spherical coordinate system:

$$\frac{1}{r'^2} \frac{d}{dr'} \left( r'^2 \frac{dA_z}{dr'} \right) + \gamma A_z = 0 \tag{387}$$

There are two independent solutions to this equation:

$$A_z = \frac{C_1}{r'} e^{-i\gamma r'}$$

$$A_z = \frac{C_2}{r'} e^{i\gamma r'} \tag{388}$$

The second solution does not behave properly, inasmuch as it becomes infinite for large values of $r'$, and so the constant $C_2$ must be zero. The only acceptable solution is the first. As the wave number, $\gamma$, becomes small, Eq. (386) reduces to Laplace's equation (or Poisson's equation, if the origin is included), and Eq. (383) applies:

$$A_z = \frac{I\,dl}{4\pi r'}; \quad \gamma \to 0$$

so:

$$C_1 = \frac{I\,dl}{4\pi r'}$$

and so:

$$A_z = \frac{I\,dl}{4\pi r'} e^{-i\gamma r'} \tag{389}$$

This last function is the retarded vector potential for the vertically directed current dipole. The electromagnetic field components are found by substituting the solution in Eq. (389) into Eqs. (378) and (385). In spherical coordinates, these operations lead to:

$$A_{r'} = A_z \cos\theta = \frac{I\,dl}{4\pi r'} e^{-i\gamma r'} \cos\theta$$

$$A_\theta = -A_z \sin\theta = \frac{I\,dl}{4\pi r'} e^{-i\gamma r'} \sin\theta$$

$$H_\varphi = \frac{1}{r'} \left[ \frac{\partial}{\partial r'} r' \left( -\frac{I\,dl}{4\pi r'} e^{-i\gamma r'} \sin\theta \right) - \frac{\partial}{\partial\theta} \left( \frac{I\,dl}{4\pi r'} e^{-i\gamma r'} \cos\theta \right) \right]$$

$$= \frac{I\,dl}{4\pi} e^{-i\gamma r'} \left( \frac{i\gamma}{r'} + \frac{1}{r'^2} \right) \sin\theta \tag{390}$$

Similarly, the electric field is:

$$E_r' = \frac{I\,dl}{2\pi}\,e^{-i\gamma r'}\left(\frac{\sqrt{(\mu/\varepsilon)}}{r'^2} + \frac{1}{i\,\omega\varepsilon r'^2}\right)\cos\theta \qquad (391)$$

and:

$$E_\theta = \frac{I\,dl}{4\pi}\,e^{-i\gamma r'}\left(\frac{i\mu\omega}{r'} + \frac{\sqrt{(u/\varepsilon)}}{r'^2} + \frac{1}{i\omega\varepsilon r'^3}\right)\sin\theta \qquad (392)$$

For small values of $r'$ and $\omega$, the equation for the tangential component of the magnetic field, $H_\varphi$, reduces to:

$$H_\varphi = \frac{I\,dl}{4\pi r'^2}\sin\theta \qquad (393)$$

which is the same as the form of Ampere's law used earlier to describe static magnetic fields. Under the same conditions, the equations for the two components of the electric intensity reduce to:

$$E_{r'} = \frac{I\,dl}{i\,4\pi\omega\varepsilon r'^3}\cos\theta$$

$$E_\theta = \frac{I\,dl}{i\,4\pi\omega\varepsilon r'^3}\sin\theta \qquad (394)$$

These equations can be identified with the equations for the field about an electrostatic dipole where $I/i\omega$ takes the place of the charge at one end of the static dipole. In regions where the product $\omega r'$ is small, the magnetic field is very nearly in phase with the current in the dipole and 90° out-of-phase with respect to the electric field.

When the product $\omega r'$ is large, Eqs. (390)–(392) reduce to:

$$H_\varphi = \frac{i\gamma I\,dl}{4\pi r'}\,e^{-i\gamma r'}\sin\theta$$

$$E_{r'} = 0$$

$$E_\theta = \frac{i\mu\omega I\,dl}{4\pi r'}\,e^{-i\gamma r'}\sin\theta \qquad (395)$$

Under these conditions, the magnetic and electric field vectors are at right angles to one another, and over short distances the spherical wave front can be treated as a plane wave. The portions of the fields which vary inversely as the first power of distance are referred to as radiation fields; they are dominant at large distances, $r'$, and represent outward propagation of energy. The portions of the fields which vary as the inverse third power of distance are referred to as electrostatic fields, since the expressions are analogous to the expressions developed in electrostatics for the fields about a distribution of static charge. The portions of the fields which vary as the inverse square of the distance are termed the induction field. Ordinarily, the only field of importance at the distances from the source at which electromagnetic measurements are made with the various induction methods is the induction field.

Next, consider an infinitely-long wire carrying current along the $z$ axis of a cylindrical coordinate system. The wave potential $A$ for this current system should have only a $z$ component, considering the symmetry of the problem, and the wave potential should be a solution of the equation:

$$\nabla^2 A_z + \gamma^2 A_z = 0$$

except at the wire, $r = 0$. The solution to this particular differential equation is known to consist of a combination of Bessel's functions known as the Hankel functions of the second kind (Harrington, 1961). As before, $C$ is a constant which must be determined using initial or boundary conditions. Taking the magnetic field vector to be the curl of the vector potential, we have:

$$A_z = C_1 H_0^{(2)}(\gamma r) \tag{396}$$

$$\boldsymbol{H} = \nabla \times \boldsymbol{A}$$

$$H_\varphi = -\frac{\partial A_z}{\partial r} = -C_1 \frac{\partial}{\partial r} [H_0^{(2)}(\gamma r)] = C_1 \gamma H_1^{(2)}(\gamma r) \tag{397}$$

For small values of its argument, the Hankel function may be replaced by the approximation:

$$H_1^{(2)}(\gamma r) = i\frac{2}{\pi \gamma r}$$

so that for small values of $\gamma r$, we have:

$$H_\varphi = i\frac{2C_1}{\pi r} \tag{398}$$

Under these conditions, Eq. (366), developed for the static magnetic field about a long current carrying wire, should apply, so:

$$H_\varphi = \frac{I}{2\pi r}$$

$$C_1 = \frac{I}{4i} \tag{399}$$

and the proper solution for the vector potential is:

$$A_z = \frac{I}{4i} H_0^{(2)}(\gamma r) \tag{400}$$

As before, the electromagnetic field components are found by substituting this wave potential into Eqs. (378) and (385):

$$E_z = \frac{-\gamma^2 I}{4\omega \varepsilon} H_0^{(2)}(\gamma r) \tag{401}$$

$$H_\varphi = \frac{\gamma^2 I}{4i} H_1^{(2)}(\gamma r) \tag{402}$$

For large distances, and so, large values of the parameter $\gamma r$, the Hankel functions can be replaced by appropriate approximations, and so, the electric and magnetic fields can be computed from the asymptotic expressions:

$$E_z = -\sqrt{\frac{\mu}{\varepsilon}}\,\gamma I\,\sqrt{\frac{i}{8\pi\gamma r}}\,e^{-i\gamma r} \tag{403}$$

and

$$H_\varphi = \gamma I\,\sqrt{\frac{i}{8\pi\gamma r}}\,e^{-i\gamma r} \tag{404}$$

These equations represent an outward travelling wave with an amplitude decreasing as the inverse square root of distance, rather than as the inverse first power. The electric and magnetic fields are inphase and represent radiative transfer of energy.

In the case of a small loop of circulating current (see Fig. 186), it has been shown previously that for distances which are large compared with the radius

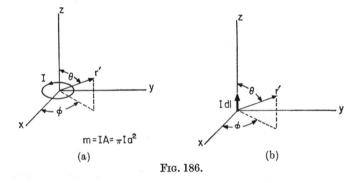

$$m = IA = \pi I a^2$$

(a)    (b)

FIG. 186.

of the loop, the loop can be represented as a magnetic dipole with magnetic moment equal to the product of current and area see (Eq. (372)). Using arguments similar to those for a short current dipole located at the origin, the vector potential for a loop of current forming a magnetic dipole at the origin can be shown to be:

$$A_\varphi = \frac{m}{4\pi}\,e^{-i\gamma r'}\left(\frac{i\gamma}{r'} + \frac{1}{r'^2}\right)\sin\theta \tag{405}$$

The electromagnetic field components about the loop may be obtained by substituting this expression for the vector potential in Eqs. (378) and (385):

$$H_{r'} = \frac{m}{2\pi}\,e^{-i\gamma r'}\left(\frac{i\gamma}{r'^2} + \frac{1}{r'^3}\right)\cos\theta \tag{406}$$

$$H_\theta = \frac{m}{4\pi}\,e^{-i\gamma r'}\left(-\frac{\gamma^2}{r'} + \frac{i\gamma}{r'^2} + \frac{1}{r'^3}\right)\sin\theta \tag{407}$$

$$E_\varphi = \frac{\sqrt{\dfrac{\mu}{\varepsilon}}\,m}{4\pi}\,e^{-i\gamma r'}\left(\frac{\gamma^2}{r'} - \frac{i\gamma}{r'^2}\right)\sin\theta \tag{408}$$

These equations have the same form as the equations for the field components about an electric dipole, except that the symbols $E$ and $H$ have been interchanged. For large values of the parameter $\omega r'$, the radiation field components may be calculated using Eqs. (395). For small values of the parameter $\omega r'$, the electric field components are negligible, and the expressions for $A$ and $H$ are:

$$A_\varphi = \frac{m}{4\pi r'^2} \sin\theta \tag{409}$$

$$H_{r'} = \frac{m}{2\pi r'^3} \cos\theta \tag{410}$$

$$H_\theta = \frac{m}{4\pi r'^3} \sin\theta \tag{411}$$

The components of the electromagnetic field observed in the $xy$ plane are:

$$H_z = \frac{m}{4\pi r'^3}\left(3\cos^2\theta - 1\right) \tag{412}$$

$$H_x = \frac{m}{4\pi r'^3}\sin\theta\,\cos\theta \tag{413}$$

These equations are useful for computing the fields about a small loop such as is used in electromagnetic surveying. The error introduced by assuming the loop to be a dipole magnetic source is less than one per cent if the distance from the loop to the observation point is more than five times the diameter of the loop.

When the distance from the loop to the observer is small and $\omega r'$ is very small, expressions for $A$ and $H$ given by Smythe (1950) may be used:

$$A_\varphi = \frac{Ih}{2\pi}\left(\frac{a}{r}\right)^{1/2}\left[\left(\frac{2}{h^2} - 1\right)\int_0^{\pi/2}\frac{d\theta}{(1 - h^2\sin^2\theta)^{1/2}} - \frac{2}{h^2}\int_0^{\pi/2}(1 - h^2\sin^2\theta)^{1/2}\,d\theta\right] \tag{414}$$

where $h^2 = 4ar[(a + r)^2 + z^2]^{-1}$

$$H_z = \frac{I}{2\pi}\frac{1}{[(a + r)^2 + z^2]^{1/2}}\left[K(h) + \frac{a^2 - r^2 - z^2}{(a - r)^2 + z^2}E(h)\right] \tag{415}$$

$$H_r = \frac{I}{2\pi}\frac{z}{r[(a + r)^2 + z^2]^{1/2}}\left[-K(h) + \frac{a^2 + r^2 + z^2}{(a - r)^2 + z^2}E(h)\right] \tag{416}$$

where the two integral expressions are actually the complete elliptic integrals of the first ($K$) and second ($E$) kinds.

## 37. MODEL STUDIES

The computation of the secondary magnetic fields associated with conductive bodies is much more difficult than the calculation of the primary field about wires and loops. Mathematical expressions for the behavior of secondary fields have been developed for only a few simple cases, such as that of a sphere or that of a horizontally stratified earth. In most cases, the formal expressions contain integrals which can be evaluated only by using numerical methods and high-speed digital computers.

Scaled model studies may be used to determine the structure of secondary magnetic fields when the mathematical approach is unreasonably difficult. Geologic conductors may be simulated by small scale models of a size which may be handled readily in a laboratory. The reduction in size is accompanied by an increase in conductivity, an increase in the frequency, or by a combination of these factors.

Since only the ratio of the secondary field amplitude to the primary field amplitude (or to the current generating the primary field) is measured in either field surveys or model studies, the results may be expressed in a way so that they are independent of the absolute dimensions of the model. Moreover, for any likely material, the electrical properties do not depend on the amplitude of the electric fields or currents used in measuring them, so it is not necessary to simulate the amplitude of the inducing magnetic field in going from the full-scale situation to the model. In scaling, it is necessary only that the dimensionless parameter, $\gamma L$ ($\gamma$ is the wave number defined previously) be the same for the model as for the full scale situation, where $L$ represents a linear dimension either full scale or in the model (Stratton, 1941). For the frequencies used in induction electromagnetic methods, the effects of displacement currents (the second term in Eq. 273) may be neglected, and the simplified expression

$$\gamma^2 = i\mu\omega\sigma$$

may be used for the wave number. In this case, the requirements for similitude between a full-scale situation and a model are that:

$$\sigma_1\mu_1\omega_1 L_1^2 = \sigma_2\mu_2\omega_2 L_2^2 \tag{417}$$

where the index 1 indicates a quantity measured in the full-scale situation and the index 2 indicates a quantity measured in the model. Since the magnetic permeability of most rocks is very nearly equal to that of free space, variations in permeability are rarely of importance, and the permeability may be assumed to be that of free space both for the full-scale situation and for the model so that:

$$\sigma_1\omega_1 L_1^2 = \sigma_2\omega_2 L_2^2 \tag{418}$$

As an example, consider that the dimensions of a model are to be one-thousandth the dimensions of the full-scale situation ($L_1 = 1000\, L_2$). The requirement for similitude is then:

$$\sigma_2\omega_2 = \sigma_1\omega_1 \times 10^6$$

It will be shown later that the principle of equivalence holds for a thin conductive sheet for the induction method, just as it applies to the galvanic resistivity method and the magneto-telluric method. At low frequencies, the induction electromagnetic response depends on the conductance, $S$, of a thin conducting sheet, rather than on the thickness or conductivity independently. This is important in the construction of models, since one conducting sheet may be substituted for another if both sheets have the same product of conductivity and thickness, $S$. This permits some latitude in selecting material for the construction of a model provided the limits required for equivalence are met.

When purely inductive methods are used in prospecting for ore bodies (current is induced in the ground, rather than driven in through galvanic contacts), the effect of eddy currents in the host rock and in the overburden is usually small, and may be neglected in model studies of such situations. In such cases, models may be constructed using metal sheets in air. Often, the frequency used in the model may be the same as the frequency used in the full-scale situation. The conductivity of a typical ore body consisting of massive sulfides should be in the range from 1 to 1000 mhos/m. Considering a model constructed of non-magnetic stainless steel having a conductivity of $1\cdot40 \times 10^6$ mhos/m and using the same frequency in the model as in the full-scale situation, a model with a linear scale reduction of 1000 : 1 would represent a geologic body with a conductivity of $1\cdot4$ mhos/m. A model constructed from aluminium having a conductivity of $3\cdot45 \times 10^7$ mhos/m with a scale reduction of 200 : 1 would simulate a geologic body having a conductivity of 853 mhos/m.

Horizontal sheets of tin or aluminium foil are sometimes placed over a model of an ore body to simulate the effect of eddy currents in a conductive over-burden. Using low frequencies and a scale reduction of 200 : 1, a sheet of aluminium foil $0\cdot1$ mm thick would simulate a 10-m thickness of overburden with a resistivity of 116 ohm-m. To best approximate actual conditions, it is necessary to provide a low resistance contact between the model overburden and the model ore body so that eddy currents can flow between the two.

If galvanic energizing is used, such as in the case of a long grounded-wire source, or if the conductivity of the host rock cannot be neglected, it is necessary to use a continuous medium in a model study. An electrolytic solution in a large tank can be used to simulate a conductive half-space, with various conductive or resistant bodies immersed in the tank serving as anomalous bodies. Solutions of sodium chloride are used most commonly, though solutions such as sulfuric acid have been used when it is necessary to obtain a higher conductivity than can be obtained readily with a salt solution. It is rarely possible to use the same frequencies in the model as in the full-scale situation when an electrolyte is used to represent a conductive half-space. For example, using a scale reduction of 1000 : 1 in the model and a brine containing 15 per cent sodium chloride by weight which has a resistivity of approximately $0\cdot06$ ohm-m, a model frequency of 60,000 c/s must be used to simulate a full scale situation where the resistivity of the host rock is 500 ohm-m and a frequency of 500 c/s is used.

Since the frequencies which must be used with electrolytic tank models are higher than the frequencies used in the full-scale situation, metals generally have too high a conductivity to be used in simulating ore bodies. Commercially available blocks of carbon and graphite, which have conductivities in the range from about $2 \times 10^4$ to $1 \cdot 25 \times 10^5$ mhos/m are usually satisfactory for use with electrolytic tank models. Unfortunately, many forms of graphite are highly anisotropic and cannot be used in models which must be isotropic.

At low frequencies, the contact resistance between the model and the electrolytic solution may be significant, and may be a function of frequency (overvoltage effects such as this are discussed in Chapter VIII). Overvoltage effects can be considered to be negligible at frequencies above about 10 kc/s. In some cases, a model may be deliberately coated with an insulating film to block current flow between the model and the electrolyte. This may be done to simulate a massive sulfide ore deposit surrounded by non-conductive gangue minerals. However, in general, current flow must be permitted between the model and the electrolyte in order to simulate field conditions.

Problems other than that of a conductive ore body surrounded by a conductive host rock may be studied using an electrolytic tank model. For example, water-proof wooden or concrete blocks can be used to simulate resistant bodies in a conductive medium. This situation may be of practical interest in surveys for ground water, and possibly in other cases.

The instrumentation used in scaled model studies is similar to the equipment used in actual field surveys. Interference from power-line harmonics and other industrial noise sources is always a problem in the laboratory. In order to obtain a favorable ratio of signal level to noise level, the transmitting loops used in model studies are made as large as possible within the limitation that they should behave as simple dipole sources at the distances involved. In practice, this limitation is sometimes violated. The power dissipated in a small transmitting coil used for model studies may be a half watt or more; heating of the coil is a limiting factor in determining the power which may be used.

When the same frequency is used in a model study as in the full-scale situation, field ratiometers and amplifiers may be used. When the model requires a higher frequency than that used in field surveys, a special high-frequency ratiometer must be constructed. In some cases, commercially available vacuum-tube voltmeters and phase angle meters may be used. This type of equipment has the advantage that a variety of frequencies may be used, but generally the precision of measurement is not as great as can be obtained with equipment constructed specifically for the purpose.

At high frequencies, capacitive coupling between the coils and the model and other stray capacitances can be troublesome. Proper shielding and grounding of the coils and connecting cables is very important.

Model apparatus must include provisions for positioning models, as well as some arrangement to support the coils so that profiles may be run at various distances and directions from the model. In some model setups, measurements can be made continuously as the coils are moved over the models, the measured parameters being recorded on strip charts.

## 38. DEPTH-SOUNDING METHODS

As in the case of galvanic resistivity methods, induction methods may be subdivided into groups containing techniques for studying variations of conductivity with depth (depth-sounding methods) and techniques for studying lateral changes in conductivity (horizontal profiling methods). Electromagnetic depth sounding methods have not yet been used extensively. However, for certain types of problems, electromagnetic depth sounding methods are more effective than galvanic sounding methods, so that it is likely that such techniques will be more widely used in the future. In the past, the lack of equipment capable of the required accuracy of measurements over a continuous frequency spectrum and the lack of theoretically-derived reference curves have held back the development of depth sounding techniques.

In making a depth sounding with an electromagnetic method, either the frequency or the spacing between the transmitter and receiver may be varied to vary the effective depth of penetration. It is usually preferable to vary the frequency rather than the spacing, so that the transmitter and receiver can be fixed in location. This reduces the effects of lateral changes in resistivity. However, there is an optimum coil separation which provides the greatest sensitivity to the presence of a boundary at a given depth. Thus, if the depths to several layers are to be determined, it may be necessary to make variable frequency measurements with several different transmitter–receiver spacings. The spacings may be chosen to emphasize the presence of each successive boundary.

Either small loop or long wire sources may be used in making soundings. Grounded wires are usually much longer than the separation between the wire and the receiver, so it is a reasonable assumption that such a source can be treated as being infinitely long. Large loops may be used rather than a grounded wire, provided measurements are made sufficiently close to one side of such a loop that the field from the other three sides is negligible.

Induction depth soundings may be interpreted by matching with theoretically derived reference curves in much the same way that galvanic soundings are interpreted. Reference curves may be obtained either by model studies or by computation. Most of the reference curves for interpreting depth soundings have been obtained by calculation. In contrast, most reference curves used in interpreting horizontal profiles have been obtained from model studies.

The effect of a very conductive earth can be determined fairly simply with the image approach used earlier in developing the galvanic resistivity equations. In this case, with a long wire situated over a perfectly conducting earth (reflection coefficient of $-1$), we can assume the presence of an image wire an equal distance below the conducting surface, carrying an equal but opposite current. The geometry of such a problem is shown in Fig. 187, where the actual and the image wires are shown running into and out of the paper. At an obser-

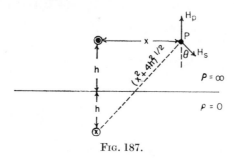

FIG. 187.

vation point, P, at a distance $x$ from the actual wire and in the same horizontal plane, the vertical component of the magnetic field, calculated using Eq. (366), is:

$$H_{p,z} = \frac{I}{2\pi x}$$

for the primary field (that due only to current flow in the actual wire) and

$$H_{s,z} = -\frac{I \cos \theta}{2\pi [x^2 + 4h^2]^{1/2}}$$

$$= -\frac{Ix}{2\pi [x^2 + 4h^2]} \tag{419}$$

for the secondary field (that due to the current apparently flowing in the image wire). The total magnetic field strength in the vertical direction at the observation point is the sum of these two fields:

$$H_z = \frac{Ix}{2\pi} \left( \frac{1}{x^2} - \frac{1}{x^2 + 4h^2} \right) \tag{420}$$

Usually, the measured field is expressed in terms of the primary field as follows:

$$\frac{H_z}{H_{z,p}} = 1 - \frac{1}{1 + 4\left(\dfrac{h}{x}\right)^2} \tag{421}$$

It should be noted that this ratio is always less than one, and approaches zero as $h/x$ approaches zero.

The horizontal component of the secondary field is:

$$H_{s,x} = \frac{Ih}{\pi[x^2 + 4h^2]} = \frac{I}{\pi}\frac{\dfrac{h}{x}}{1 + 4\left(\dfrac{h}{x}\right)^2} \tag{422}$$

The ratio of the horizontal to the vertical field is $x/2h$. Using this relationship, the depth to a perfectly conducting sheet may be calculated readily by measuring this ratio or the inclination of the magnetic field at a single known distance, $x$, from a long wire source.

The concept of a perfectly conducting earth is valid only at very high frequencies or in the case of very excellent conductors, such as flat-lying massive sulfide deposits. When the overburden resistivity is high and the frequency of the primary field is high enough that the primary field does not penetrate an appreciable distance into an ore body (all the induced current flow is at the surface of the ore body), Eqs. (421) and (422) are very good approximations. When the conductivity and the frequency are not both large, the induced currents may flow at some depth within the ore body, and the apparent depth will appear to be greater than in the case of a perfectly conducting ore body. Also, the secondary field will not be in phase with the primary field.

Another case which may be treated simply is that of a perfectly resistant earth with a magnetic permeability different than that of free space. This problem can be solved readily using the image method. With a current-carrying wire over a permeable half-space, the direction of the image wire is such that the projection of the source and of the image on the interface coincide in position and direction. The strength of the apparent image current is:

$$I' = \frac{\mu_1 - \mu_0}{\mu_1 + \mu_0}I \tag{423}$$

where $\mu_1$ and $\mu_0$ are the values for the magnetic permeability of the half-space and free space, respectively. If the wire and the point of observation are located on the same horizontal plane (see Fig. 188), the vertical component of the secondary field is:

$$H_{s,z} = \frac{I}{2\pi}\left(\frac{\mu_1 - \mu_0}{\mu_1 + \mu_0}\right)\left(\frac{x}{x^2 + 4h^2}\right) \tag{424}$$

Considering the primary field, the total field strength in the vertical direction is:

$$H_z = \frac{Ix}{2\pi}\left[\frac{1}{x^2} + \frac{\mu_1 - \mu_0}{\mu_1 + \mu_0}\left(\frac{1}{x^2 + 4h^2}\right)\right] \tag{425}$$

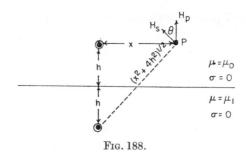

The ratio of the total vertical component to the primary field strength is:

$$\frac{H_z}{H_{p,z}} = 1 + \frac{\mu_1 - \mu_0}{\mu_1 + \mu_0} \left[ \frac{1}{1 + 4\left(\dfrac{h}{x}\right)^2} \right] = 1 + \frac{\dfrac{\mu_1}{\mu_0} - 1}{\dfrac{\mu_1}{\mu_0} + 1} \left[ \frac{1}{1 + 4\left(\dfrac{h}{x}\right)^2} \right] \quad (426)$$

In contrast to the case of a conducting earth, the ratio over a permeable earth is always greater than one, and approaches two, as a limit, as the ratio $h/x$ approaches zero, providing the ratio in permeabilities is large.

The horizontal component of the secondary field for an observation point in the same horizontal plane as the source wire is:

$$H_{s,x} = \frac{I}{2\pi} \frac{\mu_1 - \mu_0}{\mu_1 + \mu_0} \left[ \frac{2h}{x^2 + 4h^2} \right] \quad (427)$$

and the ratio of the horizontal component to the vertical component is:

$$\frac{H_{s,x}}{H_z} = \frac{(\mu_1 - \mu_0)\dfrac{h}{x}}{2(\mu_1 + \mu_0)\dfrac{h}{x} + \mu_1} = \frac{\dfrac{h}{x}\left(\dfrac{\mu_1}{\mu_0} - 1\right)}{2\left(\dfrac{\mu_1}{\mu_0} + 1\right) + \dfrac{\mu_1}{\mu_0}} \quad (428)$$

The magnetic permeability of rocks usually varies over too narrow a range to cause significant electromagnetic anomalies. As an example, the ratio $H_z/H_{p,z}$ is only 1·004 when $h/x = 0·25$ and $\mu/\mu_0 = 1·01$. An anomaly of 0·4 per cent is not usually measurable with most techniques. One might attempt to measure only the horizontal component of the secondary field in the horizontal plane containing the source wire, since there is no horizontal component of the primary field under such conditions. However, such measurements would require very precise orientation of the detector to avoid errors due to measuring a component of the vertical field.

While they provide some insight into the problem, calculations based on the concept of a perfectly conducting earth or a perfectly insulating magnetic earth are seldom useful. More general mathematical results or scale model studies are required.

A formal solution for the field about a long wire over a layered earth may be obtained, but in general, evaluation of the resulting integral equation requires the use of a high speed computer. Numerical results are available for the magnetic fields above a homogeneous earth or a thin sheet, and for the electric fields induced by an electromagnetic field at the surface of a two-layer earth.

The equation for the electromagnetic field about a long wire may be developed as follows (Wait, 1953). First, consider a plane electromagnetic wave incident on the surface of a stratified earth consisting of $M$ layers resting on a conductive half-space. The direction of the electric field vector is taken to be perpendicular to the plane of incidence (the $xz$ plane in Fig. 189), and the angle of incidence, $\theta$, is measured from the negative (upward) $z$ axis. The conductivity and di-

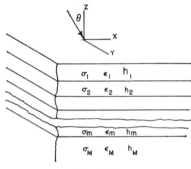

Fig. 189.

electric constant for each layer are designated by $\sigma_m$ and $\varepsilon_m$, respectively. The magnetic permeability of all layers is assumed to have the same value, $\mu_0$. As was established for the magneto-telluric problem in Chapter IV, the electric field in any of the layers is a solution of the wave equation:

$$\nabla^2 E_m - \gamma_m^2 E_m = 0 \qquad (429)$$

However, in the present problem, the possibility that displacement currents may be appreciable can be taken into account by using the complete expression for the wave number:

$$\gamma_m^2 = i\mu_0\omega\sigma_m - \mu_0\omega^2\varepsilon_m$$

In this case, in order to conform with most of the literature, the factor $e^{i\omega t}$ is used to represent a sinusoidal variation in the amplitude, rather than $e^{-i\omega t}$, as was used in the solution of the magneto-telluric problem. The sign which is chosen is a matter of convention, not having any particular significance in the solution of the problem.

Because of the way in which we chose our coordinate system, there is a $y$ component only of the electric field vector in any of the layers. Equation (429) may therefore be simplified to the form:

$$\frac{\partial^2 E_{m,y}}{\partial x^2} + \frac{\partial^2 E_{m,y}}{\partial z^2} - \gamma_m^2 E_{m,y} = 0 \tag{430}$$

Using the method of separation of variables as was done for the galvanic resistivity problem in Chapter III, two independent differential equations may be obtained:

$$\frac{d^2 F_m(x)}{dz^2} + \lambda^2 F_m(x) = 0 \tag{431}$$

$$\frac{d^2 G_m(z)}{dz^2} - u_m^2 G_m(z) = 0 \tag{432}$$

where $F_m \cdot G_m = E_{m,y}$

$$\gamma_m^2 = u_m^2 - \lambda^2$$

The solution to Eq. (429) may be formed from the product of the solutions for Eqs. (431) and (432):

$$E_{y,m} = (C_1 e^{-i\lambda x} + C_2 e^{i\lambda x}) \cdot (C_3 e^{-u_m z} + C_4 e^{u_m z}) \tag{433}$$

Considering only waves travelling in the positive $x$ direction, the constant $C_2$ must be made zero. The assumed solution may be rewritten as:

$$E_{y,m} = [a_m e^{-u_m z} + b_m e^{u_m z}] e^{-i\lambda x} \tag{434}$$

where the arbitrary constant, $\lambda$, may have any value whatsoever.

The electromagnetic field in the air may be computed using the theory for reflection of plane waves from a conducting interface:

$$E_{y,0} = a_0 \exp[-\gamma_0(z \cos \theta + x \sin \theta)] + b_0 \exp[\gamma_0(z \cos \theta - x \sin \theta)]$$
$$= [a_0 \exp(-\gamma_0 z \cos \theta) + b_0 \exp(\gamma_0 z \cos \theta)] \exp(-\gamma_0 x \sin \theta) \tag{435}$$

where $a_0 \exp[-\gamma_0(z \cos \theta + x \sin \theta)]$ represents the amplitude of the incident wave and $b_0 \exp[\gamma_0(z \cos \theta - x \sin \theta)]$ represents the amplitude of the reflected wave.

By comparison with Eq. (434), we see that the parameter $\lambda$ must have the value

$$i\lambda = \gamma_0 \sin \theta \tag{436}$$

In the lowermost layer (the conductive half-space), only waves travelling downward are permissible, and so, the coefficient $b_M$ must be zero. The rest of the coefficients ($a_0$ to $a_m$, $b_0$ to $b_m$ and $a_M$) can be evaluated by applying appropriate boundary conditions at each of the boundary planes. These conditions state that $E_{m,y}$ and $H_{m,x}$ must be continuous across each interface.

These conditions applied at the boundary between the $m$th and $m$th $+ 1$ layer (where $z = z_m$) lead to the equation:

$$a_m \exp(-u_m z_m) + b_m \exp(u_m z_m) = a_{m+1} \exp(-u_{m+1} z_m) + b_{m+1} \exp(u_{m+1} z_m)$$
(437)

The components of the magnetic field vector may be found by applying one of Maxwell's equations:

$$\nabla \times E = -\mu_0 \frac{\partial H}{\partial t}$$

$$H_{m,x} = \frac{1}{i\mu_m \omega} \frac{\partial E_{m,y}}{\partial z}$$
(438)

$$H_{m,z} = -\frac{1}{i u_m \omega} \frac{\partial E_{m,y}}{\partial x}$$
(439)

Combining Eq. (437) with Eq. (438), the value for $H_{m,x}$ may be obtained for the $m$th and $m$th $+ 1$ layers and equated as required by the condition of continuity:

$$\frac{a_m u_m}{\mu_m} \left[ a_m \exp(-u_m z_m) - b_m \exp(u_m z_m) \right]$$

$$= \frac{a_{m+1} u_{m+1}}{\mu_{m+1}} \left[ a_{m+1} \exp(-u_{m+1} z_m) - b_{m+1} \exp(u_{m+1} z_m) \right]$$
(440)

As indicated by Wait (1953) this procedure yields a set of $2m$ simultaneous algebraic equations which may be solved progressing backwards from the lowermost boundary to provide the result:

$$\frac{b_0}{a_0} = \frac{k_0 - Z_1}{k_0 + Z_1} = R(\lambda)$$
(441)

where

$$Z_1 = k_1 \frac{Z_2 + k_1 \tanh u_1 h_1}{k_1 + Z_2 \tanh u_1 h_1}$$

$$Z_2 = k_2 \frac{Z_3 + k_2 \tanh u_2 h_2}{k_2 + Z_3 \tanh u_2 h_2}$$

$$\vdots$$

$$Z_{M-1} = k_{M-1} \frac{Z_M + k_{M-1} \tanh u_{M-1} h_{M-1}}{k_{M-1} + Z_M \tanh u_{M-1} h_{M-1}}$$

$$Z_M = k_M$$

$$k_m = \frac{u_m}{i\mu_m \omega}; \quad m = 0, 1, 2 \cdots M$$
(442)

A general solution for Eq. (430) was found by Wait by taking a linear combination (sum) of special solutions for specific values of the separation parameter, $\lambda$. Since $\lambda$ may assume a continuum of values, such a linear combination in the most general form is an integral over a range in $\lambda$ from $-\infty$ to $\infty$ for solutions of the form in Eq. (434):

$$E_{y,0} = \int_{-\infty}^{\infty} a_0(\lambda) \left[ e^{-u_0 z} + R(\lambda) e^{u_0 z} \right] e^{-i\lambda x} d\lambda \qquad (443)$$

This integral is of the same form as a Fourier integral with the subject function:

$$a_0(\lambda) \left[ e^{-u_0 z} + R(\lambda) e^{u_0 z} \right]$$

Now consider the source of the electromagnetic field to be a long wire along the $y$ axis at a height $-h$, and carrying a current, $I$. In Eq. (443), the parameter $a_0$ may be evaluated by considering the field about the wire in the absence of the conductive layers. The effect of the conductive layers is removed by setting $R(\lambda)$ equal to zero. The inverse of the Fourier integral may be used in solving Eq. (443) for the parameter $a_0$:

$$a_0(\lambda) e^{-u_0 z} = \frac{1}{2\pi} \int_{-\infty}^{\infty} E_{y,0} e^{i\lambda x} d\lambda \qquad (444)$$

The value for $E_{y,0}$ should be the same as the value for $E_z$ in Eq. (401):

$$\begin{aligned} E_{y,0} &= -\frac{\gamma_0^2 I}{4\omega\varepsilon_0} H_0^{(2)}(\gamma_0 r) \\ &= \frac{I\omega\mu_0}{4} \left[ \frac{2i}{\pi} K_0(\gamma_0 r) \right] \\ &= \frac{iI\omega\mu_0}{2\pi} K_0 \left( \gamma_0 \{x^2 + (z+h)^2\}^{1/2} \right) \end{aligned} \qquad (445)$$

where $r = \{x^2 + (z+h)^2\}^{1/2}$ and where $K_0$ is the modified Bessel's function of the second kind. Substituting this last expression in Eq. (444) in place of $E_{y,0}$, we have:

$$a_0(\lambda) e^{-u_0 z} = \frac{iI\mu_0\omega}{4\pi^2} \int_{-\infty}^{\infty} K_0(\gamma_0 \{x^2 + (z+h)^2\}^{1/2}) e^{i\lambda x} d\lambda \qquad (446)$$

The definite integral in Eq. (446) may be evaluated using form 917 from Campbell's and Foster's tables:

$$a_0(\lambda) = \frac{i\mu_0\omega I \exp - h(\gamma_0^2 - \lambda^2)^{1/2}}{4\pi u_0} \qquad (447)$$

The electric field in air is found by substituting the expression from Eq. (447) in Eq. (443):

$$E_{y,0} = \frac{i\mu_0\omega I}{2\pi} \int_0^{\infty} \frac{1}{u_0} \left[ e^{-u_0(z+h)} + R(\lambda) e^{u_0(z-h)} \right] \cos \lambda x \, d\lambda \qquad (448)$$

Equation (448) is a formal solution for the electric field over a layered earth. Equivalent expressions for the magnetic field can be found using Eqs. (448) and (438) or (439). We shall now consider several special cases for which Eq. (448) has been evaluated.

First, consider a long, current-carrying wire above a homogeneous half-space which has the electrical properties $\sigma_1$, $\varepsilon_1$ and $\mu_0$. Then:

$$R(\lambda) = \frac{u_0 - u_1}{u_0 + u_1}$$

$$E_{y,0} = \frac{i\mu_0\omega I}{2\pi} \int\limits_0^\infty \frac{1}{u_0} \left[ e^{-u_0(z+h)} + \frac{u_0 - u_1}{u_0 + u_1} e^{u_0(z-h)} \right] \cos \lambda x \, d\lambda \qquad (449)$$

If the wire is on the surface of the half-space ($h = 0$), this equation becomes:

$$E_{y,0} = \frac{i\mu_0\omega I}{\pi} \int\limits_0^\infty \frac{\cos \lambda x}{u_1 + u_0} \, d\lambda. \qquad (450)$$

This last equation may be rewritten as:

$$E_{y,0} = \lim_{\alpha \to 0} \frac{i\mu_0\omega I}{\pi} \int\limits_0^\infty \frac{\cos \lambda x}{u_1 + u_0} e^{-\lambda\alpha} \, d\lambda \qquad (451)$$

or as:

$$E_{y,0} = \lim_{\alpha \to 0} \frac{i\mu_0\omega I}{\pi(\gamma_1^2 - \gamma_0^2)} \left[ \int\limits_0^\infty e^{-\lambda\alpha} u_1 \cos \lambda x \, d\lambda - \int\limits_0^\infty e^{-\lambda\alpha} \cos \lambda x \, d\lambda \right] \qquad (452)$$

Using form 918 from Campbell's and Foster's tables, it follows that:

$$\lim_{\alpha \to 0} \int\limits_0^\infty \frac{\cos \lambda x}{u_1} e^{-\lambda\alpha} \, d\lambda = K_0(\gamma_1 x) \qquad (453)$$

and also that:

$$\lim_{\alpha \to 0} \int\limits_0^\infty \cos \lambda x \, e^{-\lambda\alpha} u_1 \, d\lambda = \left( \gamma_1^2 - \frac{\partial^2}{\partial x^2} \right) K_0(\gamma_1 x) = - \frac{\gamma_1 K_1(\gamma_1 x)}{x} \qquad (454)$$

Thus, Eq. (450) becomes:

$$E_{y,0} = \frac{i\mu_0\omega I}{\pi(\gamma_1^2 - \gamma_0^2) x^2} [\gamma_0 x K_1(\gamma_0 x) - \gamma_1 x K_1(\gamma_1 x)] \qquad (455)$$

Under the same conditions, the vertical component of the magnetic field
will be:

$$H_z = -\frac{1}{i\mu_0\omega} \frac{\partial E_{y,0}}{\partial x}$$

$$= \frac{I}{\pi(\gamma_1^2 - \gamma_0^2) x^3} [2\gamma_0 x K_1(\gamma_0 x) + \gamma_0^2 x^2 K_0(\gamma_0 x) - 2\gamma_1 x K_1(\gamma_1 x) - \gamma_1^2 x^2 K_0(\gamma_1 x)]$$

$$(456)$$

If the offset distance, $x$, from the wire to the point of observation is much less
than a free space wavelength, as is usually true in the application of the induc-
tion method, the condition $|\gamma_0 x| \ll 1$ means that the Bessel's function may be
approximated as:

$$K_1(\gamma_0 x) \cong \frac{1}{\gamma_0 x}$$

so that:

$$H_z = \frac{I}{\pi\gamma_1^2 x^3} [2 - 2\gamma_1 x K_1(\gamma_1 x) - (\gamma_1 x)^2 K_0(\gamma_1 x)] \qquad (457)$$

An alternate expression for the magnetic field was given in Eq. (366) for the
case in which no conductive earth was present:

$$H_{z,p} = \frac{I}{2\pi x}$$

The ratio of the magnetic field observed at the surface of a homogeneous earth
(given by Eq. 457) to the primary field strength is then (Wait, 1953):

$$\frac{H_z}{H_{z,p}} = \frac{2}{(\gamma_1 x)^2} [2 - 2\gamma_1 x K_1(\gamma_1 x) - (\gamma_1 x)^2 K_0(\gamma_1 x)] \qquad (458)$$

where the wave number, $\gamma_1$, can be expressed as follows, neglecting displacement currents:

$$\gamma_1 = (i\sigma_1\mu_0\omega)^{1/2}$$

Rather than expanding $\sqrt{i}$ in the form used in Chapter IV, we can make use
of an expression for the wave number in terms of a combination of ker and
kei functions (McLachlan, 1955):

$$\text{ker } y + i \text{ kei } y = K_0(y\sqrt{i})$$
$$i \text{ ker}_1 y - \text{kei } y = K_0(y\sqrt{i}) \qquad (459)$$

By numerical integration of an expression equivalent to that in Eq. (448),
Price (1950) obtained numerical results for the magnetic field strength on and
above the surface of a homogeneous conducting halfspace. The numerical
results given by Price are presented graphically in Fig. 190 along with a curve
computed from Eq. (458). The data are plotted as functions of the two

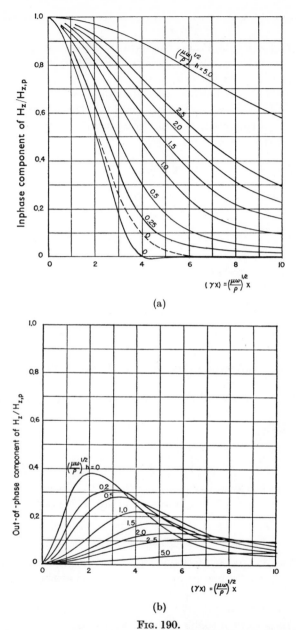

FIG. 190.

(a) Inphase component of the normalized magnetic field strength about a
long wire raised above a homogeneous earth. (Solid curves after A. T. Price,
dashed curve computed from Eq. (458).)

(b) Out-of-phase component of the normalized magnetic field strength about
a long wire raised above a homogeneous earth. (Solid curves after A. T. Price;
dashed curve computed from Eq. (458).)

parameters $\gamma_1 x$ and $\gamma_1(h + z)$. Price's results are most accurate for large values of these two parameters; the curve computed from Eq. (458) is exact for the source wire laying on the surface $(\gamma_1(h + z) = 0)$.

In the case of a very conductive earth, the quantity $\gamma_1 x$ may be quite large at distances from the wire which are still small compared to a wave length in free space. For this condition, asymptotic expressions may be used for the Bessel's functions remaining in Eq. (458):

$$K_0(\gamma_1 x) \to \sqrt{\frac{\pi}{2\gamma_1 x}}\, e^{-\gamma_1 x}$$

$$K_1(\gamma_1 x) \to \sqrt{\frac{\pi}{2\gamma_1 x}}\, e^{-\gamma_1 x}$$

$$\frac{H_z}{H_{z,p}} \cong \frac{2}{(\gamma_1 x)^2}\left[2 - \sqrt{(2\pi\gamma_1 x)}\, e^{-\gamma_1 x} - \sqrt{\frac{\pi}{2}}\gamma_1 x e^{-\gamma_1 x}\right] \tag{460}$$

and, if $|\gamma_1 x| > 8$, this equation may be written simply as:

$$\frac{H_z}{H_{z,p}} = \frac{4}{(\gamma_1 x)^2}$$

$$= -\frac{4i\varrho}{\mu_0 \omega x^2} \tag{461}$$

Thus, the magnetic field is almost exactly out of phase with the primary field for large values of the parameter $\gamma_1 x$.

The horizontal component of the magnetic field about a long wire on the surface of a conductive half-space has been obtained from Eq. (448) by Wait for the case $\gamma_0 x \ll 1$:

$$H_x = \frac{I}{2}\frac{\partial}{\partial x}\left[\frac{1}{\gamma_1 x}\{I_1(\gamma_1 x) - L_1(\gamma_0 x)\}\right] \tag{462}$$

where $I_1$ is the modified Bessel's function of the first kind and $L_t$ is Nicholson's function. For small values of the quantity $\gamma_1 x$, Eq. (462) may be simplified to:

$$H_x \cong \frac{I}{2\pi x}\left(\frac{\gamma_1 x}{\sqrt{2}}\right) \tag{463}$$

Price (1950) has computed the values for the horizontal component of the magnetic field about a long wire on or above a conducting half space (see Fig. 191).

These curves may be used to determine the resistivity of a homogeneous earth from measurements of the magnetic field strength about a long current-carrying wire. For the case in which the wire is on the surface of the earth, in principle at least, it is necessary only to measure the inphase component at one frequency or the out-of-phase component at two frequencies in order to determine the resistivity. At least two measurements of the out-of-phase component are required, since a single measurement may provide ambiguous results. In practice, the quantity $H_z/H_{z,p}$ should be measured over a range of values for the parameter $\gamma_1 x$ by varying either the frequency of the current flowing in the wire or the distance from the wire at which observations are made. Comparison of the results of such a series of measurements with the reference curves in Figs. 190 and 191 will indicate whether or not the assumption of a homogeneous earth is reasonably valid and will provide a more certain value for the resistivity if there are local lateral variations in resistivity.

One procedure for determining the resistivity from a set of measurements consists of plotting the field data to a set of bi-logarithmic coordinates, using $x/f$ as the abscissa, $f$ being the frequency in c/s. This plot of the field data is then overlain on a set of reference curves drawn to the same bi-logarithmic coordinates and moved about until the data points match simultaneously with reference curves for the inphase and out-of-phase components. If the two sets

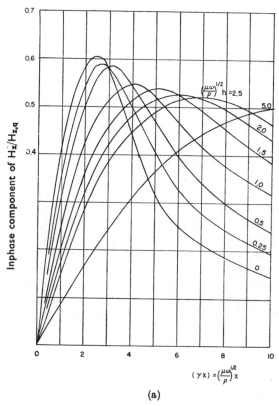

(a)

FIG. 191.

(a) Inphase component of the normalized horizontal field strength about a long wire raised above a homogeneous conducting half-space (after A. T. Price).

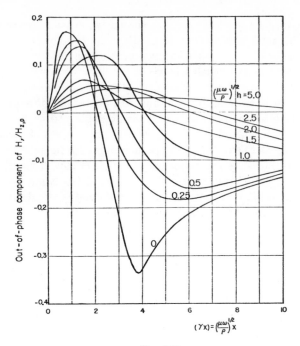

FIG. 191.

(b) Out-of-phase component of the normalized horizontal field strength about a long wire raised above a homogeneous conducting half-space (after A. T. Price).

of data cannot be matched simultaneously with the reference curves for the same set of parameters, it means that there are vertical or lateral changes in conductivity within the earth.

Once a match between the field data and the reference curves has been made, the resistivity of the earth can be determined by comparing matching coordinates on the field and reference curves. For example, the value of $x\sqrt{f}$ may be read from the field plot at a point where the parameter $\gamma_1 x = 1$ on the reference curves. The resistivity is then determined simply from the expression:

$$\varrho = 2\pi f \mu_0 x_1^2 \tag{464}$$

When logarithmic curve matching is used, it is not necessary to convert the field measurements of voltage and current into magnetic field values to obtain the ratio $H_z/H_{z,p}$. For example, the voltage, $V$, induced in the receiving coil divided by the frequency, $f$, is proportional to the ratio $H_z/H_{z,p}$:

$$\frac{V}{f} = C_1 \frac{H_z}{H_{z,p}} \tag{465}$$

If the ratio $V/f$ is plotted as a function of $x \sqrt{f}$ on the field curve, a match between the field plot and the reference curves provides values not only for $\gamma_1$ but also for $C_1$.

The amplitude and phase angle for the magnetic field may be used rather than the inphase and out-of-phase components in determining resistivity. In comparing field data with reference curves, the amplitude measurements are plotted as a function of $x \sqrt{f}$ on bi-logarithmic coordinates, while the phase angle data are plotted to semi-logarithmic coordinates. An example, showing amplitude and phase angle measurements over a thick section of the Helmut fanglomerate in southern Arizona, is given in Fig. 192. Measurements were made with a fixed frequency (500 c/s) but with a varying offset distance, $x$. The data are

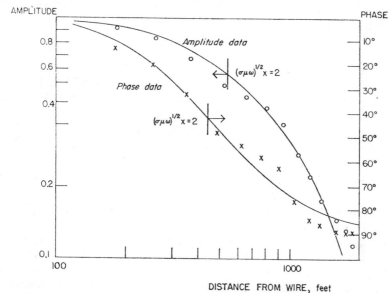

FIG. 192. Example of amplitude and phase data measured as a function of distance from a long grounded-wire carrying current at 500 c/s. Matching the data with reference curves as shown provides an apparent resistivity of 24 ohm-m for the amplitude data and of 16 ohm-m for the phase data.

shown with the reference curves which provide the best match. The correspondence between the field data and the reference curves is far from perfect, departures being caused by local variations in resistivity, but the apparent resistivity determined for the rock is a reasonable value.

In interpreting these data for the match shown, the value $x \sqrt{\omega} = 8740$ on the field plot matches with the value $\gamma_1 x = 2 \cdot 0$ on the reference curves. For the phase angle measurements, the value $x \sqrt{\omega} = 7120$ on the field plot matches with the value $\gamma_1 x = 2 \cdot 0$ on the reference curves. These two matching points provide two values for the resistivity, 24 and 16 ohm-m, respectively.

In curve matching, it is helpful if measurements can be made over a range in frequencies and spacings such that the plot of field data exhibits some diagnostic feature, such as a maximum. Considering this, the curves for the ratio $H_x/H_{z,p}$ provide better matching conditions than the curves for the ratio $H_z/H_{z,p}$, since the point where the out-of-phase component of $H_x/H_{z,p}$ crosses the zero axis can be detected easily. If it can reasonably be assumed that the earth over which measurements are being made is homogeneous, the resistivity may be determined directly by relating the value for $x\sqrt{f}$ measured in a diagnostic region of a reference curve to the appropriate value of $\gamma_1 x$ for that portion of the reference curve.

If the ratio $H_z/H_{z,p}$ is determined for spacings and frequencies such that the condition $|\gamma_1 x| > 8$ holds, the resistivity for a homogeneous half-space may be obtained from Eq. (461) without using reference curves:

$$\varrho = i\,\frac{\mu_0 \omega x^2}{4}\,\frac{H_z}{H_{z,p}} \tag{466}$$

The reference curves shown in Figs. 190 and 191 may be used to determine both the resistivity of a conducting half-space and the height of the source cable above the half-space, if the cable is elevated. One set of measurements of the inphase and out-of-phase components of $H_z/H_{z,p}$ for values of $\gamma_1 x < 3$ or of $H_x/H_{z,p}$ for values of $\gamma_1 x < 1$ would be sufficient under ideal conditions to determine both $\varrho$ and $h$. For values of $\gamma_1 x$ greater than these minimum values, the reference curves cross one another, and more sets of data points would be required to avoid possible ambiguities.

The procedure which is generally used in practice consists of making measurements over a range of values for $x\sqrt{\omega}$, following which the data are interpreted by curve matching. This is the same procedure used for a long wire resting directly on the conductivity half-space, with the exception that the data may be matched to one of a family of reference curves, rather than to a unique reference curve. The resistivity is determined from the correspondence between the coordinate systems on which the data and the reference curves are plotted, as before. The height of the cable above the half-space may be determined once the resistivity has been determined, using the parameter $\gamma_1 h$ assigned to the particular member of the family of reference curves which was selected as the best match with the field data.

The curves in Figs. 190 and 191 are plotted for parametric values of the quantity $\gamma_1 h$, and are designed for use with field measurements in which the frequency is fixed, and only the spacing, $x$, is varied in making a survey. Price's curves are replotted in Fig. 193 with parametric values of the ratio $h/x$ for use in interpreting experimental data obtained with a varying frequency but a fixed spacing.

We shall now consider another case for which reference curves are available, that of a long wire over a thin conducting sheet. In order to find the field

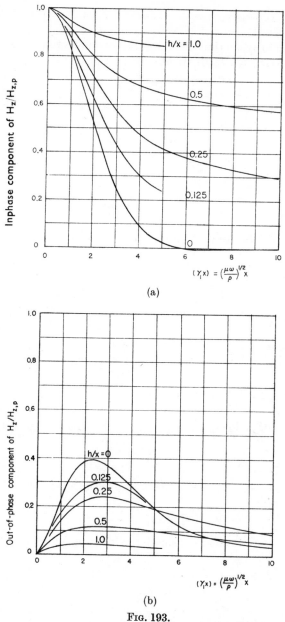

Fɪɢ. 193.

(a) Inphase component of the normalized vertical field strength about a long wire above a homogeneous conducting half-space.
(Solid curves after A. T. Price; dashed curve computed from Eq. (458).)
(b) Out-of-phase component of the normalized vertical field strength about a long wire above a homogeneous conducting half-space. (Solid curves after A. T. Price; dashed curve computed from Eq. (458).)

strength components over a thin sheet, we let the conductivity of all the layers but the first be zero in Eq. (448) (Wait, 1953):

$$M = 2$$

$$u_m = u_0$$

$$\gamma_2 = \gamma_0$$

so that:

$$R(\lambda) = \frac{u_0 - u_e}{u_0 + u_e}$$

where:

$$u_e = u_1 \frac{u_0 + u_1 \tanh u_1 h_1}{u_1 + u_0 \tanh u_1 h_1}$$

From Eq. (448), the electric field intensity in the air is:

$$E_y = \frac{i\mu_0 \omega I}{2\pi} \int_0^\infty \frac{1}{u_0} \left[ e^{-u_0(z+h)} + \frac{u_0 - u_e}{u_0 + u_e} e^{u_0(z-h)} \right] \cos \lambda x \, d\lambda \qquad (467)$$

or, rearranging terms:

$$E_y = \frac{i\mu_0 \omega I}{2\pi} \int_0^\infty \left[ \frac{e^{-u_0(z+h)}}{u_0} - \frac{e^{u_0(z-h)}}{u_0} + \frac{2}{u_e + u_0} e^{u_0(z-h)} \right] \cos \lambda x \, d\lambda \qquad (468)$$

or, using identities for Bessel's functions:

$$E_y = \frac{i\mu_0 \omega I}{2\pi} \left\{ K_0(\gamma_0 r_1) - K_0(\gamma_0 r_2) + \int_0^\infty \frac{2 e^{u_0(z-h)}}{u_e + u_0} \cos \lambda x \, d\lambda \right\} \qquad (469)$$

where $r_1^2 = (z + h)^2 + x^2$ and $r_2^2 = (z - h)^2 + x^2$.

If we again ignore the contribution of displacement currents to the wave number, $\gamma_1$:

$$\gamma_1^2 = i\sigma_1 \mu_0 \omega$$

and if the thickness of the sheet is small, the following approximations may be made:

$$\tanh u_1 h_1 = u_1 h_1 = u_1 t$$

and

$$u_1^2 h_1 = 2iq$$

where

$$q = \sigma_1 \mu_0 \omega t / 2$$

so that

$$u_e = u_0 + 2iq$$

Moreover, if the distance between the wire and the measurement point is much less than a free-space wavelength, we can let $\gamma_0 = 0$, so that $u_0 = \lambda$. Then, when $z$ is small:

$$K_0(z) \cong -(\log z + C - \log 2) \qquad (470)$$

so that a simplified expression for the electric field intensity is:

$$E_y \cong \frac{i\mu\omega I}{2\pi} \left\{ \log \frac{r_2}{r_1} + J(B) \right\} \tag{471}$$

where $B = h - z$, and

$$J(B) = \int\limits_0^\infty \frac{e^{-\lambda B}}{\lambda + iq} \cos\lambda x \, d\lambda \tag{472}$$

The $J$-integral in Eq. (472) may be integrated. First, we make use of an integral identity:

$$\frac{i}{i\lambda - q} = -i \int\limits_0^\infty e^{(i\lambda - q)} \, d^\alpha \, d\alpha$$

so that the $J$-integral may be rewritten as a double integral:

$$J(B) = -i \int\limits_0^\infty \int\limits_0^\infty e^{-(B - i\alpha)} \, e^{-q\alpha} \cos\lambda x \, d\lambda \, d\alpha \tag{473}$$

The $J$-integral in this form may be integrated with respect to $\lambda$:

$$J(B) = -i \int\limits_0^\infty \frac{e^{-qx}(B - i\alpha) \, d\alpha}{(B - i\alpha)^2 + x^2} \tag{474}$$

This integral expression may now be integrated in parts as follows:

$$dv = e^{-q\alpha} \, d\alpha$$

$$u = \frac{B - i\alpha}{(B - i\alpha)^2 + x^2}$$

$$v = -\frac{1}{q} e^{-qx}$$

$$J(B) = i \left[ \frac{1}{q} e^{-q\alpha} \left\{ \frac{B - i\alpha}{(B - i\alpha)^2 + x^2} \right\} \right]_{\alpha=0}^{\alpha=\infty} + \frac{1}{q} \int\limits_0^\infty e^{-q\alpha} \frac{\partial}{\partial\alpha} \left( \frac{B - i\alpha}{(B - i\alpha)^2 + x^2} \right) d\alpha$$

$$= \frac{i}{q} \left[ \frac{B}{B^2 + x^2} + \int\limits_0^\infty e^{-qx} \frac{\partial}{\partial\alpha} \left( \frac{B - i\alpha}{(B - i\alpha)^2 + x^2} \right) d\alpha \right] \tag{475}$$

Successive integrations by parts will generate a series (Wait, 1953):

$$J(B) = \lim_{x \to 0} i \sum_{m=0}^\infty \left( \frac{1}{q} \right)^{m+1} \frac{\partial^m}{\partial\alpha^m} \left( \frac{B - i\alpha}{(B - i\alpha)^2 + x^2} \right)$$

$$= \sum_{m=0}^\infty \left( -\frac{1}{q} \right)^{m+1} \frac{\partial^m}{\partial B^m} \left( \frac{B}{B^2 + x^2} \right) \tag{476}$$

The components of the magnetic field intensity may be found by forming the proper derivative of the expression for electric field intensity:

$$H_z = \frac{1}{i\mu_0\omega} \frac{\partial E_y}{\partial x}$$

$$H_x = \frac{1}{i\mu_0\omega} \frac{\partial E_y}{\partial z} \tag{477}$$

If the receiver is located at $z = -h$ (at the same height as the line source), we have:

$$H_z = -\frac{I}{2\pi} \frac{\partial}{\partial x} \left[ \log \frac{4h^2 + x^2}{x^2} + J(2h) \right] \tag{478}$$

Dividing this value for the magnetic field strength by the expression for the primary field strength, we have:

$$\frac{H_z}{H_{z,p}} = \left[ 1 - \frac{x^2}{4h^2 + x^2} - x \frac{\partial J(2h)}{\partial x} \right] \tag{479}$$

Expanding the expression in Eq. (477), the first four terms are:

$$\frac{H_z}{H_{z,p}} = \left[ 1 - \frac{1}{4H^2 + 1} - \frac{i}{W} \left( \frac{4H}{(4H^2 + 1)^2} \right) + \frac{1}{W^2} \left( \frac{24H^2 - 2}{(4H^2 + 1)^3} \right) \right.$$
$$\left. + \frac{i}{W^3} \left( \frac{192H^3 - 48H}{(4H^2 + 1)^4} \right) - \frac{i}{W_4} \left( \frac{1920H^4 - 960H^2 + 24}{(4H^2 + 1)^5} \right) + \cdots \right] \tag{480}$$

where $H = x/h$ and $W = xq$.

When $W$ is large, Eq. (480) simplifies to:

$$\frac{H_z}{H_{z,p}} = 1 - \frac{1}{4H^2 + 1} \tag{481}$$

This is the same as Eq. (421), which was obtained using the theory of images for a wire raised over an earth with zero resistivity.

The convergence of the series in Eq. (480) proceeds slowly for much of the range of interest, and alternate expressions that have been developed by Sundberg (1931) and others are sometimes more suitable.

The ratio $H_z/H_{z,p}$ is shown plotted as a function of $W(= xq)$ for parametric values of $H(= x/h)$ in Fig. 194. The curves representing magnetic field strength over a conducting sheet resemble the curves for magnetic field strength over a homogeneous earth in a general way, but there are significant differences. For instance, the maximum value for the out-of-phase component observed over a thin sheet is considerably larger than the maximum value for the same component observed over a homogeneous half-space.

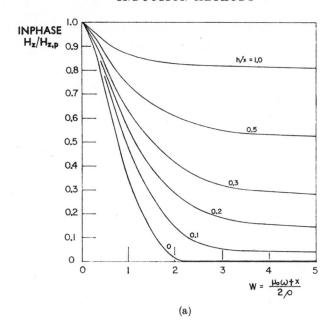

(a)

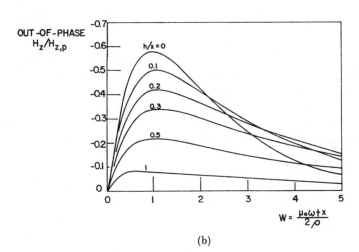

(b)

FIG. 194.

(a) Inphase component of the vertical field about a long wire above a thin sheet.

(b) Out-of-phase component of the vertical field about a long wire above a thin, conducting sheet.

From Eqs. (471) and (477), the first terms of the series expression for the horizontal component of the magnetic field strength are:

$$H_x = \frac{I}{2\pi} \left\{ - \frac{2h}{4h^2 + x^2} - \frac{i}{q} \left[ \frac{(h - z)^2 - x^2}{[(h - z)^2 + x^2]^2} \right] \right.$$
$$\left. - \frac{i}{q^2} \left[ \frac{6(h - z) x^2 - 2(h - z)^3}{[(h - z)^2 + x^2]^3} \right] + \cdots \right\} \tag{482}$$

When the observation point is in the same horizontal plane as the source wire (that is, when $z = -h$), the ratio of the horizontal component of field strength to the primary field strength is:

$$\frac{H_x}{H_{z,p}} = \left\{ - \frac{2H}{4H^2 + 1} - \frac{i}{W} \left[ \frac{4H^2 - 1}{(4H^2 + 1)^2} \right] - \frac{1}{W^2} \left[ \frac{12H - 16H^3}{(4H^2 + 1)^3} \right] + \cdots \right\} \tag{483}$$

The first term in this series is the same as the expression derived in Eq. (422) for the horizontal field intensity over a perfectly conducting earth.

The ratio $H_x/H_{z,p}$ is shown plotted as a function of the variable $W$ for parametric values of the quantity $h/x$ in Fig. 195. For small values of $h/x$, the out-of-phase component reverses sign as $W$ is increased, while the magnitude of the inphase component is less than the magnitude of the vertical component of the field. On the other hand, for large values of $h/x$, the ratio $H_x/H_{z,p}$ is much larger than the ratio $H_z/H_{z,p}$, particularly for the inphase component. Considering this behavior of the field components, it appears that observations of the ratio $H_z/H_{z,p}$ are preferable for shallow investigations while observations of the ratio $H_x/H_{x,p}$ are preferable for deep investigations.

Observations of field intensity components about a long wire over a thin, horizontal conducting sheet may be interpreted using logarithmic curve matching procedures similar to the ones already described. When plotted to logarithmic coordinates, the curves in Figs. 194 and 195 may be used to interpret observations made with a fixed offset spacing and a varying frequency. The ratio $h/x$, from which a value for $h$ may be computed, is determined either from the reference curve to which the field observations are matched or by interpolation between reference curves. The conductance of the sheet, defined as usual by the product $\sigma t$ (conductivity times thickness), may be determined from the correspondence between the abscissas for the two sets of curves (reference and field). If both the inphase and out-of-phase field curves do not match corresponding reference curves, the field data do not represent the effects of a thin horizontal sheet. The maximum thickness which a sheet may have and yet behave as a thin sheet is given by one author as:

$$t_{max} = \left( \frac{1}{\mu\omega\sigma} \right)^{1/2} \tag{484}$$

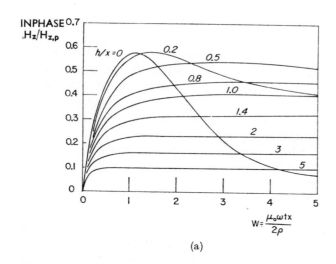

(a)

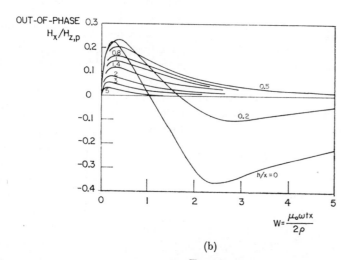

(b)

FIG. 195.

(a) Inphase component of the horizontal field about a long wire above a thin, conducting sheet.

(b) Out-of-phase component of the horizontal field about a long wire above a thin, conducting sheet.

and as:

$$t_{\max} = 300 \left(\frac{1}{\sigma f}\right)^{1/2} \tag{485}$$

by another author. These two conditions correspond to thickness equal to 0·707 and 0·60 skin-depth distances, respectively. However, the maximum thickness a thin sheet may have is a function of the ratios $h/x$ and $t/x$, as well as of $\sigma$ and $f$, being least for small values of $h/x$ and large values of $t/x$. Often the maximum thickness a thin sheet may have and still appear to be a thin sheet is less that that given by these expressions.

The equations and the curves for the field components about a long wire over a thin sheet are the basis for interpreting the data obtained with the "Sundberg" or compensator method. In the Sundberg method, measurements are made along a traverse oriented at right angles to a long grounded-wire, or at right angles to one leg of a large insulated loop. A single receiving coil and ratiometer are used to measure real and imaginary parts of the vertical and/or horizontal components of the magnetic field. The amplitudes are referred to the current flowing in the source cable. The reference signal is transmitted from the source cable to the ratiometer from a reference coil placed near the source wire over a two-wire cable. Ordinarily, two or more frequencies are used so that depth soundings can be made by varying both frequency and spacing. The number of readings and the spacing between readings along a profile are determined by the complexity of the problem and the depth to which one wishes to explore.

For many purposes it is more convenient to combine the curves in Figs. 194 and 195 into a single Argand diagram. On the Argand diagram in Fig. 196, the inphase component of $H_z/H_{z,p}$ is plotted along the abscissa and the out-of-

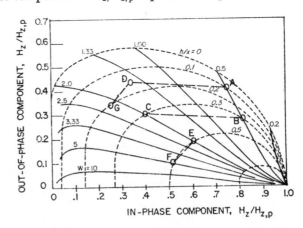

FIG. 196. Argand diagram for the vertical component of the electromagnetic field about a long wire above a thin, conducting sheet.

phase component of the same ratio is plotted along the ordinate.) Solid lines connect points for which the ratio $h/x$ is some constant parametric value, while the dashed lines connect points for which the value of $W$ assumes some constant parametric value. A similar Argand diagram for the ratio $H_x/H_{z,p}$ is given in Fig. 197. Using the Argand diagram in Fig. 196, both $h/x$ and $W$ may be evaluated using a single measurement of the real and imaginary components of the ratio, $H_z/H_{z,p}$. The quantities $h$ and $\sigma t$ may be computed from these determinations of $h/x$ and $W$ since the distance $x$ from the wire to the receiver is known. Providing the ratio $h/x$ is greater than a half, $h/x$ and $W$ may also be determined from a single measurement of the real and imaginary parts of the ratio $H_x/H_{z,p}$. If the ratio $h/x$ is less than a half, a single measurement could be interpreted as indicating either one of two possible layers having different depths and conductances but producing the same effect on the electromagnetic field; two or more sets of measurements with different values for $x$ or $W$ are required to resolve this uncertainty.

The electromagnetic field observed at a point below a thin conducting sheet remains nearly constant, regardless of where the sheet is placed, so long as it is between the source and the observation point. This phenomenon makes it possible to determine readily the electromagnetic fields in the presence of two or more parallel sheets at different depths (Sundberg and Hedström, 1933).

The field at point P in Fig. 198 contributed by currents flowing in sheet 2 is the same no matter where sheet 1 is located so long as it is located between the source and sheet 2. In the case in which sheet 1 is close to the top of sheet 2, the secondary field contributed by currents flowing in both sheets is the same as the secondary field due to a single sheet at the same depth with a conductance $\sigma t = \sigma_1 t_1 + \sigma_2 t_2$. At the observation point P, the difference between the field contributed by currents in a composite sheet at a depth $Z_2$ and the field contributed by the single sheet 1 at a depth $Z_2$ is exactly equal to the field contribution which would be observed by placing sheet 2 under sheet 1. The total field at point P contributed by both sheets is found by adding the field computed for the presence of sheet 2 to the field computed for the presence of sheet 1. This process may be repeated to determine the field contributed by several thin sheets at different depths.

As an example, consider the case of three conducting layers:

| Layer | $h/x$ | $W_i$ | $\Sigma W_i$ |
|-------|-------|-------|--------------|
| 1 | 0·1 | 0·5 | 0·5 |
| 2 | 0·3 | 1·5 | 2·0 |
| 3 | 0·5 | 3·0 | 5·0 |

The vertical field component contributed by layer 1 is given by the position of point A on Fig. 196. The field which would be contributed by layer 1 if it were placed at a depth to make $h/x = 0·3$ is given by the position of point B;

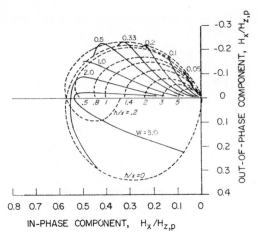

Fig. 197. Argand diagram for the horizontal component of the electromagnetic field about a long wire above a thin, conducting sheet.

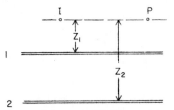

Fig. 198. Definition of quantities used in discussing the electromagnetic-field above a long wire over several conducting sheets. The source wire is I, the observer is located at point P.

the field contribution from both layer 1 and layer 2 with $h/x = 0.3$ is given by the position of point C. The field contributed by the addition of sheet 2 beneath sheet 1 is given by the difference between points B and C. The total effect of the two sheets, indicated by point D, is found by adding the phasor B–C to A. The effect of layer 3 may be added to the effect of the first two layers in a similar manner.

In practice, the problem of interpretation requires the use of the inverse to this procedure to find values for $h/x$ and $W$ for each layer using measurements of $H_z/H_{z,p}$ or $H_x/H_{x,p}$. Ideally, each set of measurements of the real and imaginary parts of the normalized field strength taken at a single offset spacing and frequency is sufficient to determine one set of values for $h/x$ and $W$. Measurement of the field component $H_z/H_{z,p}$ at three different frequencies for a single offset spacing should be sufficient to determine the depth and conductivity of three layers. The number of layers must be known or assumed before an interpretation may be made. Then, trial solutions are worked out using two sets of data, with the values for all but two of the unknown quantities being arbitrarily assumed in making each trial solution. When the solutions for the two sets of data are essentially the same, a satisfactory solution has been obtained.

Usually, interpretation by this procedure is too tedious and the resolution of the method is not good enough to warrant the assumption of more than three layers. If more than three layers are actually present, the effect in assuming the existence of only two or three layers is to represent several of the actual layers as a single fictitious layer. However, the fictitious two or three layer system interpreted from the electromagnetic measurements may give an accurate picture of significant structural or conductivity changes in the earth.

With the equipment commonly used, the maximum depth range for the Sundberg method is about 1500 ft. Before seismic methods became highly developed, the Sundberg method was used for shallow investigations of potentially oil-bearing structures. At present, the Sundberg method is used in mining exploration.

Any technique in which one or more of the components of the electromagnetic field about a long wire can be measured as a function of frequency or spacing can be used as a depth sounding method. It must be emphasized that techniques in which the frequency is the quantity varied are the most effective. Reference curves for a homogeneous earth and for a thin layer are useful in many cases, but they need to be supplemented by curves calculated for thick layers.

## Two-loop sounding methods

We shall now consider inductive sounding methods in which a small current loop is used as a source. If the separation between the source loop and the receiver loop is somewhat more than five times the diameter of either loop, both loops may be treated mathematically as dipoles (Wait, 1954). As with the long wire methods, quantities are measured in terms of normalized field intensities. However, the concept of *mutual impedance* or *mutual coupling* is used when dealing with two-loop systems, rather than using the ratio of primary plus secondary to primary field strengths. The mutual impedance, $Z$, for two loops is defined as $Z = V/I$, where $V$ is the voltage induced in the receiving loop and $I$ is the current supplied to the transmitting loop.

There are four ways in which the transmitting and receiving loops may be oriented with respect to each other which are commonly used in electromagnetic prospecting and which are suitable for depth soundings (Wait, 1955). These are (a) horizontal coplanar, (b) vertical coplanar, (c) vertical coaxial, and (d) perpendicular. Other two-loop arrangements may be used for horizontal profiling but are not commonly used for depth sounding. Since the mathematical development is essentially the same, we shall consider also the sounding method in which a small horizontal coplanar loop and wire element are used.

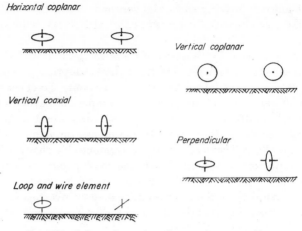

FIG. 199. Coil configurations for two-loop soundings.

The mutual coupling between loops in free space, designated by the symbol $Z_0$, may be calculated readily using Eqs. (412) and (413). For coplanar loops, the angle $\theta$ is 90°, so the field intensity at the receiving loop is:

$$H = -\frac{m}{4\pi r^3} = -\frac{n_1 A_1 I}{4\pi r^3} \tag{486}$$

where $n_1$ and $A_1$ are the number of turns and the area of the transmitting loop. Then:

$$Z_0 = \frac{V}{I} = \frac{i\mu_0 \omega n_1 n_2 A_1 A_2}{4\pi r^3} \tag{487}$$

where $n_2$ and $A_2$ are the number of turns and the area of the receiving loop. In the case of coaxial loops, the angle $\theta$ is zero, so $H$ and $Z_0$ are:

$$H = \frac{m}{2\pi r^3} = \frac{n_1 A_1 I}{2\pi r^3} \tag{488}$$

$$Z_0 = -\frac{i\mu_0 \omega n_1 n_2 A_1 A_2}{2\pi r^3} \tag{489}$$

It should be noted that the mutual coupling for coaxial loops is twice that for coplanar loops separated the same distance.

From Eq. (408) the mutual coupling between a coplanar loop and electric dipole in free space is

$$Z_0 = \frac{-i\mu_0 \omega A_1 s}{4\pi r^2} \sin B \tag{490}$$

where $s$ is the length of the dipole and $B$ is the angle between the direction of the electric dipole and the line joining the loop and dipole.

Some idea of the behavior of the different arrangements may be obtained by considering the simple model of loops above a perfectly conducting earth. The effect of a perfect conductor may be simulated by placing an image dipole at a depth $-h$ below the transmitting loops (as shown in Fig. 200).

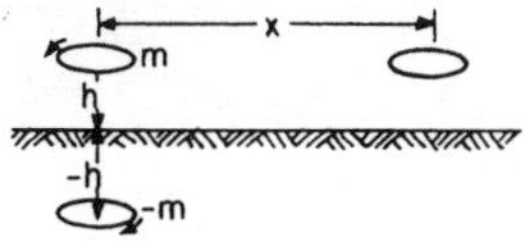

FIG. 200.

For horizontal coplanar loops, the primary field is:

$$H_{z,p} = -\frac{m}{4\pi x^3} \tag{491}$$

The vertical component of the secondary field is:

$$H_{z,s} = \frac{m}{4\pi(x^2 + 4h^2)^{3/2}}\left(1 - 3\frac{4h^2}{x^2 + 4h^2}\right) \tag{492}$$

The ratio of the total field to the primary is the same as the ratio of mutual coupling in the presence of the conductor to the mutual coupling in free space:

$$\frac{H_z}{H_{z,p}} = \frac{Z}{Z_0} = 1 - \frac{1}{\left[4\left(\dfrac{h}{x}\right)^2 + 1\right]^{3/2}} + \frac{12\left(\dfrac{h}{x}\right)^2}{\left[4\left(\dfrac{h}{x}\right)^2 + 1\right]^{5/2}} \tag{493}$$

When the ratio $h/x$ is small, the impedance ratio is less than unity. The condition to be satisfied to have $Z/Z_0 = 1$ is:

$$12\left(\frac{h}{x}\right)^2 = 4\left(\frac{h}{x}\right)^2 + 1$$

$$\frac{h}{x} = 0{\cdot}354 \tag{494}$$

For $h/x$ ratios greater than 0·354, the impedance ratio reaches a maximum value of approximately 1·2 at $h/x = 0{\cdot}61$ and then decreases toward zero as $h/x$ becomes larger (see Fig. 201).

For vertical coplanar loops having their axes in a common plane:

$$H_{p,y} = -\frac{m}{4\pi x^3}$$

$$H_{s,y} = -\frac{m}{4\pi(x^2 + 4h^2)^{3/2}} \tag{495}$$

$$\frac{Z}{Z_0} = 1 + \frac{1}{\left[4\left(\dfrac{h}{x}\right)^2 + 1\right]^{3/2}} \tag{496}$$

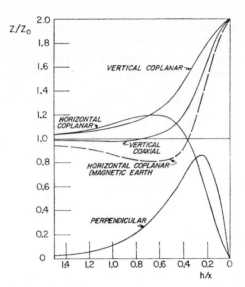

Fig. 201. Mutual coupling between loops raised over an ideally conducting earth or over an insulating but ideally magnetic earth.

The impedance ratio approaches the value of one when the ratio $h/x$ becomes very large, and the impedance ratio rises monotonically approaching the value two as the ratio $h/x$ approaches zero.

For vertical coaxial loops:

$$H_{p.y} = \frac{m}{2\pi x^3} \qquad (497)$$

$$H_{s,y} = \frac{m}{4\pi(x^2 + 4h^2)^{3/2}}\left(\frac{3x^2}{4h^2 + x^2} - 1\right) \qquad (498)$$

$$\frac{Z}{Z_0} = 1 + \frac{3}{2\left[4\left(\dfrac{h}{x}\right)^2 + 1\right]^{5/2}} - \frac{1}{2\left[4\left(\dfrac{h}{x}\right)^2 + 1\right]^{3/2}} \qquad (499)$$

For large values of $h/x$, the value for $Z/Z_0$ is slightly less than one. At $h/x = 1/\sqrt{2}$, the value for $Z/Z_0$ is one, and for smaller values of $h/x$, $Z/Z_0$ assumes values greater than unity approaching two as $h/x$ approaches zero.

For loops perpendicular to each other (the axis of one coil passes through a radial in the plane of the other coil):

$$H_{p,r} = 0 \qquad (500)$$

$$H_{p,z} = -\frac{m}{4\pi x^3} \qquad (501)$$

$$H_{s,x} = -\frac{3m}{4\pi(x^2 + 4h^2)^{3/2}}\left[\frac{x}{(x^2 + 4h^2)^{1/2}}\right]\left[\frac{2h}{(x^2 + 4h^2)^{1/2}}\right] \qquad (502)$$

$$\frac{Z}{Z_0} = \frac{6\left(\dfrac{h}{x}\right)}{\left[4\left(\dfrac{h}{x}\right)^2 + 1\right]^{5/2}} \tag{503}$$

where $Z_0$ is the value computed for coplanar loops, rather than for perpendicular loops. If the value of $Z_0$ computed for coaxial loops is used instead, this last equation is:

$$\frac{Z}{Z_0} = \frac{3\left(\dfrac{h}{x}\right)}{\left[4\left(\dfrac{h}{x}\right)^2 + 1\right]^{5/2}} \tag{504}$$

The mutual impedance ratio for perpendicular loops is zero for $h/x = 0$, rises to a maximum value of 0·86 at $h/x = 0·25$, and then approaches zero as $h/x$ becomes large.

The method of images may be used also to compute coupling ratios for loops above a homogeneous earth with zero conductivity, but with a specified magnetic permeability. As in the case of the long wire considered earlier, the image current in the loop is given by:

$$I' = \frac{\mu - \mu_0}{\mu + \mu_0} I$$

and flows in the same direction as the current in the source loop. For horizontal coplanar loops:

$$B_{z,p} = -\frac{m\mu_0}{4\pi x^3} \tag{505}$$

$$B_{z,s} = \frac{m\mu_0}{4\pi}\left(\frac{\mu - \mu_0}{\mu + \mu_0}\right)\left[\frac{1}{(x^2 + 4h^2)^{3/2}}\right]\left[3\,\frac{4h^2}{x^2 + 4h^2} - 1\right] \tag{506}$$

$$\frac{Z}{Z_0} = 1 + \frac{\mu - \mu_0}{\mu + \mu_0}\left\{\frac{1}{\left[4\left(\dfrac{h}{x}\right)^2 + 1\right]^{3/2}} - \frac{12\left(\dfrac{h}{x}\right)^2}{\left[4\left(\dfrac{h}{x}\right)^2 + 1\right]^{5/2}}\right\} \tag{507}$$

It should be noted that the magnetic field is expressed in terms of magnetic induction, $B$, rather than in terms of field intensity, $H$, since we are no longer assuming constant permeability. The coupling ratio in the presence of a magnetic earth is greater than unity for values of $h/z$ less than 0·354, and less than unity for values of $h/x$ greater than 0·354, which is opposite to the behavior of the coupling ratio in the presence of a conductive earth. For loops resting on the surface of the earth, rather than raised, we have:

$$\frac{Z}{Z_0} = 1 + \frac{\mu - \mu_0}{\mu + \mu_0} = \frac{2\mu}{\mu + \mu_0} \tag{508}$$

When the permeability is very large, the curve for mutual coupling over a magnetic earth is the mirror image of the coupling curve over a very conductive earth.

Similar expressions may be derived for other coil orientations; in all cases, the behavior of $Z/Z_0$ for a magnetic half-space is opposite to the behavior of $Z/Z_0$ for a perfectly conducting half-space.

These relationships which have been derived for a perfectly conducting earth can seldom be applied to a practical problem in a quantitative way, but they help in understanding the behavior of the mutual coupling ratios for loops above an earth with finite resistivity. In particular, it should be emphasized again that the secondary field at the receiving coil may add to or subtract from the primary field, depending on the geometry of the problem, so that the ratio $Z/Z_0$ may have values greater than or less than unity.

The relationships for a magnetic half-space are useful in interpreting results from certain airborne methods, but in general the effects of variations in magnetic permeability are seldom of much importance.

Next we shall calculate the field components and the coupling ratios for loops on or above the surface of a half-space with finite conductivity. Assume that we have placed a small current-carrying source loop at a height $h$ above a two-layer earth. The upper layer has a thickness, $d$; the lower layer is infinite in extent. The upper layer is assumed to have a conductivity, $\sigma_1$; the lower layer is assumed to have a conductivity, $\sigma_2$. The magnetic permeability of both regions is assumed to be the same as that for free space. The definition of these various dimensions is indicated in Fig. 202.

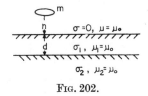

FIG. 202.

As was done in determining the primary field about a dipole source, a vector potential will be used. However, cylindrical rather than spherical coordinates will be used, and it is convenient to use a vector potential, $F$, given by:

$$E = -\nabla \times F$$

$$H = -(\sigma + i\varepsilon\omega) F + \frac{1}{\mu\omega} \nabla(\nabla \cdot F) \tag{509}$$

rather than the vector potential $A$ used earlier. The vector potentials $A$ and $F$ are known as the magnetic and the electric vector potentials, respectively, though some authors refer to $F$ as a magnetic vector potential.

By comparison with Eqs. (385) and (389) for the primary vector potential $A$ of an electric dipole, we see that the primary vector potential $F$ for a magnetic dipole is:

$$F = \frac{i\mu\omega m}{4\pi r'} e^{-\gamma r'} \tag{510}$$

where $r'^2 = r^2 + z^2$. The expression for electric vector potential, given here, is much simpler than the expression for the magnetic vector potential for a magnetic dipole source.

The potential functions $F_i$ ($i = 0, 1, 2$) for each of the three regions in our layered earth must each be a solution of the wave equation:

$$\nabla^2 F_i + \gamma_i^2 F_i = 0 \tag{511}$$

which is, when expanded for a cylindrical coordinate system:

$$\frac{1}{r} \frac{\partial}{\partial r} \left( r \frac{\partial F_i}{\partial r} \right) + \frac{1}{r^2} \frac{\partial^2 F_i}{\partial \varphi^2} + \frac{\partial^2 F_i}{\partial z^2} - \gamma_i^2 F_i = 0 \tag{512}$$

For our problem, where the fields must be finite at the origin and where the dipole is located at $r = 0$, a complete solution for the wave equation is made up of special solutions of the following type:

$$\exp \pm [(\lambda^2 + \gamma_i^2) z] J_0(\lambda r) \tag{513}$$

where $J_0$ is the Bessel function of the first kind. The sign in the exponential term must be chosen so as to reject any solutions which are not finite as $z$ tends to become infinite either in the plus or minus sense.

Following Wait (1951) and Bhattacharyya (1955) the general solution for $F$ is given by the summation of the primary vector potential and the vector potential given in Eq. (513), where the latter is to be taken for a continuum of values for the parameter $\lambda$. The vector potential functions in the three media are:

$$F_1 = C \left[ \frac{e^{-\gamma_0 r'}}{r'} + \int_0^\infty \psi_1(\lambda) \exp \left[ -(\lambda^2 + \gamma_0^2)^{1/2} z \right] J_0(\lambda r) \, d\lambda \right], \quad 0 < z$$

$$F_2 = C \left[ \frac{e^{-\gamma_0 r'}}{r'} + \int_0^\infty \psi_2(\lambda) \exp \left[ -(\lambda^2 + \gamma_0^2)^{1/2} z \right] J_0(\lambda r) \, d\lambda \right], \quad -h < z < 0$$

$$F_3 = C \left[ \int_0^\infty \{\psi_3(\lambda) \exp \left[ (\lambda^2 + \gamma_1^2)^{1/2} z \right] + \psi_4(\lambda) \exp \left[ -(\lambda^2 + \gamma_1^2)^{1/2} z \right] \} J_0(\lambda r) \, d\lambda \right],$$
$$- (h + d) < z < -h$$

$$F_4 = C \left[ \int_0^\infty \psi_5(\lambda) \exp \left[ (\lambda^2 + \gamma_2^2)^{1/2} z \right] J_0(\lambda r) \, d\lambda \right], \quad z < -(h + d) \tag{514}$$

where $C = \dfrac{i\mu\omega m}{4\pi}$ and $\psi_1,\,\psi_2,\,\psi_3,\,\psi_4$ and $\psi_5$ are functions of $\lambda$ with values to be determined from the various boundary conditions.

The exponential function representing the primary field may be transformed into an integral expression using the Sommerfeld integral:

$$\frac{e^{-\gamma_1 r'}}{r'} = \int\limits_0^\infty \frac{\lambda}{(\lambda^2 + \gamma_1^2)^{1/2}} \exp\left[-(\lambda^2 + \gamma_1^2)^{1/2} z\right] J_0(\lambda r)\,d\lambda; \quad z \geqq 0$$

$$\frac{e^{-\gamma_0 r'}}{r'} = \int\limits_0^\infty \frac{\lambda}{(\lambda^2 + \gamma_0^2)^{1/2}} \exp\left[(\lambda^2 + \gamma_0^2)^{1/2} z\right] J_0(\lambda r)\,d\lambda; \quad -h \leqq z \leqq 0 \quad (515)$$

Neglecting displacement currents, setting $\gamma_0 = 0$, and using the notation

$$(\lambda^2 + \gamma_1^2)^{1/2} = v_1, \quad (\lambda^2 + \gamma_2^2)^{1/2} = v_2, \quad \gamma_i^2 = i\sigma\mu_0\omega$$

Eq. (514) may be written as:

$$F_1 = C \int\limits_0^\infty \left[e^{-\lambda z} + \psi_1(\lambda)\,e^{-\lambda z}\right] J_0(\lambda r)\,d\lambda$$

$$F_2 = C \int\limits_0^\infty \left[e^{-\lambda z} + \psi_2(\lambda)\,e^{-\lambda z}\right] J_0(\lambda r)\,d\lambda$$

$$F_3 = C \int\limits_0^\infty \left[\psi_3(\lambda)\,e^{v_1 z} + \psi_4(\lambda)\,e^{-v_1 z}\right] J_0(\lambda r)\,d\lambda$$

$$F_4 = C \int\limits_0^\infty \psi_5(\lambda)\,e^{v_2 z} J_0(\lambda r)\,d\lambda \qquad (516)$$

Solutions for the functions $\psi_1,\,\psi_2,\,\psi_3,\,\psi_4$ and $\psi_5$ may now be obtained by using boundary conditions which require that the tangential electric and magnetic fields be continuous across the layer boundaries. The tangential field components are given by Eqs. (509) as:

$$E_{\varphi i} = \frac{\partial F_i}{\partial r}$$

$$H_{ri} = \frac{1}{i\mu\omega}\,\frac{\partial^2 F_i}{\partial r\,\partial z} \qquad (517)$$

The boundary conditions are:

$$\frac{\partial F_1}{\partial r} = \frac{\partial F_2}{\partial r},\quad \frac{\partial^2 F_1}{\partial r\,\partial z} = \frac{\partial^2 F_2}{\partial r\,\partial z} \quad \text{at}\ \ z = 0$$

$$\frac{\partial F_2}{\partial r} = \frac{\partial F_3}{\partial r},\quad \frac{\partial^2 F_2}{\partial r\,\partial z} = \frac{\partial^2 F_3}{\partial r\,\partial z} \quad \text{at}\ \ z = -h$$

$$\frac{\partial F_3}{\partial r} = \frac{\partial F_4}{\partial r},\quad \frac{\partial^2 F_3}{\partial r\,\partial z} = \frac{\partial^2 F_4}{\partial r\,\partial z} \quad \text{at}\ \ z = -(h + d) \qquad (518)$$

These equations may be integrated directly with respect to $r$. The constants of integration introduced by this process all turn out to be zero since all the expressions must have zero value when $r$ tends to infinity. Thus, we have the simpler set of expressions for the boundary conditions:

$$F_1 = F_2, \qquad \frac{\partial F_1}{\partial z} = \frac{\partial F_2}{\partial z} \quad \text{at} \quad z = 0$$

$$F_2 = F_3, \qquad \frac{\partial F_2}{\partial z} = \frac{\partial F_3}{\partial z} \quad \text{at} \quad z = -h$$

$$F_3 = F_4, \qquad \frac{\partial F_3}{\partial z} = \frac{\partial F_4}{\partial z} \quad \text{at} \quad z = -(h + d) \qquad (519)$$

Using Eqs. (516) and (519), we obtain the following equations from which the $\psi$ functions may be determined:

$$1 + \psi_1(\lambda) = 1 + \psi_2(\lambda)$$
$$e^{-\lambda h} + \psi_2(\lambda)\, e^{\lambda h} = \psi_3(\lambda)\, e^{-v_1 h} + \psi_4(\lambda) e^{-v_1 h}$$
$$\lambda e^{-\lambda h} - \lambda \psi_2(\lambda)\, e^{\lambda h} = v_1 \psi_3(\lambda)\, e^{-v_1 h} - v_1 \psi_4(\lambda)\, e^{v_1 h}$$
$$\psi_3(\lambda)\, e^{-v_1(h+d)} + \psi_4(\lambda)\, e^{v_1(h+d)} = \psi_5(\lambda)\, e^{-v_2(h+d)}$$
$$v_1 \psi_3(\lambda)\, e^{-v_1(h+d)} - v_1 \psi_4(\lambda)\, e^{v_1(h+d)} = v_2 \psi_5(\lambda)\, e^{-v_2(h+d)} \qquad (520)$$

Solving for $\psi_1(\lambda)$ and $\psi_2(\lambda)$:

$$\psi_1(\lambda) = \psi_2(\lambda) = \frac{-(v_1 - \lambda)(v_1 + v_2) + (v_1 - v_2)(v_1 + \lambda)\, e^{-2v_1 d}}{(v_1 + \lambda)(v_1 + v_2) - (v_1 - v_2)(v_1 - \lambda)\, e^{-2v_1 d}}\, e^{-2\lambda h}$$

Then:

$$F_1 = F_2 = C\left[\frac{1}{r'} + \int_0^\infty R(\lambda)\, e^{-\lambda(z+2h)}\, J_0(\lambda r)\, d\lambda\right], \quad -h < z \qquad (521)$$

where

$$R(\lambda) = \frac{-(v_1 - \lambda)(v_1 + v_2) + (v_1 - v_2)(v_1 + \lambda)\, e^{-2v_1 d}}{(v_1 + \lambda)(v_1 + v_2) - (v_1 - v_2)(v_1 - \lambda)\, e^{-2v_1 d}}$$

The solutions for $\psi_3(\lambda)$ and $\psi_4(\lambda)$ are:

$$\psi_3 = \frac{2\lambda(v_1 + v_2)\, e^{h(v_1 - \lambda)}}{(v_1 + \lambda)(v_1 + v_2) - (v_1 - \lambda)(v_1 - v_2)\, e^{-2v_1 d}}$$

$$\psi_4 = \frac{2\lambda(v_1 - v_2)\, e^{-h(v_1 + \lambda) - 2v_1 d}}{(v_1 + \lambda)(v_1 + v_2) - (v_1 - \lambda)(v_1 - v_2)\, e^{-2v_1 d}}$$

Then:

$$F_3 = 2C \int_0^\infty \frac{(v_1 + v_2)\, e^{v_1(h+z)} + (v_1 - v_2)\, e^{-v_1(h+2d+z)}}{(v_1 + v_2)(v_1 + \lambda) - (v_1 - v_2)(v_1 - \lambda)\, e^{-2v_1 d}}\, \lambda e^{-\lambda h}\, J_0(\lambda r)\, d\lambda,$$

$$-(d + h) < z \leq 0 \qquad (522)$$

Equations for the field components above the surface, at the surface, or below the surface of the earth may be obtained from these expressions for $F_2$ and $F_3$. Numerical integration using digital computers is the only practical means for evaluating the integrals in the general case, but for a dipole source located at the surface of a homogeneous earth, equations for $E$ and $H$ observed at the surface may be obtained in closed form or in terms of tabulated functions. For this special case, $h = z = 0$, $v_1 = v_2 = v$ and $F_3$ is given by:

$$F_3 = 2C \int_0^\infty \frac{\lambda}{(v + \lambda)} J_0(\lambda r)\, d\lambda \tag{523}$$

This last equation may also be written as:

$$F_3 = \frac{2C}{\gamma^2} \left[ \int_0^\infty \lambda (\lambda^2 + \gamma^2)^{1/2} J_0(\lambda r)\, d\lambda - \int_0^\infty \lambda^2 J_0(\lambda r)\, d\lambda \right] \tag{524}$$

which may be evaluated with the help of the Sommerfeld integral:

$$\int_0^\infty \frac{\exp\left[(\lambda^2 + \gamma^2)^{1/2} z\right]}{(\lambda^2 + \gamma^2)^{1/2}} \lambda J_0(\lambda r)\, d\lambda = \frac{\exp\left[-\gamma (r^2 + z^2)^{1/2}\right]}{(r^2 + z^2)^{1/2}} \tag{525}$$

Differentiating both sides of this identity twice with respect to $z$, and then setting $z = 0$, we have:

$$\int_0^\infty \lambda (\lambda^2 + \gamma^2)^{1/2} J_0(\lambda r)\, d\lambda = -\frac{(1 + \gamma r)\, e^{-\gamma r}}{r^3} \tag{526}$$

Then:

$$F = \frac{2C}{\gamma r^3} \left[ 1 - (1 + \gamma r)\, e^{-\gamma r} \right] \tag{527}$$

The electric and magnetic field strengths for this special case may now be derived. The electric field has only a tangential component given by:

$$E_\varphi = \frac{\partial F}{\partial r} = \frac{2C}{\gamma^2 r^4} \left[ (3 + 3\gamma r + \gamma^2 r^2)\, e^{-\gamma r} - 3 \right] \tag{528}$$

From Eq. (509), the vertical component of the magnetic field is given by:

$$i\mu_0 \omega H_z = -\gamma^2 F + \frac{\partial^2 F}{\partial z^2} \tag{529}$$

However, since $F$ is a solution of the wave equation (Eq. 511), and since $F$ does not vary with $\varphi$, we have:

$$i\mu_0 \omega H_z = \frac{\partial^2 F}{\partial z^2} - \gamma^2 F = -\frac{1}{r} \frac{\partial}{\partial r} \left( r \frac{\partial F}{\partial r} \right) \tag{530}$$

Carrying out the indicated differentiation, we have:

$$i\mu_0\omega H_z = \frac{2C}{\gamma^2 r^5}\left[(9 + 9\gamma r + 4\gamma^2 r^2 + \gamma^3 r^3)\,e^{-\gamma r} - 9\right] \tag{531}$$

Using Eq. (528), the mutual impedance between a single-turn loop with an area $dA$ and a short wire element with a length $ds$ is:

$$Z = \frac{dA\,ds}{2\pi\sigma r^4}\left[(3 + 3\gamma r + \gamma^2 r^2)\,e^{-\gamma r} - 3\right]\sin\beta \tag{532}$$

where $\beta$ is the angle between the direction of the wire and the line joining the wire and the loop. From Eq. (491) the mutual coupling ratio, $Z/Z_0$, is

$$\frac{Z}{Z_0} = \frac{2}{\gamma^2 r^2}\left[3 - (3 + 3\gamma r + \gamma^2 r^2)\,e^{-\gamma r}\right] \tag{533}$$

From Eq. (531), the mutual coupling, $Z$, between two single-turn loops is:

$$Z = \frac{dA_1\,dA_2}{2\pi\sigma r^5}\left[9 - (9 + 9\gamma r + 4\gamma^2 r^2 + \gamma^3 r^3)\,e^{-\gamma r}\right] \tag{534}$$

The mutual coupling ratio, $Z/Z_0$, for loops is:

$$\frac{Z}{Z_0} = \frac{2}{\gamma^2 r^2}\left[9 - (9 + 9\gamma r + 4\gamma^2 r^2 + \gamma^3 r^3)\,e^{-\gamma r}\right] \tag{535}$$

The fields about a horizontally-directed dipole source may be derived using similar procedures. The mutual impedance ratios for three other loop arrangements on the surface of a homogeneous earth are (Wait, 1955):

*Vertical coplanar loops*

$$\frac{Z}{Z_0} = 2\left[\frac{e^{-\gamma r}}{\gamma^2 r^2}(3 + 3\gamma r + \gamma^2 r^2) + 1 - \frac{3}{\gamma^2 r^2}\right] \tag{536}$$

*Vertical coaxial loops*

$$\frac{Z}{Z_0} = \frac{e^{-\gamma r}}{\gamma^2 r^2}(12 + 12\gamma r + 5\gamma^2 r^2 + \gamma^3 r^3) + 2 - \frac{12}{\gamma^2 r^2} \tag{537}$$

*Perpendicular loops*

$$\frac{Z}{Z_0} = -\frac{1}{2}\left[\gamma^2 r^2(I_1 K_1 - I_0 K_0) + 4\gamma r(I_1 K_0 - I_0 K_1) \div 16\,I_1 K_1\right] \tag{538}$$

where $I_0$, $I_1$, $K_0$ and $K_1$ are modified Bessel functions with the argument $\gamma r/2$, and where the value for $Z_0$ for perpendicular loops is taken to be that for coaxial loops.

The inphase and out-of-phase components are plotted as a function of loop separation measured in radian wave lengths in Fig. 203, and tables of values are given in the Appendix for the four different coil arrangements. The curves for a loop and a wire element are the mirror image, about the line $Z/Z_0 = 1$, of the curves for the vertical coplanar arrangement.

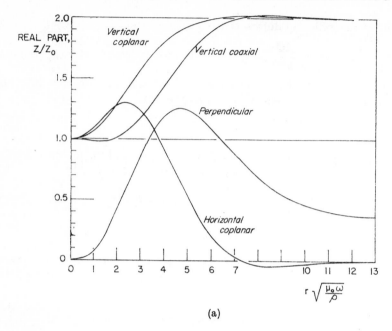

(a)

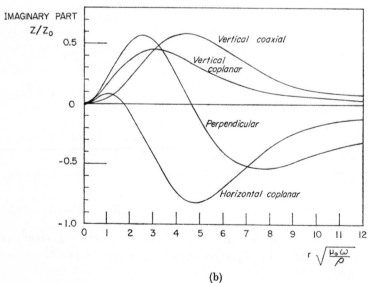

(b)

FIG. 203.

(a) Inphase components of mutual coupling between loops on the surface of a homogeneous earth.

(b) Out-of-phase components of mutual coupling between loops on the surface of a homogeneous earth.

The curves for horizontal coplanar loops are similar in character to the curves shown in Fig. 190 for the ratio $H_z/H_{z,p}$, but are a little more complicated in that maxima occur for small values of $\gamma r$. The curves for vertical coaxial loops tend to be a mirror image of the curves for horizontal coplanar loops. The coupling ratio for vertical coplanar loops shows the least complicated behavior for any of the four loop arrangements being considered. The curves for the inphase component of the ratio $Z/Z_0$ for loops on the surface of an earth with finite conductivity are similar to the curves for $Z/Z_0$ plotted as a function of $h/x$ for loops raised above the surface of a perfectly conducting earth (shown in Fig. 201). The values for $Z/Z_0$ when $\gamma r = 0$ (corresponding to $h/x = \infty$) and for $\gamma r - = \infty$ (corresponding to $h/x = 0$) are identical and the curves show equivalent maxima and minima for intermediate values of $\gamma r$ and $h/x$. This similarity is explained logically by considering that for a finitely conductive earth with $|\gamma r|$ very large, the eddy currents in the earth are concentrated near the surface, just as in the case of a perfectly conducting earth. On the other hand, for small values of $|\gamma r|$, eddy currents are distributed through a thick section of the earth, placing the effective image source of current at a considerable depth.

In order to determine the ratio $Z/Z_0$ for loops or a loop and a wire element raised over a two-layer earth, we return to Eqs. (521) and (529). Since $\gamma_0$ may be taken as zero, we have:

$$i\mu_0\omega H_z = \frac{\partial^2}{\partial z^2}\left\{ C\left[ \frac{1}{r'} + \int_0^\infty R(\lambda)\, e^{-\lambda(z+2h)}\, J_0(\lambda r)\, d\lambda \right]\right\}$$

$$H_z = \frac{3mz^2}{4\pi(r^2+z^2)^{5/2}} - \frac{m}{4\pi(r^2+z^2)^{3/2}} + \frac{m}{4\pi}\int_0^\infty \lambda^2 R(\lambda)\, e^{-\lambda(z+2h)}\, J_0(\lambda r)\, d\lambda$$

$$E_\varphi = \frac{\partial}{\partial r}\left\{ C\left[ \frac{1}{r'} + \int_0^\infty R(\lambda)\, e^{-\lambda(z+2h)}\, J_0(\lambda r)\, d\lambda \right]\right\}$$

$$E_\varphi = \frac{i\mu_0\omega m}{4\pi}\left[ -\frac{r}{(r^2+z^2)^{3/2}} + \int_0^\infty \lambda R(\lambda)\, e^{-\lambda(z+2h)}\, J_1(\lambda r)\, d\lambda \right] \tag{539}$$

where the first two terms may be recognized as being the primary field and the last term as the secondary field. The horizontal field components due to both vertical and horizontal dipole sources may by obtained in a similar manner and the results used to calculate values for the ratio $Z/Z_0$ for any coil orientation. Wait (1959) has placed these results into a form particularly suited to computation with a digital computer. In order to be consistent with the notation used by Wait, the origin of coordinates will be shifted to the ground surface, so that the dipole source will be located at $z' = h + z$ in the new notation. The substitution $\lambda = g^2/\delta^2$ will be used.

For the vertical dipole source, the primary components are:

$$H_{x,p} = \frac{3mx(z' - h)}{4\pi r''^5}$$

$$H_{y,p} = \frac{3my(z' - h)}{4\pi r''^5}$$

$$H_{z,p} = \frac{3m(z' - h)^2}{4\pi r''^5} - \frac{m}{4\pi r''^3}$$

$$E_{\varphi,p} = -\frac{i\mu_0\omega mr}{4\pi r''^3} \tag{540}$$

where $r''^2 = x^2 + y^2 + (z' - h)^2$

The secondary field components are:

$$H_{x,s} = -\frac{m}{4\pi(\delta)^3} T_1(A, B)\left(\frac{x}{r}\right)$$

$$H_{y,s} = -\frac{m}{4\pi(\delta)^3} T_1(A, B)\left(\frac{y}{r}\right)$$

$$H_{z,s} = -\frac{m}{4\pi(\delta)^3} T_0(A, B)$$

$$E_{\varphi,s} = \frac{i\mu_0\omega m}{4\pi\delta^2} T_2(A, B) \tag{541}$$

For a horizontal dipole source, the primary field components are:

$$H_{x,p} = \frac{3mxy}{4\pi r''^5}$$

$$H_{y,p} = \frac{3my^2}{4\pi r''^5} - \frac{m}{4\pi r''^3}$$

$$H_{z,p} = \frac{3my(z' - h)}{4\pi r''^5} \tag{542}$$

The secondary field components are:

$$H_{x,s} = \frac{m}{4\pi(\delta)^3}\left(\frac{xy}{r^2}\right)\left[\frac{2}{B} T_2(A, B) - T_0(A, B)\right]$$

$$H_{y,s} = \frac{m}{4\pi(\delta)^3}\left(\frac{y^2}{r^2}\right)\left[\left(1 - \frac{x^2}{y^2}\right)\frac{1}{B} T_2(A, B) - T_0(A, B)\right]$$

$$H_{z,s} = \frac{m}{4\pi(\delta)^3}\left(\frac{y}{r}\right) T_1(A, B) \tag{543}$$

**where**

$$T_0(A, B) = \int_0^\infty R(D, g)\, g^2 e^{-gA}\, J_0(g, B)\, dg$$

$$T_1(A, B) = \int_0^\infty R(D, g)\, g^2 e^{-gA}\, J_1(g, B)\, dg$$

$$T_2(A, B) = \int_0^\infty R(D, g)\, g e^{-gA}\, J_1(g, B),\, dg$$

$$R(g) = 1 - 2g \frac{(U + V) + (U - V)\, e^{-UD}}{(U + g)(U + V) - (U - g)(U - V)\, e^{-UD}}$$

$$U = (g^2 + 2i)^{1/2}$$

$$V = (g^2 + 2iK)^{1/2}$$

$$\delta = \left(\frac{2}{\sigma_1 \mu_0 \omega}\right)^{1/2}$$

$$K = \sigma_2/\sigma_1$$

$$A = \frac{Z + h}{\delta}$$

$$B = \frac{r}{\delta}$$

$$D = \frac{2d}{\delta} \tag{544}$$

Using these formulas for $H$, values may be computed for the ratio $Z/Z_0$ for any arbitrary arrangement of loops; the loops need not be in the same horizontal plane, nor do their axes need to be parallel or perpendicular to one another. The coupling ratios for four two-coil and a coil and wire element arrangement of particular interest may be computed readily:

For *horizontal coplanar loops*

$$Z/Z_0 = 1 + B^3 T_0 \tag{545}$$

For *vertical coplanar loops*

$$Z/Z_0 = 1 + B^2 T_2 \tag{546}$$

For *vertical coaxial loops*

$$Z/Z_0 = 2[1 + \tfrac{1}{2}B^2(T_2 - BT_0)] \tag{547}$$

For *perpendicular loops*

$$Z/Z_0 = B^3 T_1 \tag{548}$$

For *a horizontal coplanar loop and wire element*

$$Z/Z_0 = 1 - B^2 T_2(A, B) \tag{549}$$

For two-loop arrangements the value for $Z_0$ is taken to be that for coplanar arrangements.

The only practical means for evaluating the functions $T_0$, $T_1$ and $T_2$ in the most general case is by the use of a digital computer, but Wait (1958) has developed asymptotic expansions which can be used when $A$ and $B$ are much larger than one. The first three terms in these expansions are, respectively,

$$T_0 \cong \frac{2A^2 - B^2}{(A^2 + B^2)^{5/2}} - \left(\frac{2}{i}\right)^{1/2} Q \frac{6A^3 - 9AB^2}{(A^2 + B^2)^{7/2}} - iQ^2 \frac{24A^4 - 72A^2B^2 + 9B^4}{(A^2 + B^2)^{9/2}}$$

$$T_1 \cong \frac{3AB}{(A^2 + B^2)^{5/2}} - \left(\frac{2}{i}\right)^{1/2} Q \frac{12A^2B - 3B^3}{(A^2 + B^2)^{7/2}} - iQ^2 \frac{60A^3B - 45AB^3}{(A^2 + B^2)^{9/2}}$$

$$T_2 \cong \frac{B}{(A^2 + B^2)^{3/2}} - \left(\frac{2}{i}\right)^{1/2} Q \frac{3AB}{(A^2 + B^2)^{5/2}} - iQ^2 \frac{12A^2B - 3B^3}{(A^2 + B^2)^{7/2}}$$

where

$$Q = \frac{(1 + K^{1/2}) + (1 - K^{1/2}) \exp\left[-(2i)^{1/2} D\right]}{(1 + K^{1/2}) - (1 \cdot K^{1/2}) \exp\left[-(2i)^{1/2} D\right]} \tag{550}$$

For the special case in which both the source loop and the receiving loop are located on the earth's surface, $A = 0$ and the integrals in Eq. (544) do not converge. When $A = 0$, the functions $T_0$, $T_1$ and $T_2$ may be replaced by the following expressions:

$$\int_0^\infty [R(D, g) - R(\infty, g)] g^2 J_0(g, B) \, dg + \int_0^\infty R(\infty, g) g^2 J_0(g, B) \, dg = T_0' + T_0''$$

$$\int_0^\infty [R(D, g) - R(\infty, g)] g^2 J_1(g, B) \, dg + \int_0^\infty R(\infty, g) g^2 J_0(g, B) \, dg = T_1' + T_1''$$

$$\int_0^\infty [R(D, g) - R(\infty, g)] g J_1(g, B) \, dg + \int_0^\infty R(\infty, g) g^2 J_0(g, B) \, dg = T_2' + T_2''$$

$$\tag{551}$$

The integrals $T_0'$, $T_1'$ and $T_2'$ are convergent when

$$[R(D, g)] - [R(D, g)]_{D \to \infty} \tag{552}$$

is treated as a single quantity. The integrals $T_0''$, $T_1''$ and $T_2''$ need not be evaluated by digital computer; they represent the secondary field for a source dipole on the surface of a homogeneous earth with a conductivity $\sigma = \sigma_1$ and may be determined from the expressions in closed form which were developed for a homogeneous earth. The mutual coupling ratios, $Z/Z_0'$ for the special loop arrangements over a two-layer earth with $A = 0$ are given by the following equations:

$$\frac{Z'}{Z_0} = 1 + B^3 T'_0 + \frac{Z}{Z_0}\bigg|_{D \to \infty}$$

$$\frac{Z'}{Z_0} = 1 + B^2 T'_2 + \frac{Z}{Z_0}\bigg|_{D \to \infty} - 1$$

$$\frac{Z'}{Z_0} = 2\left[1 + \frac{1}{2} B^2 (T'_2 - BT'_0)\right] + \frac{Z}{Z_0}\bigg|_{D \to \infty}$$

$$\frac{Z'}{Z_0} = B^3 T'_1 + \frac{Z}{Z_0}\bigg|_{D \to \infty}$$

$$\frac{Z'}{Z_0} = 1 - B^2 T'_2 + \frac{Z}{Z_0}\bigg|_{D \to \infty} - 1 \qquad (553)$$

where the parameter $Z/Z_0|_{D = \infty}$ is evaluated using the appropriate equation from the set (533)–(538).

The case for loops on the surface of a two-layer earth is of greatest interest, at least from the standpoint of depth soundings. The case for loops raised above the surface of the earth is important in the interpretation of airborne surveys, which are primarily horizontal profiling measurements rather than depth-sounding measurements. However, there are some exploration problems in which it is necessary to determine the presence of conductive rock beneath a highly resistive surface layer which has a negligible effect on mutual coupling. An example of such a problem is the determination of the thickness of glacial ice or other surficial ice cover. Determining the thickness of dry sand in a desert area or of igneous rocks thrust over sedimentary rocks are other examples of practical applications.

The case of a thin conducting sheet is of practical interest; numerical results for loops above a thin sheet may be obtained by evaluating the $T$ integrals with $K$ set equal to zero and $d/r$ assigned small values. The conducting sheet need not be parallel to the plane containing the loops, so the case of a gently dipping dike may be evaluated. Such calculations will not be valid near the outcrop of the dike, however.

Expressions similar to those obtained for a long wire over a conducting sheet have been developed for the mutual coupling between loops over a conducting sheet (Wait, 1953). One of these expressions may be used to obtain a very simple equation for the ratio $Z/Z_0$ for a layered earth provided the value for $|\gamma r|$ is very small. For perpendicular loops both placed at a height $z = h$ over a thin sheet, we have (from Wait, 1962):

$$\frac{Z}{Z_0} = iWG^3 + \frac{W^2 G^2}{1 + (1 - G)^{1/2}} + \frac{iW^3 G}{1 + (1 - G)^{1/2}}$$

where $W = \dfrac{\sigma \mu_0 \omega r \, (\mathrm{d}h)}{2}$

$\mathrm{d}h$ = thickness of the sheet

$$G = \frac{r}{[4h^2 + r^2]^{1/2}} \tag{554}$$

At very low frequencies, only the first term of the three is important; this is equivalent to neglecting the self-inductance of the eddy current paths within the sheet and considering only their resistance. When this can be done, Eq. (554) is simply:

$$\frac{Z}{Z_0} = \frac{i\sigma\mu_0\omega}{2} \frac{r^4}{(4h^2 + r^2)^{3/2}} \, \mathrm{d}h \tag{555}$$

With the same assumptions and approximation, this expression may be integrated to give the response of a layered earth:

$$\frac{Z}{Z_0} = \frac{i\mu_0\omega}{2} \int_0^\infty \sigma(h) \frac{r^4}{(4h^2 + r^2)^{3/2}} \, \mathrm{d}h \tag{556}$$

where $\sigma(h)$ is the conductivity expressed as a function of depth. This integral is readily evaluated to provide values for the ratio $Z/Z_0$ for situations such as a number of uniform layers, a linear variation of conductivity with depth, or an exponential variation of conductivity with depth. For an earth having three uniform layers such that:

$$\sigma(h) = \sigma_1 \quad \text{for} \quad 0 < h < h_1$$
$$\sigma(h) = \sigma_2 \quad \text{for} \quad h_1 < h < h_2$$
$$(\sigma h) = \sigma_3 \quad \text{for} \quad h_2 < h < h_3 = \infty$$

the mutual impedance ratio is:

$$\frac{Z}{Z_0} = \left(\frac{\gamma_1 r}{2}\right)^2 \frac{1}{\left[4\left(\dfrac{h_1}{r}\right)^2 + 1\right]^{1/2}} + \left(\frac{\gamma_2 r}{2}\right)^2 \left[\frac{1}{\left[4\left(\dfrac{h_2}{r}\right)^2 + 1\right]^{1/2}} - \frac{1}{\left[4\left(\dfrac{h_1}{r}\right)^2 + 1\right]^{1/2}}\right]$$
$$+ \left(\frac{\gamma_3 r}{2}\right)^2 \left[1 - \frac{1}{\left[4\left(\dfrac{h_2}{r_1}\right)^2 + 1\right]^{1/2}}\right] \tag{557}$$

where

$$\gamma_1^2 = i\sigma_1\mu_0\omega$$
$$\gamma_2^2 = i\sigma_2\mu_0\omega$$
$$\gamma_3^2 = i\sigma_3\mu_0\omega$$

This approximation provides accurate results if the restriction that $|\gamma r| < \frac{1}{2}$ is observed for all the layers. This method of solution provides only the out-of-phase component; however, for a homogeneous earth, as an example, the in-phase component is almost negligible for $|\gamma r| < \frac{1}{2}$ (see Fig. 203 for proof).

Extensive tables of values for the ratio $Z/Z_0$ as a function of the parameters $A$, $B$, $D$ and $K$ for loops on and above the surface of a two-layer earth soon will be available in the literature. A few curves and tables are available for coupling ratios over a homogeneous earth (Wait, 1956). Families of curves for various loop configurations raised above a homogeneous earth are shown in Fig. 204. These curves are plotted for parametric values for the ratio $h/r$ to show the variation in $Z/Z_0$ as $\sigma$ or $\omega$ change. If instead we wish to consider changes in the coupling ratio, $Z/Z_0$, as $r$ is varied, $\sigma$ and $\omega$ remaining constant, the curves must be replotted so that the constant parameter for each curve is $A$ rather than $h/r$. For large values of $B$, the curves for the inphase components approach the values for $Z/Z_0$ given by Eqs. (493), (496), (499) and (503). For small values of $h/x$, the curves for loops raised above the surface have much the same shape as the curves for $h/x = 0$; as the ratio $h/x$ is increased, the curves become less complicated in form.

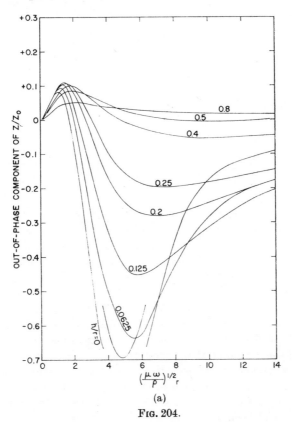

(a)

Fig. 204.

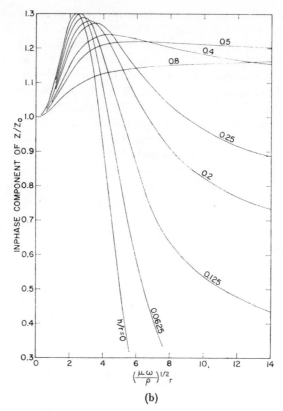

FIG. 204 (a)–(b). Mutual impedance ratio for horizontal coplanar loops raised above a uniform conducting earth.

Two-loop soundings may be interpreted by the same procedures used for interpreting long-wire soundings. Field measurements are plotted with the parameter $r\sqrt{f}$ as the abscissa; bi-logarithmic coordinates must be used if the quantity measured is $C_1 Z/Z_0$ (the situation in which the calibration constant, $C_1$, for the equipment is not known), while semi-logarithmic coordinates may be used if the field observations are reduced numerically to values for the ratio $Z/Z_0$. When bi-logarithmic coordinates are used, the field data may be transposed either vertically or horizontally over the theoretical curves, the only requirement being that the coordinate axes of both the field curves and the master curves must be parallel when a match is obtained. If measurements have been made at a sufficiently low frequency, the impedance ratio may be assumed to approach unity, so that the low-frequency end of the curve may be placed at $Z/Z_0 = 1$ on the master curves. Curve matching may then be carried out as though the impedance ratios had been measured in absolute terms.

An example of the interpretation of two-loop sounding data is shown in Fig. 205. The measurements shown in this illustration were made over thick ice floating on water in the Arctic Ocean. Two square-loops of flexible cable, 10 m on a side, were placed on the surface of the ice at separations ranging from 61 to 152 m, though for a single set of data, as shown here, only a single spacing was used. The diameters of the loops were less than one-fifth the separation

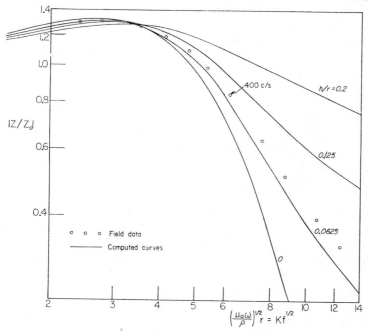

FIG. 205. Comparison of mutual coupling measurements made over ice drifting on sea water with theoretical coupling curves for loops raised above a uniform conductor.

between loop, as required by theory. The transmitting loop was energized with a battery-operated power oscillator delivering about 5 W of power to the transmitter coil. The voltage induced in the receiving coil was amplified and measured with a battery-operated a.c. voltmeter. The current flowing in the transmitter coil was monitored with the same voltmeter by measuring the voltage across a low ohmic-value resistor connected in series with the transmitter coil. A twisted, shielded pair of wires was used to couple the reference voltage from the transmitter to the receiver.

The ratio of the magnitude of the voltage induced in the receiving loop to the magnitude of the current in the transmitting loop is the magnitude of the mutual coupling, $Z$. This coupling was a function of the frequency response characteristic of the equipment as well as of the thickness of the ice. A second set of measurements was made with the loops very close together over thick ice to determine $Z_0/K$ as a function of frequency, where $K$ is a constant dependent on the coil geometry and spacing. Using these two sets of measurements, the quantity $KZ/Z_0$ was calculated, $K$ being equivalent to a calibration factor $C_1$ used above.

It is not possible to calculate a value for the calibration factor, $K$, with the desired accuracy, so the field data shown in Fig. 205 were plotted on bi-logarithmic coordinates for interpretation. The values for mutual impedance, expressed in uncalibrated instrument units, were plotted as a function of the square root of frequency, as shown by the data points on Fig. 205. These data were then compared with a family of theoretical curves (from Fig. 204) for different values of $h/r$, with $\left(\dfrac{\mu_0\omega}{\varrho}\right)^{1/2} r$ as the abscissa, also plotted on bi-logarithmic coordinates. The data suggest the best $h/r$ ratio to compare would be approximately 0·08. Since the coil separation was 200 ft, this leads to an interpretation for the ice thickness of 16 ft. A frequency of 400 c/s on the abscissa of the field curve corresponds with $\left(\dfrac{\mu_0\omega}{l}\right)^{1/2} r = 6{\cdot}18$ on the abscissa of the reference curves. From this relationship, the relationship for any other corresponding points, the resistivity of the sea water is determined to be 0·31 ohm-m.

Between June 9 and June 29, 1958, 80 such soundings were made over ice in the Arctic Ocean. The thicknesses determined from coupling measurements provided an accuracy of $\pm 10$ per cent, in comparison with other methods for determining ice thickness (observations of flexural waves and drilling).

The amount of field data which are required to determine the resistivity and thickness of the layers of a stratified earth depends on the complexity of the problem and whether $Z/Z_0$ or $C_1 Z/Z_0$ is measured. For loops on the surface of a homogeneous earth, a single measurement of both the real and the imaginary components of $Z/Z_0$ or the measurement of one component at two frequencies is sufficient to give the resistivity. When curve matching is not required, Argand diagrams such as Fig. 206 are convenient to use. Using the measured values of the real and imaginary components, the value of $B$ may be read directly from the curve for $h/r = 0$. Then

$$\varrho = \frac{\mu_0\omega r^2}{2B^2}$$

Depending on the loop configuration and the range of $B$ and $h/r$, a single measurement of both components may be sufficient for the determination of both $\varrho$ and $h$ for loops raised above homogeneous earth. However, for the horizontal coplanar and the perpendicular configurations, more than one set of measurements is often required for a unique determination of the parameters.

When the earth is layered, it is always necessary to measure one or both components of $Z/Z_0$ at a number of frequencies or spacings and to use curve match-

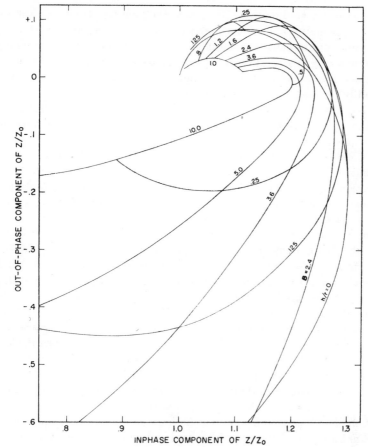

FIG. 206. Argand diagram for horizontal coplanar loops on and above the surface of a homogeneous earth.

ing or an equivalent interpretative procedure. When two components are measured, the two curves may be interpreted independently; if the interpretations are valid they will, of course, agree with each other.

A number of other techniques for interpreting depth soundings have been devised. In some of these techniques, the first step is to calculate an apparent resistivity $\varrho_a$ for each measurement, of $Z/Z_0$. This apparent resistivity is that of a fictitious homogeneous half space in which the resistivity varies as a function of the frequency and spacing. The field data are then plotted with $\varrho_a$ as the ordinate and $f^{\frac{1}{2}}$ as the abscissa and the reference curves are plotted with $\varrho_a/\varrho_1$ as the ordinate and a quantity proportional to $B$ or $D$ as the abscissa, using bi-logarithmic co-ordinates for both sets of curves. Interpretation of the induction soundings can then be carried out in essentially the same manner as for galvanic resistivity soundings. If $\varrho_1 \gg \varrho_2$ the high-frequency asymptote of the curves approaches $\varrho_1$ and the low-frequency asymptote approaches $\varrho_2$. The main difficulty in using this method of interpretation is in determining a unique value for $\varrho_a$. Over certain ranges of the parameters, where the curve is monotonically increasing or decreasing, an apparent resistivity can be determined readily by using curves for a homogeneous earth such as Figs. 203(a) or 203(b) and considering each component separately. Kozulin (1960) has described a method in which both components of $Z/Z_0$ are used to determine a complex "effective" resistivity.

In addition to methods using a long wire or a small loop as a source, a number of other induction sounding methods are sometimes employed. In the central ring induction method, the field in the center of a large circular transmitting loop is measured with reference to the current in the loop. Either the radius of the loop or the frequency may be varied to make a sounding; the maximum depth of investigation is about equal to the radius of the loop. At the frequencies used, the inphase component of the secondary field usually is negligible. Using an approximation for the out-of-phase component, Stefanescu (1936) has developed interpretative curves and procedures. Kosenkov (1963) has developed more exact expressions for the field of a large loop on the surface of a homogeneous earth.

Short grounded-wires are sometimes used as sources; in many cases the receiver is sufficiently far removed that the wire may be considered as an electric dipole. If the vertical component of the magnetic field is measured by means of a small loop, the system is equivalent to one in which the electric field about a magnetic dipole is measured. By the theorem of reciprocity, Eqs. (533), (549) and (553) can be used to obtain the response of either system.

Induction sounding methods, in which short grounded-wires equivalent to electric dipoles are used for both the source and receiver, are employed extensively by Russian investigators.

So far we have considered only those induction methods which employ quasi-steady state (sinusoidally varying) sources of current. In recent years, other induction methods have been developed which employ transient currents. The current waveform may approximate a step function or may consist of repetitive pulses having a square, trapezoidal, sawtooth, or half-sine waveform. In some of these "transient" methods, the secondary field is measured while the primary field or the time rate of change of the primary field is zero. Such transient methods are insensitive to errors in spacing and orientation of the loops, and the depth range is not necessarily a function of the spacing between the source and receiver. An analysis of the shape of the received transient waveform is equivalent to measuring the response at a number of frequencies for a harmonically varying source. In some transient induction methods which employ repetitive pulses, the integral of the transient voltage is measured. Unless the integral is measured as a function of time, such as in the INPUT method (see the section on airborne methods Chapter VI), these integral methods are not useful for making vertical soundings since all frequency information is lost.

It is somewhat more difficult to design satisfactory equipment for transient measurements than for continuous wave measurements. Current waveforms which closely approximate step functions or square pulses are difficult to produce in loops because of the inductance of the loop. When primary current waveforms having very sharp rise and fall times are generated, large voltage spikes are induced in the receiving coil which may cause ringing at the self-resonant frequency of the coil and may, if the spacing is short, damage the receiving apparatus. For these and other reasons trapezoidal or half-sine wave current waveforms are easier to employ and may be preferred to square waves even though they contain less high-frequency energy. The receiving apparatus cannot be tuned in order to reject noise because the bass-band must be wide enough to pass the frequencies of interest. Noise from power lines or other industrial sources can be reduced by injecting noise into the system in such a manner as to effect cancellation. Integral methods are inherently insensitive to non-cohe: rent noise.

The response of a transient system may be obtained directly by model experiments using the desired transient waveform, or it may be calculated. Most commonly, when calculations are made, the transient response is determined from the harmonic response of an equivalent system. If a repetitive primary field is employed, the response may be approximated by taking the sum of the responses for the fundamental and first few harmonics of the primary waveform. In general, however, it is just as easy or is necessary to consider all frequencies. If the harmonic response $f(i\omega)$ of a system is known for all frequencies, the response of a similar system for a step function may be obtained, formally at least, by means of the Fourier or Laplace transform. From the response for a step function, the response for any other waveform may be obtained by superposition.

For quasi-steady state conditions, $i\omega$ in the expressions for the harmonic response of a system may be replaced by $s$ as used in the Laplace transform so that $f(i\omega) = F(s)$. A function of time $F(t)$ is related to $F(s)$ by the complex inversion integral

$$F(t) = \frac{1}{2\pi i} \int\limits_{c-i\infty}^{c+i\infty} F(s)\, e^{st}\, ds \qquad (558)$$

where $c$ is a small, real and positive, arbitrary constant. In practice $F(t)$ can often be found directly from tables of Laplace transforms without employing the inversion integral.

The step function response for loops on the surface of a conductive homogeneous earth or over a thin sheet may be obtained in closed form from the harmonic response $Z(s)$ and Eq. (558). For a unit step function of current, $I(t) = u(t)$, the Laplace transform is $I(s) = 1/s$. Replacing $i\omega$ by $s$ in Eq. (533) for horizontal coplanar loops above a homogeneous earth, the transform of the voltage induced in the receiving loops is:

$$V(s) = -Z(s)\,I(s) = \frac{dA_1\,dA_2}{2\pi\sigma r^5}\left[ -\frac{9}{s} + \frac{9}{s}\,e^{-\alpha s^{1/2}} + 9\alpha s^{-1/2}\,e^{-\alpha s^{1/2}} \right.$$

$$\left. + 4\alpha^2 e^{-\alpha s^{1/2}} + \alpha^3\,e^{-\alpha s^{1/2}} \right]$$

where
$$\alpha = (\sigma\mu_0)^{1/2}\,r \tag{559}$$

Employing the inversion integral

$$V(t) = \left[\frac{dA_1\,dA_2}{2\pi\sigma r^5}\right]\frac{1}{2\pi i}\int_{c-i\infty}^{c+i\infty}\left[ -\frac{9}{s} + \frac{9}{s}\,e^{-\alpha s^{1/2}} + 9\alpha s^{1/2}\,e^{-\alpha s^{1/2}} + 4\alpha^2 e^{-\alpha s^{1/2}} \right.$$

$$\left. + \alpha^3 s^{1/2}\,e^{-\alpha s^{1/2}} \right] e^{st}\,ds \tag{560}$$

The inverse transforms of the first two terms of the integrand are well known and may be found in tables of transforms.

$$\frac{1}{2\pi i}\int_{c-i\infty}^{c+i\infty} -\frac{9}{s}\,e^{st}\,ds = -9u(t)$$

and
$$\frac{1}{2\pi i}\int_{c-i\infty}^{c+i\infty}\frac{9}{s}\,e^{-\alpha s^{1/2}} e^{st}\,ds = 9\left(1 - erf\frac{\alpha}{2t^{1/2}}\right)u(t) \tag{561}$$

Differentiating each side of this last equation three times with respect to $\alpha$ yields (Wait, 1951)

$$\frac{1}{2\pi i}\int_{c-i\infty}^{c+i\infty} \alpha s^{-1/2}\,e^{-\alpha s^{1/2}} e^{st}\,ds = \frac{\alpha}{2t^{1/2}}\,erf'\left(\frac{\alpha}{2t^{1/2}}\right)u(t)$$

$$\frac{1}{2\pi i}\int_{c-i\infty}^{c+i\infty} \alpha^2\,e^{-\alpha s^{1/2}} e^{st}\,ds = 2\left(\frac{\alpha}{2t^{1/2}}\right)^3 erf'\left(\frac{\alpha}{2t^{1/2}}\right)u(t)$$

$$\frac{1}{2\pi i}\int_{c-i\infty}^{c+i\infty} \alpha^3 s^{1/2}\,e^{-\alpha s^{1/2}} e^{st}\,ds = \left[-2\left(\frac{\alpha}{2t^{1/2}}\right)^4 erf''\left(\frac{\alpha}{2t^{1/2}}\right) - 2\left(\frac{\alpha}{2t^{1/2}}\right)^3 erf'\left(\frac{\alpha}{2t^{1/2}}\right)\right]u(t)$$

$$\tag{562}$$

so that

$$V(t) = \frac{dA_1\,dA_2}{2\pi\sigma r^5}\,P(t)\,u(t) = \frac{\mu_0\,dA_1\,dA_2}{2\pi\alpha^2 r^3}\,P(t)\,u(t) \qquad (563)$$

where

$$P(t) = \left\{9\,erf\left(\frac{\alpha}{2t^{1/2}}\right) + 2\left(\frac{\alpha}{2t^{1/2}}\right)^4 erf''\left(\frac{\alpha}{2t^{1/2}}\right) - \left[9\left(\frac{\alpha}{2t^{1/2}}\right) + 6\left(\frac{\alpha}{2t^{1/2}}\right)^3\right]erf'\left(\frac{\alpha}{2t^{1/2}}\right)\right\}$$

$$(564)$$

The function $P(t)$ is plotted in Fig. 207 for the parameter $t/\alpha^2$. This parameter, $t/\alpha^2$, is the step function response parameter for a homogeneous earth just as $(\sigma\mu_0\omega)t$ is the harmonic response parameter for a homogeneous earth. The

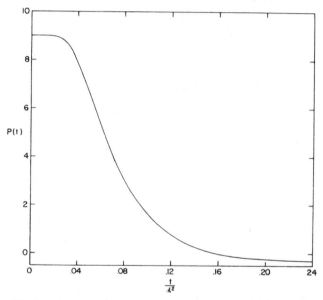

Fig. 207. Step function response for horizontal coplanar loops on surface of a homogeneous earth.

function $V(t)$ may be obtained for other loop configurations over a homogeneous earth by the same procedure as outlined above. Since displacement currents were neglected in the development of the expression for $Z(s)$, Eq. (563) and Fig. 207 are not valid for very small values of $t/\alpha^2$. Integral solutions for certain cases in which displacement currents are considered have been developed by Bhattacharyya (1959) and Wait (1960).

By the use of the Laplace transform, the step function response for a thin horizontal sheet also may be obtained in simple form (Wait, 1956). The results show that the secondary field may be obtained by assuming that the field from the sheet is the same as the field of an image dipole located a distance $h + t/q$

below the plane of the sheet. For a vertical dipole located a distance $h$ above a thin sheet

$$H_z(t) = \frac{m}{4\pi}\left[\frac{3Z^2}{r'^5} - \frac{1}{r'^3} - \frac{3(B + t/q)^2}{r_a'^5} + \frac{1}{r_a'^3}\right]u(t)$$

where

$$
\begin{aligned}
q &= \sigma\,\mu_0\,\mathrm{d}/2 \\
r_a' &= [(B + t/q)^2 + r^2]^{1/2} \\
B &= 2h + z
\end{aligned}
\qquad (565)
$$

For very small times, the sheet behaves as a perfect conductor and Eq. (565) is equivalent to Eq. (491). For very large times, the image recedes to a great distance from the sheet leaving only the field from the primary dipole. The step function response for a thin sheet for other coil configurations may be obtained by substituting $z' + 2h' + t/q$ for $z + 2h$ in Eqs. (495), (498) and (502).

The voltage induced in the receiving loop of a horizontal coplanar configuration raised above a horizontal sheet may be obtained by differentiating Eq. (565).

$$V(t) = -\mu_0\,\mathrm{d}A_2\frac{\mathrm{d}(H_z(t))}{\mathrm{d}t} \qquad (566)$$

Neglecting the impulse voltage at $t = 0$

$$V(t) = \frac{\mu_0 I_0\,\mathrm{d}A_1\,\mathrm{d}A_2}{4\pi q}\left[\frac{9(2h + t/q)}{r_a^5} - \frac{15(2h + t/q)^3}{r_a^7}\right]u(t) \qquad (567)$$

Grounded wires with electrode configurations essentially the same as are used with galvanic resistivity methods were employed in most of the early transient methods. Single or repetitive square or sawtooth waveforms were used. In some of the methods the current pulses were sufficiently long for the voltage from the receiving array to reach a steady value permitting calculation of the d.c. apparent resistivity. A variety of techniques was used to study the transients following the beginning or end of the current pulse. In some methods the transients were recorded; in other methods the transients from the earth were compared with transients generated in R–C networks. The circuit parameters required for cancellation were a measure of the amplitude and shape of the transient. Interpretation of the results of early transient surveys was primarily qualitative. In some cases, the data were complicated by induced polarization effects.

Orsinger and Van Nostrand (1954) have described a two-loop sounding method employing a repetitive square wave source current which they term "the electromagnetic reflection method". Large horizontal square loops separated by a distance of about three times the length of a side are used. In addition to the field apparatus, a scale model apparatus, using as models sheets of metal placed in air, is employed in the field. The scale model transmitting loop is placed in series with a part of the current flowing through the field loop. The voltages from the two receiving coils are mixed in opposition and their difference is displayed on an oscilloscope. The models are varied until satisfactory cancellation of the two signals is obtained. Since the frequencies in the full scale and the model setup are the same, the thickness and conductivity of the various earth strata are obtained from the parameters of the corresponding model and the relationship (see Eq. 418):

$$\sigma_1 L_1^2 = \sigma_2 L_2^2$$

Generally the sweep and triggering of the oscilloscope are adjusted so that only the first part of each square wave is displayed. Noise from power lines is largely eliminated by filtering off the noise from the raw signal in a parallel branch, shifting the phase, and then mixing the noise with the signal to effect cancellation.

In the last decade, Russian investigators have been particularly active in the development and application of transient as well as continuous wave electromagnetic sounding methods. Enenstein et al. (1959) describe equipment and interpretative techniques for a transient method which employs rectangular current pulses in conjunction with electric dipole arrays for both transmitter and receiver. Following the initiation of the current pulse, the voltage from the receiving dipole is recorded by means of an oscilloscope which is equipped with a camera. The apparent resistivity is calculated by means of Eq. (85) or (91) at various intervals of time following the beginning of the pulse. The apparent resistivity is then plotted as a function of time on bi-logarithmic paper and interpreted by matching to theoretical curves in much the same way as for conventional galvanic soundings.

Transient methods employing a horizontal loop for the receiver and horizontal loops or horizontal wire elements for transmitters also have been developed by Russian investigators. In most of these methods, only the latter part of the transient curve is studied. Koreleva and Skugarevskaya (1962) and others have developed formulae and reference curves which can be used in the interpretation of results obtained with these methods.

## 39. HORIZONTAL PROFILING

At the present time, induction methods are used primarily for horizontal profiling. Massive sulfide ore bodies or other very good conductors usually are the targets of such surveys. Many methods and techniques have been used in horizontal profiling, and, unfortunately, there is no systematic terminology or classification for the various methods. Often several different names are used for the same method; commonly, the name of the manufacturer of the equipment or of a contractor, rather than a descriptive name (acronym), is employed.

Since standard names are not in common usage, methods may be described and classified in terms of (1) the technique used in traversing, particularly whether the source is fixed or moving; (2) the size and type of source; and (3) the quantity or quantities measured.

With a fixed source method, the transmitting setup remains in a fixed location while the receiving loop is moved about to explore the immediate area. With a moving source method, both transmitter and receiver are moved about, usually with a fixed separation between the two. The fixed source induction method may be compared with galvanic resistivity methods in which half the array is moved while the other half remains fixed; moving source methods are the counterpart of resistivity profiling methods in which all of the electrodes are moved for each new station. In using a moving source method, the line between source and receiver may be in the direction of movement (in-line technique) or it may be perpendicular to the direction of movement (broadside technique).

It is important that the difference between fixed source and moving source methods be considered in selecting the best method for a particular problem and in interpreting the results obtained. For many purposes, the depth range of moving source equipment is not adequate and large fixed sources must be used. However, considerable time and effort are involved in laying out a large source; hence, when a limited depth range may be tolerated, moving source methods are often preferable, particularly for reconnaissance work.

With a fixed source method, two like conductors do not, in general, provide identical anomalies unless the anomalous conductors are symmetrically located with respect to the source, because the primary field or free-space coupling is different at each conductor. The shapes of the anomalies obtained with fixed source methods, however, are relatively simple compared with the shapes of anomalies obtained with moving source methods, because the coupling between the source and the conductor is constant for any one setup and also because the primary field at the conductor is usually quite homogeneous. With a moving source method, the free-space coupling between source and receiver is constant; therefore, in the absence of other disturbances two like conductors provide identical anomalies. In general, anomalies obtained with moving source methods are complicated because the coupling between the source and conductor varies. Anomalies obtained with moving source methods usually have more maxima and minima then those obtained with fixed source methods, particularly if the conductor is not deeply buried. Usually, several closely spaced conductors are more easily defined with fixed source methods, although it may be difficult to assess the anomalies in terms of their magnitudes. Not only does the form of the primary field vary with distance from the source, but the conductor nearest the source also may tend to screen those further away. When moving source methods are used, the presence of more than one shallow conductor within a distance comparable to or less than the separation between the source and receiver results in data which may be very difficult to interpret reasonably. When it is not necessary to go to great lengths to provide a connecting cable or line-of-sight path between the source and receiver, the broadside technique is usually preferable to the in-line technique; in following traverse lines normal to the strike of a series of closely spaced conductors, the results obtained with the broadside technique will be simpler than the results obtained with an in-line technique.

For practical reasons, a small portable loop and current generator comprise the only type of source equipment used with moving source methods. Fixed sources may consist of insulated loops placed either horizontally or vertically. Long grounded-wires may also be used. The diameter of portable source loops used in moving source methods and of vertical loops used in fixed source methods is usually less than a fifth of the separation between the source and receiver so that they may be treated as simple magnetic dipoles. Fixed horizontal loops are often so large that the fields from each side of the loop must be considered separately, and the primary field varies as some inverse power of distance between the first and third. In the case of a long grounded-wire source, measurements are usually made in the distance range where the induction field varies approximately with the inverse distance. Since a long grounded-wire develops galvanic currents in the ground as well as eddy currents, some methods in which a grounded wire source is used are designed to accentuate the effects of galvanic currents and to minimize the response from eddy currents. In such instances, the electromagnetic or induction technique used is referred to as being galvanic. The relative merits of long grounded-wires as opposed to large loops or of horizontal loops as opposed to vertical loops will be discussed under the specific profiling methods.

As was pointed out previously, the complete specification of the field at any point requires the equivalent of the measurement of three spatial components and their phase angle with respect to the field at the source. The field is never completely specified in any of the routine methods. The direction only of the projection of the field on one or two planes may be measured. Commonly the amplitude and phase angle, or the real and imaginary components of one spatial component of the field, relative to the source current, are measured. In other methods, the ratio of one component detected at two different points or the ratio of two different components detected at the same point are measured.

Methods in which the direction of the field is measured are quite sensitive and instrumentation for these methods is simple. A qualitative indication of the conductivity of a body can be obtained by making measurements at two frequencies. Usually, more information about the nature of a conductor can be obtained with methods in which the real and imaginary parts of one or more components are measured; a qualitative indication of conductivity can be derived from measurements at a single frequency. Also, for a given frequency the out-of-phase component or phase angle is more sensitive to the presence of weakly conductive areas than is the direction of the field.

When the conductivity of the body is relatively slight and the frequency is low, the shapes of the inphase and out-of-phase anomalies usually are about the same. Ordinarily, the component having the greatest amplitude is used to deduce the geometry of the conductors. For some conductors, when the conductivity and frequency are high, the shapes of the inphase and out-of-phase components are quite different; usually the inphase anomalies are simpler and easier to interpret than the out-of-phase anomalies.

Methods in which the ratio of the field observed at two different locations or the ratio of two components observed at a common point are used do not require a reference signal from the source. "Ratio" methods are quite sensitive and are sufficiently flexible that a field technique can be varied to be most suitable for a particular problem. The results obtained with ratio methods are usually somewhat more complicated and difficult to interpret than the measurements of a single component in relation to the source.

In general, the secondary fields caused by conductors such as may be found by horizontal profiling cannot readily be determined by mathematical analysis and most reference curves for the interpretation of horizontal profiling data are obtained by scaled model studies. Reference curves obtained by mathematical analysis are available for some simple shapes such as spheres and infinitely long cylinders of arbitrary conductivity and permeability, and for an infinite half-plane with zero resistivity.

Consider a sphere of radius $R$, with conductivity $\sigma_1$ and magnetic permeability $\mu_1$ placed in an extensive medium with zero conductivity and with permeability $\mu_2$. We will assume that displacement currents can be neglected. The primary field, $H_p = H_{z,p}$, will be assumed to be uniform over the volume occupied by the sphere (see Fig. 208).

The vector potential, $F_{z,p}$, for the primary field is:

$$F_{z,p} = -\sigma_1 H_{z,p} = -\frac{1}{\sigma_1} H_0 \qquad (568)$$

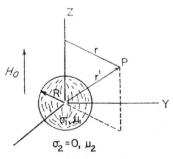

FIG. 208.

The solutions for the wave equation giving expressions for $F_1$ inside the sphere and for $F_2$ outside the sphere have been given by Wait (1951):

$$F_1 = F_{1,z} = \sum_{n=0}^{\infty} \frac{b_n \hat{I}_n(\gamma_1 r')}{r'} \, P_n(\cos\theta) \, i\mu_1\omega$$

$$F_2 = F_{2,z} = \sum_{n=0}^{\infty} \frac{a_n \hat{K}_n(0)}{r'} \, P_n(\cos\theta) \, i\mu_2\omega - \frac{1}{\sigma_2}H_{z,p} \tag{569}$$

where $\hat{I}_n(Z)$ and $\hat{K}_n(Z)$ are spherical Bessel functions and $P_n(\cos\theta)$ is the Legendre function of the first kind. The coefficients $a_n$ and $b_n$ may be determined from boundary conditions which require that the tangential magnetic field and the normal flux density be continuous at $r' = R$. The external magnetic field can be determined from the vector potential $F_{2,z}$. The results are:

$$H_{z,s} = -\frac{3}{2} R^3 H_{z,p}(M + iN)\left(-\frac{1}{r'^3} + \frac{3z^2}{r'^5}\right)$$

$$H_{r,s} = -\frac{3}{2} R^3 H_{z,p}(M + iN)\frac{3zr}{r'^5}$$

where $M + iN$

$$= -\frac{2}{3}\left[\frac{2\mu_1(\sinh\alpha - \alpha\cosh\alpha) + \mu_2(\sinh\alpha - \alpha\cosh\alpha + \alpha^2\sinh\alpha)}{2\mu_1(\sinh\alpha - \alpha\cosh\alpha) - 2\mu_2(\sinh\alpha - \alpha\cosh\alpha + \alpha^2\sinh\alpha)}\right]$$

$$\alpha = (i\mu_1\sigma_1\omega)^{1/2}R \tag{570}$$

The equation for $H_{z,s}$ may also be written as:

$$H_{z,s} = \frac{m}{4\pi r'^3}\left[-1 + \frac{3z^2}{r'^2}\right] \tag{571}$$

By comparison with Eq. (412), we see that the secondary field at the sphere is equivalent to that of a dipole with a moment, $m$, given by:

$$m = -6\pi R^3 H_{z,p}(M + iN) \tag{572}$$

Thus, the phase angle for the anomalous field observed in profiling over a buried sphere depends only on $\alpha$ and $R$. The amplitude of the anomaly depends on all the parameters, but the shape of the anomaly depends only on the position of the traverse with respect to the center of the sphere. The functions $M$ and $N$ are plotted graphically as functions of $\alpha$ for various values of the permeability ratio, $\mu_1/\mu_2$, in Fig. 209. A curve for $H_{z,s}/H_{z,p}(M + iN)$ is shown plotted as a function of $x/R$ for $y = 0$ and $R/z = 0.8$ in Fig. 210. Multiplication of the ordinates of Fig. 210 by the appropriate values for $M$ and $N$ (taken from Fig. 209) will give the anomaly in terms of the ratio $H_z/H_{z,p}$.

Ward (1959) has shown that by using Eq. (570) it is possible to determine all the parameters for a buried sphere from measurements of electromagnetic field strength in which $\omega$ is varied over a wide range. The following discussion applies to the interpretation of any field results in which the primary field is nearly homogeneous in the area of interest, and in which the vertical or horizontal component of the field is measured.

*Location.* Once the sphere has been detected, its exact location can be determined from a series of closely spaced traverses.

*Depth.* The anomaly in $H_z/H_{z,p}$ is zero at $|x| = x_0$ (for $y = 0$) when $1/r'^3 = 3z^3/r'^5$. The depth, $d$, to the center of the sphere is given by $z = d = \sqrt{2}\, x_0$. Similarly, if $H_x/H_{z,p}$ is measured, the depth to the sphere is given by $z = d = x_0$, where $x_0$ is now the horizontal distance between the maximum and minimum.

*Magnetic permeability.* The permeability contrast, $\mu_1/\mu_2$, can be found independently of the other parameters provided the field at one point is measured at a frequency sufficiently low that eddy currents are negligible and at a

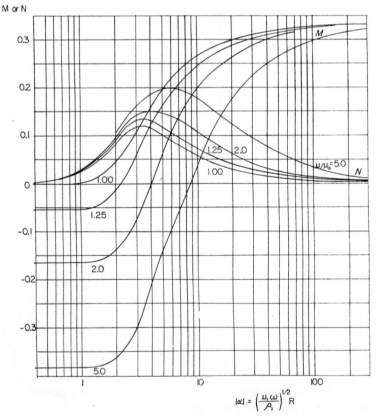

$$|\alpha| = \left(\frac{\mu_1 \omega}{\rho_1}\right)^{1/2} R$$

FIG. 209. Variation of the functions $M$ and $N$ (defined in Eq. (558)) with $|\alpha|$ for a conducting, permeable sphere.

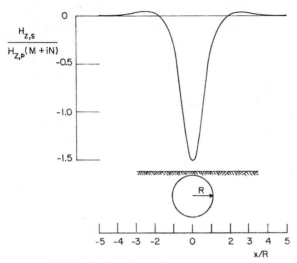

FIG. 210. Variation in electromagnetic field intensity over a buried sphere for $y = 0$ and $R/z = 0 \cdot 8$.

frequency sufficiently high that eddy current saturation is closely approached. At very low frequencies, $\alpha \to 0$ and

$$\lim_{\alpha \to 0} (M + iN) = -\frac{2}{3}\left(\frac{\mu_1 - \mu_2}{\mu_1 + 2\mu_2}\right) \tag{573}$$

At very high frequencies, $\alpha \to \infty$ and

$$\lim_{\alpha \to \infty} (M + iN) = 1/3 \tag{574}$$

The ratio of the low frequency response to the high frequency response is:

$$\frac{H_s(0)}{H_s(\infty)} = -2\left(\frac{\dfrac{\mu_1}{\mu_2} - 1}{\dfrac{\mu_1}{\mu_2} + 2}\right) \tag{575}$$

so that:

$$\frac{\mu_1}{\mu_2} = \frac{2H_s(\infty) - 2H_s(0)}{2H_s(\infty) + H_s(0)} \tag{576}$$

Ordinarily, $\mu_2$ will be nearly equal to $\mu_0$, so $\mu_1 \cong \mu_0(\mu_1/\mu_2)$. Either the vertical or horizontal field can be used in this analysis. Furthermore, it is necessary only to measure $KH_{z,s}$ or $KH_{x,s}$, where $K$ is any arbitrary calibration constant; therefore, the equipment need not be calibrated. As was pointed out in the discussion on sounding methods, this simplifies the instrumentation required for multi-frequency measurements. If remnant magnetization is negligible, the low frequency response, $H_{z,s}(0)$, can be determined from measurements of $H_z(0)$ using an ordinary vertical component magnetometer, and from a value for $H_{z,p}(0)$ which may be taken from charts giving the normal values for the vertical component of the earth's magnetic field. It is interesting to note that for large values of the permeability ratio, the magnetic or low frequency response is twice the high frequency or eddy current response.

The permeability contrast may also be found by curve matching provided that field measurements are made over a sufficiently wide range. Curve matching is facilitated by replotting Fig. 209 in terms of the amplitude $A$ $\left(= (M^2 + N^2)^{1/2}\right)$ and $\varphi \left(= \tan^{-1}\left(\dfrac{N}{M}\right)\right)$ using bi-logarithmic coordinates for $A$ and semi-logarithmic coordinates for $\varphi$ plotted as a function of $\alpha$. Once the proper super-position of the field curves over the theoretical curves has been found, $\mu_1/\mu_2$, the value of $A$ for any value of $\alpha$, and corresponding values of the field frequency and the theoretical $\alpha$ are determined. From these quantities $M$ and $N$ and $\sigma_1$ may be determined.

Regardless of which technique is used, the permeability contrast must be determined before or together with the determination of the radius and the con-ductivity. However, it should be emphasized again that for many conductors the permeability is so low that the effect on electromagnetic measurements is neg-ligible and one can proceed directly with the determination of the radius and conductivity by assuming $\mu_1 = \mu_2 = \mu_0$.

*Radius of the sphere.* Once the depth to the center of the sphere and the per-meability have been determined, it is easy to determine the radius, $R$, of the sphere using either the low frequency or the high frequency response. Using the ow frequency response, Eq. (573) yields:

$$R^3 = \frac{H_{z,s}(0)}{H_{z,p}} \left(\frac{\mu_1 + 2\mu_2}{\mu_1 - \mu_2}\right)\left(\frac{r'^5}{3z^2 - r'^2}\right) \tag{577}$$

Using the high frequency response:

$$R^3 = -2\frac{H_{z,s}(\infty)}{H_{z,p}}\left(\frac{r'^5}{3z^2 - r'^2}\right) \tag{578}$$

Similar expressions may be obtained for $H_{x,s}$, rather than $H_{z,s}$.

If neither the low frequency nor the high frequency response is determined. but if $M$ and $N$ have been determined by curve matching, the radius may be determined from the measured response and values of $M$ and $N$ for an arbitrary frequency using the following expression:

$$R^3 = \frac{H_{z,s}(\omega)}{H_{z,p}}\left(\frac{1}{M + iN}\right)\left(\frac{r'^5}{3z^2 - r'^2}\right) \tag{579}$$

*Conductivity.* If the conductivity has not been determined by curve match-ing, it may be determined from $\mu_1$, $R$ and the value of $\alpha$ at the critical frequency,

$\omega_c$, where the inphase component of $H_{z,s}$ is nil:

$$\sigma_1 = \frac{|\alpha^2|}{\mu_1 \omega_c R^2} \tag{580}$$

Solutions for a conducting sphere in a dipolar field have also been obtained (March, 1953). In such circumstances, the secondary field of the sphere is no longer dipolar, but may be expressed as an infinite series of magnetic multipoles located at the center of the sphere. The relative amplitudes of the multipoles depend on the ratio of the radius of the sphere to the distance between the source and the sphere. When this ratio is very small, only the first, or dipolar, term is important and the results given above for a sphere in a homogeneous field may be employed. Ordinarily, the results for the case in which the ratio of the radius to the separation is not small are too complicated to be of much help in most practical field problems.

The results given in the preceding paragraphs can be used only for anomalous conductive masses which approximate spheres in shape. For masses of other shapes, however, the form of the curves giving the inphase and out-of-phase anomalies as a function of frequency is much the same as the response curves for a sphere.

Solutions for the secondary field about an infinitely long, conducting circular cylinder have been obtained for both dipole and infinite line sources. The results for a dipole source are quite complicated (Wait, 1960), and therefore, the following discussion is limited to the results for an infinite line source. Consider a

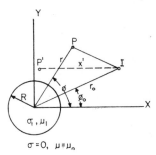

FIG. 211.

cylinder with conductivity $\sigma_1$, magnetic permeability $\mu_1$ and radius, $R$. The axis of the cylinder is placed along the $z$ axis of a cylindrical coordinate system, as shown in Fig. 211. The coordinates of the source wire, which carries a current, $I$, are $r_0$, $\varphi_0$. The solutions are (Wait, 1952):

$$H_{r,s} = -\frac{I}{2\pi} \sum_{n=1}^{\infty} q_n \frac{R^{2n}}{r_0^n r^{n+1}} \sin n(\varphi - \varphi_0)$$

$$H_{\varphi,s} = \frac{I}{2\pi} \sum_{n=1}^{\infty} q_n \frac{R^{2n}}{r_0^n r^{n+1}} \cos n(\varphi - \varphi_0) \tag{581}$$

where

$$q_n = -\frac{n\left(\dfrac{\mu_1}{\mu_0} + 1\right) I_n(\alpha) - \alpha I_{n-1}(\alpha)}{n\left(\dfrac{\mu_1}{\mu_0} - 1\right) I_n(\alpha) + \alpha I_{n-1}(\alpha)}$$

$$\alpha = (i\sigma_1\mu_1\omega)^{1/2}R$$

and $I_n(\alpha)$ and $I_{n-1}(\alpha)$ are modified Bessel functions of the first kind with argument $\alpha$.

If either of the ratios $R/r$ or $R/r_0$ is much less than one, only the first term of the series in Eq. (568) is significant, and the fields are then simply:

$$H_{r,s} = -\frac{I}{2\pi r_0} q_1 \frac{R^2}{r^2} \sin(\varphi - \varphi_0)$$

$$H_{\varphi,s} = \frac{I}{2\pi r_0} q_1 \frac{R^2}{r^2} \cos(\varphi - \varphi_0) \tag{582}$$

where

$$q_1 = -\frac{\left(\dfrac{\mu_1}{\mu_0} + 1\right) I_1(\alpha) - \alpha I_0(\alpha)}{\left(\dfrac{\mu_1}{\mu_0} - 1\right) I_1(\alpha) + \alpha I_0(\alpha)}$$

At the surface of the earth, the vertical and horizontal components of the secondary field are given by:

$$H_{x,s} = -\frac{I}{2\pi r_0} q_1 \frac{R^2}{r^2} [\sin(\varphi - \varphi_0) \cos\varphi + \cos(\varphi - \varphi_0) \sin\varphi]$$

$$H_{y,s} = -\frac{I}{2\pi r_0} q_1 \frac{R^2}{r^2} [\sin(\varphi - \varphi_0) \sin\varphi - \cos(\varphi - \varphi_0) \cos\varphi] \tag{583}$$

The ratio of the total field to the primary field is:

$$\frac{H_x}{H_{y,p}} = -\frac{x'}{r_0} q_1 \frac{R^2}{r^2} [\sin(\varphi - \varphi_0) \cos\varphi + \cos(\varphi - \varphi_0) \sin\varphi]$$

$$\frac{H_y}{H_{z,p}} = 1 - \frac{x'}{r_0} q_1 \frac{R^2}{r^2} [\sin(\varphi - \varphi_0) \sin\varphi - \cos(\varphi - \varphi_0) \cos\varphi] \tag{584}$$

where $x'$ is the horizontal distance from the wire to the observation point.

As was the case for the conducting sphere, the response for a conducting cylinder is the product of two functions, one which depends only on the properties of the cylinder, the radius of the cylinder and the frequency, and one which depends only on the geometrical relationship between the source, the observation point and the cylinder. The function $q$ for the cylinder is compared graphically with the $M$ and $N$ functions for a sphere in Fig. 212 for the one case in which the permeability ratio is unity. It is seen that the variation of the

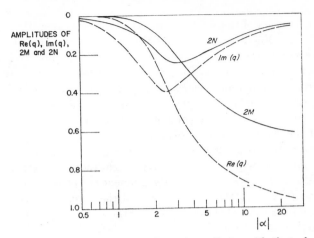

FIG. 212. Comparison of response of a cylinder with that of a sphere as a function of $|\alpha|$ for $\mu_1 = \mu_0$.

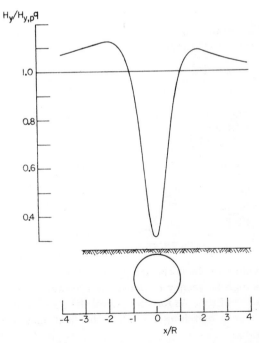

FIG. 213. Variation in electromagnetic field intensity over a buried cylinder for $x'/R = 10$ and $y/R = 1\cdot2$.

secondary field as a function of frequency is similar in the two cases. However, comparing Fig. 213 with Fig. 210, we see that the shapes of the anomalies in the two cases differ significantly; the maxima for the anomaly observed over a cylinder are much more pronounced than the maxima for the anomaly observed over a sphere. This difference occurs because the secondary field about an infinite cylinder vary as the inverse square of the distance of the observation point from the cylinder, whereas the secondary field about a sphere varies as the inverse cube of the distance to the observation point.

Just as for a sphere, the parameters of a cylinder may be determined from measurements in which the very low- and the very high-frequency response or the response over a wide range of intermediate frequencies is measured.

*Depth.* The distance between the points at which the anomaly is zero is a measure of the depth to the axis of the cylinder. For the anomaly in the vertical component to be zero

$$\sin(\varphi - \varphi_0)\sin\varphi = \cos(\varphi - \varphi_0)\cos\varphi$$

This relationship is satisfied for

$$d = \pm x \sqrt{\frac{x_0}{x_0 - 2x}} \tag{585}$$

when

$$x_0 \gg d, \quad d \approx \pm x.$$

*Permeability.* The permeability contrast can be found from the ratio of the low frequency response to the high frequency response. At very low frequencies $\alpha \to 0$ and

$$\lim_{\alpha \to 0} q = -\frac{\mu_1/\mu_0 - 1}{\mu_1/\mu_0 + 1} \tag{586}$$

At very high frequencies $\alpha \to \infty$ and

$$\lim_{\alpha \to \infty} q = 1 \tag{587}$$

The ratio of the low frequency response to the high frequency response is

$$\frac{H_s(0)}{H_s(\infty)} = -\frac{\mu_1/\mu_0 - 1}{\mu_1/\mu_0 + 1} \tag{588}$$

so that

$$\mu_1/\mu_0 = \frac{H_s(\infty) - H_s(0)}{H_s(\infty) + H_s(0)} \tag{589}$$

For very large values of the permeability contrast, the magnitude of the low frequency and the high frequency response are the same, whereas for a sphere they differ by a factor of two. As for the sphere, the permeability contrast may be found also by curve matching provided the field measurements cover a sufficiently wide range.

*Radius.* Having obtained the depth and permeability contrast, the radius may be obtained from the low frequency response, or it may be obtained from the depth and the high frequency response. Using the vertical component

$$R^2 = \frac{H_{y,s}}{H_{y,p}} \left[ \frac{d^2 r_0}{x_0} \right] \left[ \frac{1}{\sin(\varphi - \varphi_0)\sin\varphi - \cos(\varphi - \varphi_0)\cos\varphi} \right] \quad (590)$$

It is convenient to use a measurement from directly above the cylinder so that $\varphi = 90°$.

Then

$$R^2 = \frac{H_{y,s}}{H_{y,p}} \left( \frac{r_0}{x_0} \right)^2 \frac{d^2}{q} \quad (591)$$

*Conductivity.* Having obtained $\mu_1$ and $R$, $\sigma_1$ can be calculated from the value of $\alpha$ at the critical frequency $\omega_c$, where the inphase component is nil, or from the value of $\alpha$ at some other frequency as obtained by curve matching.

A semi-infinite half plane of vanishing thickness and infinite conductivity is one other important mathematical model for which computations readily can be made (Wesley, 1958 and West, 1960). The method of images may be applied to this problem by finding a Green's function in a space of two windings. Wesley (1958) has given expression for the primary and secondary fields at $P$ (Fig. 214$m$) due to an arbitrarily oriented dipole at $P_0$ above a vertical half-plane. The $z$ axis runs along the upper edge of the half plane.

$$\boldsymbol{H}_p = [3\boldsymbol{m} \cdot \boldsymbol{R}_0 \boldsymbol{R}_0 - \boldsymbol{m} \, R_0^2][\pi + 2\tan^{-1} g_0/R_0 + 2g_0 R_0/X^2]/8\pi R_0^5$$
$$+ \, \boldsymbol{R}_0 \boldsymbol{m} \cdot [g_0 \boldsymbol{R}_0 + R_0^2 \, \nabla^0 g_0]/2\pi^2 R_0^2 X^4$$
$$- \, \boldsymbol{m} \cdot [\boldsymbol{R}_0 - g_0 \, \nabla^0 g_0] \, \nabla g_0/2\pi^2 X^4 - \nabla[\boldsymbol{m} \cdot \nabla^0 g_0]/4\pi X^2$$

$$\boldsymbol{H}_s = [3(\boldsymbol{m} \cdot \boldsymbol{R}_1 - 2(x + x_0) m_x) \, \boldsymbol{R}_1 - \boldsymbol{m} R_1^2 + 2m_x R_1^2 \, \boldsymbol{e}_x][\pi + 2\tan^{-1} g_1/R_1 +$$
$$+ \, 2g_1 R_1/X^2]/8\pi^2 R^5 + \boldsymbol{R}_1 \boldsymbol{m} \cdot [g_1(\boldsymbol{R}_1 - 2(x + x_0) \, \boldsymbol{e}_x + R_1^2 \, \nabla^0 g_1]/2\pi^2 R_1^2 X^4$$
$$- \, \boldsymbol{m} \cdot [\boldsymbol{R}_1 - 2(x + x_0) \boldsymbol{e}_x - g_1 \, \nabla^0 g_1] \, \nabla g_1/2\pi X^4 - \nabla[\boldsymbol{m} \cdot \nabla^0 g_1]/4\pi^2 X^2 \quad (592)$$

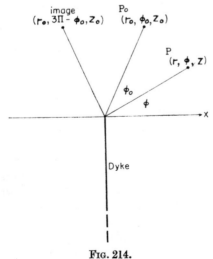

Fig. 214.

where

$$\boldsymbol{H}_p = \text{field due to source}$$
$$\boldsymbol{H}_s = \text{field due to image}$$
$$\boldsymbol{e}_x, \boldsymbol{e}_y, \boldsymbol{e}_z = \text{unit vectors}$$
$$\boldsymbol{m} = \text{moment of the dipole}$$
$$R_0^2 = (x - x_0)^2 + (y - y_0)^2 + (z - z_0)^2$$
$$R_1^2 = (x + x_0)^2 + (y - y_0)^2 + (z - z_0)^2$$
$$g_0 = 2 \; \big/(r r_0) \cos \tfrac{1}{2}(\varphi - \varphi_0)$$
$$g_1 = 2 \; \big/(r r_0) \cos \tfrac{1}{2}(\varphi + \varphi_0 - 3\pi)$$
$$X = (r + r_0)^2 + (z - z_0)^2$$
$$\nabla^0 g_0 = (h_0 \boldsymbol{e}_x - g_1 \boldsymbol{e}_y)/2 r_0$$
$$\nabla^0 g_1 = (h_1 \boldsymbol{e}_x - g_0 \boldsymbol{e}_y)/2 r_0$$
$$h_0 = 2 \; \big/(r r_0) \cos \tfrac{1}{2}(\varphi + \varphi_0)$$
$$h_1 = 2 \; \big/(r r_0) \cos \tfrac{1}{2}(\varphi - \varphi_0 - 3\pi)$$
$$\nabla g_0 = (h_0 \boldsymbol{e}_x - g_1 \boldsymbol{e}_y)/2 r$$
$$\nabla g_1 = -(h_1 \boldsymbol{e}_x + g_0 \boldsymbol{e}_y)/2 r$$
$$\nabla[\boldsymbol{m} \cdot \nabla^0 g_0] = [-h_1 \boldsymbol{e}_z \times \boldsymbol{m} + g_0(m_x \boldsymbol{e}_x + m_y \boldsymbol{e}_y)]/4 r r_0$$
$$\nabla[\boldsymbol{m} \cdot \nabla^0 g_1] = [h_0 \boldsymbol{e}_z \times (\boldsymbol{m} - 2m_x \boldsymbol{e}_x) + g_1(-m_x \boldsymbol{e}_x + m_y \boldsymbol{e}_y)]/4 r r_0 \tag{593}$$

For any particular coil configuration, these equations become much simpler; curves which have been calculated for specific methods are given in later sections.

### 39a. Dip angle method

The most common method in which the direction of the field is measured is the dip angle method (sometimes the dip angle method is also referred to as the vertical loop method). The source loop is usually placed in the vertical plane and may be either fixed or moving. Measurements are made in the plane of the transmitting loop, where, in the absence of conductors, the field is horizontal and normal to the plane of the source loop. The measuring apparatus consists of a single receiver loop equipped with an amplifier and headphones which serve as a null detector, and an inclinometer which is used to measure the angle between the plane of the loop and the horizontal plane. For some work, the loop may be equipped with a device to measure the azimuth with respect to a known bearing when the loop is oriented vertically.

Let us choose a coordinate system as shown in Fig. 215, with the source loop placed at point P′ in the $y'z'$ plane and with the observation point P placed as shown. The usual technique for making a measurement consists of

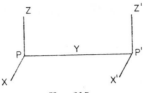

FIG. 215.

placing the receiver loop in the vertical plane and rotating it about the vertical axis until a minimum signal is noted. The angle between the plane in which this minimum signal is observed and the $xz$ plane is called the strike angle (S.A.), and is sometimes measured and recorded if means are available. After finding the strike angle, the receiving loop is rotated back and forth through the horizontal plane about a horizontal axis normal to the strike-angle plane until a minimum signal is observed. The angle between this plane and the horizontal is the dip angle (D.A.) and is measured with the inclinometer. If the operator knows the bearing to the transmitter coil, an alternate method consists of measuring the dip angle by rotating the receiving loop about a horizontal axis in the plane of the transmitting loop. Unless the strike angle is unusually large, approximately the same dip angle will be measured with either technique.

The loop orientation which provides a maximum received signal could be determined rather than the orientation for minimum signal; however, the position for minimum signal can be located much more accurately than the position for a maximum signal. Assuming that the out-of-phase component of the received signal is negligible when the receiver coil is misoriented from the plane in which a maximum signal would be observed, the signal varies as the cosine of the misorientation angle, whereas when the coil is misoriented from the plane in which a minimum signal would be observed, the signal varies as the sine of the misorientation angle.

When the dip angle is measured in the plane of the transmitting loop, the total field intensity at the receiving loop is:

$$H = |H_x| \sin \text{D.A.} \, (\cos\varphi_x + i \sin \varphi_x) - |H_z| \cos \text{D.A.} \, (\cos\varphi_z + i \sin \varphi_z) \quad (594)$$

where $\varphi_z$ and $\varphi_x$ are the phase angles for $H_z$ and $H_x$, and $H_x = H_{x,p} + H_{x,s}$.

If the difference between the two phase angles is zero, the dip angle is found by setting $H$ equal to zero:

$$\text{D.A.} = \arctan \frac{|H_z|}{|H_x|} \quad (595)$$

If the two phase angles are not the same, there is no angle at which $H$ is zero, but there will be an angle at which a minimum $H$ is observed. This angle is:

$$\text{D.A.} = \arctan \frac{|H_z|}{|H_x|} \cos (\varphi_z - \varphi_x) \quad (596)$$

The corresponding expression for the strike angle is:

$$\text{S.A.} = \arctan \frac{|H_y|}{|H_x|} \cos (\varphi_y - \varphi_x) \tag{597}$$

When a fixed transmitter is used, measurements are made along traverse lines laid out across the regional strike, or, less commonly, along radials from the transmitter. Unless radial traverses are used, it is necessary to reorient for each receiver coil location. In wooded areas, this can be done by plotting the positions of all the stations on a map with the transmitter loop at the center. The mast supporting the loop is placed through a hole in the center of the map, and by use of a pointer attached to the mast the loop is oriented so that its plane passes through each receiving station. Coordination between the observer and the operator at the source loop is necessary. Radios may be used or measurements may be made on a time schedule. When a schedule is used, the transmitter current may be interrupted while the loop is being reoriented, so that the observer may keep track of the schedule.

When a moving source is used, either the in-line or the broadside technique may be employed. With the in-line technique, the transmitting and receiving loops are moved together at a fixed spacing along traverses laid out at an angle of about 15° to 45° to the regional strike; if the angle between the traverse directions and the regional strike is too large, the coupling between the source and a sheet-like conductor is poor, and if the angle is too small, short narrow conductive zones may be missed unless the traverse lines are spaced close together.

For the broadside technique, the loops are moved simultaneously along two parallel traverses which are perpendicular to the regional strike. Measurements are made with the plane of the transmitting loop parallel to the strike. When a moving source is used, voice communication between the transmitter and receiver stations is usually possible.

With the fixed source method, the data are plotted at the receiving station location. This is customary also for the moving-source methods, but it can be preferable to plot the data at the point midway between the transmitter and receiver.

Most of the earlier dip angle surveys, whether for reconnaissance or for detailed exploration, were made using a fixed source. More recently, with the development of light-weight portable equipment, a larger proportion of dip angle surveys are being made with moving source methods, particularly in instances where a large depth range is not necessary. The maximum usable separation ranges from about 1200 to 5000 ft, using fixed sources, and is about 400 to 800 ft with moving sources, depending on the precision required and the skill of the operator. In most instances, an error of two or three degrees is allowable at the maximum separation.

The frequencies used range from 400 to 5000 c/s; often two frequencies are used, with one being three to five times the other.

The comments made at the beginning of the discussion on profiling methods concerning the relative merits of fixed source as opposed to moving source methods applies to the dip angle method as well as to the other profiling methods. However, it should be noted that even with the fixed source method, in dip angle surveys, the source is not truly fixed in position but is rotated, causing the coupling between the source and conductors to vary somewhat.

In an area where there are several closely spaced conductors, the fixed source method may be used advantageously in delineating the individual conducting zones. In such a circumstance, the source is placed directly over and parallel to the strike of a single conductive zone to maximize the coupling between the source and this conductor and to minimize the coupling with all other conductors. By so doing, it is possible to determine the characteristics of a weakly conductive zone which is surrounded by other, more highly conductive zones.

The in-line moving source method is usually preferred in reconnaissance work since only one traverse line is required. Because the exact spacing between coils is not critical, it is sometimes satisfactory to measure distances by pacing. The broadside method is usually preferred for detailed studies over conductive zones which have already been located, or in areas where brushed lines have been established for other purposes.

The theoretical development for the secondary field about a sphere as a function of frequency can, at least in part, be applied to the interpretation of dip angle anomalies, provided that the conductive zone is small compared with the distance to the source.

In using Eqs. (558), it should be remembered that in the dip angle method, the primary field is horizontally directed at the point of observation rather than vertically directed. With the dip angle method, $H_{z,s}$ provides the horizontal component of the secondary field, which is parallel to the direction of the primary field, and $H_{r,s} \cos \varphi$ and $H_{r,s} \sin \varphi$ provide the vertical component and the horizontal component normal to the primary field, respectively.

The interpretation of field results is simplified if the secondary field is small compared with the primary field, so that it may be written:

$$D.A. = \arctan \frac{H_{z,s}}{H_p} \tag{598}$$

Commercial dip angle equipment does not usually provide a wide enough range of frequencies to permit the determination of all the parameters for a sphere. The position of the sphere and its depth can be determined from measurements at a single frequency. If $\alpha$ is very small, the permeability contrast and the radius of the sphere can also be determined.

The dip angle responses for a perfectly conducting half-plane can be calculated from Eqs. (592) and (593). For conductors other than a sphere or a perfectly conducting half-plane, it is generally necessary to use scaled model studies for reference curves. However, there are a few simple interpretation guides which are based on the assumption that eddy currents are concentrated in the edges of a conductive zone which are sometimes useful.

Consider a thin vertical sheet in the plane of the source and assume that the frequency is sufficiently high that the out-of-phase component is negligible. There will be a concentration of eddy currents flowing along the edges of the sheet. If the depth and strike extent of the sheet are very large, only the current concentration in the upper edge of the sheet need be considered. The primary field has only a horizontal component, $H_{x,p}$.

At point P (on the surface in Fig. 216), the vertical component of the field due to the current concentration, $I'$, is:

$$H_z = - \frac{I'x}{2\pi(x^2 + z^2)} \tag{599}$$

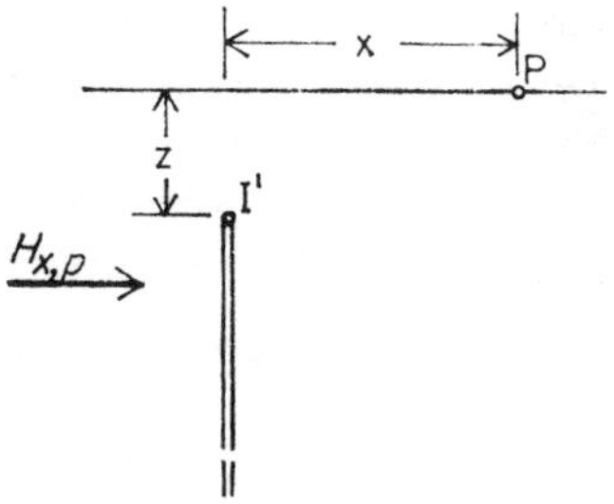

FIG. 216.

and the total horizontal field is:

$$H_x = H_{x,p} + \frac{I'z}{2\pi(x^2 + z^2)} \tag{600}$$

In general, $I'$ must be determined by model experiments, though in some instances, it may be computed. For our purposes, we will assume that the horizontal component of the secondary field at the surface for $x = 0$ is related to the primary field at the surface by a direct proportionality with a constant, $K$:

$$K = \frac{I'}{2\pi z H_{x,p}} \tag{601}$$

Having made this assumption, we may write:

$$H_z = - \frac{KxzH_{x,p}}{(x^2 + z^2)} \tag{602}$$

$$H_x = \left(1 + \frac{Kz^2}{x^2 + z^2}\right)H_{x,p} \tag{603}$$

The dip angle is then:

$$\text{D.A.} = \arctan\left(-\frac{Kxz}{x^2 + z^2(1 + K)}\right) \tag{604}$$

A curve showing the dip angle as a function of $x/z$ for $K = 1$ and for $K = 2$ is shown in Fig. 217.

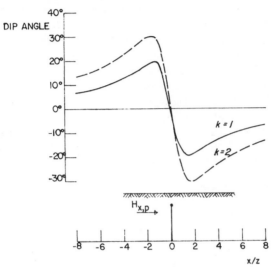

FIG. 217. Calculated dip angle curves for $K = 1$ and $K = 2$.

The distance, $x$, from the sub-outcrop of the sheet to the locations where the maximum and minimum dip angles are observed may be found by differentiating Eq. (604) and setting the derivative equal to zero:

$$x = \pm z(1 + K)^{1/2} \tag{605}$$

Using dip angle observations, one may determine the horizontal position, the approximate depth and an estimate of the magnitude of $K$ for a single current concentration. The horizontal position of the conductive zone is found from the position along the traverse where the dip angle is zero. This location is commonly referred to as the "crossover", or reversal, since the dip angle changes from positive values to negative. For small values of $K$, the maximum and minimum values for the dip angle are observed at distances $x = z$. The depth, $z$, to the current concentration is then about equal to the distance between the crossover and the maximum or minimum; however, it may be preferable in some cases to take the depth as one half the distance between the maximum and the minimum. At the maximum and the minimum, the magnitude of the dip angle is:

$$|\text{D.A.}| = \arctan\left|\frac{K}{2(1 + K)^{1/2}}\right| \tag{606}$$

For small values of $K$, the dip angle is:

$$|\text{D.A.}| \cong \arctan \frac{K}{2} \tag{607}$$

The discussion in the preceding paragraphs applies to the fixed source method for cases in which the conductive zone is roughly in the plane of the transmitting loop and the depth to the conductor is small compared with the separation between the transmitter and receiver. In most cases, the assumptions that the primary field is homogeneous and that the eddy currents may be represented by a single current filament are not valid. Also, the factor $K$ changes as the transmitting loop is rotated or moved, so it is usually necessary to use model curves in interpreting field surveys.

A few curves for the dip angles observed over a highly conductive half plane using the fixed source method are shown in Figs. 218, 219 and 220. When the transmitter is located directly over a vertical sheet, the left and right branches

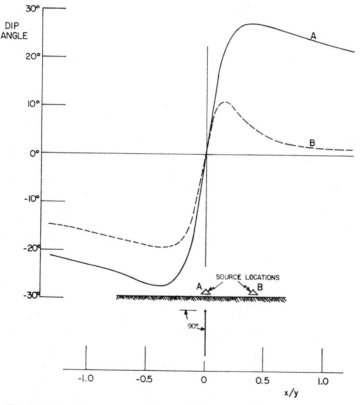

Fig. 218. Fixed-source dip angle anomalies over a half-plane with $d/y = 0.12$.

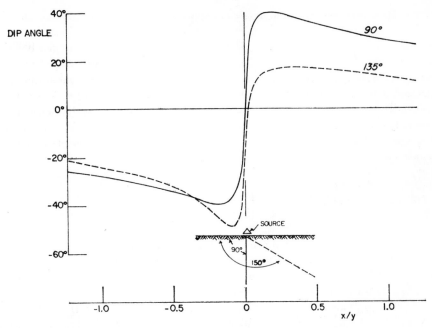

FIG. 219. Fixed-source dip angle anomalies over a half-plane with $d/y = 0.02$.

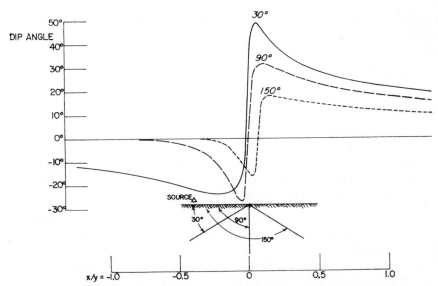

FIG. 220. Fixed source dip angle anomalies over a half plane with $d/y = 0.02$.

of the curve relating dip angle to traverse position have exactly the same shape and amplitude. When the transmitter is offset, or when the conductive sheet dips at some angle other than 90°, the maximum and minimum portions of the curves differ. For the vertical sheet shown in Fig. 218, the depth to the top of the sheet is only one third the distance from the crossover to the maximum or minimum point, indicating that $K$ (from Eq. 601) is quite large. When the transmitter is placed directly over the conductive sheet, the flanks of the anomaly approach zero very slowly.

Some of the parameters for a conductive sheet which may be determined from fixed-source dip angle measurements are:

*Horizontal location.* For thin, steeply dipping sheets, the crossover position locates what is commonly called the axis of the conductor, which is nearly coincident with the upper edge of the conductor. For a gently dipping sheet, the crossover is observed slightly downdip from the upper edge. If a wide conductor has considerable depth extent, it may be desirable, when using the fixed source method, to repeat some of the traverses with the source on the opposite side of the conductive zone, since the indication of the far edge from the dip angle anomaly curve may be indistinct. If two or more conductive zones are present, and are seperated by a distance equal to or less than their depth of burial, only a single crossover will be observed. When a weakly conductive zone occurs close to a more highly conductive zone, the weak conductor may not give rise to a crossover, but only to an inflection in the dip angle anomaly curve related to the presence of the stronger conductor.

*Strike.* Variations in the strike of a conductive zone will have some effect on the dip angle anomaly curve, but it is not usually possible to determine the strike direction from data obtained along a single traverse. Location of the axis of a conductive zone along several traverses is required for a precise determination of the strike direction.

*Dip.* If the conductive zone is dipping, the magnitudes of the two branches of the dip angle anomaly curve and the shape of the curves will not be the same on the two sides of the crossover point. Dip can be estimated by comparison of field data with model curves. If the dip is less than 45°, the effect of dip is slight and cannot always be recognized in field measurements. The dip is most easily estimated from measurements made with the transmitter located directly over the upper edge of the conductor.

*Depth.* The distance between the crossover point and the maximum and minimum points on the dip angle anomaly curve, as well as the slope of the curve at the crossover, are directly related to the depth. In making a depth estimate, the effect of dip, strike and transmitter offset also must be considered. For a thin vertical sheet with great depth extent, the distance between the crossover point and the maximum or minimum point ranges from one to three times the depth, depending on the conductivity and the length of the conductor. Measurements made at two frequencies with the transmitter located directly over the half plane are sufficient to determine both the depth and the conductance of the half plane as mentioned below.

*Depth extent.* If the conductive zone is limited in depth extent, the current filaments in the lower edge of the conductive zone will have the effect of decreasing the amplitudes of the flanks of the anomaly. This effect is difficult to recognize if anomalies from more than one conductive zone are superimposed on one another.

*Length.* For conductive zones which are no longer than one and a half times the loop spacing, the magnitude of the anomaly is a function of the ratio of this length to the loop separation. However, the most practical means for determining the length of a conductive zone is to run a series of parallel traverses so that one or more of the traverses is near or beyond each end of the conductive zone.

*Conductivity.* Below the saturation frequency, the amplitude of the dip angle anomaly depends on the frequency, so that measurements at two frequencies provide at least a qualitative indication of the conductivity, or for thin sheet-like conductors, of the conductance. In a given area, conductors may be graded in terms of their relative conductivities by considering the ratio of the magnitudes of the anomalies at two different frequencies. Since the sharpness with which the null position may be determined depends on the ratio of the out-of-phase component to the inphase component, the operator may estimate the width of the null and use this as a qualitative indication of conductivity.

If the dip angle anomaly is sufficiently easy to interpret that the other parameters can be determined, the conductivity may be estimated by comparison of the field data with model curves. If, for example, it has been determined that the conductive zone is a thin vertical sheet which can be represented by a half plane, the conductivity and the depth to the upper edge may be estimated from the amplitudes measured at two frequencies.

*Anisotropy.* When traverse lines cross a wide, flat-topped, massive conductive zone, the dip angle will be nearly zero so long as neither loop is near one edge of the conductive zone. However, if such an extensive conductive zone is "schistose" (that is, if the zone consists of a great number of thin conducting sheets separated by non-conducting zones), an anomaly in observed dip angle will persist so long as the line between the source and the receiver is not exactly parallel to the strike (Swanson, 1961). Also, in the case of a schistose conductor, the direction of the dip angle reverses when the positions of the source and the receiver are reversed. These phenomena serve to distinguish highly anisotropic conductors from massive conductors. The types of rocks which commonly display this sort of extreme anisotropy are usually metamorphosed sediments which retain much of their original bedding; carbon or graphite is usually the main conductive mineral. Metamorphosed iron formation may be highly anisotropic due to alternating bands of magnetite or specular hematite with barren rock. Some metamorphic rocks bearing syngenetic sulfides are highly anisotropic.

The interpretation of data obtained with moving source dip angle methods differs in many respects from the interpretation of data obtained with a fixed source method. Some typical interpretation curves for the broadside technique over a half-plane are shown in Figs. 221 and 222. If the dip and strike relative to the traverse line are both 90°, the left- and right-hand branches of the dip angle anomaly curve are identical except for sign. If the sheet is not vertical, the crossover is shifted in the down dip direction and the left- and right-hand branches differ in magnitude and shape. Variation of the strike direction from 90° has a similar but more marked effect on the anomaly (see Fig. 222).

If the strike is approximately 90° from the traverse direction, the flanks of the anomalies obtained with a moving source method are much sharper than the flanks of the anomalies obtained with a fixed source method. In general, the magnitude and shape, as well as the sign of moving-source anomalies change if the positions of the source and receiver are interchanged. For this reason, it is desirable to keep the relative positions of the source and receiver the same when a detailed survey is being run. As a generalization, the crossover may be said to be located almost directly above the axis of a thin sheet, provided the dip and strike are large, no matter which technique is used. It is difficult to formulate other generalizations which apply to all of the many dip angle measurement techniques.

## 39b. AFMAG Method

Variations in the intensity of the earth's magnetic field in the audio frequency range (see the discussion in Chapter IV) provide the source of energy for the *AFMAG method* (Ward, 1959), which is commonly grouped with the induction profiling methods. Most of the energy contained in the earth's natural electromagnetic field at audio frequencies is initiated by atmospheric electric discharges, or *sferics*. At large distances from a lightning stroke, the electromagnetic field is essentially a plane wave, with the electric field tilted slightly forward (in the direction of propagation) from the vertical and with the magnetic field being nearly horizontal, provided the earth in the neighborhood of the observation point consists of uniform horizontal layers. The azimuth of the magnetic field is more or less random in time since the sources tend to be widely distributed. However, near a lateral change in resistivity in the earth, and particularly in the vicinity of a highly conductive zone, the plane of polarization of the electromagnetic field will be tilted from the horizontal and the azimuth will be less random, showing a tendency to be directed normal to the surface of the

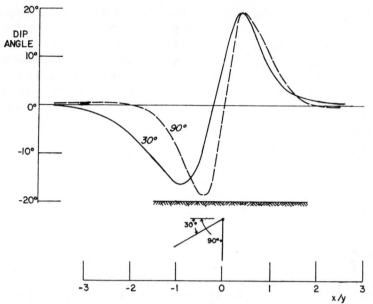

Fig. 221. Moving-source (broadside) dip angle anomaly curves over a conductive half-plane with $d/y = 0.12$, and with the strike direction relative to the traverse direction being 90°.

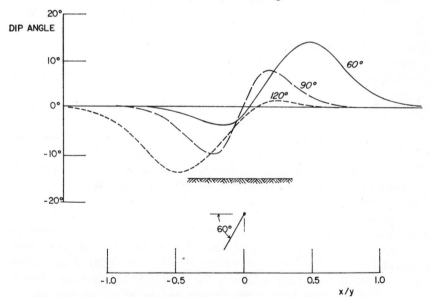

Fig. 222. Moving-source (broadside) dip angle anomaly curves over a conductive half-plane for $d/y = 0.25$ and for strike angles relative to the traverse direction of 60°, 90° and 120°.

conductive zone because the secondary field associated with eddy currents flowing in the conductive zone reinforce fields so polarized.

In the AFMAG method, the inclination or tilt angle and, in most cases, the azimuth of the natural field are measured. The equipment consists of two detecting coils with ferromagnetic cores fastened together and suspended by a frame so that they may be rotated about a horizontal axis (see Fig. 223), amplifiers, a correlation circuit or ratiometer, and a measuring meter. In

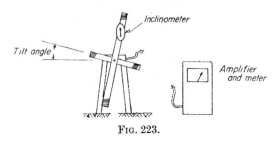

FIG. 223.

making an azimuth measurement, the amplifier and meter are connected to one of the coils (the larger one, if they are not of equal size). This coil is placed in a horizontal position and then rotated about the vertical axis until the mean azimuth of the signal (the direction for maximum signal) is found.

To find the tilt angle, the larger coil is connected to the meter through an amplifier and signal correlation circuit, using the smaller coil to provide a reference voltage, and then the larger coil is rotated about the horizontal axis normal to the mean azimuth until a minimum signal level is noted. Alternate equipment using coils of equal size may be employed, but the parameters measured are the same. Systems with but a single detection coil may be used, but two-coil systems are more sensitive. If the measured azimuth is not approximately normal to the strike of the conductive zone, it may be preferable to measure the tilt angle in a plane normal to the strike rather than in the plane of the measured azimuth. Otherwise, variations in the azimuth of the incident signal with time may cause some variation in the observed tilt angle.

The coils and amplifiers are tuned sharply to the chosen operating frequency; ordinarily, two frequencies which differ as much as possible from harmonics of power distribution frequencies (such as 150 and 510 c/s) are chosen. Low-noise amplifiers are used in order that the instrumental noise level will be determined by the level of thermal noises generated in the coils themselves. Depending on the location and the time of day, the natural signal level may drop to a value comparable with the instrumental noise level, making AFMAG measurements impossible. In the northern hemisphere, the least favorable times (those times when the natural noise level is the lowest) are the winter months and the daylight hours. At mid-latitudes, the natural field is usually strong enough for measurements to be made at any time of the day during the summer. However, intense signals from local thunderstorms sometimes interfere with AFMAG measurements. The tilt angle for signals from local sources may be appreciable even in the absence of conductors, and the azimuth will not generally coincide with the mean azimuth for signals from distant sources.

In making an AFMAG survey, readings are taken along traverses laid out perpendicular to the assumed geologic strike. Because conductors can be detected at greater distances with AFMAG than with other induction methods (such as the dip angle method), the interval between stations may be fairly large, in the range from 100 to 200 ft. Detailed surveys may be made in the immediate vicinity of conductive zones.

AFMAG tilt angles may be plotted and interpreted in much the same way as conventional dip angle data. However, it is sometimes useful to consider the azimuth data as well as the tilt angles. To do so, a planimetric presentation of the data may be used, with an arrow or vector representing each reading. The tilt is indicated by the magnitude of each vector, the head of the arrow indicating the direction of tilt; the azimuth is indicated by the azimuth of the arrow on the map. By convention, the direction of tilt is defined so that the arrow points toward the conductive zone giving rise to the anomaly. The planimetric or vector presentation of results is particularly useful when the traverse lines are not approximately normal to the strike of the conductive zones, or when the survey is of the reconnaissance type. The vector plot may indicate the presence of conductors even though none of the traverses intersect the conductive zone and no tilt angle crossovers are observed. This is indicated in an idealized survey example in Fig. 224.

AFMAG is sometimes regarded as a dip angle technique in which the source is at infinity. Since the source is far removed, the AFMAG method has a much greater vertical and horizontal range for the detection of conductive zones than do methods using a local source. In the case of a very large conductive zone, the entire zone is energized by the natural field, whereas, with other methods, only the part of the zone near the source is energized. This makes the AFMAG method more responsive than other methods to large masses with a small conductivity contrast with respect to the surrounding rocks.

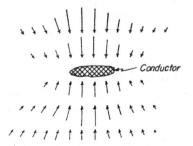

FIG. 224. Idealized vector plot of AFMAG results.

Inasmuch as all conductive zones are essentially the same distance from the source, AFMAG anomalies tend to be a more accurate representation of the size and conductivity of a conductive zone than anomalies observed with other fixed source methods. This can sometimes be a disadvantage inasmuch as anomalies caused by relatively small, highly conductive bodies of potential economic value may be obscured by much larger anomalies caused by large, low conductivity masses (shear zones or beds of conductive slate, for example).

The depth to the effective current concentration in the upper edge of a conductive sheet is greater for the AFMAG method than for the dip angle method. If the conductor is dipping, the AFMAG crossover is shifted in the downdip direction further than is the dip angle crossover, providing a means for determining the dip of a conductive zone when both types of surveys are run. Also, the low-frequency crossover will occur further downdip than the high frequency crossover. The slope of the AFMAG anomaly curve will be less than the slope of the dip angle anomaly curve at the crossover, suggesting a greater depth to the conductive zone. If the conductive zone has a finite depth extent, the flank of the AFMAG curves will drop off in amplitude faster than the flanks of the corresponding dip angle curves.

AFMAG, because of its greater range, is a valuable tool in exploring for large deep-seated conductive zones. AFMAG is also valuable in mapping features of geologic interest, such as faults and shear zones, where contrasts in resistivity may exist. For detailed studies of shallow conductive zones, other inductive methods should be used with the AFMAG method to provide the best results.

### 39c. Slingram method

The *slingram* (or *loop frame*) method (Werner, 1958) uses two small portable loops similar to those used for moving-source dip angle measurements. One of the loops is a transmitter, the other a receiver. A ratiometer such as is shown in Fig. 183 is used in measuring the inphase and out-of-phase components of the voltage induced in the receiving loop. Most commonly, the two loops are oriented to be horizontal and coplanar; for this reason, the slingram method is sometimes referred to as the *horizontal loop method* to distinguish it from dip angle methods which usually employ a vertical loop. However, other orientations for the loops, such as vertical coaxial or vertical coplanar configurations, may also be used if the coils are equipped with the necessary level bubbles and sighting devices.

Usually, one or two fixed frequencies between 400 and 4000 c/s are used, although frequencies as high as 20,000 c/s have been employed in special applications. The loop spacing is held constant for each set of measurements, and may range from about 50 ft to a maximum of about 300 ft. Since a connecting cable is required to transmit a reference signal between the transmitter and receiver, it is almost always necessary to use the in-line profiling technique for traversing. The broadside technique can be used with small loop separations in flat areas where trees and brush pose no problem. For most reconnaissance work, an interval between stations of one-half the loop spacing is satisfactory. More detailed measurements may be required in the vicinity of conductors. Ordinarily, traverses are laid out perpendicular to the assumed regional strike.

The mutual coupling ratio, $Z/Z_0$, may be measured directly with a ratiometer if the equipment is adjusted to read 100 per cent and 0 per cent for the real and imaginary components, respectively, over an area where the conductivity is known to be slight. Measurements are usually plotted in the form of profiles assuming that the point of measurement is midway between the two loops. If the equipment is properly designed and constructed, the theorem of reciprocity is valid and the positions of the transmitter and receiver may be interchanged without affecting the measurements. When traverses are placed sufficiently close together, the measurements may, for special purposes, be contoured.

Errors in the measurements caused by errors in the loop spacing or orientations must be considered. When no conductive zones are present, the mutual coupling has only an inphase component, and errors may be calculated readily. Consider the case in which a ratiometer is adjusted to read 100 per cent at a coil spacing, $r$, and through some surveying error, the actual coil spacing is $r_a$. The ratiometer will provide a reading $(r/r_a)^3$ 100 per cent. Thus, an error of 2 per cent in the coil separation will cause an error of about 6 per cent in the measurement of the inphase component. If one of the coils is misoriented from its proper position through an angle $\delta$, the ratiometer measurement will be $(\cos \delta)$ 100 per cent. Usually, a fairly large error in orientation of one of the coils can be tolerated; if the angle $\delta$ is 8°, the error in measurement will be only about 1 per cent. Serious errors can occur, however, when traversing rugged terrain with the horizontal coplanar or vertical coaxial loop arrangements. Consider the case in which both coils in a horizontal coplanar arrangement are held level, but the line connecting the coils makes an angle $\delta$ with the horizontal. Equation (412) may be rewritten for this situation as:

$$H_{z,p} = \frac{m}{4\pi r^3} (3 \sin^2 \delta - 1) \tag{608}$$

Since $Z/Z_0 = H_z/H_{z,p}$, the error in the measurement is $(3 \sin^2 \delta)$ 100 per cent. For a slope of 8°, the error is 5·8 per cent. When the horizontal coplanar or vertical coaxial arrangements are used in hilly terrain, it is necessary to orient the coils parallel to the slope, or to make corrections to the measurements when the inphase component is considered in interpretation. In practice, the inphase component may be ignored in making reconnaissance surveys in rough terrain. When this is done, there is some risk that deeply buried, highly conductive bodies which do not produce significant anomalies in the out-of-phase component may be missed. Measurements made with the vertical coplanar arrangement are not affected by differences in elevation between the coils, though there are other disadvantages which limit the utility of this coil arrangement.

The error contributed by the misorientation of one coil may be more significant in the presence of conductors than in their absence. As an example, assume that the receiving coil of a horizontal coplanar arrangement is misoriented through an angle of 8° and that the vertical and horizontal components of the secondary field are −50 per cent and +50 per cent of the primary field, respectively. The measured field is then (cos 8°) 50 per cent + (sin 8°) 50 per cent, and is in error by 6·5 per cent. In general, the magnitudes of such errors can be determined only by model experiments.

The theory presented in a preceding section for the mutual coupling between loops over a homogeneous or horizontally stratified earth may be applied directly to slingram measurements. If the earth is known to be homogeneous, the conductivity may be determined using the curves given in Fig. 203. If the earth is layered, slingram profiles can be used to estimate variations of the thickness of near surface layers between locations where induction or resistivity soundings have been made. The slingram method has been used to map variations in conductivity and thickness of layers and structures such as buried faults in flat-lying sedimentary rocks where conductivity contrasts are relatively small. However, the method is used most frequently to locate highly conductive zones such as are associated with massive sulfide deposits.

Equation (570) for the response of a buried sphere is not particularly useful in calculating theoretical slingram anomalies since the assumption that the primary field is homogeneous over the volume occupied by the sphere is not reasonable for any sphere large enough to be detected readily by the slingram method. Equations for a sphere in a dipolar field have been developed, but they are quite complicated. An idea of the response over steeply dipping sheets may be obtained by assuming that eddy currents are concentrated in the upper edge of a sheet. However, this approximation is difficult to apply to a moving source method since the magnitude of the eddy current is different for each station. Equations for the response of a perfectly conducting half-plane can be used to calculate slingram anomalies when the out-of-phase component is negligible. In most cases, model curves are used in interpreting slingram data.

The response observed over a perfectly conducting vertical half-plane is shown by the curves in Fig. 225 for various ratios of the depth, $d$, to the loop separation, $r$. Except for differences in amplitude, the inphase and out-of-phase curves for a sheet with finite conductivity are nearly identical to the curves in Fig. 225. (Traverses are assumed to be normal to the strike of the conductive sheet and the coils are assumed to be horizontal and coplanar.) Variations in the value of the ratio $d/r$ cause large changes in the amplitudes of these curves but only subtle changes in the shapes of the curves. The inphase and out-of-phase curves cross through values of 100 per cent and 0 per cent, respectively, at the

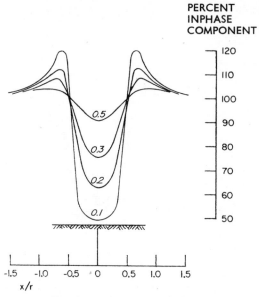

FIG. 225. Slingram anomalies observed over a conducting, vertical half-plane for various depths of cover, $d/r$.

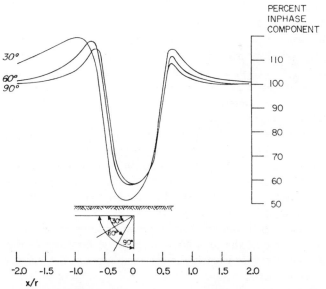

FIG. 226. Slingram anomalies observed over a dipping, conducting half-plane for a depth of burial, $d/r = 0·15$.

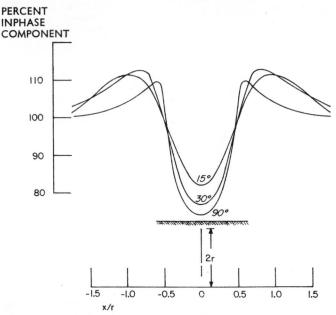

Fɪɢ. 227. Slingram anomaly curves observed over a thin vertical sheet buried at a depth $d/r = 0.1$, with $W = 5.3$ for strike directions making angles of $90°$, $30°$ and $15°$ with the traverse direction.

points $x/r = \pm0.5$, corresponding to the positions where one or the other of the loops is directly above the sheet.

If the sheet dips at an angle other than vertical, the anomalies are non-symmetric (see Fig. 226); for any but very small values of $d/r$, the maximum observed on the downdip side of the sheet is of greater amplitude than the maximum observed on the updip side, and the minimum is shifted toward the downdip side. Unless the dip is quite shallow, the 100 per cent and 0 per cent crossing points are almost the same as for a vertical sheet.

For a vertical sheet, the effect of crossing the strike at an angle other than $90°$ is to decrease the minimum and to increase the amplitude and the width of the maxima (see Fig. 227). The separation between the points at which the maxima are observed varies significantly with changes in the traverse direction in relation to the strike, but the 100 per cent and 0 per cent crossing points remain at the positions $x/r = \pm0.5$.

Variations of both the strike direction and steepness of dip cause an increase in the amplitude of the maxima and in the asymmetry of the anomaly. The quantity which varies least is the distance between the 100 per cent and 0 per cent crossing points.

The effect of finite depth extent in the case of a vertical sheet is to reduce the magnitude of the anomalies and, for very small depth extents, to change the

shape of the anomaly curve (see Fig. 228). In the cases in which the sheet dips
at an angle other than 90°, the effect of a finite depth extent is more pronounced; for shallow dips, a second minimum develops over the downdip edge
of the sheet.

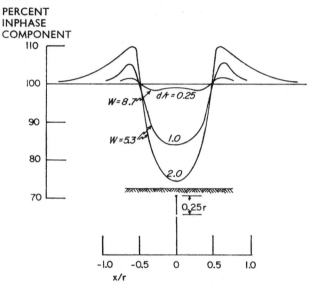

FIG. 228. Slingram anomaly curves observed over a vertical sheet of limited
depth extent, for depth extents of 0·25 r, 1·0 r and 2·0 r.

There are minor variations in the shape of the anomaly curves observed over
a thin sheet with variations in the value of the parameter $W$ ($W = \sigma\mu\omega tr$) and
there are small differences in shape between the inphase and out-of-phase
anomaly curves. These effects may be attributed to the fact that the phase angle
of the current within the sheet varies as a function of distance from the edge of
the sheet and of the conductivity in the sheet. If the actual current distribution
within the sheet is approximated by two current concentrations, one of which
is inphase and the other out-of-phase, the out-of-phase current component
flows nearer the edge of the sheet than the inphase current, while both current
concentrations tend to flow nearer the edge of the sheet as either the conductivity or the frequency, or both, are increased. Differences between the shapes of
the inphase and the out-of-phase anomaly curves are most pronounced for
gently dipping sheets; in the case of a vertical sheet, there is scarcely any
difference inasmuch as, as has already been mentioned, the shape of the
anomaly curve is nearly independent of depth for a vertical sheet.

Over the interior part of a large horizontal sheet, the inphase component can
be greater or less than 100 per cent, and the out-of-phase component can be
either positive or negative, depending on the values for the parameters $d/r$ and $W$.

The anomaly observed over the interior part of a sheet of finite thickness can be calculated from Eqs. (545)–(548). Near the edge of a horizontal sheet, the anomaly curve shows a non-symmetric variation through a maximum point outside the edge and a minimum over and inside the edge (see Fig. 229). For fairly large values of the parameter, $W$, the out-of-phase anomaly curve is complicated by the presence of a second maximum point inside the edge of the sheet (see curve for $d/r = 0{\cdot}1$, $W = 2{\cdot}7$ in Fig. 229). In the case of a horizontal sheet, the 100 per cent and 0 per cent crossing points are located a distance

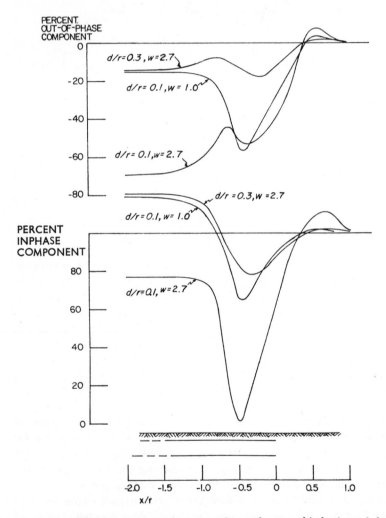

FIG. 229. Slingram anomaly curves observed over a thin horizontal sheet for depths of burial, $d/r = 0{\cdot}1$ and $0{\cdot}3$, and for $W = 1{\cdot}0$ and $2{\cdot}7$.

$x/r \approx 0\cdot3$–$0\cdot5$ from the edge of the sheet, reflecting the fact that the mean current concentrations are a short distance inside the edge of the sheet.

If the width of a horizontal sheet is less than about three times the coil spacing, the anomaly curves for the two edges overlap. In most cases, the anomaly curve has three maxima and two minima, although for large values of $W$ the out-of-phase anomaly curve may have four maxima and three minima.

The anomaly curve observed over a dike or sheet of finite thickness differs somewhat in character from the anomaly curve observed over a sheet of vanishing thickness. In the first place, as was pointed out in discussing the field of a long wire source over a horizontal sheet, the response as a function of frequency depends on $W$ and $t$ as two independent variables if the value of $W$

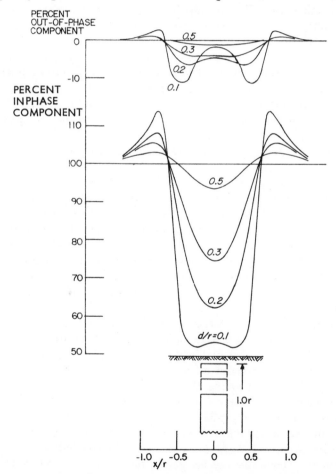

Fig. 230. Slingram anomaly curves observed over a dike for $B = 37$, and for various depths of burial.

is larger than some threshold value. In the second place, the shape of the anomaly curve differs from that observed over a thin sheet (Fig. 230) if the thickness of the sheet is an appreciable fraction of the coil separation. For the curves in Fig. 230, the response parameter $B$, defined by Eq. (544) using $\sigma$ as the conductivity of the dike, serves as a parameter describing the complete set. In traversing normal to the strike of a vertical dike, maxima are observed while both coils are outside the edges of the dike. If the ratio $t/r$ is small, or if $d/r$ is large, a minimum will be observed over the dike, just as in the case of a thin sheet. If the ratio $t/r$ is large, or if $d/r$ is small, both maxima and minima will be observed, just as is the case for a horizontal sheet. The 100 per cent and 0 per cent crossing points are displaced from the edges slightly toward the interior of the dike. Variations in strike and dip from $90°$ affect the anomaly curve observed over a dike in much the same way similar variations affect the anomaly curves observed over thin sheets.

The anomaly curve observed over a series of parallel, closely spaced, highly conductive thin sheets, when traversed at a large angle with respect to their strike direction, is almost the same as the anomaly curve observed over a dike with the same overall width and a lower conductivity (see Fig. 231). If the separation between the sheets is increased to about one-third the coil spacing, the effects of each individual sheet may be recognized, providing $d/r$ is small. However, for certain spacings of the sheets, the anomaly curve may still resemble the anomaly caused by a wide dike. When a series of closely spaced parallel vertical sheets is traversed at a small angle with respect to the strike direction, the anomaly in the inphase response is everywhere greater than 100 per cent and the anomaly in out-of-phase response will be positive. Over an isotropic dike with the same overall width, minima with values less than 100 per cent for the inphase response and negative values for the out-of-phase component are always observed, regardless of the strike direction. As was the case for the dip angle method, this behavior of the anomaly curves makes it an easy matter to distinguish between a wide schistose conductor and a wide massive or isotropic conductor.

In order to make a detailed interpretation of a field anomaly curve, the interpreter may need to refer to a large number of model curves. When large numbers of reference curves are available, it is convenient to condense the data by plotting auxiliary graphs based on one or more key parameters. As an example, an Argand diagram giving the minimum values for the real and imaginary components observed over a half plane as a function of the parameters $d/r$ and $W$ is shown in Fig. 232.

Figure 233 shows curves for the ratio of the smaller to the larger maximum value as a function of dip and depth for a highly conductive half-plane.

To summarize the above discussions, guides for determining some of the parameters for a conductor when using the horizontal coplanar loop arrangement are as follows:

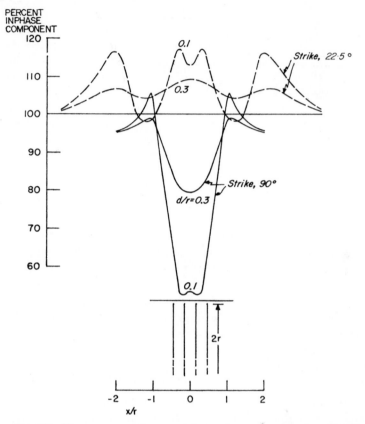

FIG. 231. Slingram anomaly curves observed over a series of vertical sheets
for $W = 5\cdot3$ and for strike directions of 90° and 22·5°.

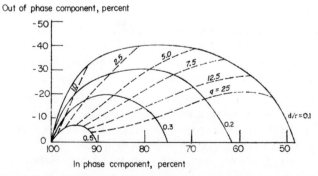

FIG. 232. Argand diagram for the minimum in the slingram anomaly curves
observed over a thin vertical sheet for various values of $W$ and $d/r$.

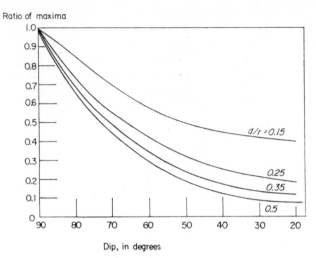

FIG. 233. Ratio of maxima observed on a slingram anomaly curve over a thin
sheet as a function of dip.

*Horizontal location.* A vertical sheet is located directly under the point
where the minimum on the anomaly curve is observed. If the sheet is dipping
other than vertically, the horizontal position of the upper edge will be displaced
slightly in the updip direction from the point where the minimum is observed.
The edges of a vertical dike or a horizontal sheet are located at $x/r = 0.3$ to
$0.5$, taking the 100 per cent and 0 per cent crossing points as $x/r = 0$.

*Strike.* The width of the flanking maxima and the ratio of their amplitudes
to the amplitude of the minimum can be used to estimate the strike of a
vertical sheet or dike, except for ambiguity in assigning the quadrant. Of
course, two or more traverses over a conductor will also establish the strike
direction.

*Dip.* A dip other than 90° causes the anomaly curve to be non-symmetric;
in most cases, the larger of the maxima is observed in the downdip direction.
The dip of a highly conductive half plane may be estimated from the curves in
Fig. 233. Similar curves may be used for less conductive sheets; variations in
the ratio of the maxima with variations in conductivity are rather small.

*Depth.* If the shape of the anomaly curve indicates that the conductive zone
is a thin vertical sheet, the depth to the upper edge may be determined readily
from the values of the inphase and out-of-phase minima and interpolation from
Fig. 232. The depth to the upper edge of a dipping sheet may be determined in
a similar manner, provided that sufficient model information is available.
Measurements at two or more frequencies or spacings are, in general, required
to obtain the depth to a horizontal sheet. The depth to the edges of a wide

horizontal sheet or the depth to the top of a wide dike may be estimated by making measurements with two or more spacings and comparing the shapes of the anomaly curves with model curves. The horizontal position of the minimum which is associated with the edge of a horizontal sheet depends on $d/r$ and, to a lesser extent, on $W$. In the case of a wide dike, the depth is most easily obtained by determining the critical value of $d/r$ below which a maximum is developed over the center of the dike (see Fig. 230).

*Depth extent.* A sheet with finite depth extent may be recognized by the shape of the anomaly curve if the depth extent is small or if the sheet is gently dipping. If measurements over a vertical sheet at two or more frequencies and spacings do not yield consistent results when interpreted (using Fig. 232), a finite depth extent is indicated. In principle, a set of Argand diagrams, each prepared for a different ratio of depth extent to loop spacing, could be used to interpret the results of measurements made over sheets with finite dimensions.

*Length.* The most practical means for determining the length of a conductor is to run a sufficient number of traverses to locate the ends of the conductive zone. If the traverse is near one end of the conductive zone, the amplitude of the anomaly curve is diminished in much the same way as is observed when the conductor has a finite depth extent.

*Conductivity.* For thin sheets, $W$ is determined from the Argand diagram, together with the depth. The conductivity cannot be determined unless the thickness is measured by an independent means. For thick dikes and for other shapes, the conductivity can be estimated provided the field curves can be matched with a model curve. From Eq. (418), the conductivity $\sigma_1$ is

$$\sigma_1 = \frac{\sigma_2 \omega_2 L_2^2}{\omega_1 L_1^2}$$

In using this equation, the physical dimensions of the model must be known. If the value for $B$ is given, rather than the dimensions, we may use the equation:

$$\sigma_1 = \frac{2B^2}{\mu_0 \omega_1 L_1^2}$$

*Anisotropy.* Measurements made along traverses which intersect the strike direction at a small angle indicate whether or not a wide conductor is anisotropic.

The horizontal coplanar loop arrangement is usually more sensitive and more convenient to use than the other loop arrangements. However, in some cases there is an advantage to be gained in using one of the other loop configurations, particularly if a detailed survey of a conductor whose position has already been determined is to be made.

Examples of the anomaly curves obtained with three coil configurations are shown in Figs. 234 and 235. As a generalization, we may say that the horizontal coplanar arrangement is the most sensitive and provides the most complicated anomaly curves, and that the vertical coplanar arrangement is the least sensitive and provides the simplest anomaly curves. The vertical coplanar arrangement is sometimes useful when working in steep terrain inasmuch as differences in elevation between the loops do not cause errors in the inphase component. The chief disadvantage of this configuration is that it is not very sensitive to steeply dipping sheet-like conductors unless the traverse is made at a small angle relative to the strike direction. For wide horizontal sheets, the vertical coplanar arrangement is often as sensitive as the horizontal coplanar arrangement and the anomaly curves are less complex. Over the center of the sheet, the anomaly is always greater than 100 per cent and positive, and the anomaly associated with the edge of the sheet is small. Inphase values of less than 100 per cent and negative values for the out-of-phase component are observed over a thin sheet when the strike is near 90°. The converse is true for wide dikes or for vertical sheets whose strike direction makes a small angle with the traverse direction. Therefore, traverses made at right angles to the strike direction with the vertical coplanar arrangement can be used to distinguish between a wide isotropic conductor and an anisotropic conductive zone.

With the exception of horizontal sheets having a width several times the coil spacing, the vertical coaxial arrangement usually is more sensitive than the vertical coplanar arrangement. For the range of $d/r$ values encountered

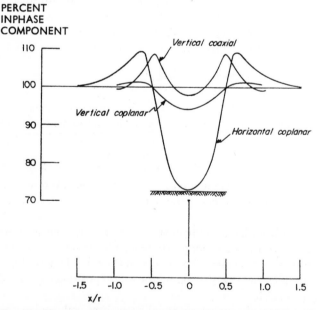

Fig. 234. Slingram anomaly curves observed over a vertical sheet for $W = 5\cdot3$ and $d/r = 0\cdot1$ for various coil arrangements.

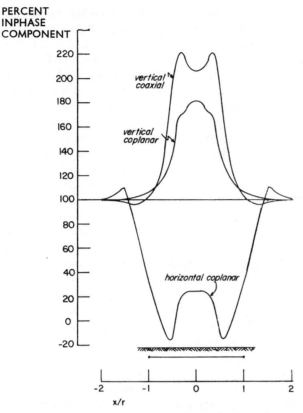

FIG. 235. Slingram anomaly curves observed over a horizontal sheet for $d/r = 0\cdot1$ and $W = 5\cdot3$ for various loop configurations.

in ground surveys, the sensitivity of the configuration is rather poor for vertical sheets. Over a dipping sheet, the maximum on the downdip side may be several times as large as the maximum for a vertical sheet. As was the case with the horizontal coplanar arrangement, the dip may be estimated from the ratio of the two maxima. The total width of the anomaly observed over any particular conductor is usually less for the vertical coil configurations than for the horizontal configuration. For this reason, in addition to the fact that the anomalies tend to have a simple form, the vertical configurations are particularly useful in delineating individual conductors in areas where several conductors occur close to one another.

The perpendicular coil configuration is, in most cases, more sensitive than the vertical configurations for ground measurements. It is seldom used because of the difficulty in maintaining proper coil orientation.

In discussing the interpretation of induction horizontal profiling measurements made over highly conductive zones, we have until now neglected eddy currents induced in the host rock and overburden. This assumption is not valid unless the resistivity of the host rock and overburden is high and the distance between the source and the receiver is small. The surrounding rock may cause three effects: (1) attenuation and distortion of the primary field between the source and the target conductive zone, along with attenuation and distortion of the secondary field between the target conductive zone and the receiver, (2) current flow from the host rock through the conductive zone, and (3) anomalies caused by variations in the conductivity of the host rock. All three effects are inter-related to some extent, but the last of the three effects introduces the problem of distinguishing conductive zones of economic interest from conductive zones which are not significant.

Shielding of a good conductor by a conductive layer tends to decrease the amplitude of the anomaly curve whereas the effect of current flow between the host rock and a target conductive zone may increase the amplitude of the anomaly. The extent of such shielding depends on the conductivity and thickness of the layer above the target zone and on the frequency. By reducing the frequency, the extent of shielding can be reduced and the anomaly accentuated, providing the reduction in frequency does not reduce the response for the target conductor by a like amount. In effect then, when the optimum frequency is chosen, the shielding will be a function of the conductivity contrast between the target conductor and the overlying rock or overburden. The orientation of the source, the size and shape of the conductive zone and the position of the conductive zone relative to the source and receiver are all factors which determine the attenuation of an anomaly. Estimation of shielding effects based on the attenuation factor for a plane wave is done occasionally, but the results are likely to be misleading.

Currents flowing from the surrounding rock through a conductive zone are likely to be important whenever conditions are such that an appreciable amount of shielding takes place. One exception is the situation in which the conductive zone does not have a sub-outcrop surface in contact with the overburden and in which the host rock is highly resistive; another exception is the case of a massive sulfide body inside a shell of highly resistive gangue minerals. The currents flowing from the host rock into a conductive zone may be regarded as normal eddy currents, induced in the host rock by the source, which pass through the conductive zone as part of their paths. The currents will tend to concentrate in the conductive zone, thereby generating an anomaly. This effect is most pronounced for conductive zones which are elongate in the direction of normal eddy current flow in the earth. Extreme examples of this phenomenon are sometimes observed in making measurements near buried pipes. A pipe which, by itself, would not produce an anomaly in free space may concentrate a sufficient amount of current to cause a strong anomaly immediately around the pipe. Such anomalies are seen principally in the out-of-phase component inasmuch as the eddy currents in a poorly conducting medium are almost 90° out-of-phase.

As one might expect, the response of a conductive body located in a poorly conducting half-space is much more difficult to calculate than the response of the same body in free space. D'yaakanov (1959) and Negi (1962) have developed formulae which may be applicable in some cases for calculating the response of a buried sphere energized by a homogeneous primary field. Using an approximate method, Schaub (1962) has included the effect of an induced electric dipole in the calculation of AFMAG anomaly curves for a buried sphere. The effect of the electric dipole is to decrease the tilt angle and to increase the phase shift between the secondary and primary fields. Calculations for other cases will be mentioned in discussing the turam method. At the present time, model experiments are in general the only satisfactory approach to determining the effects of a poorly conducting host rock and overburden.

In ground surveys, the anomaly produced by a uniform conductive overburden and homogeneous country rock is constant everywhere for the slingram method and zero for the dip angle method. However, the overburden is seldom uniform in characteristics over large areas and lateral variations in the conductivity or thickness produce anomalies unless the response parameter for the overburden is low everywhere. Anomalies caused by variations in the characteristics of a conductive overburden are similar to those caused by a flat sheet and are usually observed primarily in the out-of-phase component. They are readily distinguishable from anomalies caused by near-surface, highly conductive zones but their existence may tend to obscure anomalies from deeply buried conductors.

In ore prospecting, conductors which are of no economic value are likely to have the same dip and strike as conductors which are of potential value. The anomalies from these non-economic conductors may have many or all of the characteristics of anomalies from potential ore bodies. In general, carbonaceous, graphitic or sulfidic slates and schists, as well as water-saturated shear zones, are less conductive than massive sulfide deposits so that in a given area the ratio of the inphase to the out-of-phase component may be a useful criterion in selecting targets worthy of further investigation. Beds of slate and schist are likely to have a greater strike extent and to be more anisotropic than massive sulfide deposits. There are, however, carbonaceous and graphitic beds which are as conductive as most massive sulfide deposits, and there are ore bodies which are not very conductive and which are anisotropic. When prospecting is done in areas where there are many non-economic conductive zones, other prospecting methods are used to help select conductive zones which are worthwhile drilling. Magnetic methods, gravity measurements and geochemical surveys often are useful in distinguishing between sulfides and graphitic zones or shear zones. Other electrical methods, such as induced polarization and self potential surveys, are useful in differentiating between conductive clays or shear zones and mineralized zones.

In addition to large and well-defined anomalies, variations in the conductivity of the host rock may produce smaller anomalies of all sizes and shapes. The irregular pattern of small, overlapping anomalies commonly seen on profile data is sometimes called "geologic noise". Geologic noise may be characteristic of a rock type, being greater, for example, over slates than over graywackes. Small anomalies caused by deeply buried conductors which are of interest may be obscured by geologic noise, and the flanks of large anomalies may be obscured by geologic noise and distorted to such a degree that a quantitative interpretation of the anomaly may not be made.

### 39 d. Turam Method

The *turam* method (Hedström, 1940; Bosschart, 1914) is very similar to the Sundberg or compensator method described earlier in this chapter. In both methods, large fixed sources are used, but the turam method differs from the Sundberg method in that the ratio of field intensities at two points is measured rather than the ratio of field intensity at a single point to the current in the source. The measuring equipment consists of two loops, a ratiometer and a null-meter. Since ratios are measured rather than absolute field intensities, variations in the source current are not important. Most commonly, only the ratio of vertical field components is measured, although ratios of horizontal field intensities or the ratio of horizontal to vertical components are sometimes measured for special purposes. The frequencies used range from 100 to 800 c/s; often measurements are made at two frequencies which differ by a ratio of 3 or 4 (for example, the two frequencies, 200 and 800 c/s, may be used).

In a typical turam survey, a cable several kilometers long is used for a source. The cable is laid out parallel to the probable strike of the ore deposits or conductive zones being sought, and grounded at both ends. Measurements are made with the coils placed in line along traverses perpendicular to the cable. When there is no conductive material present, accurate measurements can be made at distances up to one kilometer from a long cable carrying one ampere of current. Ordinarily, the spacing between the coils is fixed at a distance of approximately 25 m; however, when near-surface conducting zones are to be studied in detail, the spacing may be made smaller, or in some cases, the separation between the coils may be varied from 5 to 25 m by leaving the lagging coil in one position while the leading coil is moved away from it. When this is done, the lagging coil is moved up to the farthermost position occupied by the leading coil in advancing the coil pair, to make the computation of the field easy.

Inphase and out-of-phase ratios may be measured, as well as amplitude ratios and phase angle differences; the latter are more convenient. The measured amplitude ratios or the real and imaginary components are usually normalized by dividing them by ratios computed from the expression for the primary field. When there is no conductive material present, the normalized ratio is always one. As a first approximation, the normalized ratios of vertical field components are given by:

$$\frac{H_{i+1,n}}{H_{i,n}} = \frac{1 + \dfrac{H_s}{H_p} + \dfrac{L}{H_p}\dfrac{\mathrm{d}H_s}{\mathrm{d}x}}{1 + \dfrac{H_s}{H_p}} \tag{609}$$

where $H_{i,n}$ is the normalized vertical field at station $i$, $L$ is the coil spacing, $H_s/H_p$ is the ratio of the secondary to the primary field intensities at station $i$, and $\mathrm{d}H_s/\mathrm{d}x$ is the derivative of the secondary field with respect to the horizontal position. Thus, variations in the value for the ratio $H_{i+1,n}/H_{i,n}$ are a measure of the horizontal gradient of the secondary field strength.

To facilitate interpretation of turam measurements, field curves are often computed using the normalized ratios. The field $H_i$ at some station (usually the one nearest the cable) is measured with respect to the current in the cable. This field value is then multiplied successively by the normalized amplitude ratios, $H_{i,n}/H_{i-1,n}$; to yield the normalized field strength at the $i$th station:

$$H_{i,n} = H_{1,n}\left(\frac{H_{2,n}}{H_{1,n}}\right)\left(\frac{H_{3,n}}{H_{2,n}}\right)\cdots\left(\frac{H_{i,n}}{H_{i-1,n}}\right) \tag{610}$$

In order to find the difference in phase angle between the first and the $i$th stations, the measured phase angle differences are summed:

$$(\alpha_i - \alpha_1) = (\alpha_2 - \alpha_1) + (\alpha_3 - \alpha_2) + \cdots (\alpha_i - \alpha_{i-1}) \tag{611}$$

The real and imaginary components of the normalized field may be calculated from the amplitude and the phase angle.

When an insulated loop is used as the source, the horizontal component of the primary field is zero in the plane of the loop. If the ratios of the horizontal to the vertical field intensities are measured as well as the vertical field ratios, it is possible to calculate $H_{x,s}/H_{z,p}$. When a grounded cable is used for a source, as in the turam method, the return current through the earth establishes a horizontal field which may be regarded as the primary field. When the earth is homogeneous, and providing both the frequency and the conductivity are very low, this field is given by Eq. (373). This field component is usually small compared to the other field components, except in the area around the electrodes, and its existence is usually ignored.

When there are no conductive zones, errors in measurement caused by coil misorientation or by variations in coil spacing may be calculated readily. The primary field intensity about a long-wire source decreases approximately as the first power of the distance. Therefore, the error in the observed amplitude of the vertical field intensity ratio will be almost directly proportional to the error made in coil spacing. The error caused by misorientation of one of the coils is proportional to $(1 - \cos \delta)$, $\delta$ being the angle between the plane of the coil and the plane containing the wire and the center of the coil. In calculating the primary field, it is sometimes necessary to consider changes in elevation along the cable. This is tedious, but can be done by considering the cable to consist of a number of straight segments. When there are conductive zones present, the error caused by misorientation of the coils can be surprisingly large as a result of the existence of a large horizontal field component. In the case of measurements of the horizontal field component, large errors may be caused by the existence of the primary vertical field.

Highly conductive zones near the surface can cause such large anomalies that the field intensity may approach zero, or even in some cases pass through zero and reverse-phase by 180°. When this occurs, the turam ratios may assume extreme values, zero or infinity in the positive or negative sense, depending on the positions of the coils. Since extreme values for the ratios cannot be measured accurately, the normalized field cannot be computed. In order to avoid this problem, when extreme values are measured for the ratios, it is common practice to bridge the anomalous area by choosing other sets of coil stations which provide reasonable values for the ratios.

Sometimes, observations are also made along a base line parallel to the cable and connecting the ends of the traverses running normal to the cable, so that loop positions coincide with the positions occupied on the normal traverses. Using measurements along this base line as a reference, any serious errors in the measurements made along the traverses, or errors in calculations can be detected and corrected to improve the overall accuracy of the survey.

Since turam ratios may be computed from observations of field intensities, as well as the inverse, the turam and Sundberg methods are equivalent. In recent years, the reference cable required in the Sundberg method has been replaced with a radio link. The real and imaginary components of the field relative to the current flowing in the source cable are measured using a radio signal for phase comparison. Because this is a more accurate method, the Sundberg system is to be preferred. The field-intensity curves computed from ratios measured with the turam method are not as accurate as the field intensity curves measured directly with the Sundberg method, using either a wire link or radio link for a reference signal. However, errors between adjacent stations are smaller than those obtained with the Sundberg method using a radio link. The Sundberg method with a reference cable is the slowest, with a radio-transmitted reference signal the fastest. Instrumentation for the turam method is the simplest; instrumentation for the radio reference method is the most complicated.

The following discussion of interpretive techniques for turam profiling applies for the most part to measurements made with the Sundberg method or the radio reference method. However, turam ratios are rarely calculated from field intensity measurements made by the other methods.

Equations (570) and (584) can be used to calculate turam anomaly curves which would be observed over a buried sphere or over a long cylinder oriented parallel to the source, provided the radius of the sphere or cylinder is small compared with the distance from the cable. Since the field about a long wire source is more nearly homogeneous than the field about a dipole source, Eqs. (570) for a sphere are more reasonable approximations for measurements made with the turam method than for measurements made with the dip angle method.

With the equipment which is commercially available, measurements may be made with at most three frequencies, so that it may not be possible to obtain enough information to determine the permeability, radius and conductivity of a spherical conducting zone. However, the horizontal location and the depth of such a conducting zone may be determined readily using the procedures previously outlined.

The assumption that eddy currents are concentrated along the edges of a conductive zone is more useful in the interpretation of turam data than in the interpretation of dip angle data. As before, let us consider a thin, steeply, dipping sheet. If both the strike extent and the dip extent are large, we need consider only the current concentrations in the upper edge. We may also assume that the source cable is very long in comparison with the distance from the sheet to the cable and that

$$I' = - (U + iV) I$$

where $I'$ is the current in the sheet and $I$ is the source current. The normalized vertical and horizontal field intensities may be found as follows:

$$H_z = H_{z,p} + H_{z,s}$$

$$H_z = \frac{I}{2\pi x_1} - \frac{(U + iV) Ix}{2\pi(x^2 + d^2)}$$

$$\frac{H_z}{H_{z,p}} = 1 - \frac{Uxx_1}{x^2 + d^2} - \frac{iVxx_1}{x^2 + d^2} \tag{612}$$

where $x$ is measured from the sheet and $x_1$ is measured from the cable. Curves for the field intensity and the turam ratio calculated from the real part of this last expression are shown in Fig. 236 for $x_1/d = 50$ at $x = 0$, $L/d = 0.5$ and $U = 0.01$. The current concentration is located directly under the point where $H_z/H_{z,p} = 1$, and very nearly coincides with the inflection point of this curve. The horizontal distance, $x_m$, between the current concentrations and the maximum or minimum of the field intensity curve may be found by setting the derivative of Eq. (612) equal to zero. Provided $x_1 \gg x$, $X_m$ will be about equal to 1. The left-hand branch of the field intensity curve approaches the limit one at the cable and the right-hand branch approaches an asymptotic value of less than one, with the value depending on $U$. If $x_1 \gg x$, the minimum of the ratio curve is almost directly over the current concentration and the

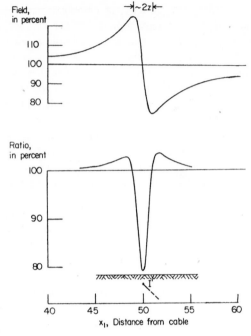

FIG. 236. Vertical component turam anomaly curves observed over a line current for $d = 1$, $L/d = 0.5$ and $U = 0.01$.

distance between the points where the curve passes through 1 is almost exactly twice the depth. If we assume that the equation for the ratio curve is given by $dH_{z,s}/dx$, the distance between the maxima is $2\sqrt{3}$ times the depth and the amplitude of the maxima is one-eighth the amplitude of the minimum. The exact shape and amplitude depend on the spacing between the receiving loops and on the locations of the stations as well as on other parameters. Out-of-phase anomalies computed from Eq. (612) are similar in appearance to the inphase anomalies shown in Fig. 236, except that they are symmetrical. The pragmatic rules given previously for determining the depth are exact in the case of the out-of-phase field intensity curves, and are nearly correct for the out-of-phase ratio curve.

When the anomaly curve is simple in character, the ratio anomaly can be used for interpretation without need to calculate field intensity anomaly curves. Since the ratio curve is a measure of the gradient of field intensity, ratio profiles tend to emphasize small features which may be scarcely discernible on the field intensity profiles. If an anomaly curve is complicated in character, and particularly if there appears to be more than one conductive zone present, it is easier to use the field intensity curves for interpretation.

If the conductive sheet dips under the energizing cable, rather than away from it as in Fig. 236, the eddy currents flow in the opposite sense, and the current is given by

$$I' = (U + iV) I$$

and Eq. (612) becomes

$$\frac{H_z}{H_{z,\,p}} = 1 + \frac{U x x_1}{x^2 + d^2} + \frac{i V x x_1}{x^2 + d^2} \tag{613}$$

The field intensity and ratio curves for this case are very nearly mirror images of the curves shown in Fig. 236.

When it is assumed, as it was in the preceding discussion, that the effect of a thin, conductive sheet with great depth extent can be represented by a single line current, the shape of the anomaly curve does not depend on the magnitude of the dip. However, the amplitude of the anomaly curve depends on the amount of primary flux linking the sheet, and hence the amplitude depends strongly on the dip angle, being almost zero for a dip of 90°.

The horizontal field intensity about a current concentration in the upper edge of a sheet dipping away from the energizing cable is:

$$H_x = H_{x,\,s} = \frac{(U + iV)\, I x}{4\pi (x^2 + d^2)}$$

$$\frac{H_x}{H_{z,\,p}} = \frac{U I\, dx_1}{x^2 + d^2} + \frac{i V I\, dx_1}{x^2 + d^2} \tag{614}$$

Curves for field intensity and turam ratio calculated from these expressions for $x_1/d = 50$ at $x = 0$, $L = 0\cdot5\, d$ and $U = 0\cdot01$ or $V = 0\cdot01$ are shown in Fig. 237. The shape of the ratio curve for the horizontal component is similar to the shape of the curve for the vertical component of field intensity for a sheet dipping away from the cable, with maximum and minimum points occurring at about the same locations. Except for the two small maxima on the curves, the field intensity curves for the horizontal component are similar to the ratio curves for the vertical component for a sheet dipping towards the cable. If $x_1 \gg d$, the points at which $H_x/H_{z,\,p}$ is equal to half its maximum value are located at $x = \pm d$.

For a conductive zone which is located at a considerable distance from the cable, the depth range of the turam method is greater than the depth range for the dip angle or slingram methods. Consequently, a conductive zone is less likely to appear to have an infinite depth and strike extent when the turam method is used than when the other methods are used. The eddy currents in a thin sheet with a finite depth extent may be approximated with current concentrations flowing in opposite directions in the upper and lower edges. When measurements are made over a dipping sheet with limited dimensions, the anomaly curves are asymmetric and the positions of the maxima and minima are shifted from the locations where they are observed for an infinite sheet.

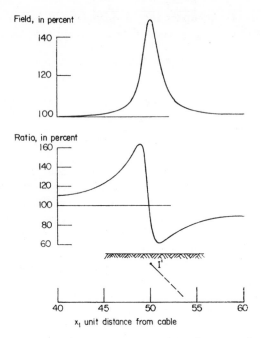

FIG. 237. Horizontal component turam anomaly curves above a line current
for $d = 1$, $L/d = 0.5$ and $U = 0.01$.

In attempting to synthesize an anomaly, it may be desirable to assume that
there are several line currents flowing at various depths (Werner, 1958). The
individual magnitudes of these currents need not be the same, but the algebraic
sum of the currents in the two directions must be zero, if the anomaly is due
only to eddy currents induced in the sheet.

Scale models consisting of metal sheets placed in air are useful in interpreting
turam data, particularly when the resistivity of the overburden and host rock
is high, and when the conductive zones are not extremely elongate in the direction parallel to the energizing cable. Curves obtained in such model studies in
which a large horizontal loop was used as the source are shown in Fig. 238 for
two dipping models with a limited depth extent. The response parameter chosen
for these model studies was $W'$ ($= \sigma\mu_0\omega t x_1$), where $t$ was the thickness of the
sheet, and $x_1$ was the horizontal distance from the near leg of the loop to the
upper edge of the sheet. In Fig. 238, the dimensions are given in arbitrary
units of distance rather than in ratios of distances. The lower edge of the sheet
had an effect on the results, particularly for the sheet which was dipping
at 20°. When the depth was small, the distance between the minimum and the
horizontal upper edge of the sheet was observed to be greater than the depth.

Inphase component, percent

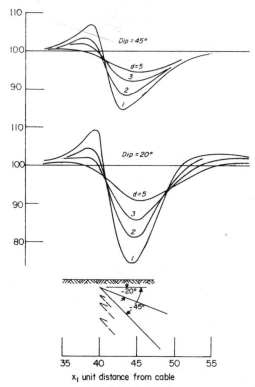

Fig. 238. Normalized vertical field over a dipping sheet with limited depth extent for $W = 21$, $d = 1, 2, 3$ and $5$ units and for dips of $-20°$ and $-45°$.

Both the vertical and the horizontal field intensities over a dipping sheet are shown in Fig. 239 for two values of $W'$. The sheet was placed at the center of a large horizontal loop in obtaining these curves. Because the sheet had a greater depth extent than the one used in obtaining the curves in Fig. 238, the effect of the lower edge was found to be less pronounced than in the preceding example. Using the pragmatic rules for finding the horizontal position and depth to a current concentration, it was found that the mean current concentration was nearer the upper edge for the larger value of $W'$, and also that the out-of-phase current was nearer the upper edge than the inphase current concentration.

Model data such as these, in which conduction in the overburden and host rock is neglected, are useful in interpreting turam measurements but are, in general, not as valuable for the turam method as for the dip angle method, or slingram method. In addition to the secondary effects in the host rock and over-burden which were discussed under these other methods, in the case of the

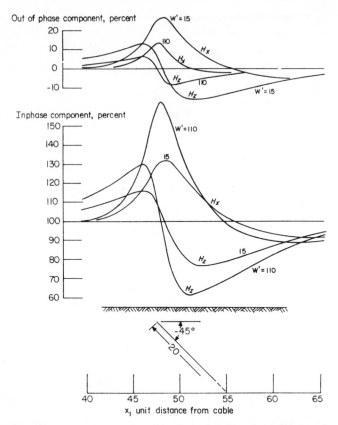

Fig. 239. Normalized vertical and horizontal fields over a dipping sheet of limited extent for $W' = 15$ and 110, $d = 5$ units and a dip of $-45°$.

turam method, secondary effects may be caused by galvanic or earth-return currents flowing in a conductive zone when a grounded cable is used as a source.

Grounded cables are usually preferred over insulated loops as excitation sources, particularly in routine exploration, because they are easier to lay out, and because secondary effects from the earth-return current tend to make the method more sensitive. This increased sensitivity is due to the concentration of return currents in zones of high conductivity. In the case of long, cylindrical bodies which are more or less oriented parallel to the cable, and particularly if such bodies are not very highly conductive, the fields due to galvanic currents may exceed the secondary fields due to eddy currents. The anomalous inphase response due to a concentration of galvanic current is given by the real part of the expression in Eq. (612) corresponding to the inphase anomaly caused by eddy currents in a sheet which dips away from the cable. However, the out-of-phase anomaly when galvanic currents are important sometimes will be that given by the imaginary part of the expression in Eq. (613) so that a negative anomaly in the out-of-phase response corresponds to a positive anomaly in the inphase response, and a positive out-of-phase anomaly corresponds to a minimum in the inphase response. This behavior apparently is caused by the ratio of inductance to resistance being greater for the cable than for the earth-return portion of the current path, so that the galvanic current in the conductive zone has a positive phase angle with respect to the current in the cable. The positions of the ends of the cable relative to the location of the conductive zone are very important; in some cases, tracing the buried extension of a conductive zone is made easy by grounding one end of the source cable in an outcrop or in a drill hole which intersects part of the conductor.

Because anomalies caused by eddy currents alone are less complicated and easier to interpret than anomalies caused by a combination of causes, insulated loops are sometimes used in place of grounded wires when complicated problems are being studied. Most commonly, measurements are made outside such a loop source only, since this is the most favorable area for detecting steeply dipping conductors. The area inside the loop is favorable for the detection of flat-lying conductors, inasmuch as the primary field is almost vertical everywhere beneath the center of a loop. Under some circumstances, a small loop, with a width as small as 20 m, may be used in making inductive measurements along a few short profiles.

When the conductivity of the country rock and overburden is appreciable, some or all of the effects discussed previously for the dip angle and slingram method may be significant. The tendency for eddy currents induced in the host rock to concentrate in highly conductive bodies is most pronounced in the case of long conductors oriented parallel to a grounded cable or located outside of and parallel to one leg of an insulated loop, and least pronounced for equidimensional conductive zones beneath the center of a loop. Usually this type of response is more important in the turam method than in the dip angle or slingram methods.

Wait (1961) has shown that the secondary fields from a conducting sphere, when excited by an electric dipole source, are frequency independent and are inphase with the source current provided the free space wavelength and the resistivity of the country rock are large. However, if the resistivity of the country rock is low the secondary fields may have an out-of-phase component.

In addition to these effects which are common to all three methods, the primary electric field of a large loop may have a significant influence on turam results when the conductivity of the host rock is high. In calculating the fields of a large, un-shielded, circular loop over a homogeneous earth, Kosenkov (1963) has included the secondary magnetic and electric fields caused by the primary electric field. The phase angle of the secondary magnetic field of electric origin is close to the phase of the primary electric field; for highly conductive rock, this causes the phase of measured field outside the loop to be less than that calculated if the electric field is ignored. Inside the loop the phase is low over half of the area and high over the other half.

Approximate computations by Svetov (1961), based on the assumption that the source is a plane wave, indicate that the anomaly due to eddy currents induced in the host rock but flowing through the conductive zone ("electrical excitation"), may be much larger than the anomaly due to currents induced directly in the conductor, in the case of a long cylinder placed parallel to the electric field. The relative importance of magnetic induction increases as the conductivity contrast between the cylinder and the medium increases.

Ideally, the frequencies used in horizontal profiling should be low enough that the response parameter for the medium ($|\gamma_1 x|$) is small at the maximum spacing used. If the conductivity contrast between the conductive zones and the surrounding medium is not sufficiently large, $|\gamma_1 x|$ will be appreciable at any frequency high enough to produce a response from the conductive zone. When $|\gamma_1 x|$ is significant, field intensity profiles will resemble the curves shown in Figs. 192 and 191 or 195, depending on whether the medium is homogeneous or layered. When the response of the medium appears to be appreciable, and the conductors which are the object of exploration are deeply buried, an attempt should be made to match the field data with theoretical curves for a homogeneous or layered half-space using the usual techniques for interpretation of depth soundings. Sharp deviations between the best fitting theoretical curve and the field data represent lateral variations in the properties of the ground or the presence of conductive zones.

Another important consideration is that in the case in which $|\gamma_1 x|$ for the ground is appreciable, the inducing field for a conductor is the sum of the primary field and the secondary fields from eddy currents flowing in the earth. At large distances from the source, the horizontal component of the inducing field near the surface may be larger than the vertical component, and both components may have a large phase shift with respect to the primary field. This phenomenon may be regarded as one in which the earth acts as a shield in the horizontal direction.

D'yakonov (1960) has developed a general solution and asymptotic expressions for the response of a cylinder buried in a conductive half-space. His results indicate that the electric field at a point on the surface directly over a deeply buried cylinder is increased by the presence of a conducting half-space, whereas the horizontal component of the magnetic field is decreased by the presence of a conducting half-space. Also, for points on the surface at large horizontal distances from the cylinder, the electric field and the horizontal component of the magnetic field vary as $1/x^2$, whereas the vertical component of the magnetic field varies as $1/x^3$. Analytical results such as these are limited in usefulness because of the approximations which must be used in obtaining numerical results and because only a few problems have, or can be, treated.

Very few turam model data for which the presence of a host environment is considered have been published. Without such data, the interpreter cannot hope to make precise interpretations when the response of the medium or the interaction with conductive zones is appreciable. The interpreter must recognize the presence of such effects, and allow for them at least in a qualitative way. Alternately, the type of source and its orientation may be chosen to minimize such effects; for example, a loop source may be used in place of a grounded cable to eliminate galvanic currents. When adequate model data are not available, synthesis of observed anomaly curves from calculations based on a number of current filaments provides a means for approximately determining some of the parameters for a conductor. If galvanic currents and eddy currents from the surrounding rock are considered, the sum of the assumed currents need not be zero.

Based on these ideas, guides for determining some of the parameters for a conductive zone when using the turam, Sundberg or radio-reference method are:

*Horizontal location.* The upper edge of a steeply-dipping thin sheet is located beneath the point where the minimum is observed on the vertical field ratio curve, the point where the maximum of the horizontal field curve is observed and at the point where the inflection on the vertical field intensity curve is observed. If the sheet does not dip steeply, and particularly if the sheet has a limited depth extent, these points are displaced in the downdip direction from the edge of the sheet. Usually, the edge may be located most precisely from the out-of-phase anomaly. If the sheet is lying horizontally, or if it has a very shallow dip, both edges may be located, approximately, using these rules. Depending on the conductivity of the sheet, the edge actually may be closer to the point where the maximum vertical field intensity is observed than to the point where the maximum horizontal field intensity is observed. If the conductive zone is a thick dike-like structure, the edge which is farthest from the cable will not show up distinctly unless the dike is near the center of an energizing loop. When a grounded wire is used, the far edge of a dike may be located by making a second set of measurements with the source cable located on the far side of the conductor. When dealing with a conductive zone made up of a number of individual conductors, each inflection on the vertical field intensity curve, no matter how small, represents a change in conductivity, or a separate conductor. For delineating individual conductors in a complicated area, the turam method is superior to the other methods which have been discussed.

*Dip.* If the maximum point on the curve for vertical field intensity is located nearer the cable than the minimum point, the direction of dip for a thin sheet is away from the cable. The dip is towards the cable, if the minimum point is closer to the cable than the maximum point, provided the anomaly is due primarily to eddy currents induced in the sheet. If galvanic currents are strong enough, the maximum point may be observed closer to the cable than the minimum point even though the sheet is dipping under the cable. Often the direction of dip can be determined from a comparison of the inphase and out-of-phase component anomalies or by a comparison of the anomalies observed at two different frequencies. Characteristic features on the inphase anomaly curve or on the anomaly curve obtained at a low frequency are displaced in the down dip direction from the corresponding features on the out-of-phase curve or the high frequency curve. If the direction of dip cannot be determined with certainty because of the effects of galvanic currents or because the conductive zone is a wide dike, a second set of measurements should be made with the source cable placed on the opposite side of the conductor. In cases in which only eddy currents in a conducting sheet need be considered, the magnitude of the dip in either direction can be estimated from a comparison with model data or computations based on current filaments. Estimates of the dip are most reliable when the depth extent of the conductor is limited.

*Depth.* The depth to the upper edge of a conductor which is parallel to the cable is equal to or slightly less than one-half the distance between the maximum and minimum points on the anomaly curve. The amplitude of the anomaly curve depends on the depth of burial, as well as on the dip, strike and response parameter. In principle, Argand diagrams which give the depth and conductance of a thin sheet as functions of the maximum amplitude of the anomalies in the inphase and out-of-phase response could be constructed for various parametric values of dip and depth extent, using model data. These diagrams could be used to determine the depth and conductance in the same manner as was described for the slingram method.

*Depth extent.* If the dip can be determined by comparison of data with model curves, it may be possible also to estimate the depth extent. If the depth extent is not large compared with the depth to the upper edge of a conductive sheet, the determinations of dip and depth extent are interdependent.

*Strike and length.* The only practical way for determining the strike direction and the length of a conductive zone is by running enough profiles to outline the upper edge or surface of the conductive zone. If the strike of the conductor is not parallel to the direction of the cable, the anomalies will be broadened and the depth estimates will be in error by the amount of broadening. The effects of galvanic currents are most likely to be important when the conductive zone is very long.

*Conductivity.* The conductivity, or, in the case of thin sheets, the conductance, can be calculated if the field data can be matched with model curves. If, for example, the observed data match one of the model curves shown in Fig. 238 and 239, we have:

$$\sigma_1 t_1 = \frac{W'}{\mu_0 \omega_1 x_1} \tag{615}$$

*Anisotropy.* The effects of anisotropy on turam measurements have not been studied in detail. However, the response of wide isotropic and wide anisotropic conductive zones is probably, quite different. For example, the horizontal field intensity has a maximum value over one edge of an isotropic dike and a minimum value over the other edge, whereas one would expect to observe a single broad maximum over a highly anisotropic dike.

### 39 e. Other Horizontal Profiling Methods

Although a great variety of techniques have been employed for induction profiling at one time or another, the three methods which we have discussed in detail (dip angle, slingram and turam) are the most widely used at present and have largely supplanted some of the older methods. Except for instrumental details, many of the other techniques which are employed are quite similar to one or more of these three methods. We shall discuss briefly a few other techniques which do differ somewhat.

The "Enslin" (Enslin, 1955) method employs a straight grounded-wire as the source. The horizontal component of the field which is tangential is measured along semi-circles about one of the electrodes. Measurements are not made near the wire in order to avoid the region of strong primary fields. The measurements can be made with conventional turam or radio reference equipment, or with other specialized equipment.

As indicated by Eq. (373), the tangential field is constant along a circle about an electrode provided the earth is homogeneous. However, if a zone of higher conductivity occurs in the vicinity of the electrode, galvanic currents tend to concentrate in this zone and an anomaly is measured. Under conditions for which the method was designed, horizontal induction fields are negligible. The method has been used successfully for locating water filled zones which occur in a dense, highly resistive, host rock.

Methods employing transient waveforms are being developed for use in horizontal profiling, as well as for depth sounding. A big advantage of transient methods in profiling is that the depth range is not dependent on the separation between transmitter and receiver, provided the configuration of the transmitter and receiver is suitable for the conductor being sought. Transient methods in which the shape of the waveform of the secondary field is measured can be used in place of multi-frequency measurements for the determination of the parameters of a conductor. One feature of transient methods is that they do not, as contrasted with harmonic methods, respond to bodies having low electrical conductivity and a high magnetic permeability.

A transient method employed in ore prospecting has been described by Kovalenko (1962). Measurements are made inside a large square source loop. The primary current is a d.c. current which is interrupted to approximate a negative step function. The transient voltage is recorded at various times ranging from 3 to 40 miliseconds following the interruption of the primary current. Noise from power lines is reduced by cancellation techniques as discussed earlier. The transient method is quite effective in distinguishing between highly conductive ore bodies and overburden of intermediate conductivity.

A number of so called "bore hole electromagnetic methods" have been devised. The distinction between these methods and the induction logging method described in Chapter II is that the former employ a fixed transmitter on the surface and one or more receivers in the hole, and are intended to detect conductive bodies which occur at a considerable distance from the bore hole, rather than to measure the resistivity of the wall rock.

In the dip angle bore hole method (Ward and Harvey, 1954), a large vertical transmitting loop is placed directly over the hole. A small coil and pre-amplifier are lowered into the hole and readings are taken at the desired depths. The received signal is amplified and fed to headphones or an indicating meter. To take a reading, the operator tilts a small vernier coil, fed in series from the main loop, until minimum signal is observed. The tilt angle of the vernier coil is read by means of an inclinometer. If the hole is vertical and if there are no conductors in the vicinity, the dip angle is always zero. If a conductor occurs within range of the apparatus, anomalous dip angles will be measured.

Other bore hole methods employ a large horizontal loop laid out with the center of the loop at the hole. In one technique, which is analagous to the turam method, the ratio or difference between the voltages induced in two receiving coils is measured. A constant separation is maintained between the coils for any one log. Alternatively, the voltage induced in a single receiving coil can be measured with reference to the current in the transmitting coil. The primary field of the transmitting coil can be calculated and the results normalized just as for the turam or Sundberg method. Interpretation of the results is similar to the interpretation of surface turam measurements except that source and receiver are co-axial.

Deviations of the hole from vertical can yield erroneous results in bore hole induction surveying. However, anomalies caused by changes in direction of the hole affect the inphase component only and are likely to have shapes different from those of anomalies caused by conductive bodies. If a directional survey of the hole is made, the effects of changes in direction can be allowed for in interpreting electromagnetic surveys.

## 40. AIRBORNE METHODS

Airborne electromagnetic methods are widely used in prospecting for conductive ore bodies. Location of ore bodies is usually the primary purpose, but airborne electromagnetic data often can be used in mapping conductive marker beds for geological purposes. Airborne induction methods are unique among the aerial surveying methods inasmuch as an artificial field is used with an appreciable depth of investigation.

The construction of a practical airborne electromagnetic system is a much more difficult problem than the construction of a counterpart ground system for a number of reasons. One principle difficulty is that the separation between source and receiver which can be obtained is very limited. The lowest safe and practical altitude for surveying usually is from 200 to 500 ft above the ground surface, depending on the terrain and the type of aircraft. At this distance above a conducting half-space, the ratio of secondary field from the earth to the primary field is very small, and an airborne system must be more sensitive than a comparable ground system. Most aircraft are of all-metal construction, contributing large secondary fields which must be discriminated against. Noise introduced into the receiving system by motion of the receiving in the earth's field loop is also a problem, though of less importance. Noise is also contributed by the aircraft's electrical system. A continuous measurement and recording system is required. Procedures for establishing the true flight path of the aircraft have been well developed for other types of aerial surveying and will not be discussed here. Since airborne induction measurements are strongly dependent on altitude, the aircraft should be equipped with a precise radar altimeter.

The problem of obtaining adequate sensitivity in an airborne system has been solved in at least five basically different ways:

1. Perhaps the most direct approach, though not the first historically (Pemberton, 1962) is the use of loops attached rigidly to the extremities of the airframe. The electronic system is made extremely sensitive to overcome the difficulty of small loop separation.

2. A second approach is the use of a receiving loop in a "bird" which is towed at the end of a long cable or by a second aircraft. In a towed bird system, the magnitude of the mutual coupling between the loops varies with motion of the bird relative to the aircraft. A number of different schemes are used in which the quantities measured are essentially independent of these variations in mutual coupling.

3. One of the most recently developed airborne systems uses a repetitive transient primary field. The secondary field is measured during the intervals when the primary field is zero.

4. Another method, which is not completely airborne, uses a long grounded cable as the source; the horizontal component of the field about the wire is measured with a loop and associated recording equipment attached to a fixed-wing aircraft or a helicopter.

5. The AFMAG method is a fifth means for making airborne electromagnetic measurements.

## 40a. Airborne slingram

A number of fixed-loop systems have been devised for use with helicopters and fixed-wing aircraft. Basically, these are extremely sensitive slingram systems in which the real and imaginary parts of the mutual coupling are measured at a single frequency which may range from 320 to 4000 c/s. Vertical loop arrangements are used in preference to horizontal loop arrangements since, for large ratios of height to separation, vertical loop configurations are more sensitive to steeply dipping conductors and less sensitive to flat-lying conductors. When the vertical coplanar loop configuration is used, coils with ferromagnetic cores are placed in pods attached to either wingtip. For vertical coaxial loop arrangements, the transmitting coil is attached to brackets on the nose of the aircraft and the receiving coil is placed in a boom or "stinger" extending from the tail. Vertical coaxial arrangements are used with helicopters as well as on fixed-wing aircraft; in one system installed on a small helicopter, the coils are placed at either end of a long, light-weight detachable boom. In other systems, the coils are placed at either end of a large bird which is towed by the helicopter.

Servo-systems can be used to make ratiometers such as the one shown in Fig. 183 self balancing and continuously recording. However, it is difficult to obtain the required sensitivity in such a system when the ratio $Z/Z_0$ is measured directly. The block diagram in Fig. 240 is an example of one of the types of measuring systems used for airborne measurements. The balancing network is

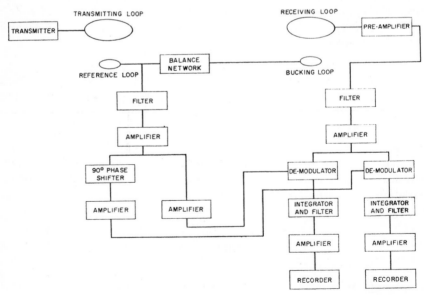

Fig. 240.

adjusted so that the reference signal supplied to the bucking coil is such that the primary field is cancelled at the receiving coil. The secondary field arising from eddy currents in the airframe is also cancelled, provided it is constant. When the system is flown within range of a conductive zone, a voltage proportional to the secondary field from the conductor is induced in the receiving coil. This signal is amplified, filtered and fed to two phase-sensitive demodulators. The output of these demodulators is integrated, amplified and recorded. Reference voltages for the demodulators are provided from a reference winding coupled to the transmitting loop. The reference circuits must be adjusted for phase shifts so that one channel gives the in-phase component of the secondary field from the earth and the other channel gives the out-of-phase component. By calibrating such a system properly, the parameter recorded is $Z_s/Z_1$ $= Z/Z_0 - 1$, commonly expressed as the secondary field in parts per million of the primary field.

Since it is not necessary to measure $Z_s/Z_1$ with a high degree of precision, the active components of this system, such as amplifiers and demodulators, need not be highly stable. The stability and noise level of the system depend primarily on the mechanical stability of the coils and the electrical stability of both the coils and the balancing network.

Proper mounting of the loops is essential. Vibrations and flexing of the airframe or coil mountings cause variations in the orientation and separation of the loops with respect to one another and with respect to the aircraft, introducing noise into the system. Also, vibration of the coils in the earth's magnetic field causes noise. Shock mounting of the receiver coil eliminates much of the noise contributed by this last mechanism, but tends to increase the amount of low frequency noise caused by changing orientations of the loops. Coplanar loops are suspended somewhat below the wing tips to minimize variations in separation of the coils as the wings flex. A change in separation of one part in a hundred thousand causes a change in the free-space mutual coupling of about 30 ppm, which is greater than the internal noise level of some of the more sensitive fixed-coil systems. For coaxial coil configurations mounted in a bird or on a boom, the structure is designed and supported in such a fashion that the loops tend to remain parallel to each other during flexures of the structure.

Variations of the intensity of the secondary field at the aircraft caused by the movement of control surfaces such as ailerons are sometimes a source of noise. Variations in the contact resistance between various parts of the airframe can cause changes in the eddy current flow pattern and the associated secondary field at the aircraft. Careful electrical bonding of the various portions of the airframe largely eliminates this source of noise.

Low pass filters placed between the demodulators and the recorder reject most of the noise which is of high enough frequency that it cannot be confused with lower-frequency earth-return signals. Some systems also have high pass filters to eliminate very low frequency noise and drift such as may be caused by variation of coil properties with temperature. Such systems are responsive only to vertical conductors and the edges of horizontal conductors. This type of response is acceptable for most prospecting but is not desirable if data are to be used in geologic mapping.

Response data for airborne slingram systems may be obtained from calculations and model studies using the same equations and techniques which have been described in previous sections. Since the ratio of height to loop separation ranges from two to five for an airborne system, most reference data for ground slingram surveys do not cover the range of interest for airborne slingram. Because the resolution of airborne measurements is inferior to that of ground measurements, fewer reference data are needed in the interpretation of airborne surveys. In some cases, calculation of the anomaly curve for a particular conductor geometry is easier for an airborne system than for a ground system; for example, Eq. (570) giving the response for a sphere may be valid for a sphere of a certain size for airborne measurements but not for ground measurements because the ratio $h/r$ is greater for the airborne equipment. Model experiments are somewhat more difficult to make for airborne measurements than for ground measurements due to the greater sensitivity which is required.

Response curves for a homogeneous earth for $h/r = 3{\cdot}0$ are shown for four different slingram configurations in Fig. 241. The vertical coaxial and perpendicular configurations are least sensitive, giving them some advantage over the other configurations in distinguishing between vertical conductors and conductive overburden and country rock. It is interesting to note that for $h/r = 3$, the response of these two configurations is about the same in magnitude; for smaller values of $h/r$, the perpendicular arrangement is more sensitive and for larger values of $h/r$, the vertical coaxial arrangement is the more sensitive.

Model studies for a vertical coplanar arrangement using high-pass filtering have been made by Boyd and Roberts (1961) for tabular bodies. Assuming that an anomaly must have an amplitude of at least 80 ppm to be definitely recognizable, their results indicate that a tabular body having dimensions of $150 \times 50 \times 60$ m can be detected from heights up to 120 m ($h/r = 4{\cdot}8$), provided the flight line is directly over the body.

Wieduwilt (1962) has compiled complete sets of response curves similar to those in Fig. 232 for a vertical coaxial system over thin horizontal and vertical sheets. These curves can be used to obtain the conductance and depth of thin vertical dikes. For dikes having a width which is large compared with $h/r$, the peak response is positive rather than negative.

Because of the large ratio of height to coil separation, airborne slingram anomaly curves over individual conductors have relatively simple shapes and they reflect only the gross character of conductive zones. Anomaly curves over a series of conductors may, of course, be complicated. Geologic and instrumental noise levels usually are high. For these reasons, the interpretation

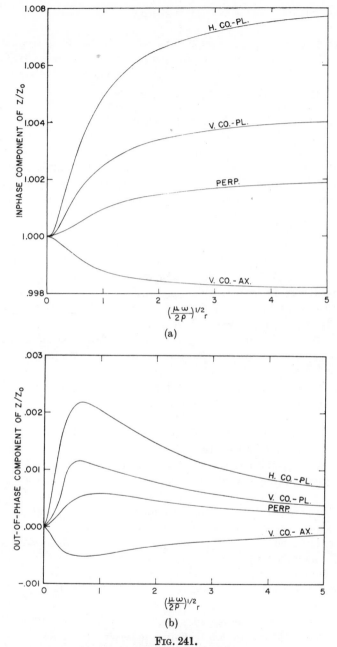

Fig. 241.

(a) Inphase components of mutual coupling for loops above a homogeneous earth for $h/r = 3 \cdot 0$.

(b) Out-of-phase component of mutual coupling for loops above a homogeneous earth for $h/r = 3 \cdot 0$.

of airborne measurements is seldom as detailed as is the interpretation of ground measurements. In airborne prospecting, the primary objective in interpretation is to distinguish between conductors of potential economic interest and conductive zones which are of no economic value. In using electromagnetic surveys for more general geologic purposes, interpretation is usually limited to tracing and correlating the sub-outcrops of conductive zones.

If the shape of an anomaly curve indicates the presence of a wide conductive horizontal sheet or a wide zone of uniformly conductive earth, the conductance or the conductivity may be estimated roughly using the approach outlined in the section on depth sounding. In many cases, airborne slingram measurements made at a single frequency do not provide enough information to permit the interpreter to distinguish between a single conductive layer, several layers or a homogeneous earth.

If the shape of the anomaly curve indicates a steeply dipping thin sheet or dike, the depth and the conductance may be estimated from the peak amplitudes of the inphase and out-of-phase components using curves such as those given by Wieduwilt. In areas in which the thickness of the overburden is known to be uniform, and if this thickness is known, the depth can be assumed to be the sum of the thickness of the overburden and the height of the aircraft above the ground. When the depth is known, a more accurate estimate of the conductance may be made. An alternate method for evaluating anomalies when the depth is known consists of normalizing the anomalies in terms of the percentage of the anomaly which would have been observed over a perfectly conducting half plane having its upper edge at the same depth as the upper edge of the actual conductor (Fig. 242).

In the case of wide dikes and other three-dimensional bodies, the width can be determined from the half-width of the anomaly curve or from other criteria, depending on the particular coil configuration used. If the depth to the top of the body can be estimated, the anomalies can be normalized using the response of a perfectly conducting half plane or other suitable conductor geometry. The ratio of the peak amplitudes of the inphase and out-of-phase anomalies often is used as a qualitative measure of the relative conductivity. In matching airborne curves with model curves, it is difficult to make a unique selection of one model curve which best fits the observed data because of the lack of resolution in airborne measurements. As a result, estimates of actual conductivity or other interpretive parameters are likely to be crude.

Model curves obtained using metal sheets in air are less useful in the interpretation of airborne measurements than in the interpretation of ground surveys because of the greater effect that a conductive overburden may have on airborne measurements. In the case of a homogeneous earth, the maximum value of the out-of-phase component and the asymptotic value for the inphase component are reached at smaller value of $B$ for large values of the ratio $h/r$ than for small values of the ratio $h/r$. In effect, an airborne electromagnetic system is responsive to a greater volume of earth than a ground system with the same spacing. Not only is the response from the overburden and the host rock relatively more important, but the effects of current flow from the surrounding medium into highly conductive bodies are more pronounced.

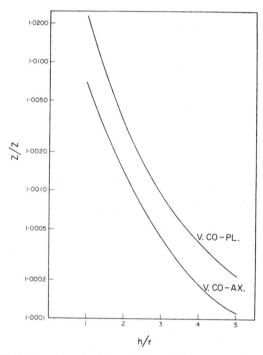

Fig. 242. Variation of maximum anomaly with $h/r$ for various coil configurations above a perfectly conducting vertical half-plane.

In interpreting airborne electromagnetic results, maximum use should be made of any other geologic or geophysical information which may be available. Magnetic field and natural gamma radiation measurements are usually made along with electromagnetic measurements in a typical airborne survey. Many times, massive sulfide bodies will be somewhat magnetic, while conductors of no economic interest, such as graphitic slate or water-filled shear zones are non-magnetic. When there is a correspondence between electromagnetic and magnetic anomaly curves, additional information about the shape of the body can be derived from the magnetic survey data. Swamps and lakes, which are likely to be conductive, are observed as lows in natural radioactivity. In many cases, the geology of the survey area may be sufficiently well known that ambiguities in interpretation may be readily resolved.

## 40b. The quadrature system

One of the most widely used towed-bird systems is known as the *quadrature system* or the *dual-frequency phase shift system*. In such a system, the phase shift of the electromagnetic field at two frequencies is observed at the receiver. The transmitting loop is a cable stretched around the aircraft between wingtips and tail in such a manner that the axis of the loop is nearly vertical. The loop is powered with several hundred watts at two frequencies, typically 400 and 2300 c/s. A horizontal-axis receiving coil is towed in a bird at the end of a long cable. In a typical installation in an amphibious aircraft, the bird maintains a position about 420 ft behind and 220 ft below the aircraft. At this location, the direction of the primary field is about 25° from the axis of the coil. There is a small out-of-phase secondary field at the receiving coil due to currents induced in the aircraft. Motion of the bird in this field is a source of noise. To reduce noise from this source, an auxiliary horizontal-axis loop, powered by a current 90° out of phase with respect to the current in the main transmitter loop, is used to cancel the secondary field of the aircraft. Narrowband circuits are used to measure and record the phase shift of the field at the receiving loop relative to the current in the transmitter loop. Except under conditions of excessive turbulence, or when there are nearby electrical storms, the noise level is less than 0·05°.

At normal flying heights of 350–500 ft, the coil configuration used with the phase shift system is sensitive to conductors of almost any shape or attitude; in comparison with vertical slingram configurations, there is less tendency to discriminate against horizontal sheets and to accentuate vertical sheets. Since the inphase component of the secondary field is not measured, there is some possibility of not detecting highly conductive ore bodies which do not cause significant out-of-phase fields. In practice, the possibility of missing a large, highly conductive near-surface conductive body is slight because out-of-phase currents flow between the body and the surrounding medium and because there is usually a halo of disseminated mineralization around highly conductive zones.

The response of the system to horizontally layered media may be calculated from Eqs. (540)–(544). A number of response curves, compiled from model studies for thin sheets and from calculations for an infinite half plane, showing the effects of variations in depth, depth extent, dip, strike and conductivity are given by Patterson (1961). In the case of a vertical sheet, the peak response is observed when the receiving coil is almost directly over the sheet. For some dipping sheets, a secondary maximum is observed as the transmitter passes over the edge of the sheet. Generally, the anomalies obtained with this system are asymmetrical. The anomalies at either edge of a wide horizontal sheet differ, causing a herringbone pattern when flight lines flown alternately in opposite directions are used to prepare a contour map of electromagnetic response.

Gaur (1963) has conducted a series of model experiments for the dual-frequency phase shift system in which the models were placed in a tank full of brine, simulating a conductive host rock and overburden. In the case of horizontal sheets, the presence of the brine environment changes the shape of the anomaly curves and increases the peak amplitude by as much as a factor of three. Similar results are observed for vertical sheets, except that there is less change in the shape of the anomaly curve.

The procedures used in interpreting data obtained with the dual-frequency phase shift system are much the same as those used in interpreting slingram data. Anomalies caused by a conductive overburden are characterized by low ratios of the low frequency to the high frequency response, wide anomalies and a change in the form of the curves between adjacent traverses. Anomaly curves for steeply dipping bodies have high ratios of low frequency to high frequency response and are narrow. The depth and the conductance of steeply dipping thin sheets may be determined from the peak responses at the two frequencies by means of theoretical curves such as those given by Patterson.

## 40 c. The rotating field method

Another means for obtaining adequate sensitivity in an airborne electromagnetic system using a towed bird is the use of two or more sets of loops which respond differently to a conductor. The ratio of responses or the difference in response is observed and recorded. Ideally in such a system, the differences between free-space mutual couplings for each set of loops should be constant as the bird moves relative to the aircraft.

The *rotating field method* (Törnquist, 1958) uses two transmitting loops which are attached to the aircraft and two receiving loops which are placed in a bird. One of the transmitting loops is vertical with its plane passing through the center line of the aircraft; the second transmitter loop is orthogonal to the first and is approximately horizontal. A similar arrangement of receiving loops is towed in a bird as nearly directly behind the aircraft as is possible or is towed at the end of a short cable by a second aircraft. The inclinations of the two nominally horizontal loops are adjusted for the height of the bird so that the loops are as nearly coplanar as possible.

As indicated in Fig. 243, the transmitting coils are powered by two currents at the same frequency but differing in phase by 90°. The amplitudes of the currents are adjusted so that the fields produced by the two coils are equal in magnitude. The result is a rotating magnetic field such as would be generated by rotating a bar magnet about an axis normal to its magnetic axis. The signals from the receiving coils are amplified, shifted in phase from each other by 90° and then the inphase and out-of-phase components of the difference between the two signals, measured in terms of the primary field at the nominal separation, are determined and recorded.

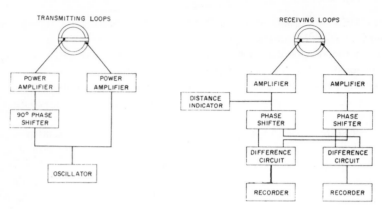

FIG. 243.

In the absence of secondary fields, and as long as the sets of transmitting and receiving coils are coplanar, the difference in signals detected by the receiving coils is zero despite variations in the separation between the sets of coils or rotations of either set about their common axis. When the system is within range of a conductor, the signals from the two receiving coils are, in general, not equal, and an anomaly is recorded. The size and shape of the anomaly does vary with the separation between the sets of coils.

Rotations of the receiving coils about any axis other than the axis which is common to the two sets of coils causes noise in the inphase component of the recorded signal. Also, the secondary field of the aircraft causes noise in both components as the bird rotates relative to the aircraft. The high noise level which is characteristic of the rotating field method is offset by the fact that the ratio of height to coil separation is small, particularly when two aircraft are used.

Frequencies ranging from 300 to 5000 c/s have been used with the rotary field method (Schaub, 1962). The flying height varies from 40 to 180 m, depending on the aircraft and the terrain. Spacings varying from 150 to 330 m are used, depending on the height, as well as on the resolution and sensitivity required, and the signal to noise ratio of the equipment. The lead aircraft tows the receiving coils by means of a short cable at a distance of 17 to 20 m below the aircraft. The second aircraft, carrying the transmitting equipment, flies directly behind the bird. A system of blinking lights actuated by the signal amplitude from one of the receiving coils is used to indicate to the pilot of the second aircraft whether the spacing is correct, too long or too short.

The voltage induced in the horizontal receiving loop is proportional to $H_{z,p,h} + H_{z,s,h} + H_{z,s,v}$ and the voltage induced in the vertical coil is proportional to $H_{y,p,v} + H_{y,s,v} + H_{y,s,h}$, where the subscript $h$ is used to indicate fields associated with the horizontal transmitting loop and the subscript $v$ is used for fields associated with the vertical transmitting loop (Hedstrom and Parasnis, 1959). The magnitudes of $H_{z,s,v}$ and $H_{y,s,h}$ are equal. When flying normal to the strike of two dimensional conductors or when flying over a homogeneous or horizontally stratified earth, $H_{z,s,v}$ and $H_{y,s,h}$ are zero. In such cases, the quantity $R$ measured by the system is

$$R = \frac{Z_v - Z_h}{Z_{oh}} \tag{616}$$

where $Z_h$ and $Z_v$ are the mutual couplings between the horizontal coplanar and the vertical coplanar configurations, respectively, and $Z_{oh}$ is the free space coupling for coplanar coils at the nominal loop separation. When flying over three dimensional conductors or at an angle other than 90° to the strike of a two dimensional conductor, $H_{z,s,v}$ and $H_{y,s,h}$ are, in general, not zero.

Anomaly curves for the rotating field method can be calculated by combining data for the horizontal coplanar, vertical coplanar and null perpendicular arrangements, or curves may be obtained directly by scaled model experiments. Torquist (1958) and Hedstrom and Parasnis (1959a, b) have obtained scale model results for various geometries of thin sheets. Schaub (1961) has calculated anomaly curves for a sphere, infinite half plane and homogeneous half space, and has confirmed the results with model studies. For flights made normal to the strike of a vertical sheet, the response of the vertical coil system is very small, and, as a result, the anomaly obtained with the rotary field method is very nearly the same as that which would be obtained with a slingram system using a horizontal coplanar coil configuration. For a homogeneous earth, the response observed with a rotating field is, except for small values of the ratio $h/r$ and large values of $B$, less than the response observed with either a vertical or horizontal coplanar slingram system. At critical values for $h/r$ and $B$, one component of the response is zero (Schaub, 1961).

For simple conditions, the interpretation of results obtained with the rotary field method is not too different from the interpretation of slingram results. Since the ratio $h/r$ is quite small, a detailed interpretation can be made for some conductor geometries. When the geological environment is complex and the flight lines are not normal to all of the conductors, rotary field results are very complicated inasmuch as they are, essentially, the combined results obtained with three different coil configurations.

## 40d. Other dual transmitter systems

A variety of other techniques have been proposed in which the ratio or difference between dual transmitter fields is measured. The simplest of such techniques uses two parallel transmitting coils rigidly fastened together and a similar arrangement of receiving coils placed in a bird. One coil system operates at a frequency suitable to detect the conductive zones being sought and the other operates at a very low frequency, preferably low enough that the response is slight, even for large, highly conductive bodies. The two signals from the receiving coils are amplified by selective circuits, rectified and the difference recorded. When no conductive zones are present, the difference between the two signals is zero or constant, independent of movements of the bird relative to the aircraft. When a conductive zone is present, the mutual coupling between coils changes more at the high frequency than at the low frequency, and an anomaly is recorded.

This technique has not proved to be very useful. The weight and power requirements for a system using a very low frequency as a reference are excessive. Also, variations in the amplitude of either of the transmitted signals or changes in the gains of either of the receiving channels cause drift or noise.

Another technique (Slichter, 1955) uses a set of orthogonal transmitting coils attached to the aircraft and a set of orthogonal receiving coils carried in a towed bird in such a fashion that one coil is coplanar and the other is coaxial with respect to its counterpart on the aircraft. The transmitting coils are powered with two frequencies which differ enough to permit separation of the signals in the receiver circuitry. The amplitudes of the transmitted signals are made the same by electronic regulators. The signal from each of the coils is filtered to remove the unwanted frequency and to eliminate noise. After filtering, the signals are mixed and amplified by a common amplifier so that the gain in each channel will be the same. After amplification, the two signals again are separated, after which they are rectified and the difference recorded. In free space, this difference will be zero despite changes in coil separation. Rotation of the bird about one of its axes causes an error signal which varies as the cosine of the angle. These error signals may be detected by means of a third receiving coil which is orthogonal to the other two, and amplified and recorded. The error signal also may be used to actuate a servomechanism which rotates the transmitting coils to help compensate for misalignment between transmitting and receiving coils as the bird moves about relative to the aircraft. As in the rotating field method, conductors are detected because they generate unequal changes in the mutual coupling in the two coil systems. Vertical coplanar and vertical coaxial coil configurations are preferred in this method although the horizontal coplanar arrangement could also be used for one of the coil pairs.

By using a third set of coils parallel to one of the other sets but operating at a substantially different frequency, a second difference signal can be obtained. A comparison of the amplitudes of the two signals provides an indication of the conductivity in a conductive zone. One such system which has been used extensively (Pemberton, 1961) employs a combination of coils and frequencies such that three differences and one error signal are recorded. To minimize noise caused by the aircraft itself and to facilitate the use of vertical transmitting coils, an aircraft constructed mostly of wood is used with this system.

The response of a *difference system* may be calculated by taking the differences between the response for each coil configuration considered separately. There is no coupling between the various coil configurations, such as can occur in the rotary field method. Results obtained with this method are likely to be more complicated than those obtained with a slingram system, but they are less complicated than the results obtained with the rotary field method.

## 40 e. Transient response systems

Fundamentally, the problem in achieving adequate sensitivity in an airborne electromagnetic system is that of detecting very small secondary fields from the earth in the presence of a large primary field. This problem may be avoided by using a pulsed primary field and by measuring the transient secondary field from the earth while the primary field is zero. To be practical, an airborne transient system must use a repetitive primary field, so that the recorded parameters will be essentially continuous.

While one serious problem in making airborne measurements is eliminated by using transient response measurements, other problems become more severe and new problems arise. In most cases, only a small part of the energy in the secondary field remains after the end of the energizing pulse. As a result, a more intense primary field must be used in a transient method than in a continuous wave method in order to measure a secondary response of the same magnitude. It is more difficult to design circuitry for transient signals than for sinusoidal signals. Inasmuch as many of the circuit elements must have a wide band pass, there is more problem in rejecting atmospheric and other noise.

The only airborne transient system which has been reported to date is the INPUT system (Barringer, 1963), which uses a half sine wave primary pulse, with alternating polarity. A large, horizontal transmitting loop is stretched around the aircraft. A vertical receiving coil with its axis aligned with the flight direction is towed in a bird at the end of a long cable. The primary purpose for using a long cable is to remove the receiving coil as far as possible from secondary fields induced in the aircraft. The signal seen by the receiving coil is blocked from the amplifiers during the primary pulse. During the interval between pulses, the received signal is amplified and fed to a series of electronic gates which sample the signal at four different times. The signal from each gate consists of a series of pulses with constant width and with an amplitude which is a function of the amplitude of the secondary field during the time the gate was open. These pulse series are integrated using circuits with appropriate time constants and recorded on four separate channels.

The delay of the gate for the first channel is selected so that this channel will be responsive to poor conductors, including conductive overburden. Delays in the other channels are successively longer so that channel four is sensitive only to conductors having a large value for the response parameter. A comparison of the anomalies recorded on the four channels serves to indicate the conductivity of the body in the same way that comparison of measurements made at several frequencies with continuous wave methods does.

Because the coil configurations are much the same, the shapes of anomaly curves obtained with the INPUT system are similar to the shapes of the anomaly curves observed with the dual frequency phase shift system. To the extent that the shape of the anomaly curve is independent of frequency, model experiments in which the coil configuration is simulated and in which continuous waves are used can be employed to obtain the shapes for INPUT anomaly curves. The exact shape and amplitude of INPUT anomalies can be determined from model experiments in which airborne circuitry is simulated or by calculation from the continuous wave response as discussed in previous sections.

## 40f. The AFMAG system

Instrumentation for the AFMAG method can be adapted to airborne measurements. Two orthogonal receiving coils of equal sensitivity are towed in a bird behind a fixed-wing aircraft or helicopter. The length of the tow cable is determined by the magnitude of the secondary fields induced in the airframe and the electrical noise generated by the aircraft. The coils are suspended so that their axes are at 45° from the horizontal and aligned with the flight direction. The signals from the two coils are compared in such a fashion that the output is approximately proportional to the tilt angle (Ward, 1959). As with a ground system, measurements are made at two frequencies. Also, a continuous record of the signal strength at each frequency is made to help evaluate the quality of the data. In some airborne systems, the phase difference between signals from the two coils is measured.

Inasmuch as it is not possible to determine the mean azimuth of the field and then measure the tilt angle in that direction as is done in making ground surveys, flight lines are laid out perpendicular to the regional strike. As pointed out in the discussion on ground AFMAG surveys, the natural field tends to become aligned perpendicular to the regional strike. The field strength must be somewhat higher for airborne measurements, so in many parts of the world, AFMAG methods can be used only rarely. For large targets, the AFMAG method provides a greater depth of penetration than any of the other airborne electromagnetic methods.

## 40g. Airborne radio reference method

The radio reference method can be adapted for use with an aircraft. A long grounded cable is used as the source. In general, more current and a longer cable, up to 20 km or more in length, are used for airborne measurements than for ground surveys. A vertical receiving coil with its axis in the line of flight is attached to a light aircraft or is towed in a bird at the end of a short cable. The inphase and out-of-phase components of the signal induced in the coil are measured with phase sensitive detectors. A reference signal for phase detection is transmitted from the source by radio. Under favorable conditions, flight lines may be extended on either side of the cable to about twice the length of the cable (Mizyuk, 1960). The area immediately around the cable is not surveyed because of the large gradients and intense vertical field in this area.

# REFERENCES

BARRINGER, A. R. (1963) The use of audio and radio frequency pulses for terrain sensing, *Proc. Second Symp. on Remote Sensing of Enviromeṇt*, The University of Michigan, 201–214.

BHATTACHARYYA, B. K. (1955) Electromagnetic induction in a two-layer earth, *J. Geophys. Research* **60** (3), 279–288.

BHATTACHARYYA, B. K. (1959) Electromagnetic fields of a transient magnetic dipole on the earth's surface, *Geophysics* **24**, 89–108.

BHATTACHARYYA, B. K. (1963) Electromagnetic fields of a vertical magnetic dipole placed above the earth's surface, *Geophysics* **28**, 408–425.

BOSSCHART, R. A. (1964) *Analytical interpretation of fixed source data*, Uitgeverei Waltman-Delft, Holland.

BOYD, D. and ROBERTS, B. C. (1961) Model experiments and survey results from a wing tip-mounted electromagnetic prospecting system, *Geophys. Prospecting* **9**.

CAMPBELL, G. A. and FOSTER, R. M. (1948) *Fourier integrals for practical applications*, D. Van Nostrand, New York.

D'YAKANOV, B. P. (1959) The diffraction of electromagnetic waves by a sphere located in a half space, *Bull. Acad. Sci. U.S.S.R., Ser. Geophys.* **11**.

D'YAKANOV, B. P. (1960) Asymptotic expressions for electromagnetic fields caused by a cylindrical in homogeneity, *Bull. Acad. Sci. U.S.S.R., Ser. Geophys.*, 954–958.

ENSLIN, J. F. (1955) A new electromagnetic field technique, *Geophysics* **20** (2), 318–334.

ENENSTEIN, B. S., SKUGAREVSKAYA, O. A. and RYBAKOVA, E. V. (1959) Some data on soundings by the method of pulsing an electric current in the earth, *Bull. Acad. Sci. U.S.S.R., Ser. Geophys.* 1486–1491.

GAUR, V. K. (1963) Electromagnetic model experiments simulating an airborne method of prospecting, *Bull. Natl. Geophys. Research Inst.* **1**, 167–174.

HARRINGTON, R. F. (1961) *Time Harmonic Electromagnetic Fields*, McGraw-Hill, New York.

HEDSTRÖM, E. H. (1940) Phase measurements in electrical prospecting, *Trans. Am. Inst. Mining Met. Petrol. Engrs.* **138**, 456–571.

HEDSTRÖM, E. H. and PARASNIS, D. S. (1959a) Some model experiments relating to electromagnetic prospecting with special reference to airborne work, *Geophys. Prospecting* **6**, 435–447.

HEDSTRÖM, E. H. and PARASNIS, D. S. (1959b) Reply to comments by N. R. Paterson, *Geophys. Prospecting* **7**, 448–470.

KOSENKOV, O. M. (1963) The electromagnetic field of a circular loop at the earth–air boundary, *Bull. Acad. Sci. U.S.S.R. Ser. Geophys.* 1845–1851.

KOROLEVA, K. P. and SKUGARESVSKAYA, O. A. (1962) The later stage of the establishment of a magnetic field in a layered medium, *Bull. Acad. Sci. U.S.S.R., Ser. Geophys.*, 506–513.

KOVALENKO, V. F. (1962) Recording transitional processes in induction resistance measurements in south Ural copper pyrite deposits, *Int. Geol. Res.*, vol. 4, no. 12.

KOZULIN, YU. N. (1960) On the theory of electromagnetic frequency sounding of multi-layered structures, *Bull. Acad. Sci. U.S.S.R., Ser. Geophys.*, 1204–1212.

MARCH, H. W. (1953) The field of a magnetic dipole in the presence of a conducting sphere, *Geophysics* 18, 671–684.

McLACHLAN, N. W. 1959 *Bessel Functions for Engineers*, Clarendon Press, Oxford.

MIZYUK, L. YA. (1960) Methods and apparatus for airborne prospecting, *Bull. Acad. Sci. U.S.S.R., Ser. Geophys.*, 789–797.

NEGI, J. G. (1962) Diffraction of electromagnetic waves by an inhomogeneous sphere, *Geophysics* 27, 480–492.

ORSINGER, A. O. and VAN NOSTRAND, R. (1954) A field evaluation of the electromagnetic reflection method, *Geophysics* 19, 478–489.

PATTERSON, N. R. (1961) Experimental and field data for the dual frequency phase-shift method of airborne electromagnetic prospecting, *Geophysics* 26, 601–617.

PEMBERTON, R. H. (1962) Airborne electromagnetics in review, *Geophysics* 27, 691–713.

PRICE, A. T. (1950) Electromagnetic induction in a semi-infinite conductor with a plane boundary, *Quart. J. Mech. and Appl. Math.* 3, 385–410.

SCHAUB, YU. B. (1962) The influence of the specific resistance of the surrounding medium on the form of anomaly curves obtained in aerial electrical prospecting, *Bull. Acad. Sci. U.S.S.R., Ser. Geophys.*, 652–658.

SCHAUB, YU. B. (1961) An experimental test of the rotating magnetic field method, *Bull. Acad. Sci., U.S.S.R., Ser. Geophys.*, 1015–1021.

SCHAUB, YU. B. (1962) Prospecting by the rotating magnetic field method using LI-2 and AN-2 airplanes, *Bull. Acad. Sci. U.S.S.R., Ser. Geophys.*, 925–933.

SLICHTER, L. B. (1955) Geophysics applied to prospecting for ores, *Econ. Geol., Fiftieth Anniv. Vol.*, 885–969.

SMYTHE, W. R. (1950) *Static and Dynamic Electricity*, McGraw-Hill, New York.

STEFANESCU, S. (1936) *Beitr. angew. Geophys.* 6, 168–201.

STRATTON, J. A. (1941) *Electromagnetic Theory*, McGraw-Hill, New York.

SWANSON, H. E. (1961) Model studies of an apparatus for electromagnetic prospecting, *Trans. Am. Inst. Mining Met. Petrol. Engrs.* 220, 234–238.

SUNDBERG, K. (1931) Principles of the Swedish geoelectrical methods, *Ergänzungshefte angew. Geophys.* I.

SVETOV, B. S. (1961) Role of the procedure used for exciting the field in electrical prospecting by the low frequency inductive method, *Bull. Acad. Sci. U.S.S.R., Ser. Geophys.*, 1826–1831.

SUNDBERG, K. and HEDSTRÖM, F. H. (1934) Structural investigations by electromagnetic methods, *Proc. World Petrol. Congr.* V. B, part 4, 102–110, Inst. of Petroleum Technologists.

TÖRNQUIST, G. (1958) Some practical results of airborne electromagnetic prospecting in Sweden, *Geophys. Prospecting* 6, 112–126.

WAIT, J. R. (1951) A conducting sphere in a time varying magnetic field. *Geophysics* 16, 666.

WAIT, J. R. (1951) Oscillating magnetic dipole over an horizontally stratified earth, *Can. J. Phys.* 29, 577–592.

WAIT, J. R. (1952) The cylindrical ore body in the presence of a cable carrying an oscillating current, *Geophysics* 17, 378–386.

WAIT, J. R. (1953) The fields of a line source of current over a stratified conductor, *Appl. Sci. Research* B3, 279–292.

WAIT, J. R. (1953) Induction in a conducting sheet by a small current-carrying loop, *Appl. Sci. Research* B3, 230–235.

WAIT, J. R. (1954) Mutual coupling of loops lying on the ground, *Geophysics* 19 (2), 290–296.

WAIT, J. R. (1955) Mutual electromagnetic coupling of loops over a homogeneous ground, *Geophysics* **20**, 630–637.

WAIT, J. R. (1956) Shielding of a transient electromagnetic dipole field by a conductive sheet, *Can. J. Phys.* **34**, 890–893.

WAIT, J. R. (1956) Mutual electromagnetic coupling of loops over a homogeneous ground, *Geophysics* **21** (2), 479–484.

WAIT, J. R. (1958) Induction by an oscillating magnetic dipole over a two-layer ground, *Appl. Sci. Research* B7, 73–80.

WAIT, J. R. (1960). Propagation of electromagnetic pulses in a homogeneous conducting earth, *Appl. Sci. Research* B8, 213–253.

WAIT, J. R. (1960) Some solutions for electromagnetic problems involving spheroidal, spherical, and cylindrical bodies. *J. Research Natl. Bur. Standards*, B64, 15–32.

WAIT, J. R. (1960) On the electromagnetic response of a conducting sphere to a dipole field, *Geophysics* **25** (3), 649–658.

WAIT, J. R. (1962) A note on the electromagnetic response of a stratified earth, *Geophysics* **27**, 382–385.

WAIT, J. R. (1962) *Electromagnetic Waves in Stratified Media*, Pergamon Press, Oxford.

WARD, S. H. and HARVEY, H. A. (1954) Electromagnetic surveying of diamond drill holes, *Canadian Mining Manual*, Northern Miner Press.

WARD, S. H. (1959) AFMAG—airborne and ground, *Geophysics* **24**.

WARD, S. H. (1959) Unique determination of conductivity, susceptibility, size and depth in multi-frequency electromagnetic exploration, *Geophysics* **26**, 531–546.

WERNER, S. (1955) Geophysical history of a deep-seated pyritic ore body in northern Sweden, *Geophys. Surv. in Min., Hydro. and Eng. Proj.* E.A.E.G., E. J. Brill, Leiden, Holland.

WESLEY, J. P. (1958) Response of dyke to oscillating dipole, *Geophysics* **23**, 128–133.

WEST, G. F. (1960) Quantitative interpretation of electromagnetic prospecting measurements, Ph. D. Thesis, Univ. of Toronto.

WIEDUWILT, W. G. (1962) Interpretation techniques for a single frequency airborne electromagnetic device, *Geophysics*, **27**, 493–506.

CHAPTER 7

# RADIO WAVE METHODS

THE electromagnetic waves transmitted from radio broadcast stations may be used as an energy source in determining the electrical properties of the earth. The use of radio-frequency electromagnetic fields can be considered as an essentially different method than the inductive methods discussed in the preceding chapter, in as much as the instrumental methods used are very different.

The frequencies used in radio-wave transmission are generally much higher than those used in the inductive methods. If existing radio transmission stations are used as a signal source, only limited ranges in frequency are available. Standard broadcast stations cover the frequency range from 540 to 1640 kc/s (in the United States). A great many commercial and amateur stations have broadcast frequencies from 1640 kc/s to 30 Mc/s and above, but for the most part, transmissions are intermittent, and many of these stations cannot be considered a reliable source. There are also a few stations operating in the frequency range from 100 to 540 kc/s, and a very few stations broadcasting in the range 10 to 100 kc/s.

Considering that the depth to which radio waves can penetrate into a conductive material such as the earth is very limited, it is usually preferable that measurements be made at as low a frequency as is possible. For the frequencies used by the standard broadcast stations, the skin depth (depth at which the signal strength is reduced by the ratio $1/e$) ranges from a few tens of feet in normally conductive rock to a few hundreds of feet in resistive rock. For this reason, radio-wave measurements using standard broadcast stations as an energy source are best used for studying the electrical properties of the soil cover and overburden, and in areas where the soil is reasonably thin, the bedrock geology.

The transmitting antenna normally used in radio stations consists of a vertical wire supplied with an oscillatory current, so that the antenna can be treated as a current dipole source, as was done in Chapter VI. At the relatively large distances at which electromagnetic field measurements are normally used, the equations for the electric and magnetic field components (Eqs. (390), (391) and 392)) simplify considerably in that the only terms which are of significant size are the terms in $1/r$:

$$H_\varphi = i\,\frac{I\gamma e^{-i\gamma r}\,dl}{4\pi r}\sin\theta = i\,\frac{I\gamma e^{-i\gamma r}\,dl}{4\pi r} \qquad (617)$$

$$E_r = 0 \qquad (618)$$

$$E_\theta = i\,\frac{I\mu\omega e^{-i\gamma r}\,dl}{4\pi r}\sin\theta = i\,\frac{I\mu\omega e^{-i\gamma r}\,dl}{4\pi r} \qquad (619)$$

These equations are applicable only in a completely uniform space. If the presence of a conductive earth, as well as the effect of the ionosphere are considered, the problem is very much more complicated. Since the wavelengths used are short compared to the distance travelled by the waves, the concepts of ray optics may be applied in a qualitative way. The propagation from a transmitter to a receiver location can take place along a multiplicity of paths, as indicated very roughly in Fig. 244. The various paths which may be followed are:

1. A direct line-of-sight path from the transmitter to the receiver;
2. A path in which the ray is reflected from the surface of the earth;
3. A path (or many paths) in which the ray is reflected from the lower surface of the ionosphere;
4. A path in which energy is continually re-radiated by currents induced in the ground (the surface wave).

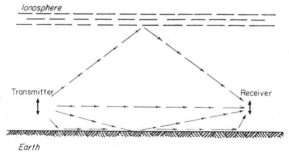

FIG. 244. Paths which may be followed by a radio wave in travelling from a transmitting antenna to a receiving antenna. Paths include direct transmission, reflection from the earth and ionosphere, and a surface wave.

In addition, energy may follow reflected ray paths from irregularities on the earth's surface, such as mountains. The electrical properties of the earth may be computed from the rate at which the amplitude of the ground wave decreases with distance from the transmitting antenna, but it must be distinguished from the other varieties of waves which may arrive at the receiver.

If both the transmitting antenna and receiving antenna are close to the surface of the earth (antenna heights less than 5 per cent of the antenna separation), the wave reflected from the earth's surface almost exactly cancels the direct arriving wave, since the travel distances along the two paths are almost equal, and since the phase of the reflected wave is inverted with respect to the phase of the direct-arriving wave as a result of the reflection. This means the total signal strength will consist only of the ground wave and ionosphere reflections. Ordinarily, the ionosphere reflected waves are important only at distances greater than a few tens of miles from the transmitter. Therefore, at distances of 10 or 20 miles, the only signal observed at the earth's surface is the ground wave.

Norton (1941) has presented a set of master curves for the decay of ground wave amplitude with distance which may be used in interpreting such measurements. The set of curves are shown in Fig. 245. The amplitude of wave intensity is expressed as a ratio to the wave intensity which would be observed at a unit numerical distance, $p$, if the field intensity decayed purely as the inverse distance. The numerical distance is defined in terms of the frequency and the electrical properties of the earth:

$$p = \frac{\pi}{x} \frac{R}{\lambda} \frac{\cos^2 b''}{\cos b'} \cong \frac{\pi}{x} \frac{R}{\lambda} \cos b \tag{620}$$

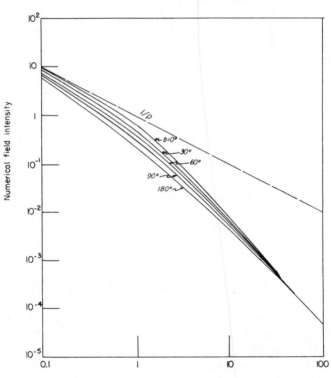

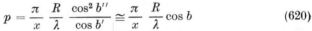

FIG. 245. Decay of ground wave intensity with propagation distance. Numerical distance is defined in the text.

where $R$ is the distance from the transmitter to the receiver, $\lambda$ is the full wavelength, measured in air, and the parameters $x$ and $b$ are computed from values for the electrical properties:

$$x = \frac{1 \cdot 80 \times 10^4}{f \varrho} \tag{621}$$

$$b = 2b'' - b' \cong \arctan \frac{\varepsilon - 1}{x} \tag{622}$$

$$\tan b' = \frac{\varepsilon - 1}{x} \; .$$

$$\tan b'' = \frac{\varepsilon}{x}$$

The frequency is expressed in Mc/s, the resistivity in ohm-m, and the dielectric constant as the ratio to that of free space. The dielectric ratio normally ranges from values as low as 4 over dry quartizitic rocks to as much as 20 over rocks with a high water-content, and to 81 for water (see Section 8, Chapter I). The approximation in which $b'$ and $b''$ can be considered equal usually applies with a reasonable precision.

The full wavelength in air may be computed simply from the expression:

$$\lambda = \frac{\text{Velocity of light}}{\text{Frequency}} \tag{623}$$

or, in the mks system:

$$\lambda = \frac{3 \cdot 0 \times 10^8}{f}$$

At 1 Mc/s, the wavelength is 300 m.

The curves given in Fig. 246 may be used in a logarithmic curve-matching procedure similar to that used for other methods. The field measurements of electric field intensity, made with some sort of radio receiver having a calibrated antenna for measuring field strength, are plotted as a function of transmitter–receiver separation on a bi-logarithmic graph having the same dimensions for the grid. These data are superimposed over the theoretical curves shown in Fig. 246 such that the data points fall along one of the theoretical curves. When this match is obtained, three parameters are read from the graphs; the numerical distance, $p_1$ for which the dashed line on the theoretical graph intersects the unit real distance axis (one meter, one kilometer, etc.), the ratio $R/\lambda$ for the unit real distance (wavelengths per unit real distance), and the angle $b$.

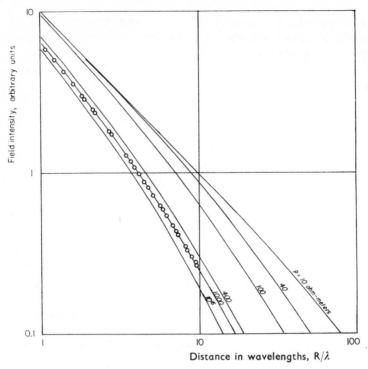

Fig. 246. Radio wave field intensity curves computed for a frequency of 3·5 Mc/s and a dielectric constant of 15. The data points were observed along a traverse from a transmitter located over glacial till in Minnesota.

Once these three parameters have been determined, the electrical properties may be computed from the relations:

$$x \cong \frac{\pi}{P_1}\left(\frac{R}{\lambda}\right)_1 \cos b \quad . \tag{624}$$

$$\varrho = \frac{1\cdot 80 \times 10^4}{x f_{Mc}} \tag{625}$$

$$\varepsilon \cong x \tan b - 1 \tag{626}$$

There is very little spread between the curves for various values of the quantity, $b$, and high-quality data must be obtained in order that a selection may be made between the various curves. An alternate approach consists of computing families of curves for the particular frequency at which measurements are being made, using values of dielectric constant and electrical resistivity for the curve parameters. A family of such curves computed for a frequency of 3·5 Mc/s and a dielectric constant ratio of 15, and for resistivities ranging from 10 to $10^6$ ohm-m is shown in Fig. 246.

The curves shown in Fig. 246 were constructed for the interpretation of field intensity measurements made over a thick section of glacial till in Minnesota. Since the ratio of transmitter–receiver distance to wavelength is known, the data may be shifted only vertically in obtaining a match with these curves. The data shown in the illustration match well with the curve for an earth resistivity of 1000 ohm-m, though the uncertainty may be as great as a factor of 2, considering how closely spaced the curves for various earth resistivity values are about the one which was actually chosen.

Higher resolution might be obtained if the earth were more conductive or if a lower frequency had been used. However, the frequencies which may be used for such surveys are severely restricted by the Federal Communications Commission, unless very low transmitter powers are used. An example of a field intensity curve conducted in an area where the rock is more conductive is shown in Fig. 247. This survey had been made at the same frequency as the preceding example, except the location was over sedimentary rocks in western

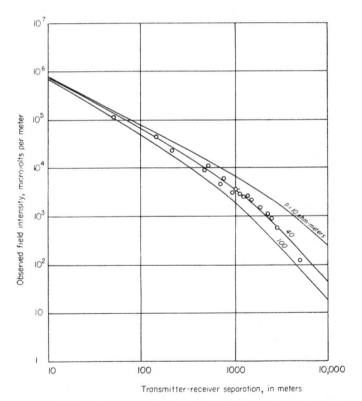

FIG. 247. Field intensity measurements made in western Colorado at a frequency of 3·5 Mc/s.

Colorado. The data match well with a curve for an earth resistivity of 40 ohm-m, a value which corresponds with resistivities determined with the galvanic method. In this case, the uncertainty of the resistivity value determined with the radio-wave field strength measurements is about 25 per cent, considerably less than in the first example, where measurements were made over a high resistivity earth.

The theoretical curves shown in Fig. 245 through 247 are computed for a uniform earth. In the case of a layered earth, the same curves may be used by considering the resistivity determined from the field strength measurements to be an "effective resistivity". If values for the effective resistivity are determined at several frequencies, these measurements may then be compared with the theoretical curves for the magneto-telluric method (Fig. 142) to determine the true resistivities of the various layers. In so doing, it must be remembered that the value of wave number, $\gamma_1$, for the first layer is that for a frequency at which displacement currents are important, and the complete expression for wave number given in Eq. (273) must be used.

It would be possible to conduct a depth sounding using radio-wave field strength profiles for a variety of frequencies, but such a technique is not used in practice for several reasons:

1. Instrumentally, it is difficult to transmit the range of frequencies required. Each frequency requires a different antenna length to obtain good radiation efficiency. Also, permission must be obtained from the Federal Communications commission for each frequency used, unless very low radiation power is used. Such permission cannot be obtained readily for a range of frequencies covering several decades, as is desired if a sounding is to be made.

2. In order to determine a value for the effective resistivity with a reasonable certainty, it is necessary to make field strength measurements over a range of distances corresponding to ten wavelengths at the lowest frequency used. In practice, this might be a distance of 10 km. In so doing, the layering must be assumed to be uniform over such a lateral distance, and this is rarely the case for the depths to which radio waves normally penetrate.

3. It is usually easier to determine the resistivity as a function of depth using the galvanic method at the depths normally reached by radio waves.

The calculation of earth resistivity from radio-wave decay rates is not as accurate as determinations made by other methods, nor as convenient, unless an existing transmitter can be used as a power source. The advantage of the method is that the effective resistivity is determined over a relatively large area.

It has been suggested many times in the literature that local variations in radio-wave field intensity which occur within a distance of a few meters or tens of meters may be used to locate locally conductive areas, such as fault traces or shallow ore bodies. The intensity of the radio-wave field will normally tend to decrease less rapidly than normal over a small conductive area, and may actually increase locally. Frequently, there may be a zone of anomalously low intensity on the side of the conductive zone away from the transmitter.

Local variations in field strength may be studied, using a distant radio station as a source, providing some corrections may be made for variations in the level of power transmitted. Broadcast stations normally have a power variation of 10–20 per cent, though this averages out if the signal level is observed and averaged over a period of several minutes. It is also desirable to measure radio-wave intensities from several stations operating at about the same frequency, but for which the signals arrive in the survey area from different directions. The anomaly in field strength may commonly depend on the orientation of the conductive zone relative to the direction of transmission of the radio waves, and the certainty of detecting a conductive zone is increased when radio waves arriving from several directions are observed.

The use of a distant transmitter to detect the presence of conductive zones is attractive, in that a survey may be conducted rapidly. However, precautions are necessary to discriminate between local anomalies in field intensity caused by conductive zones in the earth, and the anomalies associated with such things as pipes, fences, etc.

## REFERENCES

BAYS, C. A. *et al.* (1954) Study of the propagation of electromagnetic waves through lithologic formations, Signal Corps contract DA 36–039–SO 5454.

DAY, J. P. and TROLESE, L. G. (1950) Propagation of short radio waves over desert terrain, *Proc. I. R. E.* **38**, 165–175.

Federal Communications Commission (1940) Standards of good engineering practice concerning standard broadcast stations, 550–1600 kc, U. S. Gov. Printing Office, Washington, D. C.

MILLINGTON, G. (1949) Ground-wave propagation over a land-sea boundary, *Nature* **164**, 114.

MILLINGTON, G. and A. G. ISTED (1949–1950) Ground-wave propagation over an inhomogeneous smooth earth, *Proc. Inst. Elec. Engrs.* (London) **96**-III, 53–64; **97**-III, 209–222.

MOONEY, H. M. (1954) The status of (non-direct-current) electrical prospecting, with special reference to uranium prospecting, U. S. Atomic Energy Commission, Contract AT-(49-1)-900.

NORTON, K. A. (1941) The calculation of ground-wave field intensity over a finitely conducting spherical earth, *Proc. I. R. E.* **29**, 623–639.

WAIT, J. R. (1953) Propagation of radio waves over stratified ground, *Geophysics* **18**, 416–422.

WAIT, J. R. (1964) Electromagnetic surface waves, in *Advances in Radio Research*, vol. 1, 157–217, Academic Press, London.

# CHAPTER 8

# INDUCED POLARIZATION

In all the methods for measuring the electrical properties of earth materials discussed in the preceding chapters, it has been assumed that the electrical properties do not depend on frequency. Electrical properties do depend on frequency to some extent, as was pointed out in the first chapter, but the degree of dependence in most cases is small enough that the effect can be ignored in applying the standard measurement methods which have been described. On the other hand, it is possible to design measurement methods intentionally to detect the dispersion of electrical properties as a function of frequency. One such set of methods is known by the name of *induced polarization measurements*, or *overvoltage* measurements. These methods have been discussed in detail by Wait (1959). Early work has been described by Schlumberger (1920), though modern development of the method stems largely from work done by Bleil (1953).

In the induced polarization methods, variations in either electrical resistivity or in dielectric constant which occur at extremely low frequencies are studied. These variations cannot be explained in terms of atomic or molecular structure of the material, but appear to depend on the texture or macro-structure of a rock.

The nature of induced polarization (using the term in the geophysical sense) can best be described with a hypothetical experiment. Consider a sample of rock through which a direct current has been circulating for a prolonged period of time. The ratio of electric field to current density determines the d.c. or static resistivity of the sample. If the current flow is abruptly interrupted, the electric field will decay as shown in Fig. 248. It will quickly drop to a value $E_0$, which is usually a small fraction of the static field, $E_s$. After this initial drop, the field will decay slowly over a long period of time, and may have a measurable size for time periods ranging from minutes to hours (Hartshorn, 1926).

This behavior is anomalous in that it differs from the behavior of a synthetic circuit consisting of a condenser and resistor in parallel, a circuit which is the analog used in defining the electrical properties of a rock at higher frequencies. If the same experiment were to be performed on such an R-C circuit, in which the resistance and capacitance are not dependent on time, the voltage after the interruption of current flow would decrease uniformly from the static value, with the amplitude following a simple exponential decay. The time constant

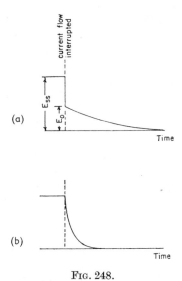

FIG. 248.
(a) Transient decay of electric field strength in a rock sample.
(b) Transient decay of voltage in an R-C circuit.

for the decay would be the product of the dielectric constant and the resistivity for the material. The peculiar form of the decay voltage actually observed in rocks indicates that a simple R-C circuit is inadequate to explain the behavior of current flow at very low frequencies. However, as was pointed out before, the departure from simple behavior is usually negligible unless measurements are specially designed to detect such departure.

The nature of induced polarization may also be seen in a hypothetical experiment using alternating current, rather than direct current. If low frequency alternating current circulates through a rock sample, the current flow can be divided into three parts; an ohmic current which flows inphase with the applied voltage, and is independent of frequency; a displacement current, which is 90° out-of-phase with the applied voltage, and is directly proportional to frequency; and an *anomalous charging current* neither inphase nor out-of-phase with the applied voltage and which is not quite proportional to the frequency (see Fig. 249). If this behavior is to be represented by an equivalent resistor–condenser circuit, as is done when constant values for resistivity and dielectric constant are assigned to a material, the resistance and capacitance of the circuit will have to be varied with the frequency. Even when this is done, the equivalent circuit will not produce the transient response shown in Fig. 248 (a), but rather, a response not too different than that shown in Fig. 248 (b).

Either type of experiment may be used to describe the induced polarization response of a rock. If alternating currents are used in determining the specific

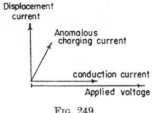

FIG. 249.

impedance of a rock sample as a function of frequency, the measurements are said to be made in the *frequency domain*. If instead, induced polarization is studied by measuring the transient decay of voltage following the interruption of steady current flow, the measurements are said to be made in the *time domain*. Measurements made in the time domain may readily be transformed into equivalent data for the frequency domain, though the computations may not be simple. It is not always possible to transform measurements made in the frequency domain into the time domain, since usually not enough data are obtained with measurements in the frequency domain.

Induced polarization, such as is observed in rocks at very low frequencies cannot be explained in terms of atomic or molecular structure, as is polarization at higher frequencies (see Section 8, Chapter I). It results from a variety of phenomena which occur more or less on a macroscopic scale when a material is built up from a mixture of mineral grains and an electrolyte. The important phenomena contributing to induced polarization appear to be electrode reactions between electrolytes and metallic mineral grains in rocks which contain such minerals, variation of electrolyte mobility in fine-grained rocks, and related phenomena.

## 41. ELECTRODE POLARIZATION AND OVERVOLTAGE

A source of induced polarization which is of considerable practical importance in geophysical prospecting involves chemical processes which take place when current flows from an electrolyte into a metal, or vice versa. Such processes can take place in rocks containing grains of conductive minerals (sulfides, native metals) dispersed through the rock. Providing these metallic grains are distributed through the rock rather than occurring as continuous filaments, current must flow part of the time through the electrolyte filling the pores of the rock and part of the time through the metallic grains. At the boundary between the electrolyte and the metal, charge is transferred from ions in solution to the metal, or from the metal to ions in solution. If the charge gained by an ion happens to be an electron, the process constitutes *reduction*; if the ion loses an electron, the process is *oxidation*.

Even when no current is flowing through such a contact, there is a potential difference between the metal grain and the solution which is caused by the tendency for the metal to go into solution. In the case of pyrite (see Fig. 250(a)), a few ferrous ions will leave the pyrite grain, leaving the pyrite with an excess negative charge and giving the solution an excess positive charge. The variation of potential about such a grain is shown in Fig. 250(b). The difference between the potential in the metal and the potential in the solution far away from the metal is the *electrode potential, $E_e$,* and has characteristic values for different metals. It also depends on temperature and on the composition of the electrolyte.

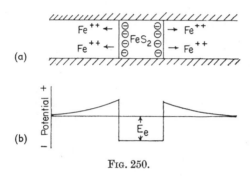

Fig. 250.

If a mineral is relatively insoluble in the electrolyte with which it is in contact, the tendency for ions to go into solution from the mineral is quickly balanced by the tendency for the same ions to precipitate out of solution. An equilibrium condition is established in which the number of ions going into solution is the same as the number of ions returning to the metal from solution.

When an electrical current flows through such a grain, there is an increased number of ions going into solution at one face and a decreased number going into solution at the other face. By unbalancing the equilibrium condition in this way, charge is transferred through the pyrite grain, and so, we have current flow.

In practice, the transfer of charge through a metallic grain is never so simple. Oxidation or reduction of the metallic atoms in the mineral grain requires somewhat more energy than the reduction of chlorine or the oxidation of hydrogen from the solution. So, superimposed on the normal exchange of ferrous ion between the pyrite and the electrolyte, there are the two other reactions indicated in Fig. 251:

1. The reduction of chlorine ion to gaseous chlorine on the anodic side of the pyrite grain, so that there is a net transfer of electrons from the electrolyte to the pyrite

2. The oxidation of gaseous hydrogen on the cathodic side of the grain, with a net transfer of electrons from the metal to the electrolyte.

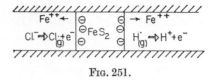

FIG. 251.

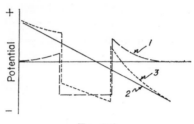

FIG. 252.

These two reactions appear to be the favored ones in most rocks, since most pore waters naturally occurring in rocks are rich in chlorine ions. The reactions result in the formation of hydrochloric acid on one side of the metallic grain, and this acid usually reacts with the sulfide to etch it. This etching may be observed on one side of a pyrite grain after a direct current has flowed through it.

These oxidation–reduction reactions affect current flow in three ways:

1. The unbalance of reaction rates at opposite faces of a grain during current flow causes concentration potentials which are not equal at the two faces (see Fig. 252). Since the regions close to the grain surfaces contain a higher concentration of ions than other parts of the solution, an electromotive force is set up between these regions and the rest of the solution according to the Nernst equation:

$$\text{e.m.f.} = \frac{RT}{nF} \ln \frac{c_1}{c_2} \tag{627}$$

where $R$ is the gas constant, $T$ is absolute temperature, $n$ is valency, $F$ is Faraday's constant and $c_1$ and $c_2$ are the ion concentrations between which the e.m.f. is developed. When added to the normal potential gradient (curve 2 in Fig. 252), these concentration voltages cause a lessening of the gradient or a reversal. In order to drive a current through these regions, the applied potential gradient must be increased. Thus, a concentration e.m.f. acts in the same manner as an added resistance.

2. The oxidation process at one face and the reduction process at the other face usually will not require the same energy. One process will tend to go faster than the other. Since, in the long run, as much current must come out one side of the grain as goes in the other, the potential gradient must be increased at

the face where the slower reaction takes place. The areas where the potential gradient must be increased to maintain current flow can be considered to be zones of added resistance.

3. If a mineral grain has a resistant surface coating, such as may be produced by oxidation or corrosion, this too will contribute excess resistance to current flow.

The extra potential gradient required to drive a current from an electrolyte into a metallic conductor is called *overvoltage*. If none of the three phenomena just described took place, it would be necessary only to apply the voltage $E_e$ to drive current into one face of a sulfide grain, and at the opposite face, the same voltage would be recovered as the current left the mineral grain. The voltage required to overcome the three processes described above increases as the current density is increased, as one would expect. The relation between overvoltage and current density is given by Tafel's law:

$$\eta = a - b \log_{10} j \qquad (628)$$

where $\eta$ is the overvoltage, $j$ is the current density, and $a$ and $b$ are experimentally determined constants. The constant $b$ is usually within the limits 0·09 and 0·13 at room temperature. The parameter $a$ varies over a wide range, with the value depending on the particular metal being used and on the properties of the electrolytic solution. Values for a few metals are:

| | |
|---|---|
| Hg in 0·2 N $H_2SO_4$ | $a = 6 \times 10^{-12}$ A/cm$^2$ |
| Hg in 1·0 N $H_2SO_4$ | $4·9 \times 10^{-13}$ |
| Hg in 10 N $H_2SO_4$ | $1·6 \times 10^{-11}$ |
| Pb in 1·0 N HCl | $2·0 \times 10^{-13}$ |
| Ag in 1·0 N HCl | $2·0 \times 10^{-4}$ |
| Cu in 1·0 N HCl | $1 \times 10^{-9}$ |
| Au in 1·0 N HCl | $1 \times 10^{-5}$ |

(All at 20°C. Values from Butler, 1951.)

Very few measurements have been reported for overvoltage at the surface of sulfide minerals, but for the values that have been measured, it appears that the parameter $a$ falls in the range from $10^{-6}$ to $10^{-3}$ A/cm$^2$.

If the overvoltage, $n$, were directly proportional to current density, its effect could be treated simply as an added resistance to current flow. However, this is not true for a material following Tafel's law, so the effective contact resistance at an electrode surface must be considered to be a function of current density. Few measurements have been made, but the curve shown in Fig. 253 is typical of the curves for surface impedance as a function of current density for natural sulfide minerals. In this case, which is the case of pyrite, the contact resistance for low current densities (less than $10^{-5}$ A/cm$^2$) is large, falling between 1000 and 10,000 ohm-m$^2$. Surface resistance is much less at high current densities.

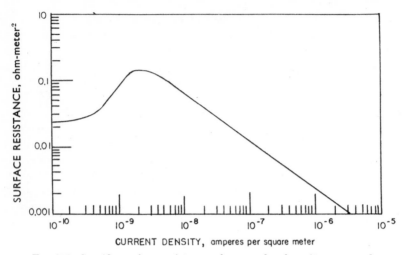

Fig. 253. Specific surface resistance of a sample of pyrite measured as a function of current density.

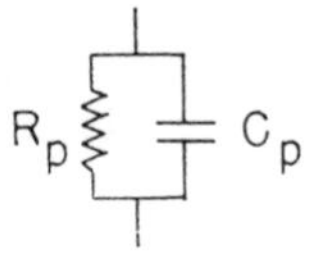

Fig. 254.

The effect that surface resistance may have on the overall properties of a rock containing disseminated metallic minerals can best be discussed by considering a synthetic circuit built up from resistors and condensers. The effect of electrode polarization may be represented crudely with a parallel combination of a resistor and condenser, as shown in Fig. 254. The resistor, $R_p$, represents the contact resistance, and so, *is a function of current density*. It also depends on frequency, decreasing rapidly with increasing frequency. The capacity, $C_p$, represents the charge that is stored at the ion concentrations which build up at the metallic surfaces before equilibrium current flow is established. This capacity is of the order of 10 to 20 pF/cm², in the case of concentration overvoltage. The capacity decreases with increasing frequency.

When sulfide grains are distributed in a porous rock, the extent to which the electrical properties are affected depends on several factors. A rock might be considered to look something like the sketch in Fig. 255(a); a certain fraction of the pores are blocked by sulfide grains, while other, parallel pores are not. This situation might be represented by the equivalent circuit in Fig. 255(b). The resistance through pores not blocked with sulfide grains is $R_{w1}$; in the sulfide-blocked pores, the resistance consists of three parts; that of the water

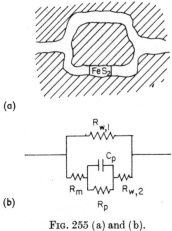

(a)

(b)

FIG. 255 (a) and (b).

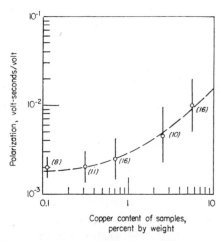

Copper content of samples,
percent by weight

FIG. 256. Induced polarization observed in samples of bornite-bearing shale from the Nonesuch Formation (Precambrian). The measure of polarization is the integral of the transient voltage recorded following the interruption of current flow through a sample. Each vertical bar represents the range of values recorded for a group of samples with an average copper content indicated by the circle at the middle of each bar.

in the pores, $R_w$ , that through the mineral grain, $R_m$, and the contact resistance, $R_p$.

This equivalent circuit may be used to discuss *qualitatively* the effect various factors have on the overall resistance of a mineralized rock:

1. The greater the number of pores which are blocked by conducting grains, the greater will be the effect of the polarization resistance, $R_p$, and the polarization capacity, $C_p$, on the overall impedance of the rock. The effect of electrode polarization should be increased as the sulfide content of the rock is increased, other things being equal. An example of the correlation between polarization effects and the sulfide content of a rock is given in Fig. 256. The measure of polarization used in this graph is the integral of transient voltage (per unit applied voltage) observed following the abrupt cessation of current flow through a sample. The samples were taken in the state of Michigan from the Precambrian Nonesuch shale, a dense, uniform shale with a porosity of about 4 per cent. The uniformity of the rock assures that the observed variations in induced polarization are caused only by the variation in the amount of sulfides (primarily bornite and native copper) in the rock. Each point plotted on the graph represents an average value for a number of samples, the number being shown in parentheses.

2. The effect of induced polarization will depend on the size of the sulfide grains. A single large sulfide grain will provide a low resistance path, so that a large amount of current will flow through it, compared to the current flowing in barren pores. However, the ratio of surface area to grain volume is small, and since polarization is a function of the amount of grain surface exposed to the electrolytic solution, the total effect of polarization will also be small. As the grain size is reduced, more surface area is exposed for the same volume of sulfides in the rock, and the effect of polarization will increase. In very fine-grained rocks, the surface resistance, $R_p$, through each sulfide grain may be so large that much of the current will bypass the pores blocked by sulfide grains, favoring the barren pores, and the effect of polarization will decrease. So, we would expect to find the greatest polarization effect for a given concentration of sulfide minerals in a rock for the case in which the sulfide grain size has an intermediate value. This is borne out by measurements made on mixtures of sand and pyrite, as shown by the experimental data in Fig. 257. In this case, maximum polarization effects were observed for pyrite grain sizes around 0·2 mm. The optimum grain size for producing polarization will depend on the bulk resistivity of a rock, and will differ from rock to rock. Generally, one would expect smaller grain sizes to provide maximum polarization in dense, non-porous rocks, since there are relatively fewer barren pores through which the current may pass than in a porous rock.

3. For the same metal content, rocks with low total porosity will polarize to a greater extent than will rocks with a high porosity. The reason is that in

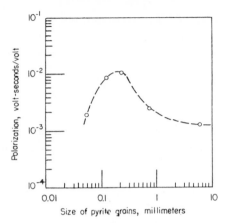

Fig. 257. Induced polarization observed in mixtures of pyrite and quartz sand, for various pyrite grain sizes.

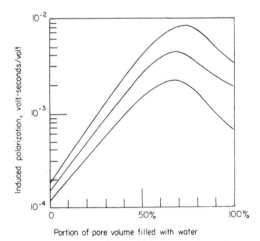

Fig. 258. Induced polarization observed for three samples of sulfide-bearing (molybdenite) granite, as a function of the water content of the samples.

low porosity rocks, a greater fraction of the total current is forced to flow through the metallic grains than in high porosity rocks. A small sulfide concentration in a dense igneous rock will cause much more noticeable polarization effects than will a similar sulfide concentration in a porous rock, such as sandstone.

4. The amount of polarization which will take place in a rock also depends on current density. Since the contact resistance between the conducting mineral grains and the electrolyte filling the pores is greater at low current densities, more of the current will bypass the sulfide grains and the overall polarization

may be less than that observed with high current-densities. However, the polarization capacity, $C_p$, increases with decreasing current density, so it is difficult to predict the range in current density which will provide the greatest amount of polarization.

5. If the electrical properties of a rock are measured with a steady state alternating current, the effects of polarization will vary with frequency. At higher frequencies, the impedance through the sulfide-blocked pores decreases, both because of the presence of the capacity, $C_p$, which shunts the resistance, $R_p$, and also because the contact resistance itself decreases with increasing frequency. The impedance of a rock will vary a few per cent per decade of frequency change in the case of a relatively barren rock to as much as 10 or 20 per cent per decade in the case of mineralized rocks. Some examples of this "frequency effect" are shown in Fig. 258.

6. The effect of polarization will depend on the degree to which the pore structure in a rock is filled with water. When a rock is partially desaturated, the resistances $R_{w1}$ and $R_{w2}$ will both be increased but the polarization resistance, $R_p$, and the resistance through the conductive mineral grain, $R_m$, will remain the same. As a result, the total resistance through a sulfide-blocked pore will decrease less rapidly as the water content of the rock is reduced than will the resistance through the barren pores. This means a greater fraction of the total current will flow through the blocked pores and the amount of polarization will increase. A series of experimentally determined curves showing how polarization varies with water content in sulfide-mineralized granite are given in Fig. 259. Here, it appears that polarization increases with decreasing water content only so long as the water content is not reduced below about 75 per cent of the available pore volume. At lower saturations, the amount of polarization is decreased markedly. It may be that at low water contents, the surfaces of the sulfide grains have a lower affinity for water than the neighboring grain surfaces, and the sulfide surfaces will dry first.


## 42. ELECTROLYTIC POLARIZATION

Polarization which is virtually indistinguishable from electrode polarization occurs in most rocks even if they do not contain any metallic minerals. This type of polarization has come to be known as a *background effect* in mineral exploration applications of induced polarization methods (see Wait, 1959). In the case of electrode polarization, we have seen that the phenomenon of induced polarization arises from the formation of ion accumulations at the surfaces of sulfide grains. Similar ion accumulations may occur in any electrolytic conductor in which the ionic mobility varies from spot to spot. Ions will tend to pile up at the boundaries between regions where mobility varies.

One mechanism by which the mobility of an ion can be changed is by increasing the viscous drag as it moves through water. Many layers of water

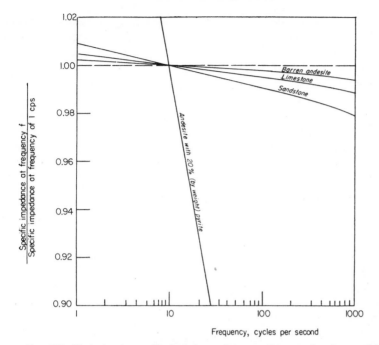

Frequency, cycles per second

FIG. 259. Variation in specific impedance for several types of rock caused by induced polarization (after Wait). The values of specific impedance (resistivity) have been normalized by dividing each value by the value determined at 10 c/s.

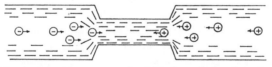

FIG. 260. Illustration of how water adsorbed along the borders of a constriction in a pore may reduce the mobility of ions in solution.

molecules are normally attached loosely to the walls of a pore. When an ion migrates through a fine pore, the forces holding these layers of water to the rock increase the viscosity, which in turn decreases the mobility of ions travelling through the pore. There will be a tendency for ionic charge to pile up on the upstream side of such constrictions, forming anion concentrations on one side and cation concentrations on the other side. Figure 260 illustrates the mechanism of ion concentration by variation in viscosity through a pore structure.

A second mechanism which changes the velocity with which ions move without actually changing their mobility may take place in rocks with fixed electrical charges distributed through the pores. This may take place when the pore structure is lined with a coating of minerals having ion-exchange capacities. When such minerals electrolyze (ionize), the exchangeable ions go into solution, leaving behind a mineral particle, usually clay, which carries a net charge. The common minerals showing exchange properties in rocks are clays, which usually have cations in ion-exchange positions in the lattice. After ionization, the clay particle forms a highly charged, immobile anion, blocking free ion flow through the pore in which it is situated (see Fig. 261). The electric field about such a fixed charge (curve 1 in Fig. 261) is added to the normal potential gradient causing current flow (curve 2), so that there will be areas of abnor-

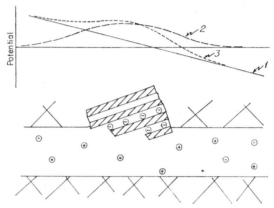

FIG. 261. Illustration of the effect of a charged clay particle. The normal potential gradient along a straight pore is shown by curve 1. The added potential caused by the presence of the clay anion is shown by curve 2. The net potential is given by curve 3.

mally low gradient on one side of a clay anion, and areas of abnormally high gradient on the other side. Ions will move more slowly in areas of low gradient, and in order to maintain continuity of current flow through such zones, the potential gradient must be steepened, and this behavior of the electric field is the same as though a resistance had been added.

At first thought, we might expect the amount of polarization in clay-bearing rocks to increase in direct proportion to the ion exchange capacity. This is not true; clay-rich rocks such as shale show relatively little ability to polarize in comparison with siltstone, which has a lower content of clay minerals. In a rock containing a large proportion of clay, almost all the negative charges may be fixed in exchange positions, with virtually no anions existing in solution. With no anions, the amount of charge which can accumulate at a potential barrier in a rock is small.

The relation between induced polarization and clay content may be predicted in a qualitative way rather simply. As an approximation, we may say that the number of polarization centers (fixed clay anions) will be proportional to the density of fixed charges on the rock framework, $a_0$. This density of fixed charges may be considered to be the ion exchange capacity of the rock, expressed in chemical equivalents per unit pore volume, rather than per unit rock weight, as was done in Chapter I. The constant of proportionality between number of clay particles and total exchange capacity will depend on the type of clay and the manner in which it is distributed in the rock. For example, if clay occurs in small clumps in the pores, it will form fewer potential barriers than if the clay lines the pore structure uniformly. We may also assume that the polarization is approximately proportional to the number of anions in solution which are free to accumulate at the potential barriers, $(a - a_0)/a$.

Using the cation exchange capacities listed in Chapter I for various clay minerals, we may estimate the clay content which will cause the maximum amount of induced polarization to take place in a rock. The amount of polarization should be proportional to the number of potential barriers formed by the clay and to the fraction of the anions free to collect at such barriers:

$$\text{Polarization} \propto \frac{a_0(a - a_0)}{a^2} \qquad (629)$$

Consider a rock which is saturated with water normally containing 100 ppm dissolved solids, and in which one of three clay minerals may occur, illite, montmorillonite or kaolinite. Curves showing the relationship between the amount of one of these clays (for 20 per cent porosity) and the factor $a_0(a - a_0)/a^2$ are given in Fig. 262. Maximum polarization would be observed for small amounts of montmorillonite clay in a rock, from 0·1 to 0·4 per cent, while in the case of kaolinite, which has a much lower exchange capacity than mont-

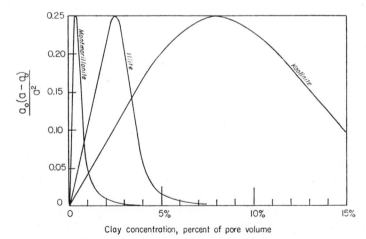

Fig. 262. Anion trap number, $a_0(a - a_0)/a^2$, as a function of clay concentration for an electrolyte with a salinity of 100 ppm. The clay content is expressed as a volume percent of the pore volume.

morillonite, maximum polarization would be observed for a clay content of
3 to 12 per cent. These figures must be considered to be only qualitative, and
in particular, if a higher water salinity were assumed, larger amounts of clay
would be required to cause maximum polarization.

The discussion in the preceding paragraph is based on a highly idealized
model of a clay-bearing rock, but the induced polarization observed in such
rocks does appear to correlate with the "ion-trap ratio" defined in Eq. (629).
The ratio may be computed for a given rock and a given water salinity, $a$,
if the cation exchange capacity, $a_0$, of the rock is known. This quantity may
be measured in the laboratory by flowing a fluid through a rock sample containing some ion which will be preferentially absorbed by the clay and by measuring the amount of absorbed material lost from the flow stream, or cation-
exchange capacity may be estimated from measurements of diffusion potential
across a rock sample. The Nernst equation gives the potential developed between
two solutions when one is diffusing into the other under ideal conditions (the
mobility of none of the ions is hindered):

$$E_0 = \frac{KT}{cF}\left(\frac{\mu_+ - \mu_-}{\mu_+ + \mu_-}\right)\ln\frac{a_c}{a_a} \tag{630}$$

where $K$ is Boltzman's constant, $T$ is the absolute temperature, $c$ is the net
charge per ion, $\mu_+$ and $\mu_-$ are the cation and anion mobilities, respectively,
and $a$ and $a_a$ are the concentrations of the more concentrated and more dilute
solutions, respectively. If the diffusion from the concentrated solution to
the dilute solution takes place through a membrane (such as a sample of rock
might form) which inhibits the mobility of the ions, the actual potential difference between the two solutions differs from the Nernst potential by an amount:

$$E_{PB} = \frac{KT}{cF}\ln\left[\frac{a_0}{a} + \sqrt{\left\{1 + \left(\frac{a_0}{2a}\right)^2\right\}}\right] \tag{631}$$

This quantity, called the phase boundary potential by McCardell and Winsauer
(1953) may be used to determine the cation exchange capacity of a rock. It
is necessary to place a rock sample as a barrier between two brine reservoirs
for which the salinity is known, and measure the potential which develops
across the rock. Equation (631) may then be solved for the anion exchange capacity, $a_0$. The value determined in this manner is more closely related to the
number of ion barriers contributing to induced polarization than is the ion
exchange capacity determined by other means, since the ion migration is
along the same pore channels in both cases, and only the exchange capacity
of clay particles exposed along these channels is measured.

Figure 263 shows the correlation between two sets of data measured on
sandstone and siltstone samples. The ion exchange capacities were deduced
from membrane potential measurements made in the manner described above,
and the ion-trap number given by Eq. (629) was calculated. This ion-trap

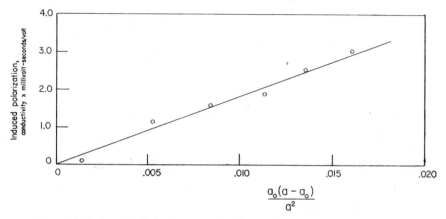

Fig. 263. Induced polarization measured on sandstone samples (taken from the Triassic Chinle Formation in northern Arizona) plotted as a function of the cation exchange capacity, $a_0$, and the total salinity of the pore water, $a$. Each plotted point represents an average of five to ten measurements.

number correlates closely with the induced polarization measured on the same samples, suggesting the idealized model of a clay-bearing rock considered earlier is reasonable, at least for these samples.

The diffusion of ions away from zones where they become concentrated during current flow obeys Fick's law, the application of which has been discussed in detail by Jost (1952). Diffusion of ions through the pores of a rock should approach closely the simple mathematical model of diffusion in a straight capillary tube, since the distance of diffusion is probably short. It is not likely that a significantly large accumulation of ions will develop during current flow, considering the low velocity with which ions in an electrical current flow (see Chapter I). If ions travel only a short distance during the development of the zones of concentration, they will need to travel only a short distance in the opposite direction during recovery, and the pore structure around each zone of accumulation should look reasonably like a capillary tube.

According to Jost, the equation describing the diffusion of ions away from an accumulation in a simple one-dimensional geometry, such as is involved in a capillary tube, is:

$$C(x, t) = \frac{C_0}{2}\left[1 - \operatorname{erf}\left(\frac{x}{2\sqrt{(Dt)}}\right)\right] \tag{632}$$

where $C(x, t)$ is the increase in concentration above normal as a function of $x$, the distance along the capillary from the zone of accumulation, and of $t$, the time following the start of diffusion; $D$ is a diffusion coefficient depending on the species of ions diffusing and on the properties of the electrolyte and $C_0$ is the initial excess concentration of the ions in the zone of accumulation.

The time rate of change of the ion concentration, divided by Faraday's number, is the diffusion expressed as an electrical current flow:

$$I = \frac{C_0}{4F} \cdot \frac{x}{\sqrt{(\pi D t^3)}} \exp\left(-\frac{x^2}{4Dt}\right) \tag{633}$$

The electrical current, I, flowing past a point at a given distance, $x$, from the zone of accumulation varies with $x$. This means the apparent current is not constant along the capillary tube, but is greatest close to the starting point for diffusion. On a megascopic scale, each ion barrier will develop a small current dipole when current flow in the ground is interrupted, and it is the net effect of all such current dipole sources that is observed as the bulk-induced polarization of a rock. The dipole moment of each current source might be found by integrating the current given in Eq. (633) along the distance $x$ through which it arises. The equation is a definite integral which may be evaluated readily if the limits are taken from $x = 0$ to $x = \infty$. Although these limits seem unreasonably large, most of the integral value will be contributed for small values of $x$, so the extension of the range in integration in $x$ to $x = \infty$ does not affect the value of the integral, but makes its evaluation possible:

$$\int_0^\infty I \, dx = \frac{C_0}{4Ft\sqrt{\gamma\pi}} \tag{634}$$

Equation (634) predicts that the source current contributing to induced polarization will decrease as the inverse first power of time. Generally, this is not observed to be the case, but the divergence from this simple theoretical model may result from the fact that the pores cannot be thought of as simple capillary tubes.

## 43. ELECTROKINETIC POLARIZATION

Charge may be separated mechanically when there is motion of an electrolyte through a pore structure. The free liquid in the center of rock pores is usually enriched in ions of one sign, while the water adsorbed to the surfaces of the pores is enriched in ions of the opposite sign (see Fig. 264). Anions are usually adsorbed on the solid surfaces in silicate rocks, and so, the free pore water is enriched in cations and carries an excess negative charge. If current passes through the pore structure, the mobile cations in the center of the pore structures carry along many molecules of water, so that electrical current flow causes a simultaneous fluid flow. This phenomenon is known as *electrokinesis*. When the current flow is interrupted externally, inertia will cause the fluid flow to continue until it is damped out by frictional forces. Since the moving fluid carries an

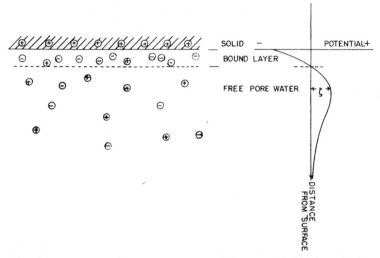

FIG. 264. Nature of the zeta-potential established by the preferential adsorption of anions on the surface of a pore.

excess negative charge, it also represents a transient decrease in electrical current following the interruption of external current flow.

A relation between the electrokinetic phenomena and rock texture may be deduced by considering von Karmen's equation for electro-osmotic velocity and Darcy's equation for hydraulic velocity (see Pirson, 1947). When a voltage is applied across a bundle of capillaries, such as a rock pore structure, the terminal velocity for the fluid flow set up by the migration of ions is:

$$v_0 = E \frac{\varepsilon \zeta}{4\pi\eta} \tag{635}$$

where $\varepsilon$ is the specific capacity of the liquid, $\eta$, is the viscosity of the liquid, $E$ is the applied potential gradient along the capillary, and $\zeta$ is the zeta potential, defined as the voltage between the adsorbed ions on the capillary surfaces and the ions in the free part of the liquid far removed from the adsorption zone. If the external electric field is removed at the instant following the interruption, the fluid will continue at the same velocity. This flow velocity (given by Eq. (635)) also constitutes an electrical current equal to:

$$I_0 = E \frac{a_0 \varepsilon \zeta}{4\pi F \eta} \tag{636}$$

where $F$ is the Faraday constant, and $a_0$ is the concentration of ions adsorbed on the solid surface. This current results in a voltage according to Ohm's law. Since the pore structure follows a devious route through the rock, the voltage gradient along a bulk sample will be reduced by a factor equal to the tortuosity of the pore paths, tortuosity being defined analytically as the ratio of the length along a pore divided by the straight line length of a sample. Thus, Eq. (636) may be used to calculate the transient electric field accompanying inertial fluid flow:

$$e_0 = E \frac{a_0 \varepsilon \zeta \varrho}{4\pi T F \eta} \tag{637}$$

According to Archie's law (discussed in Chapter I), we have the empirical rules for a granular rock:

$$T \cong \varphi^{-1}, \quad \varphi \text{ being fractional pore volume}$$

$$\varrho \cong \varrho_w \varphi^{-2} \tag{638}$$

so that the transient electric field may be expressed as:

$$e_0 \cong E \frac{a_0 \varepsilon \zeta \varrho_w}{4\pi \varphi F \eta} \tag{639}$$

$\varrho_w$ being the resistivity of the pore water.

To obtain the equation for the velocity of fluid flow after the exciting electric field has been removed, a viscous force proportional to fluid velocity is equated to the product of fluid density and fluid acceleration. The solution to such a differential equation is:

$$v = v_0 \exp\left(\frac{-\eta}{K} t\right) \tag{640}$$

where $t$ is the length of time following the interruption of the exciting voltage and $K$ is the coefficient of hydraulic permeability of a rock, defined by Darcy's law. Since the electrokinetic current is proportional to fluid velocity, the transient voltage decays according to the equation:

$$e = E \frac{\varrho_w a_0 \varepsilon \zeta}{4\pi \varphi F \eta} \exp\left(\frac{-\eta}{K} t\right) \tag{641}$$

In terms of the function $\int\limits_0^\infty \dfrac{e}{E_{ss}} \, dt$ used to represent the effects of induced polarization, we have:

$$\int\limits_0^\infty \frac{e}{E_{ss}} \, dt = \frac{\varrho_w a_0 \varepsilon \zeta K}{4\varphi \pi F \eta^2} \tag{642}$$

No induced polarization effects attributable to electrokinetic phenomens have been reported, though Marshall and Madden (1959) present some indirect measurements which suggest that ordinarily the induced polarization due to electrokinesis would be small compared to the induced polarization in the same types of rocks caused by the ion sieving process described in the preceding section. Mayper (in Wait, 1959) states that the time constants involved in the inertial decay of electrokinetic effects are far too long to explain observed data, but offers no proof.

By examining Eq. (642), we may establish the conditions for which electro-kinesis may produce significant induced polarization. The effect would be largest in rocks with a high cation adsorption capacity, high permeability and saturated with high resistivity pore water (this assures that the zeta potential will be large). The rocks most likely to meet these requirements are silt-stones, the rock type which is expected also to show the greatest amount of induced polarization by ion sieving. It is unlikely that polarization due to elec-trokinesis can be identified readily, but the phenomenon must contribute a portion of observed induced polarization effects.

Mayper suggests a second way in which electrokinesis can contribute to induced polarization. If the capillary pores through which electrokinetic flow takes place have bulges, or variations in diameter, the velocity of flow will vary from point to point. Since the moving stream is charged, a variation in flow velocity will cause an accumulation of ions at the boundary between a zone of low flow-velocity and a zone of high flow-velocity. When flow is interrupted, ions will diffuse away from such zones of accumulation, just as they would in any of the polarization mechanisms described previously. It would not be possible to distinguish polarization caused by such a phenomenon from the polarization due to ion sieving.

## 44. MEASUREMENT OF INDUCED POLARIZATION ON ROCK SAMPLES

There is no uniformity of opinion among people investigating the pheno-menon of induced polarization about the units in which the data should be expressed. If measurements are made of the change in specific impedance as a function of frequency, the magnitude of induced polarization may be expressed in terms of a percentage change in impedance per decade change in frequency, or some similar expression. If the transient decay of voltage following the inter-ruption of current flow is measured, the magnitude of induced polarization may be expressed in a great many ways; the ratio of transient voltage to steady state voltage may be taken at some specific time interval following the end of cur-rent flow, the transient voltage may be extrapolated backward to determine the ratio to steady state voltage instantaneously after current flow was inter-rupted, or the transient voltage may be integrated and compared with the steady state voltage. All measure essentially the same parameter of a rock, but the variety of units used makes it difficult to compare the meanings of the various measurements made on the same rock.

In addition to the problem of deciding the best manner in which to express induced polarization data, measurements made on samples in the laboratory are subject to some serious experimental errors, and care must be taken to eliminate these errors. In particular, the induced polarization which always takes place at the electrodes supplying current to the sample must be separated from the induced polarization which takes place within the volume of the sample, polarization at the measuring electrodes must be avoided, and if measurements are being made on mineralized samples, the fact that induced polarization is dependent on current density must be taken into account.

In order to minimize the amount of electrode polarization at the external contacts to the sample which is confused with internal polarization, four-terminal electrode systems must be used in making measurements on samples. Moreover, the terminals must be arranged so that no current flows through the contacts used for measuring voltage when current is being supplied to the sample. An elegant sample holder and electrode system for measuring induced polarization has been described by Mayper (in Wait, 1959, and see Fig. 265). A sample of the rock to be tested is cut in the form of a disc and mounted in a plastic plate, which in turn is placed between two brine reservoirs, which are used to provide a path for current flow to the sample. The current electrodes are placed in separate electrolyte reservoirs connected to the reservoirs in contact with the sample through a small canal, so that contamination of the brine in contact with the samples is minimized. The voltage drop across the sample is measured with two non-polarizing electrodes inserted in the electrolyte reservoirs in contact with the samples.

The reason for separating the current electrode chambers from the sample through a small tube is to obtain a reduction in current flow through the sample when the polarization at the current electrodes discharges. During current flow, polarization takes place at the faces of the current electrodes. When current flow is interrupted, this polarization can dissipate in two ways; charged ions in front of an electrode can combine with electronic charge in the electrode (oxidation or reduction), or the charged ions can migrate away from the electrodes through the volume of the rock. If the external circuit between the current electrodes is broken completely after current flow, the first process is favored, since the second requires that there be a complementary current flow through the external circuit (see Fig. 266).

Although elegant, a system such as that shown in Fig. 265 leads to some difficulties in instrumentation. The circuit resistance between the measuring electrodes can be large, and it may be difficult to measure the small voltages involved in induced polarization in the presence of the noise levels commonly existing in laboratories. The electrode system shown in Fig. 267 is somewhat more subject to errors from electrode polarization, but is less subject to noise problems in the laboratory. A sample is prepared in the form of a thin disc, as before, though with as large a surface area as possible. The disc is inserted between two plate electrodes, each of which has a small area in the center electrically isolated from the rest of the plate. The main portion of the plate electrodes provides current to the sample, while the small insulated sections serve as measuring electrodes.

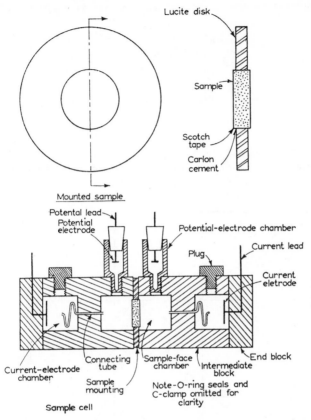

Lucite disk

Sample

Scotch tape

Carlon cement

Mounted sample

Potental lead
Potential electrode

Potential-electrode chamber

Plug

Current lead

Current eletrode

Current-electrode chamber

Connecting tube

Sample mounting

Sample-face chamber

Intermediate block

End block

Note-O-ring seals and C-clamp omitted for clarity

Sample cell

FIG. 265. Sample holder for making measurements of induced polarization (from Mayper, in Wait, 1959).

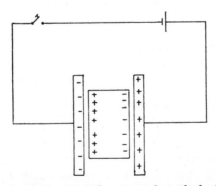

FIG. 266. Electrode polarization at the current electrodes leads to an accumulation of charge as shown. After current flow is interrupted, this accumulation discharges either by recombination across the interface between the electrodes and the sample, or by current flow within the sample accompanied by a complementary flow through the external circuit.

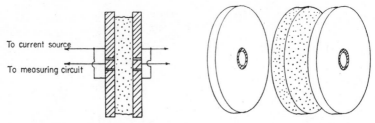

To current source

To measuring circuit

FIG. 267. Four-terminal sample holder which may be used with high-resistance samples. Insulated buttons in the face of the main current electrodes serve as measuring electrodes.

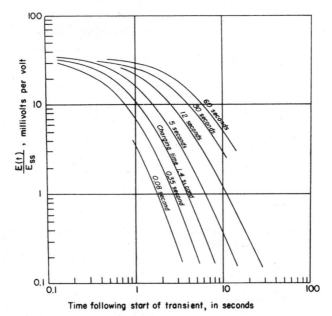

FIG. 268. Transient voltage decay observed on a sample of sandstone, as a function of the duration of charging current.

## 44 a. Pulse-transient measurements

One method for measuring the effects of induced polarization is to record the transient decay of voltage following the interruption of current flow through a sample. The general character of this transient voltage has already been described (see Fig. 248 (a)); the transient voltage starts from some initial value, $E_0$, which is less than the steady state voltage, $E_{ss}$, observed between the measuring electrodes during current flow, and decreases from this initial value gradually, but in a way that apparently cannot be described with a simple mathematical expression.

The shape of the decay curve for the transient voltage will depend on the length of time that current had been circulating in a sample before its flow was interrupted. A typical set of voltage transient curves for a variety of charging or excitation times is shown in Fig. 268, for pulse-transient type measurements made on a sandstone sample. Charging intervals ranged from less than a tenth second to a minute, with the observation that the longer the charging time, the longer was the transient voltage observable. These discharge curves differ from the discharge curve for the voltage across a simple condenser, in that the rate of discharge for long time intervals after the start of the transient is much less than that for an exponential decay.

The initial voltage from which the transient decay starts appears not to be dependent on the charging time for a sample. For example, all the curves shown in Fig. 268 appear to start from an initial voltage of 35 millivolts per volt of steady state condition. This parameter, then, might be a useful one in describing the induced polarization response of a rock. Typical values for this parameter, the initial level of the transient voltage, are as follows:

| Rock | $E_0/E_{ss}$ (mV/V) |
|---|---|
| Mineralized rock, more than 20% sulfide content | over 200 |
| Mineralized rock, between 8 and 20% sulfide content | 100 to 200 |
| Mineralized rock, between 2 and 8% sulfide content | 50 to 100 |
| Volcanic tuffs | 30 to 80 |
| Sandstone, siltstone | 10 to 50 |
| Greenstone, trap, dense volcanic rocks | 10 to 50 |
| Shale | 5 to 10 |
| Granite, granodiorite | 1 to 5 |
| Limestone, dolomite | 1 to 5 |

Although this parameter appears to be useful in describing induced polarization effects, it has certain disadvantages, particularly if the measurement is to be made in the field. In order to obtain an accurate value for the initial transient voltage, the voltage must be measured immediately after current flow is interrupted. This can be accomplished within a matter of 0·01 second in the laboratory, this small delay being caused by the finite time required to switch circuits, and by the finite decay time for polarization voltages due to atomic and molecular phenomenon, such as were described briefly in Section 8, Chapter I. In the field, transients caused by capacitive coupling between cables may mean that the transient voltage cannot be measured until some tenths of seconds following its start. In order to avoid this problem some field measurements are made at a specified time following the start of the transient. The voltage so measured is less than the initial transient voltage, $E_0$, by an amount depending on the charging pulse duration, and on the shape of the transient discharge curve. If measurements are always made consistently, this is no problem. However, it is difficult to compare such data with data obtained in other ways.

There has been a search for an analytical expression describing the shape of the voltage decay curve to be used in describing the effects of induced polarization. If induced polarization is indeed caused by ion accumulations within a rock, then according to Eq. (633), the transient decay curve should be a simple exponential curve. However, the time constant for exponential decay depends on the diffusion coefficient, $D$, and on the dimensions of the pore structures, so it may be that we need to consider a system in which there are many individual time constants controlling the transient current flow. Each individual time constant for diffusion will be determined by the properties of the pore structure at the point where the ionic charge has accumulated during current flow. The subject of transient voltage decay in materials characterized by a variety of discharge time constants has been discussed in detail by Hartshorn (1926).

Consider a system in which there are a number of sources of current contributing to the transient voltage decay, each being a simple exponential decay form. The total transient voltage decay, if all the currents are electrically in series, would be:

$$E(t) = E_0 \sum_{n=1}^{\infty} A_n e^{-u_n t} \tag{643}$$

where $A_n$ is the amplitude coefficient of the $n$th term and $u_n$ is the reciprocal of the time constant for that term. This summation can readily be extended to the case of a large number of individual transient current sources by considering a continuous spectrum of time constants:

$$E(t) = E_0 \int_0^{\infty} G(u) e^{-ut} dt \tag{644}$$

where $G(u)$ is a function describing the fraction of all the time constants which fall in a range $u$ to $u + du$.

Wagner (1913) evaluated the integral in Eq. (644) using a lognormal distribution of time constants. The lognormal distribution is one frequently encountered in studying natural phenomena, and may be suitable for describing the distribution of time constants for the movement of ions away from zones of accumulation. The expression for the lognormal distribution used by Wagner was:

$$G(u) = \frac{b u_0}{\gamma \pi} \exp\left[ -b^2 \left( \ln \frac{u}{u_0} \right)^2 \right] \tag{645}$$

where $u_0$ is the reciprocal of the mean time constant in the distribution, and the parameter $b$ is given by:

$$b = \frac{1}{\sqrt{(2\sigma^2)}} \tag{646}$$

$\sigma$ being the standard deviation of the values of $\ln(u)$. Transient voltage decay curves computed for a variety of distribution parameters, $b$, are shown in Fig. 269. The curve for an exponential decay with a single time constant is shown for comparison.

In comparing measured transient voltage curves with the curves plotted in Fig. 269, it is usually found that a curve with a distribution parameter in the range $b = 0 \cdot 3$ to $b = 0 \cdot 6$ will describe the data. Apparently there is little range in the shape of a voltage transient curve from one rock to another, and

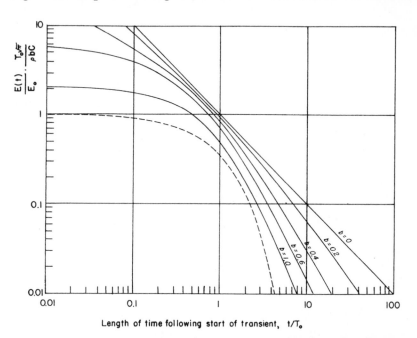

Length of time following start of transient, $t/T_0$

FIG. 269. Transient voltage curves for a system with a lognormal distribution of time constants. $T_0$ is the principle time constant in the distribution, $\varrho$ is the resistivity, $b$ is the distribution parameter for time constants, defined in the text, and $C$ is the apparent d.c. capacity of the sample. A simple exponential curve (dashed line) is shown for comparison.

for this reason, the parameter is of little use in distinguishing between rock types.

In summary, it appears that the transient voltage observed following the interruption of current flow in a rock can be described in terms of three parameters:

1. The ratio of initial transient voltage, $E_0$, to the steady-state voltage, $E_{ss}$. This parameter appears to be a sensitive index to the amount of polarization taking place in a rock.

2. The principle time constant, $T_0$, for the assumed lognormal distribution of relaxation times for the polarization. This time constant ranges from a few seconds to several hundred seconds, but does not relate to any readily observable property of a rock.

3. The distribution parameter, $b$. Rocks nearly all have about the same value for the distribution parameter, and so it does not provide information about differences between rocks.

The observation of the transient voltage decay used in determining these three parameters requires the use of a graphic recording system which is both expensive and cumbersome. For this reason, some investigators make use of the integrated transient voltage, performing the integration electronically, so that the integral can be read from a meter.

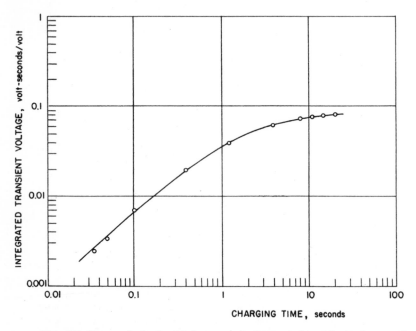

Fig. 270. Observed relationship between the integrated transient voltage and charging time for a sandstone sample. The integral is expressed in volt-sec per volt.

Considering the curves in Fig. 268, it is apparent that the time integral of the transient voltage will also be a function of the charging time for the sample. A typical charging curve—plot of the integrated transient voltage as a function of charging time—is shown in Fig. 270. For long charging times (tens of seconds), it is found that the integrated transient voltage approaches a saturation voltage, which may be used to indicate the extent of polarization in a rock. Typical values for the integrated transient voltage, per unit applied steady-state voltage, are as follows:

| Rock | $\displaystyle\int_0^\infty \frac{E(t)\,dt}{E_{ss}}$ , volt-sec per volt |
|------|----------------|
| Mineralized rock, more than 20% sulfide content | 2 to 5 |
| Mineralized rock, between 8 and 20% sulfide content | 1 to 2 |
| Mineralized rock, between 2 and 8% sulfide content | 0·5 to 1 |
| Volcanic tuffs | 0·3 to 0·8 |
| Sandstone, siltstone | 0·1 to 0·5 |
| Dense volcanic rocks | 0·1 to 0·5 |
| Shale | 0·05 to 0·1 |
| Granite, granodiorite | 0·01 to 0·05 |
| Limestone, dolomite | 0·01 to 0·02 |

The relation between the charging curve, shown in Fig. 270, and the transient discharge curves, shown in Fig. 268 and 269, can be found readily by integrating the expression for transient voltage in Eq. (644):

$$\int_0^\infty \frac{E(t)}{E_{ss}}\,dt = \frac{E_0}{E_{ss}}\left[ \int_{t=0}^\infty \int_0^\infty G(u)\,e^{-ut}\,du\,dt - \int_{t=\Delta}^\infty \int_0^\infty G(u)\,e^{-ut}\,du\,dt \right]$$

$$= \frac{E_0}{E_{ss}} \int_{t=0}^\Delta \int_0^\infty G(u)\,e^{-ut}\,du\,dt \qquad (647)$$

where $\Delta$ is the charging interval, and the first term represents the contribution to the total integral from the transient following the end of the current pulse, and the second term represents the contribution from the transient charging curve associated with the leading edge of the current pulse. Curves for the transient integral as a function of charging time and of distribution parameter, $b$, are given in Fig. 271.

If it can be considered that the transient voltage represents a transient current flow in a rock, the time-integral of the transient voltage can be identified with a capacity for storing electrical charge:

$$\int_0^\infty \frac{E(t)\,dt}{E_{ss}} = \int_0^\infty \frac{i(t)\,R\,dt}{E_{ss}} \qquad (648)$$

where $i(t)$ is the transient current in the rock, and $R$ is the resistance encountered by this current flow. If the excitation pulse is long enough to assure satura-

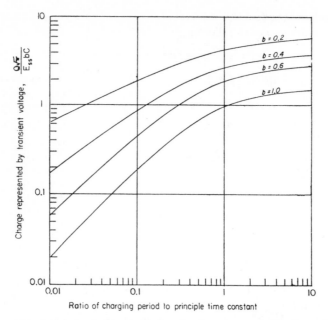

FIG. 271. Integral of the expression for transient voltage for a system consisting of a lognormal distribution of exponential time constants.

tion of the storage capacity of the rock, the time-integral of transient current can be equated with the specific capacity of the rock:

$$\int_0^\infty \frac{E(t)}{E_{ss}}\, \mathrm{d}t = \frac{QR}{E_{ss}}$$

$$= C_a R$$

$$= \varepsilon\varrho \qquad (649)$$

where $Q = \int_0^\infty i(t)\, \mathrm{d}t$, the charge represented by the transient current, and $C_a$ is the apparent specific capacity of the rock for very long charging times.

It is interesting to note that the ratio of integrated transient voltage to steady state voltage is a measure of the electrical properties of a sample which is independent of the sample dimensions. The dielectric constant which appears in Eq. (270) is inversely proportional to the resistivity of a sample, so long as a restricted suite of samples is being considered. If the integrated transient is divided by the resistivity of the sample to obtain a value for the dielectric constant in Eq. (649), the value so found may be extremely large, perhaps as great as $10^8$ times as large as the dielectric constant for air. The concept of expressing overvoltage effects as dielectric constants is the subject of discussion following a paper by Frische and von Buttlar (1957).

### 44b. Variable frequency measurements

An alternative method of studying the phenomenon of induced polarization in rock samples consists of measuring the specific impedance at a variety of frequencies, all low (see Wait, 1959). The same precautions must be taken to avoid electrode polarization when variable frequency measurements are made, as when the pulse-transient technique is used.

When resistivity is measured with an alternating current, it appears to decrease gradually with increasing frequency (see Fig. 8–12). This rate of decrease is slight in most rocks, amounting to one per cent or less decrease in resistivity per decade increase in frequency. In rocks exhibiting anomalously large induced polarization effects, the rate of decrease may be as large a 10 to 20 per cent per decade. In order to use this behavior of effective a.c. resistivity as a measure of induced polarization, it is necessary to measure the resistivity at several frequencies with sufficient accuracy that these small percentage changes will be significantly larger than errors of measurement. The range of frequencies which can be used without encountering trouble with capacitive coupling between current and measuring circuits is a few hundred cycles per second.

Measurements made with the variable frequency technique may be expressed as a percentage change in resistivity per decade frequency change:

$$P = \frac{\varrho_2 - \varrho_1}{\sqrt{(\varrho_2 \varrho_1)}} \times 100\% \tag{650}$$

where $\varrho_2$ is the resistivity measured at one frequency and $\varrho_1$ is the resistivity measured at a frequency 10 times higher. Variable frequency data may be used also to compute the "metal factor" defined by Marshall and Madden (1959):

$$MF = \frac{P \times 10^5}{\sqrt{\varrho_2 \varrho_1}} \tag{651}$$

Typical values for the per cent frequency effect per decade and for the metal factor (from Marshall and Madden, 1959) are as follows:

| Rock | Frequency effect (% per decade) | Metal factor (mhos/m) |
|---|---|---|
| Rock with concentrated sulfides | > 10 | > $10^3$ |
| High-magnetic iron ores | | 20 to 10,000 |
| Porphyry copper ores (2–10% sulfides) | 5–10 | 30 to 3000 |
| Rocks with a trace of sulfide mineralization, less than 2% | 2–5 | 1 to 100 |
| Volcanic tuff | 2–4 | 1 to 300 |
| Sandstone, siltstone | 1–3 | 1 to 30 |
| Basalt | 1–2 | $\frac{1}{2}$ to 7 |
| Granite | 0·1 to 0·5 | < $\frac{1}{2}$ |

Neither the pulse transient method nor the variable frequency method using a four-terminal sample holder will provide usable information on the effect of induced polarization at audio frequencies. Ordinarily, induced polarization measurements in the field are made at extremely low frequencies, generally under 1 c/s. However, it would be of interest to determine whether the effect of induced polarization would be detectable in induction measurements of resistivity (Chapter VI) made at audio frequencies, and in particular, if the degree of induced polarization could be measured with an induction method, thus avoiding many of the problems of electrode polarization. The techniques described so far for measuring induced polarization are limited to use at low frequencies in order that separate current and measuring circuits may be used. At high frequencies, it is difficult to separate the ground reference point of two such circuits.

The variable frequency method may be used with a two terminal sample holder if current flow can be induced in the sample capacitatively, rather than by galvanic contact. This is done by placing a disc of a rock sample to be tested between the plates of an air-gap condenser, but with a layer of some dielectric material covering the surface of the condenser plates, so that moisture from the sample does not contact the metal of the plates (see Fig. 272). The effective

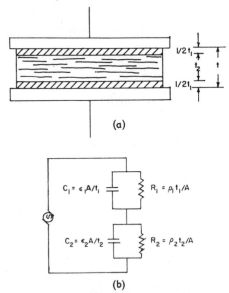

Fig. 272.

(a) Rock sample inserted between the plates of a condenser, with insulating sheets separating the moisture in the sample from contact with the metal plates.

(b) Equivalent circuit for the condenser shown in (a). The area of the sample is $A$.

capacity and a.c. resistivity of the sample along with the dielectric separators can then be measured with any accurate capacitance bridge equipment. The measured properties are those for a composite lossy dielectric, with an equivalent circuit as shown in Fig. 272(b). If the electrical properties of the dielectric separators are known, the electrical properties of the rock sample can be computed. Even with this precaution, the two-electrode method provides less satisfactory data than the four-electrode method.

The relationship between the measured characteristics of the composite condenser and the electrical properties of the two materials in it is given by von Hippel (1954). The composite material appears to have a dielectric constant given by:

$$\frac{\varepsilon}{\varepsilon_0} = \frac{t(\tau_1 + \tau_2 - \tau + \omega^2\tau_1\tau_2)}{A\varepsilon_0(R_1 + R_2)(1 + \omega^2\tau^2)} \tag{652}$$

where
$$\tau_1 = \varepsilon_1\varrho_1; \quad \tau_2 = \varepsilon_2\varrho_2$$

$$\tau = \frac{R_1\tau_2 + R_2\tau}{R_1 + R_2}$$

The dissipation factor is given by:

$$\tan \delta = \frac{\omega\varepsilon}{\varrho} = \frac{t}{A\varepsilon_0} \cdot \frac{C_1C_2}{C_1 + C_2} \cdot \left[1 + \frac{k}{1 + \omega^2\tau^2}\right] \tag{653}$$

where
$$k = \frac{(\tau_1 + \tau_2 - \tau)\,\tau - \tau_1\tau_2}{\tau_1\tau_2}$$

## 45. INDUCED POLARIZATION MEASUREMENTS IN DRILL HOLES

Measurements of induced polarization effects made on samples are subject to a variety of errors in addition to the errors contributed by electrode polarization. In particular, if the properties of mineralized rocks are being studied, alteration of the sulfide mineral surfaces by high density current flow, or merely by contact with oxygen-laden solutions may so alter the properties of a sample that they have no meaning. Inasmuch as the induced polarization measurements made in the field are usually made in the search for such ore-bearing rocks, this is a serious problem. Induced polarization measurements made in drill holes penetrating mineralized rocks offer an approach which avoids many of these problems. Induced polarization logging was not included in Chapter II, since the technique is not currently provided by the logging service companies. Induced polarization logging has been described in detail by Dakhnov and his colleagues (1952).

Either the pulse transient method or the variable frequency method might be used in measurements in bore holes, but a technique employing a repetitious pulse-transient measurement is the only one which has been used extensively.

The instrumentation required for induced polarization logging consists essentially of a switching device that alternately connects and disconnects a voltage supply to a pair of current electrodes in a well. The same system is used also to connect a pair of measuring, electrodes synchronously to a recording system only during the time intervals when current is not flowing in the well.

Any of the conventional electric logging arrays may be used in induced polarization logging, though usually, a short normal-spacing is used. A short spacing is desirable, in that the transient voltage which is to be measured is only a small percentage of the steady-state voltage. It is possible to use a zero spacing in theory, since the mud within the well contributes no induced polarization effects, and a maximum transient voltage is observed when the spacing from current electrode to measuring electrode is less than the radius of the well. If a very short spacing is used, care must be taken that none of the polarizing current flows through the measuring electrodes, polarizing them.

The wave form of the current supplied to the current electrodes in the repeating pulse-transient method is shown in Fig. 273 (a). It consists of a series of pulses of alternating polarity, separated by equal time intervals during which no current flows. The voltage detected at the measuring electrodes is distorted in waveform from that of the current as a result of induced polarization effects

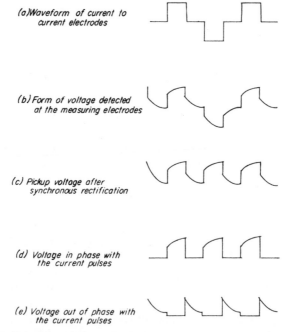

(a) Waveform of current to current electrodes

(b) Form of voltage detected at the measuring electrodes

(c) Pickup voltage after synchronous rectification

(d) Voltage in phase with the current pulses

(e) Voltage out of phase with the current pulses

FIG. 273. Principles of the repeating pulse method for measuring induced polarization.

(see the waveform in Fig. 273 (b)). Induced polarization causes a rounding of the leading edge of each current pulse, and a transient decay following the trailing edge of each pulse. Using relays or an electronic switching system, the portion of the pickup voltage which is detected during each current pulse is separated from the transient voltage detected between the current pulses. Next, each alternate half-cycle of both the "inphase" and "out-of-phase" voltages is reversed in polarity, so that all have the same sense. This reversal procedure, or synchronous rectification, serves to convert the self potential at the pickup electrodes to an alternating voltage which may be filtered from the total signal. Using the voltage detected during the inphase intervals as a reference, the amount of current to the current electrodes is varied in such a way as to maintain the inphase voltage at the measuring electrodes constant. The out-of-phase voltage, then, is the average transient voltage for that value of steady-state voltage.

Repetitious pulses can be treated as an alternating series of step functions, and the transient voltage following each pulse can be computed by taking an infinite series of transient voltages, each obeying Eq. (645):

$$E(t) = E_0 \left[ \int_0^\infty G(u)\, e^{-ut}\, du - \int_0^\infty G(u)\, e^{-u(t+\Delta)}\, du \right.$$

$$\left. + \int_0^\infty G(u)\, e^{-u(t+2\Delta)}\, du - \cdots + \int_0^\infty (-1)^{n+1}\, G(u)\, e^{-u[t+(n-1)\Delta]}\, du \right] \quad (654)$$

or ·

$$E(t) = E_0 \int_0^\infty G(u)\, [1 - e^{-u\Delta} + e^{-2u\Delta} - \cdots]\, e^{-ut}\, du \quad (655)$$

where $\Delta$ is the duration of a single current pulse. When the out-of-phase voltages are averaged, the process is equivalent to integration over the interval $2\Delta$. Therefore, the average out-of-phase voltage detected by a logging system operating at $n$ pulses per second will be:

$$V = \frac{nE_0}{2} \int_0^\Delta \int_0^\infty G(u)\, [1 - e^{-u\Delta} + e^{-2u\Delta} - \cdots]\, e^{-ut}\, du\, dt \quad (656)$$

Numerical values for this function have been computed, with the results shown in Fig. 274.

For rapid pulsing rates, the response of the pulse logging system is independent of both the pulse duration and the distribution parameter, $b$. The response depends only on the ratio $E_0/E_{ss}$. Since the unit period used in Fig. 274 is the

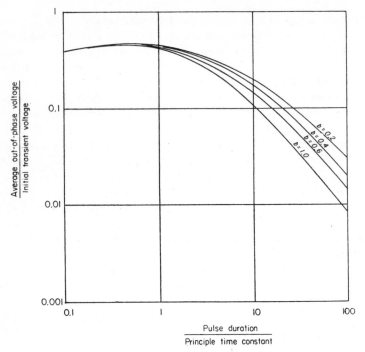

Fig. 274. Output of a repeating pulse-transient logging system, as a function of pulse duration and distribution parameter, $b$.

principle relaxation time for the induced polarization in the rock around the well, which is usually measured in tens of seconds, any reasonable pulsing rate can be considered a rapid pulsing rate. The pulsing rates normally used range from one to ten pulses per second. Higher pulsing rates lead to difficulty with capacitive coupling between the current and measuring circuits, while lower pulsing rates are not practical for a continuous logging device.

Examples of induced polarization logs are shown in Figs. 275 and 276. The logs shown in Fig. 275 were run in a drill hole penetrating highly-altered sandstone and limestone with up to 10 per cent sulfide mineral content. The minerals are disseminated through the rock so that their presence has no effect on the bulk resistivity of the rock. However, the induced polarization detected with a repeating pulse-transient logging system is closely related to the sulfide mineral content. The second example shows induced polarization measurements made in a drill hole penetrating limestone and dolomite, with a section of limestone bearing several per cent sphalerite, zinc sulfide. In this case, there is no anomalous induced polarization response observed in the mineral-bearing rocks. None is observed since sphalerite is one of the few sulfide minerals which is not a conductor. The resistivity of the ore-bearing zone, though, is much lower than that of the rest of the section. This decrease in resistivity is caused by an increased porosity and water content in the ore zone.

## 46. INDUCED POLARIZATION FIELD SURVEYS

Induced polarization surveys are used primarily in the search for sulfide ore deposits, particularly disseminated deposits of such minerals. Induced polarization surveys may be useful in the search for ground water supplies and in detecting beds with fluid permeability, but at present, not enough is known about the use of induced polarization measurements to make such applications feasible. Similarly, the induced polarization method is not used to solve structural problems, since the other geoelectrical methods (galvanic and induction resistivity methods, in particular) are better developed and provide more reliable results. However, the unique response of the induced polarization method over disseminated deposits of conductive minerals makes it extremely

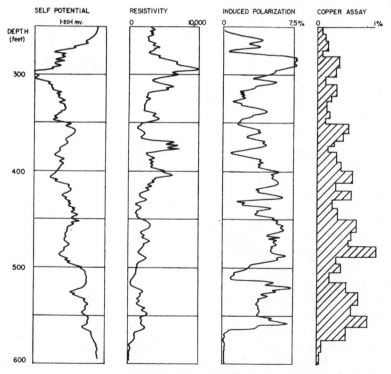

Fig. 275. Electric and induced polarization logs from a drill hole penetrating rocks containing finely disseminated copper sulfides. The host rock is calcareous sandstone.

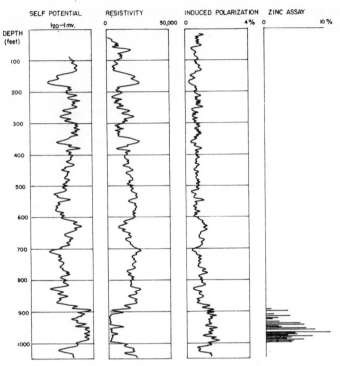

Fig. 276. Electric and induced polarization logs from a drill hole penetrating sphalerite-bearing dolomite in eastern Tennessee.

valuable in a mineral exploration program, since no other geophysical method can locate such ores.

The procedure for measuring induced polarization response in the field is very similar to the procedure used in measuring resistivity. Polarization is excited by a galvanic current driven into the earth through two current electrodes, and the effects of polarization are measured by noting the potential difference developed between two measuring electrodes. As with the galvanic resistivity method, measurements may be made by profiling, to detect changes laterally in the character of the earth, or by sounding, to detect vertical changes in the amount of polarization excited in the rock. Since induced polarization measurements are conducted primarily to find metallic ores, the profiling method is much more commonly used than the sounding method. A second reason that soundings are not commonly made is that induced polarization measurements are subject to a variety of experimental errors, so the data are not amenable to interpretation by sophisticated techniques which require accurate measurements. The induced polarization field survey is used primarily as a qualitative reconnaissance method.

A variety of instrumental techniques may be used to detect the effects of induced polarization in a field survey. Three basically different instrumental methods may be used:

1. *Dual frequency resistivity method:* The application of the variable frequency method described earlier uses two different low-frequency current sources for measuring two values of apparent resistivity for a particular electrode array.

2. *Pulse–transient method:* A pulse of steady direct-current is used to excite polarization in the earth, with the transient voltage decay following the end of the pulse being recorded.

3. *Repeating pulse–transient method:* A steady series of direct-current pulses is used to excite polarization, with the voltage persisting between the pulses being averaged and measured.

*Dual frequency resistivity method.* When resistivity is measured with an alternating current, it appears to decrease slightly with increasing frequency, as described earlier. Field measurements are usually made at two frequencies a decade apart below 10 c/s; that is, 0·1 and 1 c/s, for example. Standard methods for measuring a.c. voltages are difficult to apply with a required accuracy of 1 part in 10,000 in this frequency range. Frequently, a square wave current form may be used rather than a sinusoidally varying voltage, with the current being reversed with mechanical relays or controlled rectifiers. The current may then be measured before commutation (or "chopping"), using d.c. potentiometric methods with a high degree of measurement accuracy. The voltage between the measuring electrodes is rectified using a synchronous set of relays, so that it, too, may be measured with a potentiometer as a d.c. voltage. Chopper rectification also has the advantage that it converts the d.c. level of natural potential which normally exists between the measuring electrodes to an a.c. square-wave voltage, which does not affect the measurement. A typical dual-frequency system is shown in Fig. 277.

*Pulse-transient method.* When direct current flow in the ground is interrupted abruptly, a transient voltage may be observed to persist for many seconds or minutes. A technique for field surveys consists of energizing the ground with a single pulse of current, and observing the characteristics of this decay voltage. In practice, a fairly long current pulse is driven through the ground between a pair of current electrodes, and the voltage between a pair of measuring electrodes is recorded graphically. The current may be measured with an ammeter, rather than with a potentiometer, since small errors are not as important in the pulse transient method as they are in the dual frequency method. The length of the current pulse must be controlled with a reasonable precision ($\pm 5$ per cent), since the amplitude of the transient decay depends on the pulse length for all except pulses of very long duration. This requirement arises since there is a transient associated with the leading edge of the pulse which is of the opposite polarity to the transient following the trailing edge of the pulse. The amplitude (integrated voltage) of the trailing transient is nearly proportional to the duration of the excitation pulse for pulse lengths up to about 10 sec, and becomes independent of the pulse duration only for durations of several

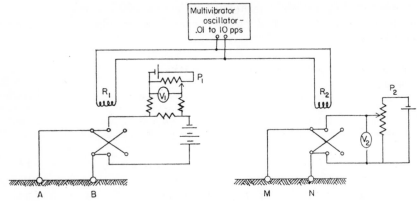

Fig. 277. Typical equipment for measuring the effects of induced polarization using the dual-frequency method. A multivibrator pulse oscillator is used to operate relays $R_1$ and $R_2$ at repetition rates between 0·01 and 10 times per sec. The potentiometer $P_1$ is used to measure the current supplied to the ground, with voltmeter $V_1$ serving to indicate potentiometer unbalance, or percentage deviation in current between measurements made at two frequencies. The potentiometer $P_2$ is used to measure the pickup electrode voltage after it has been rectified with relay $R_2$. Voltmeter $V_2$ serves to indicate potentiometer unbalance or percentage deviation between measurements made at two frequencies.

minutes. Excitation intervals of 1–5 min are commonly used in the "single-shot" pulse transient method, in order that a maximum transient voltage is recorded.

Pulses of this length may be generated in several ways. One method is to use a precision oscillator (for example, 1000 c/s) with an electronic counter which operates a relay when a preset count (or time) is reached. An alternate method is the use of a low rpm d.c. motor which operates a cam to activate a relay, starting and ending the current pulse with the desired duration.

The characteristic of the transient voltage which it would be most desirable to measure would be the initial voltage, $E_0$. Unfortunately, transient voltages which arise from capacitive and inductive coupling mask the electrochemical transient for periods approaching a second after the current is switched off. Either the transient voltage at some specified time following the end of the excitation pulse or the integrated voltage–time product during the transient discharge may be measured instead of the initial voltage. Since the shape of the transient decay curve is very similar from rock to rock, either of these two measurements should show the same behavior as the initial voltage.

If the transient voltage is recorded graphically, the voltage at a specified time following the end of the excitation pulse may be read directly from such a record. The time integral of the transient may be determined by planimeter-ing the area between the transient voltage curve and the constant level of natural potential recorded between the measuring electrodes. Integration may be carried out with an electronic circuit, eliminating the need for a graphic recorder in the field. A simple resistor–capacitor circuit such as that shown in Fig. 278 makes an effective integrating circuit for time durations up to about 1/3 the RC time constant of the circuit. The voltage across the capacitor is proportional to the time integral of the voltage applied to the circuit, providing the input resistance of the meter used to measure the voltage across the capacitor is large compared to the integrating resistance. Such a circuit integrates the total voltage applied, which is normally the sum of a transient voltage and

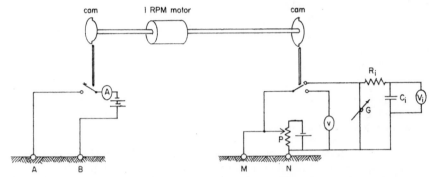

FIG. 278. Typical instrumentation used in the single-shot pulse-transient method for measuring the effects of induced polarization. The length of the excitation pulse is controlled by motor driven cams which open and close the switches as shown. Current to the ground is measured with an ammeter. The self potential level between electrodes M and N is cancelled with potentiometer, P. The steady state voltage is measured with voltmeter, v. The transient voltage is recorded with a graphic recorder, G, or integrated with the circuit $R_i C_i$, the value of the integral being read from the meter $V_i$.

a self-potential level. The self-potential level may be cancelled with a potentiometer which applies a constant voltage of opposite polarity to the measuring electrodes. The RC integrater provides excellent results for integration times up to about 10 sec, and usually, integration of a transient is terminated after that interval. The major part of the voltage–time integral has been covered after 10 sec, in most cases, and prolonged integration leads to errors from two sources:

1. Non-linearity of the integrating circuit when thc integrating time is of the same order as the time constant of the integrating circuit.

2. Inaccuracies in cancelling the self-potential level are more important relatively as the transient voltage decays to small values. The error contributed by an incompletely cancelled self potential is proportional to the integration time.

Fluxmeters may also be used to integrate a transient voltage. A fluxmeter is a ballistic galvanometer (one whose deflection is proportional to the total charge which flows through the coil) without a restoring torque. Since the fluxmeter measures total charge rather than current, it integrates the voltage applied to it. The fluxmeter has an important advantage over an RC integrater, in that the time constant of a fluxmeter is essentially infinite. The integral of a voltage can be stored on a fluxmeter until it is convenient to read it (within reason, since the balance position of a fluxmeter may drift slowly with time), while an RC integrator must be read immediately, or the charge on the condensor begins to leak off. This storage of a transient voltage integral by the fluxmeter is a convenient feature for eliminating the effect of self-potential level. A transient may be excited by current flowing in one direction between the current electrodes, with the sum of the transient voltage and the self potential level being integrated and stored by the fluxmeter, after which an identical excitation current is caused to flow in the opposite direction between the current electrodes, and the sum of the transient voltage and self-potential level is applied with the opposite polarity to the fluxmeter. If there has been no change in the self-potential level, the two contributions due to self-potential level should cancel and the resultant reading on the fluxmeter is just twice the voltage-time integral for one transient.

*Repeating pulse-transient method.* The repeating pulse method described in the preceding section may be used for surface surveys, as well as for bore-hole measurements of induced polarization.

Measurement of induced polarization effects in field surveys is subject to the same sorts of errors that make laboratory measurements of the phenomenon difficult. Fortunately, errors caused by polarization of the measuring electrodes are less important in the field than in laboratory studies, but errors caused by inductive or capacitive coupling are much more serious in the field than in the laboratory. Both capacitive coupling and electromagnetic coupling can lead to results that might be confused with the effects of induced polarization unless precautions are taken to avoid such coupling.

Electromagnetic coupling between current and measuring circuits is the same sort of coupling used in measuring earth resistivity with the inductive methods (see Chapter VI). A time-varying magnetic field generates currents in the earth whose amplitude depends on the frequency of variation, and which may be delayed in phase relative to the primary magnetic field. The measuring circuit used in detecting induced polarization effects also detects the voltage gradient set up by these induction currents. If a variable frequency method is being used, the dependence on frequency of the induction currents leads to an apparent variation in the resistivity of the earth, even if no induced polarization were taking place. In the case of a pulse-transient method, the fact that there may be a phase delay between the induction currents and the primary magnetic field leads to the observation of a transient voltage following the end of the excitation pulse, even if no induced polarization is occurring.

Wait (1959) has outlined the basic theory governing inductive coupling for circuits such as are used in induced polarization surveys, but the theory is no different than that involved in predicting the behavior of induction measurements, discussed in Chapter VI. In particular, if an electrode array is considered in which the cables to the current electrodes and to the measuring electrodes are laid out parallel to one another on the surface of the earth, as shown in Fig. 279, the coupling between the two circuits caused by induction is:

$$Z = \frac{\varrho}{2\pi a} + \frac{i\mu\omega L}{\pi\gamma^2 y^2} \left[1 - \gamma y K_1(\gamma y)\right] \tag{657}$$

where the mutual impedance, $Z$, is the voltage developed in the measuring circuit per unit current flowing in the current circuit, $a$ is the distance between the current electrode and the near measuring electrode, $L$ is the distance from the measuring electrode to the center of the array, $y$ is the separation between the two cables laying on the ground, $\gamma$ is the wave number, as previously defined, and $K_1$ denotes the modified Bessel function of the second type of order 1.

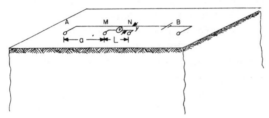

FIG. 279.

Equation (657) may be evaluated for a particular electrode array, but inasmuch as we wish the coupling to be negligible, a numerical evaluation is less important than a consideration of what variables can be used to minimize the coupling. The first term may be recognized as the normal resistive coupling between the current and measuring circuits due to current flow in the ground. Since it is frequency independent, it does not contribute to a frequency effect when a variable frequency technique is employed, unless the resistivity actually varies with frequency. No transient is observed due to the first term when a pulse-transient system is used, inasmuch as there is no phase delay, the term being only real.

The second term in Eq. (657) represents the portion of the observed voltage due to inductive coupling. The contribution depends on frequency, and so contributes to erroneous results when a variable frequency method for measuring induced polarization is used. The term is complex, so there is a phase shift which leads to erroneous transient response when a pulse-transient method is used. The most important single parameter which may be varied to minimize the contribution from this term is.$y$, the spacing between the two cables laying on the ground. The coupling varies as the square of this separation, so that coupling may be minimized by having these two cables as far apart as is possible. The coupling is proportional to the length of the electrode array, $L$, but usually, little can be done about this factor. The spacing between the measuring electrodes usually must be made large enough that there is a reasonable large voltage for the measuring circuit. The coupling is inversely proportional to the resistivity of the earth. Although this parameter is not under the control of the survey team, it does mean that more stringent precautions must be taken to avoid inductive coupling if measurements are to be made over conductive rock.

The coupling predicted by Eq. (657) will be observed if the lines connecting the various electrodes to the recording circuit and the current source are simply straight pieces of wire. In practice, very large amounts of inductive coupling may be observed if a portion of the wire used in the survey is wound on a reel. Serious coupling is observed if the unused wire for the current and measuring circuits is left wound on reels, which are placed side by side. The flux linkage between such reels may be very efficient, resembling that in a transformer if steel reels are used. Good field practice requires that the reel be moved with the outer electrodes in the current circuits, rather than being located at the center of the array with the rest of the equipment. The measuring electrodes should be connected to the recording equipment with a single length of wire of the correct length, rather than having an excess wound on a reel.

Coupling may be reduced and practically eliminated if the dipole–dipole array of electrodes is used (see Chapter III). Referring to Eq. (364) (Chapter VI); it is apparent that there is no induction field outwards along the polar axis from a current dipole. For this reason, the dipole–dipole method has been used widely for induced polarization surveys. However, since short dipoles require a very large current to produce a measurable voltage at the measuring dipole, a system using three equally spaced electrodes is commonly used (see Fig. 280).

Capacitive coupling can lead to erroneous results in induced polarization surveys in much the same way as inductive coupling (Wait, 1959). The capacity between two wires laid on the earth's surface is proportional to the length through which they are common, and decreases with the separation between the wires. This capacitance leads to erroneous detection of voltage in the measuring circuit when the measuring circuit and current circuit have finite grounding resistances. The coupling increases with increasing grounding resistance. In

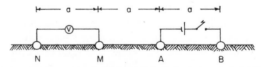

Fig. 280. The equally-spaced dipole–dipole or eltran array of electrodes used to minimize inductive coupling in induced polarization surveys.

order to minimize capacitive coupling, the cables to the current and measuring circuits should be separated as far apart as possible. Capacitive coupling is most serious when measurements are being made over high-resistivity rocks, and can be reduced by taking care to reduce grounding resistance at the electrodes to as low values as is possible, preferably under 100 ohms.

## 47. INTERPRETATION OF INDUCED POLARIZATION SURVEYS

Generally, no other interpretation of induced polarization surveys is attempted than the construction of profiles of the level of activity along traverses, or contour maps of similar data. Usually, such a presentation of the data is adequate to indicate areas of anomalous polarization activity which warrant further investigation. Occasionally, it may be worthwhile to interpret field data to determine relative levels of induced polarization activity, and relate these to the probable variations in conductive mineral content in the rock. Since ore bodies are limited in extent, such an interpretation requires the use of curves computed for induced polarization response over such things as dike-like structures, spheres and horizontal sheets.

The problem of computing the induced polarization response observed over an earth which has a uniform resistivity, but in which layers have different degrees of induced polarization activity can be treated simply. Consider an earth model which is completely uniform in resistivity, in which a surface layer of thickness $h_1$ has no induced polarization activity, and in which the space beneath the surface layer has an induced polarization activity with an amplitude denoted by $A_2$. The effect of induced polarization, when the transient response to a step wave of current is being considered, can be represented by a time-variant polarization current in the earth opposed to the normal direction of current flow during the current step wave, and a depolarization current flowing in the same direction as the original polarizing current following the end of the current step wave. The initial amplitude of the polarization or depolarization current relative to the amplitude of the exciting current is a characteristic of the material called the "induced polarization activity coefficient", and is equal to the ratio $E_0/E_{ss}$ considered earlier.

Considering for the moment, a uniform, fully infinite medium, with a single current source of strength, $I$, the polarization current density, $j_{IP}$, in a volume element, $dv$, is:

$$j_{IP} = Aj_{ss} \tag{658}$$

where $j_{ss}$ is the steady-state current density. On discharge, the sign is reversed. Since we have assumed a uniform medium, the steady-state current density may be expressed simply in terms of the total current, $I$, and the distance, $r$, from the current source to the volume element, $dv$, under consideration:

$$j_{IP} = -\frac{IA_2}{4\pi r^2} \tag{659}$$

The total polarization current in the volume $dv$ is $j_{IP}\,dS$ ($dS$ being an element of area), and forms a current dipole with moment $j_{IP}\,dS\,dr = j_{IP}\,dv$. The potential, $\delta U$, at a point $M$ due to a current dipole of this sort is:

$$\delta U_{IP} = \frac{\varrho j_{IP}\,dv\cos\alpha}{4\pi r_m^2} \tag{660}$$

where $r_m$ is the distance from the current dipole to the observation point, and $\alpha$ is the angle between $r_m$ and the dipole axis.

The total potential due to induced polarization may be found for an observation point at point $M$ by summing the contributions from volume elements such as $dv$. In doing this, a spherical coordinate system centered on the current source is chosen. The angle $\theta$ is taken in the vertical plane between the vector, $r$, and the horizontal plane, and the angle $\psi$ is formed in the horizontal plane between the line connecting points $I$ and $M$ (assuming the observation point is in the same horizontal plane as the current source), and the projection of the vector, $r$. The element of volume is then:

$$dv = r^2\cos\theta\,d\psi\,d\theta\,dr$$
$$r_m = r^2 + a^2 - ra\cos\theta$$

where $a$ is the distance between points $I$ and $M$, and

$$\cos\alpha = \frac{r - a\cos\theta}{r_m} \tag{661}$$

The total potential contributed by induced polarization is:

$$U_{IP} = -\int\limits_0^\infty \int\limits_{-\pi/2}^{\pi/2} \int\limits_0^{2\pi} \frac{\varrho IA_2}{16\pi^2}\,\frac{(r - a\cos\theta)\cos\theta\,d\theta\,d\psi\,dr}{[r^2 + a^2 - ra\cos\theta]^{3/2}} \tag{662}$$

Consider now that space is divided into three regions by two horizontal planes, one a distance $h_1$ above the current source and the other a distance $h_1$ below the current source. The region between the planes is taken to be non-polarizing, while the two regions outside the planes are characterized by an activity coefficient $A_2$. The I.P. potential along the central plane is obtained by integrating Eq. (662) through the limits:

$$\psi:\ 0\ \text{to}\ 2\pi$$
$$\theta:\ -\pi/2\ \text{to}\ \pi/2$$
$$r:\ h_0/\sin\theta\ \text{to}\ \infty$$

The integration in $\psi$ is carried out first:

$$U_{I.P.} = - \int_{-\pi/2}^{\pi/2} \int_{h/\sin\theta}^{\infty} \frac{\varrho I A_2}{8\pi} \frac{(r - a\cos\theta)\cos\theta \, d\theta \, dr}{[r^2 + a^2 - ra\cos\theta]^{3/2}} \tag{663}$$

The integration in $r$ may be carried out using standard forms in tables of integrals:

$$U_{I.P.} = - \int_{-\pi/2}^{\pi/2} \frac{\varrho I A_2}{8\pi h} \frac{\sin\theta\cos\theta \, d\theta}{\left[1 + \left(\dfrac{a}{h}\right)^2 \sin^2\theta - \dfrac{a}{h}\sin 2\theta\right]^{1/2}} \tag{664}$$

This equation may be evaluated numerically. The I. P. potential may be rewritten in the form:

$$U_{I.P.} = - \frac{\varrho I A_2}{8\pi h} F\left(\frac{a}{h}\right)$$

where

$$F\left(\frac{a}{h}\right) = \int_{-\pi/2}^{\pi/2} \frac{\sin\theta\cos\theta \, d\theta}{\left[1 + \left(\dfrac{a}{h}\right)^2 \sin^2\theta - \dfrac{a}{h}\sin 2\theta\right]^{1/2}} \tag{665}$$

For small values of $(a/h)$, $F(a/h)$ approaches the limit:

$$\lim_{a/h \to 0} F\left(\frac{a}{h}\right) = \int_{-\pi/2}^{\pi/2} \sin\theta\cos\theta \left[1 + \frac{a}{2h}\sin 2\theta\right] d\theta$$

$$= \frac{\pi a}{8h} \tag{666}$$

For large values of $a/h$, $F(a/h)$ approaches the limit:

$$\lim_{a/h \to \infty} F\left(\frac{a}{h}\right) = \int_{-\pi/2}^{\pi/2} \frac{h}{a}\cos\theta \, d\theta$$

$$= \frac{2h}{a} \tag{667}$$

In the field surveys, the I.P. potential is compared with the steady-state potential. The curve relating this ratio ($U_{I.P.}/U_{ss}$) to the separation between the current source and the measuring point is shown in Fig. 281. Curves for electrode arrays other than the one potential—one current electrode array considered here can readily be compiled by forming the required differences in potential which are measured with four-electrode methods.

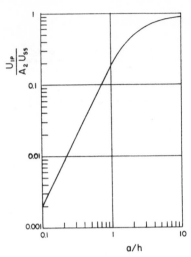

FIG. 281. Variation in induced polarization response for the case of a non-polarizing layer of thickness, $h$, resting on a polarizing substratum with activity coefficient, $A_2$. The curve has been computed for the half-Wenner electrode array.

This relatively simple problem can be expanded to more complex situations without any difficulty. For example, it has been assumed that the surface layer has no induced polarization activity whatsoever. If in fact the surface layer does have some induced polarization activity, this can be considered to contribute a uniform polarization and depolarization current throughout the medium, and it can be taken into account merely by adding a constant level of activity of that computed for the single-layer problem discussed above.

A set of curves showing the variation of observed induced polarization when a constant level of activity is added to that already included in the curve in Fig. 281 is shown in Fig. 282. Similar curves for the case in which the lower medium has a lower activity coefficient than the upper layer can be constructed in the same way, and the curves will be symmetric to those shown in Fig. 282, except the observed induced polarization will decrease with increasing spacing factor, $a$.

The curves in Fig. 282 indicate that it is necessary to use an electrode separation at least several times as great as the overburden thickness to have the observed induced polarization come reasonably close to being the true activity for the second layer. On the other hand, if the polarization contrast between the layer and the lower medium is large, the presence of the polarizing lower medium will be evident even when the electrode spacing is less than half the overburden thickness.

The response which would be measured over a section consisting of a thin layer with a high polarization activity between two zones with no activity can

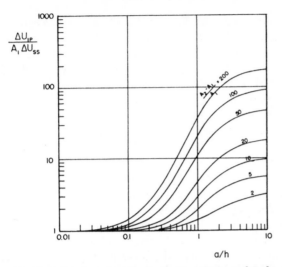

FIG. 282. Variation in induced polarization response for the case of a layer with polarization activity $A_1$ covering a substratum with activity coefficient $A_2$. Curve is computed for the half-Wenner electrode array.

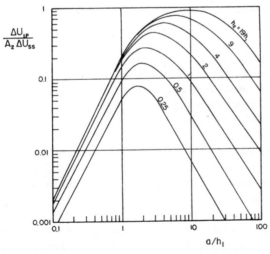

FIG. 283. Variation in induced polarization response for the case of a layer with thickness $h_2$ and activity coefficient $A_2$ lying beneath a layer with thickness $h_1$ and zero activity coefficient. Both layers cover a substratum which also has a zero activity coefficient. Curves are computed for the half-Wenner electrode array.

be obtained in a similar manner. First, the response for the second and third layers is computed (Fig. 281) assuming the third layer has the same activity coefficient as the second layer. Then, it is assumed that the third layer has a negative activity coefficient which just cancels the positive coefficient assumed in the preceding step, and the polarization due to this negative activity is subtracted from that found in the preceding step. The polarization due to the negative activity will be given by the curve in Fig. 281 if the thickness of the first two layers is assumed to be the thickness of the single non-polarizing layer used in computing that curve. The curves shown in Fig. 283 were obtained by performing such a process.

The curves in Fig. 283 show the induced polarization which would be measured for polarizing beds of varying thickness. It is apparent that the thickness of the polarizing bed must be more than ten times the overburden thickness before the observed induced polarization can approach the correct value for the activity coefficient of the polarizing bed for any electrode separation that may be used. If the polarizing layer is only a tenth as thick as the overburden, the peak level of observed polarization will be only about a twentieth the actual activity of the polarizing layer. For layer thicknesses less than the overburden thickness, the shape of the curve relating observed polarization to spacing becomes constant, and it is not possible to distinguish between a very thin bed with a high activity coefficient and a somewhat thicker bed with a lower activity coefficient.

A constant level of polarization activity could be added to these curves to represent the case in which the overburden and substratum also have some activity, as is usually the case in a practical problem. However, the results of such an addition can readily be seen from the curves in Fig. 283. The presence of a constant level of induced polarization activity will serve to limit the minimum activity for which the presence of a thin polarizing layer may be detected. For example, consider the case in which the activity of the polarizing layer is ten times as great as that of the overburden and the underlying medium. Usually, in field surveys, the observed induced polarization must be increased severalfold before the variation can be called an anomaly. Referring to Fig. 283, we see that the thin layer would have to have a thickness of at least half the overburden thickness for this to be true.

A similar approach might be used to compute the induced polarization which would be observed over a dike-like structure or over a buried sphere, both of which are geometrical shapes representing idealized ore bodies which may be the object of a field survey. Neither has been done at the present time.

The computations might also be extended to the case in which there are resistivity contrasts present as well as contrasts in the levels of polarization activity. Such computations would be tedious, inasmuch as it would be necessary to have an analytical expression for the current density for the excitation pulse valid for all space, and then the effect of resistivity changes on the flow pattern of the polarization current would have to be computed. One might predict, however, that if the rock with a high polarization activity were also more conductive than the rest of the rock, a higher induced polarization would be observed than in the case of homogeneous resistivity, since the primary current inducing polarization would have higher than normal density in the polarizing region, leading to higher than normal depolarization current. Similarly, if the polarizing regions have relatively high resistivity, the observed polarization should be less than in the case of homogeneous resistivity. Exact computations for cases in which there are resistivity contrasts have been given by Wait (1959), Frische and von Buttlar (1957) and Seigel (1959).

Under ideal conditions it may be possible to estimate the induced polarization activity coefficient for a rock from the data obtained during a survey. This, of course, is only the first part of a quantitative interpretation. Knowledge of the actual activity coefficient can be useful only if the relation between this coefficient and the ore content of a rock has been established through measurements on samples. In practice, this may be the most uncertain step in interpretation, considering the sources of error in laboratory investigations of induced polarization.

# REFERENCES

ANDERSON, L. A. (1960) Electrical properties of sulfide ores in igneous and metamorphic rocks near East Union, Maine, in *Geological Survey Research*, 1960—*short papers in the geological sciences*, pp. B125–B128.

ANDERSON, L. A. and KELLER, G. V. (1964) A study in induced polarization, *Geophysics* **29** (5), 848–864.

BELLUIGI, A. (1956) Ricerche elettro-idrologiche con polarizzazioni indotte, *Ann. geofis.* (*Rome*), vol. 9, no. 4.

BLEIL, D. F. (1953) Induced polarization: a method of geophysical prospecting, *Geophysics* **18**, (3), 636.

BULASHEVICH, U. P. (1956) Computed field of induced polarization for ore particles of spherical shape, *Izvest. Akad. Nauk S.S.S.R.*, Ser. Geofiz., no. 5, pp. 504–512.

BUTLER, J. A. V. (1951) Electrical Phenomena at Interfaces, Macmillan, New York.

DAKHNOV, V. N. (1959) Geophysical well logging, translated in English in *Quart. Colo. School of Mines*, vol. 57, no. 2, April, 1962.

DAKHNOV, V. N., LATISHOVA, M. G. and RYAPOLOV, V. A. (1952) Well logging by means of induced polarization (electrolytic logging), *Promislovaya Geofizika*, pp. 46–82.

FRASER, D. C., KEEVIL, N. B., JR. and WARD, S. H. (1964) Conductivity spectra of rocks from the Craigmont ore environment, *Geophysics*, **29** (5), 832–947.

FRISCHE, R. H. and VON BUTTLAR, H. (1957) Induced electrical polarization, *Geophysics* **22**, (3), 688 (see also, discussions by J. R. Wait in *Geophysics* **23**, 144–153).

HARTSHORN, L. (1926) A critical resume of recent work on dielectrics. *J. Inst. Elec. Engrs.* (*London*), **64**, 1152–1190.

HENKEL, J. H. (1958) Some theoretical considerations on induced polarization. *Geophysics* **23**, (2), 299.

JOST, W. (1952) *Diffusion in Solids, Liquids and Gases*, Academic Press, New York, 558 pp.

KEEVIL, N. B., JR. and WARD, S. H. (1962) Electrolyte activity; its effect on induced polarization, *Geophysics* **27** (5), 677–690.

KELLER, G. V. (1960) Pulse-transient behavior of brine-saturated sandstones, *U. S. Geol. Survey Bull.*, 1083-D, *pp.* 111–129.

KELLER, G. V. and LICASTRO, P. H. (1959) Dielectric constant and electrical resistivity of natural state cores, *U. S. Geol. Survey Bull.*, 1052-H, pp. 257–285.

MARSHALL, D. J. and MADDEN, T. R. (1959) Induced polarization, a study of its causes, *Geophysics vol.* **24** (4), 790–816.

MAUNG, TAN U. (1960) Measurement of induced polarization with variation of water saturation in core samples, M. Sc. Thesis, Colo. School of Mines.

McCARDELL, W. M., WINSAUER, W. O. and WILLIAMS, M. (1953) Origin of electric potentials observed in wells, *Trans. Am. Inst. Mining Met. Petrol. Engrs.* **198**, 41–50.

SEIGEL, H. O. (1959) Mathematical formulation and theoretical curves for induced polarization, *Geophysics* **24** (3), 547.

SEMINOV, A., FERTSHEV, M. and MALSHEVSKY, V. (1948) Concerning the use of the parameter IP in geophysical exploration, *Materiali TsNIGRI, Geofizika*, vol. 8. Gostoptekizdat.

SCHLUMBERGER, C. (1920) *Etude sur la Prospection electrique du sous-sol*, Gauthier-Villars et Cie., Paris.

SCHUFLE, J. A. (1959) Cation exchange and induced electrical polarization, *Geophysics*, **24** (1), 164.

WAGNER, K. W. (1913) The theory of a heterogeneous dielectric, *Ann. Physik* **40**, 817–855.

WAIT, J. R. (Editor) (1959) *Overvoltage Research and Geophysical Applications*, Pergamon Press, London, 158 pp.

VON HIPPEL, A. R. (1954) *Dielectrics and Waves*, John Wiley, New York, 284 pp.

# APPENDIXES

## APPENDIX A

Apparent resistivity measured with the Wenner array for a single overburden, the substratum being more conductive

| Spacing, $a/h_1$ | $\rho_2/\rho_1 = 0.818$ | 0.667 | 0.539 | 0.429 | 0.333 | 0.250 | 0.176 | 0.111 | 0.0527 | 0.000 |
|---|---|---|---|---|---|---|---|---|---|---|
| 0.8 | 0.9768 | 0.9543 | 0.9326 | 0.9114 | 0.8909 | 0.8710 | 0.8516 | 0.8326 | 0.8141 | 0.7960 |
| 1.0 | 0.9633 | 0.9279 | 0.8938 | 0.8609 | 0.8290 | 0.7981 | 0.7682 | 0.7391 | 0.7108 | 0.6833 |
| 1.2 | 0.9490 | 0.9002 | 0.8534 | 0.8084 | 0.7650 | 0.7233 | 0.6839 | 0.6439 | 0.6061 | 0.5696 |
| 1.4 | 0.9350 | 0.8732 | 0.8142 | 0.7579 | 0.7039 | 0.6522 | 0.6024 | 0.5546 | 0.5085 | 0.4640 |
| 1.6 | 0.9220 | 0.8482 | 0.7783 | 0.7118 | 0.6485 | 0.5881 | 0.5304 | 0.4751 | 0.4222 | 0.3714 |
| 1.8 | 0.9101 | 0.8257 | 0.7462 | 0.6711 | 0.5999 | 0.5324 | 0.4682 | 0.4071 | 0.3488 | 0.2932 |
| 2.0 | 0.8996 | 0.8059 | 0.7182 | 0.6358 | 0.5582 | 0.4850 | 0.4157 | 0.3502 | 0.2880 | 0.2289 |
| 2.2 | 0.8904 | 0.7887 | 0.6940 | 0.6056 | 0.5229 | 0.4452 | 0.3723 | 0.3036 | 0.2387 | 0.1773 |
| 2.4 | 0.8823 | 0.7738 | 0.6734 | 0.5801 | 0.4934 | 0.4124 | 0.3367 | 0.2657 | 0.1991 | 0.1364 |
| 2.6 | 0.8753 | 0.7610 | 0.6558 | 0.5587 | 0.4689 | 0.3854 | 0.3078 | 0.2354 | 0.1677 | 0.1044 |
| 3.0 | 0.8639 | 0.7406 | 0.6283 | 0.5256 | 0.4316 | 0.3452 | 0.2655 | 0.1918 | 0.1236 | 0.0604 |
| 3.2 | 0.8592 | 0.7324 | 0.6175 | 0.5130 | 0.4176 | 0.3304 | 0.2503 | 0.1766 | 0.1085 | 0.0457 |
| 3.6 | 0.8519 | 0.7195 | 0.6005 | 0.4934 | 0.3965 | 0.3083 | 0.2282 | 0.1548 | 0.0876 | 0.0259 |
| 4.0 | 0.8461 | 0.7097 | 0.5882 | 0.4794 | 0.3817 | 0.2935 | 0.2136 | 0.1412 | 0.0751 | 0.0146 |
| 4.4 | 0.8416 | 0.7022 | 0.5790 | 0.4693 | 0.3712 | 0.2833 | 0.2041 | 0.1323 | 0.0674 | 0.0082 |
| 5.0 | 0.8366 | 0.6941 | 0.5691 | 0.4587 | 0.3607 | 0.2733 | 0.1950 | 0.1245 | 0.0610 | 0.0034 |
| 6.0 | 0.8311 | 0.6856 | 0.5590 | 0.4483 | 0.3508 | 0.2644 | 0.1874 | 0.1186 | 0.0567 | 0.0088 |
| 7 | 0.8278 | 0.6804 | 0.5532 | 0.4425 | 0.3455 | 0.2598 | 0.1838 | 0.1159 | 0.0550 | 0.0002 |
| 8 | 0.8255 | 0.6771 | 0.5496 | 0.4390 | 0.3423 | 0.2572 | 0.1818 | 0.1146 | 0.0543 | |
| 10 | 0.8229 | 0.6733 | 0.5454 | 0.4350 | 0.3388 | 0.2544 | 0.1797 | 0.1132 | 0.0536 | |
| 12 | 0.8215 | 0.6712 | 0.5432 | 0.4330 | 0.3371 | 0.2530 | 0.1787 | 0.1125 | 0.0533 | |
| 16 | 0.8200 | 0.6692 | 0.5411 | 0.4310 | 0.3354 | 0.2516 | 0.1777 | 0.1119 | 0.0530 | |
| 20 | 0.8194 | 0.6683 | 0.5402 | 0.4301 | 0.3346 | 0.2510 | 0.1772 | 0.1116 | 0.0528 | |
| 30 | 0.8187 | 0.6674 | 0.5392 | 0.4293 | 0.3339 | 0.2505 | 0.1768 | 0.1113 | 0.0527 | |
| 50 | 0.8184 | 0.6669 | 0.5387 | 0.4288 | 0.3335 | 0.2502 | 0.1766 | 0.1112 | 0.0527 | |

Apparent resistivity measured with the Wenner array for a single overburden, the substratum being more resistant

| Spacing, $a/h_1$ | $\varrho_2/\varrho_1 = 1.2222$ | 1·5000 | 1·8571 | 2·3333 | 3·0000 | 4·0000 | 5·6667 | 9·0000 | 19·000 | large |
|---|---|---|---|---|---|---|---|---|---|---|
| 0·8 | 1·0241 | 1·0490 | 1·0750 | 1·1022 | 1·1307 | 1·1607 | 1·1926 | 1·2268 | 1·2640 | 1·3062 |
| 1·0 | 1·0383 | 1·0782 | 1·1200 | 1·1639 | 1·2104 | 1·2596 | 1·3123 | 1·3694 | 1·4322 | 1·5044 |
| 1·2 | 1·0534 | 1·1095 | 1·1686 | 1·2312 | 1·2977 | 1·3689 | 1·4458 | 1·5298 | 1·6235 | 1·7329 |
| 1·4 | 1·0685 | 1·1408 | 1·2176 | 1·2995 | 1·3872 | 1·4819 | 1·5851 | 1·6991 | 1·8278 | 1·9810 |
| 1·6 | 1·0828 | 1·1708 | 1·2649 | 1·3660 | 1·4753 | 1·5942 | 1·7251 | 1·8713 | 2·0385 | 2·2410 |
| 1·8 | 1·0959 | 1·1987 | 1·3094 | 1·4292 | 1·5597 | 1·7031 | 1·8624 | 2·0424 | 2·2510 | 2·5083 |
| 2·0 | 1·1078 | 1·2243 | 1·3505 | 1·4883 | 1·6395 | 1·8072 | 1·9954 | 2·2103 | 2·4627 | 2·7799 |
| 2·2 | 1·1186 | 1·2475 | 1·3882 | 1·5429 | 1·7143 | 1·9059 | 2·1230 | 2·3737 | 2·6721 | 3·0540 |
| 2·4 | 1·1281 | 1·2683 | 1·4225 | 1·5932 | 1·7838 | 1·9989 | 2·2450 | 2·5321 | 2·8782 | 3·3294 |
| 2·6 | 1·1367 | 1·2871 | 1·4536 | 1·6395 | 1·8486 | 2·0865 | 2·3613 | 2·6852 | 3·0808 | 3·6056 |
| 3·0 | 1·1508 | 1·3190 | 1·5077 | 1·7210 | 1·9646 | 2·2462 | 2·5772 | 2·9757 | 3·4743 | 4·1593 |
| 3·2 | 1·1567 | 1·3325 | 1·5309 | 1·7567 | 2·0164 | 2·3189 | 2·6776 | 3·1133 | 3·6653 | 4·4364 |
| 3·6 | 1·1668 | 1·3558 | 1·5715 | 1·8200 | 2·1096 | 2·4518 | 2·8640 | 3·3746 | 4·0360 | 4·9907 |
| 4·0 | 1·1748 | 1·3747 | 1·6053 | 1·8738 | 2·1905 | 2·5698 | 3·0338 | 3·6184 | 4·3922 | 5·5452 |
| 4·4 | 1·1813 | 1·3903 | 1·6336 | 1·9199 | 2·2613 | 2·6751 | 3·1886 | 3·8464 | 4·7348 | 6·0997 |
| 5 | 1·1888 | 1·4091 | 1·6683 | 1·9775 | 2·3517 | 2·8130 | 3·3965 | 4·1614 | 5·2250 | 6·9315 |
| 6 | 1·1976 | 1·4313 | 1·7110 | 2·0506 | 2·4702 | 3·0003 | 3·6898 | 4·6252 | 5·9839 | 8·3177 |
| 7 | 1·2034 | 1·4465 | 1·7411 | 2·1039 | 2·5599 | 3·1475 | 3·9306 | 5·0251 | 6·6787 | 9·7041 |
| 8 | 1·2074 | 1·4573 | 1·7630 | 2·1439 | 2·6292 | 3·2653 | 4·1308 | 5·3728 | 7·3171 | 11·090 |
| 10 | 1·2124 | 1·4711 | 1·7919 | 2·1985 | 2·7275 | 3·4396 | 4·4418 | 5·9452 | 8·4495 | 13·863 |
| 12 | 1·2153 | 1·4792 | 1·8095 | 2·2329 | 2·7922 | 3·5599 | 4·6691 | 6·3941 | 9·4421 | 16·636 |
| 16 | 1·2182 | 1·4879 | 1·8288 | 2·2720 | 2·8688 | 3·7101 | 4·9720 | 7·0439 | 11·002 | 22·181 |
| 20 | 1·2196 | 1·4921 | 1·8384 | 2·2923 | 2·9103 | 3·7961 | 5·1580 | 7·4827 | 12·223 | 27·726 |
| 30 | 1·2211 | 1·4964 | 1·8486 | 2·3141 | 2·9569 | 3·8980 | 5·3959 | 8·1091 | 14·296 | 41·589 |
| 50 | 1·2218 | 1·4987 | 1·8540 | 2·3262 | 2·9837 | 3·9602 | 5·5554 | 8·5961 | 16·387 | 69·315 |
| 100 | 1·2221 | 1·4997 | 1·8563 | 2·3315 | 2·9958 | 3·9897 | 5·6367 | 8·8825 | 18·060 | 138·63 |

# APPENDIX B

Apparent resistivities measured with the Schlumberger electrode array or the equatorial dipole array for the case of a single overburden

(a) Substratum more resistant than the overburden

| $AB/2h_1$ | $K = 0 \cdot 1$ | 0·2 | 0·3 | 0·4 | 0·5 | 0·6 | 0·7 | 0·8 | 0·9 | 1·0 |
|---|---|---|---|---|---|---|---|---|---|---|
| 0·20 | 1·0002 | 1·0004 | 1·0006 | 1·0008 | 1·0010 | 1·0013 | 1·0016 | 1·0018 | 1·0020 | 1·0024 |
| 0·40 | 1·0015 | 1·0031 | 1·0046 | 1·0064 | 1·0081 | 1·0099 | 1·0118 | 1·0138 | 1·0159 | 1·0183 |
| 0·60 | 1·0048 | 1·0136 | 1·0149 | 1·0202 | 1·0256 | 1·0244 | 1·0291 | 1·0340 | 1·0395 | 1·0451 |
| 0·80 | 1·0104 | 1·0211 | 1·0322 | 1·0437 | 1·0557 | 1·0682 | 1·0813 | 1·0955 | 1·110 | 1·127 |
| 1·0 | 1·0182 | 1·0370 | 1·0565 | 1·0767 | 1·0980 | 1·120 | 1·144 | 1·169 | 1·196 | 1·217 |
| 1·2 | 1·0275 | 1·0565 | 1·0855 | 1·117 | 1·151 | 1·185 | 1·222 | 1·261 | 1·303 | 1·352 |
| 1·4 | 1·0384 | 1·0781 | 1·120 | 1·164 | 1·210 | 1·248 | 1·299 | 1·353 | 1·411 | 1·480 |
| 2·0 | 1·0726 | 1·149 | 1·230 | 1·316 | 1·408 | 1·507 | 1·615 | 1·734 | 1·868 | 2·02 |
| 3·0 | 1·119 | 1·249 | 1·390 | 1·545 | 1·715 | 1·903 | 2·11 | 2·36 | 2·64 | 3·00 |
| 5·0 | 1·170 | 1·364 | 1·584 | 1·837 | 2·15 | 2·48 | 3·29 | 3·41 | 4·07 | 5·00 |
| 7 | 1·192 | 1·417 | 1·681 | 1·996 | 2·38 | 2·85 | 3·45 | 4·22 | 5·30 | 7·00 |
| 10 | 1·206 | 1·453 | 1·754 | 2·12 | 2·60 | 3·20 | 4·02 | 5·15 | 6·88 | 10·00 |
| 12 | 1·211 | 1·466 | 1·785 | 2·17 | 2·68 | 3·36 | 4·28 | 5·63 | 7·78 | 12·00 |
| 15 | 1·214 | 1·477 | 1·806 | 2·23 | 2·77 | 3·52 | 4·58 | 6·20 | 8·97 | 15·00 |
| 20 | 1·218 | 1·487 | 1·826 | 2·26 | 2·86 | 3·69 | 4·91 | 6·88 | 10·54 | 20·00 |
| 30 | | | | | | 3·83 | 5·24 | 7·68 | 12·74 | 30·00 |
| 50 | | | | | | 3·94 | 5·49 | 8·37 | 15·28 | 50·00 |
| Large | 1·22 | 1·50 | 1·86 | 2·34 | 3·00 | 4·00 | 5·67 | 9·00 | 19·00 | Large |

(b) Substratum more conductive than the overburden

| $AB/2h_1$ | $K = -0{\cdot}1$ | $-0{\cdot}2$ | $-0{\cdot}3$ | $-0{\cdot}4$ | $-0{\cdot}5$ | $-0{\cdot}6$ | $-0{\cdot}7$ | $-0{\cdot}8$ | $-0{\cdot}9$ | $-1{\cdot}0$ |
|---|---|---|---|---|---|---|---|---|---|---|
| 0·20 | 0·9998 | 0·9996 | 0·9993 | 0·9991 | 0·9989 | 0·9987 | 0·9986 | 0·9984 | 0·9882 | 0·9980 |
| 0·40 | 0·9985 | 0·9971 | 0·9956 | 0·9943 | 0·9928 | 0·9916 | 0·9903 | 0·9890 | 0·9877 | 0·9865 |
| 0·60 | 0·9953 | 0·9908 | 0·9863 | 0·9820 | 0·9777 | 0·9736 | 0·9695 | 0·9655 | 0·9616 | 0·9577 |
| 0·80 | 0·9899 | 0·9801 | 0·9705 | 0·9692 | 0·9521 | 0·9432 | 0·9345 | 0·9260 | 0·9176 | 0·9094 |
| 1·00 | 0·9824 | 0·9653 | 0·9487 | 0·9325 | 0·9167 | 0·9013 | 0·886 | 0·872 | 0·857 | 0·844 |
| 1·20 | 0·9732 | 0·9437 | 0·9222 | 0·898 | 0·873 | 0·851 | 0·829 | 0·807 | 0·785 | 0·764 |
| 1·40 | 0·9630 | 0·9274 | 0·893 | 0·860 | 0·827 | 0·796 | 0·765 | 0·735 | 0·706 | 0·678 |
| 2·0 | 0·9310 | 0·865 | 0·803 | 0·742 | 0·685 | 0·629 | 0·576 | 0·525 | 0·475 | 0·427 |
| 3·0 | 0·889 | 0·785 | 0·689 | 0·599 | 0·514 | 0·435 | 0·359 | 0·287 | 0·220 | 0·159 |
| 5·0 | 0·849 | 0·714 | 0·593 | 0·484 | 0·385 | 0·296 | 0·215 | 0·131 | 0·075 | −0·0160 |
| 7 | 0·834 | 0·690 | 0·565 | 0·454 | 0·355 | 0·267 | 0·189 | 0·120 | 0·0575 | 0·0033 |
| 10 | 0·826 | 0·678 | 0·551 | 0·440 | 0·342 | 0·257 | 0·182 | 0·114 | 0·0542 | |
| 12 | 0·823 | 0·675 | 0·547 | 0·435 | 0·337 | 0·253 | 0·179 | 0·112 | 0·0536 | |
| 15 | 0·821 | 0·672 | 0·544 | 0·432 | 0·335 | 0·250 | 0·177 | 0·111 | 0·0526 | |
| Large | 0·818 | 0·667 | 0·538 | 0·428 | 0·333 | 0·250 | 0·177 | 0·111 | 0·0526 | |

# APPENDIX C

## Apparent resistivities measured with the polar-dipole array for the case of a single overburden

### (a) Substratum more resistant than the overburden

| $OP/h_1$ | $K = 0.1$ | 0.2 | 0.3 | 0.4 | 0.5 | 0.6 | 0.7 | 0.8 | 0.9 | 1.0 |
|---|---|---|---|---|---|---|---|---|---|---|
| 0.40 | 0.9993 | 0.9986 | 0.9979 | 0.9972 | 0.9964 | 0.9956 | 0.9947 | 0.9938 | 0.9928 | 0.9918 |
| 0.80 | 0.9969 | 0.9937 | 0.9903 | 0.9851 | 0.9829 | 0.9790 | 0.9746 | 0.9699 | 0.9662 | 0.9586 |
| 1.20 | 0.9970 | 0.9936 | 0.9898 | 0.9854 | 0.9804 | 0.9747 | 0.9682 | 0.9604 | 0.9512 | 0.9393 |
| 1.60 | 1.0038 | 1.0070 | 1.0093 | 1.0107 | 1.0110 | 1.0099 | 1.0070 | 1.0018 | 0.9935 | 0.9795 |
| 2.00 | 1.017 | 1.034 | 1.049 | 1.063 | 1.075 | 1.085 | 1.094 | 1.098 | 1.096 | 1.085 |
| 2.40 | 1.034 | 1.069 | 1.100 | 1.132 | 1.162 | 1.189 | 1.213 | 1.231 | 1.242 | 1.237 |
| 2.80 | 1.053 | 1.106 | 1.158 | 1.206 | 1.259 | 1.306 | 1.348 | 1.386 | 1.411 | 1.416 |
| 4.0 | 1.101 | 1.207 | 1.326 | 1.427 | 1.540 | 1.653 | 1.765 | 1.869 | 1.956 | 2.008 |
| 6.0 | 1.152 | 1.317 | 1.496 | 1.691 | 1.902 | 2.13 | 2.37 | 2.62 | 2.85 | 3.00 |
| 10.0 | 1.191 | 1.412 | 1.672 | 1.976 | 2.33 | 2.76 | 3.25 | 3.84 | 4.49 | 5.00 |
| 20 | 1.214 | 1.474 | 1.798 | 2.21 | 2.74 | 3.44 | 4.41 | 5.77 | 8.00 | 10.00 |
| 24 | 1.216 | 1.483 | 1.818 | 2.24 | 2.81 | 3.58 | 4.66 | 6.28 | 9.82 | 12.00 |
| 30 | 1.219 | 1.488 | 1.830 | 2.27 | 2.87 | 3.69 | 4.84 | 6.95 | 10.10 | 15.00 |
| 40 | 1.220 | 1.494 | 1.841 | 2.30 | 2.92 | 3.81 | 5.18 | 7.47 | 11.82 | 20.00 |
| 60 | 1.221 | 1.498 | 1.852 | 2.32 | 2.96 | 3.92 | 5.42 | 7.83 | 14.05 | 30.00 |
| Large | 1.222 | 1.500 | 1.858 | 2.33 | 3.00 | 4.00 | 5.67 | 9.00 | 19.00 | Large |

(b) Substratum more conductive than the overburden

| $OP/h_1$ | $K = -0.1$ | $-0.2$ | $-0.3$ | $-0.4$ | $-0.5$ | $-0.6$ | $-0.7$ | $-0.8$ | $-0.9$ | $-1.0$ |
|---|---|---|---|---|---|---|---|---|---|---|
| 0·40 | 1·0006 | 1·0013 | 1·0019 | 1·0025 | 1·031 | 1·0037 | 1·0043 | 1·0049 | 1·0054 | 1·0059 |
| 0·80 | 1·0029 | 1·0058 | 1·0085 | 1·0111 | 1·0136 | 1·0160 | 1·0183 | 1·0206 | 1·0228 | 1·0261 |
| 1·20 | 1·0026 | 1·0050 | 1·0070 | 1·0087 | 1·0103 | 1·0116 | 1·0126 | 1·0135 | 1·0143 | 1·0146 |
| 1·60 | 0·9955 | 0·9909 | 0·9851 | 0·9793 | 0·9731 | 0·9666 | 0·9597 | 0·9526 | 0·9453 | 0·9377 |
| 2·00 | 0·982 | 0·963 | 0·944 | 0·925 | 0·905 | 0·885 | 0·864 | 0·844 | 0·823 | 0·802 |
| 2·40 | 0·965 | 0·929 | 0·893 | 0·857 | 0·821 | 0·785 | 0·749 | 0·715 | 0·677 | 0·642 |
| 2·80 | 0·947 | 0·894 | 0·841 | 0·788 | 0·736 | 0·685 | 0·634 | 0·583 | 0·535 | 0·484 |
| 3·20 | 0·930 | 0·861 | 0·793 | 0·726 | 0·660 | 0·596 | 0·533 | 0·472 | 0·415 | 0·353 |
| 3·60 | 0·915 | 0·832 | 0·751 | 0·673 | 0·597 | 0·522 | 0·451 | 0·381 | 0·313 | 0·246 |
| 4·00 | 0·902 | 0·807 | 0·716 | 0·628 | 0·544 | 0·462 | 0·385 | 0·310 | 0·239 | 0·169 |
| 5 | 0·877 | 0·761 | 0·653 | 0·551 | 0·455 | 0·365 | 0·283 | 0·205 | 0·128 | 0·062 |
| 6 | 0·861 | 0·733 | 0·616 | 0·508 | 0·408 | 0·317 | 0·233 | 0·155 | 0·085 | 0·019 |
| 10 | 0·834 | 0·690 | 0·563 | 0·452 | 0·353 | 0·265 | 0·188 | 0·119 | 0·056 | |
| 20 | 0·822 | 0·672 | 0·544 | 0·434 | 0·337 | 0·253 | 0·179 | 0·112 | 0·053 | |
| 40 | 0·819 | 0·668 | 0·540 | 0·430 | 0·333 | 0·250 | 0·177 | 0·111 | 0·053 | |
| Large | 0·818 | 0·667 | 0·539 | 0·429 | 0·333 | 0·250 | 0·177 | 0·111 | 0·053 | |

# APPENDIX D

## Auxiliary Curves

Auxiliary curves for resistivity interpretation
(type H)

| $h_2/h_1$ | $\varrho_2/\varrho_1 =$ | 2/3 | 1/2 | 1/3 | 1/5 | 1/10 | 1/20 |
|---|---|---|---|---|---|---|---|
| 0·10 | $h_f =$ | 1·10 | 1·10 | 1·10 | 1·10 | 1·10 | 1·10 |
|  | $\varrho_f =$ | 0·955 | 0·917 | 0·847 | 0·733 | 0·500 | 0·367 |
| 0·20 | $h_f =$ | 1·20 | 1·20 | 1·20 | 1·20 | 1·20 | 1·20 |
|  | $\varrho_f =$ | 0·923 | 0·857 | 0·750 | 0·600 | 0·400 | 0·240 |
| 0·33 | $h_f =$ | 1·33 | 1·33 | 1·33 | 1·33 | 1·33 | 1·33 |
|  | $\varrho_f =$ | 0·887 | 0·860 | 0·665 | 0·498 | 0·308 | 0·174 |
| 0·50 | $h_f =$ | 1·50 | 1·50 | 1·50 | 1·50 | 1·50 | 1·50 |
|  | $\varrho_f =$ | 0·857 | 0·750 | 0·600 | 0·429 | 0·250 | 0·136 |
| 1·0 | $h_f =$ | 2·0 | 2·0 | 2·0 | 2·0 | 2·0 | 2·0 |
|  | $\varrho_f =$ | 0·800 | 0·667 | 0·500 | 0·333 | 0·182 | 0·095 |
| 2·0 | $h_f =$ | 3·0 | 3·0 | 3·0 | 3·0 | 3·0 | 3·0 |
|  | $\varrho_f =$ | 0·750 | 0·600 | 0·429 | 0·273 | 0·143 | 0·073 |
| 3·0 | $h_f =$ | 4·0 | 4·0 | 4·0 | 4·0 | 4·0 | 4·0 |
|  | $\varrho_f =$ | 0·728 | 0·572 | 0·400 | 0·250 | 0·129 | 0·066 |
| 5·0 | $h_f =$ | 6·0 | 6·0 | 6·0 | 6·0 | 6·0 | 6·0 |
|  | $\varrho_f =$ | 0·706 | 0·545 | 0·375 | 0·231 | 0·118 | 0·059 |
| 10 | $h_f =$ | 11·0 | 11·0 | 11·0 | 11·0 | 11·0 | 11·0 |
|  | $\varrho_f =$ | 0·688 | 0·524 | 0·355 | 0·216 | 0·108 | 0·055 |
| 20 | $h_f =$ | 21·0 | 21·0 | 21·0 | 21·0 | 21·0 | 21·0 |
|  | $\varrho_f =$ | 0·677 | 0·512 | 0·344 | 0·208 | 0·104 | 0·052 |
| 30 | $h_f =$ | 31·0 | 31·0 | 31·0 | 31·0 | 31·0 | 31·0 |
|  | $\varrho_f =$ | 0·673 | 0·508 | 0·341 | 0·205 | 0·103 | 0·052 |
| 50 | $h_f =$ | 51·0 | 51·0 | 51·0 | 51·0 | 51·0 | 51·0 |
|  | $\varrho_f =$ | 0·672 | 0·505 | 0·338 | 0·203 | 0·102 | 0·051 |

Auxiliary curves for resistivity interpretation
(type A)

| $h_2/h_1$ | $\varrho_2/\varrho_1 =$ | 1·5 | 2 | 3 | 5 | 10 | 20 | 30 | 50 |
|---|---|---|---|---|---|---|---|---|---|
| 0·10 | $h_f =$ | 1·11 | 1·12 | 1·16 | 1·24 | 1·42 | 1·73 | 2·00 | 2·45 |
|  | $\varrho_f =$ | 1·04 | 1·07 | 1·12 | 1·21 | 1·40 | 1·73 | 2·00 | 2·45 |
| 0·20 | $h_f =$ | 1·21 | 1·24 | 1·31 | 1·44 | 1·74 | 2·22 | 2·65 | 3·50 |
|  | $\varrho_f =$ | 1·07 | 1·13 | 1·22 | 1·38 | 1·72 | 2·22 | 2·65 | 3·50 |
| 0·33 | $h_f =$ | 1·35 | 1·40 | 1·49 | 1·68 | 2·11 | 2·80 | 3·33 | 4·20 |
|  | $\varrho_f =$ | 1·11 | 1·19 | 1·34 | 1·58 | 2·05 | 2·75 | 3·30 | 4·20 |
| 0·50 | $h_f =$ | 1·52 | 1·58 | 1·70 | 1·96 | 2·51 | 3·35 | 4·04 | 5·11 |
|  | $\varrho_f =$ | 1·15 | 1·27 | 1·46 | 1·78 | 2·39 | 3·28 | 4·04 | 5·11 |
| 1·0 | $h_f =$ | 2·04 | 2·12 | 2·31 | 2·68 | 3·48 | 4·60 | 5·65 | 7·21 |
|  | $\varrho_f =$ | 1·22 | 1·41 | 1·73 | 2·24 | 3·16 | 4·48 | 5·48 | 7·07 |
| 2·0 | $h_f =$ | 3·05 | 3·16 | 3·42 | 3·92 | 5·03 | 6·72 | 8·09 | 10·2 |
|  | $\varrho_f =$ | 1·31 | 1·58 | 2·05 | 2·80 | 4·18 | 6·11 | 7·58 | 9·87 |
| 3·0 | $h_f =$ | 4·06 | 4·18 | 4·47 | 5·07 | 6·42 | 8·45 | 10·0 | 12·6 |
|  | $\varrho_f =$ | 1·35 | 1·67 | 2·24 | 3·16 | 4·83 | 7·25 | 9·10 | 11·9 |
| 5·0 | $h_f =$ | 6·07 | 6·21 | 6·55 | 7·20 | 8·76 | 11·2 | 13·1 | 16·6 |
|  | $\varrho_f =$ | 1·40 | 1·77 | 2·45 | 3·60 | 5·84 | 9·00 | 11·4 | 15·1 |
| 10 | $h_f =$ | 11·1 | 11·2 | 11·6 | 12·4 | 14·2 | 17·3 | 20·0 | 24·5 |
|  | $\varrho_f =$ | 1·44 | 1·87 | 2·68 | 4·12 | 7·11 | 11·6 | 15·0 | 20·4 |
| 20 | $h_f =$ | 21·1 | 21·2 | 21·6 | 22·5 | 24·5 | 28·3 | 31·6 | 37·4 |
|  | $\varrho_f =$ | 1·47 | 1·93 | 2·82 | 4·50 | 8·20 | 14·1 | 18·9 | 26·7 |
| 30 | $h_f =$ | 31·1 | 31·2 | 31·6 | 32·6 | 34·6 | 38·8 | 42·5 | 49·0 |
|  | $\varrho_f =$ | 1·48 | 1·95 | 2·88 | 4·65 | 8·65 | 15·5 | 21·2 | 30·6 |
| 50 | $h_f =$ | 51·1 | 51·2 | 51·7 | 52·5 | 54·8 | 59·2˙ | 63·2 | 70·8 |
|  | $\varrho_f =$ | 1·49 | 1·97 | 2·92 | 4·78 | 9·15 | 16·9 | 23·7 | 35·3 |

Auxiliary curves for resistivity interpretation
(type K)

| $h_2/h_1$ | $\varrho_2/\varrho_1 =$ | 1·5 | 2 | 3 | 5 | 10 | 20 | 30 | 50 |
|---|---|---|---|---|---|---|---|---|---|
| 0·10 | $h_f =$ | 1·12 | 1·15 | 1·22 | 1·37 | 1·72 | 2·28 | 2·76 | 3·52 |
|  | $\varrho_f =$ | 1·04 | 1·07 | 1·12 | 1·21 | 1·40 | 1·73 | 2·00 | 2·45 |
| 0·20 | $h_f =$ | 1·23 | 1·28 | 1·42 | 1·66 | 2·28 | 3·12 | 3·78 | 4·81 |
|  | $\varrho_f =$ | 1·07 | 1·13 | 1·22 | 1·38 | 1·72 | 2·22 | 2·65 | 3·50 |
| 0·30 | $h_f =$ | 1·34 | 1·41 | 1·59 | 1·95 | 2·68 | 3·68 | 4·60 | 5·81 |
|  | $\varrho_f =$ | 1·10 | 1·18 | 1·31 | 1·54 | 1·97 | 2·63 | 3·15 | 4·00 |
| 0·50 | $h_f =$ | 1·55 | 1·66 | 1·91 | 2·39 | 3·39 | 4·78 | 5·80 | 7·39 |
|  | $\varrho_f =$ | 1·15 | 1·27 | 1·46 | 1·78 | 2·39 | 3·28 | 4·00 | 5·11 |
| 1·0 | $h_f =$ | 2·08 | 2·25 | 2·61 | 3·33 | 4·73 | 6·58 | 8·08 | 10·2 |
|  | $\varrho_f =$ | 1·22 | 1·41 | 1·73 | 2·24 | 3·16 | 4·48 | 5·48 | 7·07 |
| 2·0 | $h_f =$ | 3·11 | 3·33 | 3·83 | 4·83 | 6·73 | 9·61 | 11·7 | 14·8 |
|  | $\varrho_f =$ | 1·31 | 1·58 | 2·05 | 2·80 | 4·18 | 6·11 | 7·58 | 9·87 |
| 3·0 | $h_f =$ | 4·13 | 4·36 | 4·96 | 6·09 | 8·40 | 11·9 | 14·5 | 18·3 |
|  | $\varrho_f =$ | 1·35 | 1·67 | 2·24 | 3·16 | 4·83 | 7·25 | 9·10 | 11·9 |
| 5·0 | $h_f =$ | 6·12 | 6·40 | 7·12 | 8·22 | 11·3 | 15·6 | 18·8 | 24·1 |
|  | $\varrho_f =$ | 1·40 | 1·77 | 2·45 | 3·60 | 5·84 | 9·00 | 11·4 | 15·1 |
| 10 | $h_f =$ | 11·2 | 11·5 | 12·1 | 13·7 | 17·0 | 22·9 | 27·7 | 35·1 |
|  | $\varrho_f =$ | 1·44 | 1·87 | 2·68 | 4·12 | 7·11 | 11·6 | 15·0 | 20·4 |
| 20 | $h_f =$ | 21·2 | 21·4 | 22·2 | 24·0 | 28·0 | 35·5 | 41·4 | 51·6 |
|  | $\varrho_f =$ | 1·47 | 1·93 | 2·82 | 4·50 | 8·20 | 14·1 | 18·9 | 26·7 |
| 30 | $h_f =$ | 31·2 | 31·4 | 32·2 | 34·2 | 38·8 | 46·5 | 53·5 | 64·9 |
|  | $\varrho_f =$ | 1·48 | 1·95 | 2·88 | 4·65 | 8·65 | 15·5 | 21·2 | 30·6 |
| 50 | $h_f =$ | 51·1 | 51·4 | 52·2 | 54·0 | 58·6 | 66·3 | 74·7 | 89·1 |
|  | $\varrho_f =$ | 1·49 | 1·97 | 2·92 | 4·78 | 9·15 | 16·9 | 23·7 | 35·3 |

Auxiliary curves for resistivity interpretation
(type Q)

| $h_2/h_1$ | $\varrho_2/\varrho_1 =$ | 2/3 | 1/2 | 1/3 | 1/5 | 1/10 | 1/20 |
|---|---|---|---|---|---|---|---|
| 0·10 | $h_f =$ | 1·10 | 1·09 | 1·08 | 1·08 | 1·08 | 1·08 |
|  | $\varrho_f =$ | 0·955 | 0·909 | 0·832 | 0·719 | 0·540 | 0·360 |
| 0·20 | $h_f =$ | 1·18 | 1·17 | 1·16 | 1·15 | 1·15 | 1·15 |
|  | $\varrho_f =$ | 0·907 | 0·840 | 0·725 | 0·575 | 0·383 | 0·230 |
| 0·33 | $h_f =$ | 1·32 | 1·30 | 1·29 | 1·27 | 1·25 | 1·24 |
|  | $\varrho_f =$ | 0·867 | 0·778 | 0·645 | 0·476 | 0·294 | 0·162 |
| 0·50 | $h_f =$ | 1·47 | 1·43 | 1·40 | 1·31 | 1·29 | 1·27 |
|  | $\varrho_f =$ | 0·838 | 0·715 | 0·560 | 0·374 | 0·215 | 0·115 |
| 1·0 | $h_f =$ | 1·93 | 1·82 | 1·65 | 1·62 | 1·62 | 1·62 |
|  | $\varrho_f =$ | 0·773 | 0·610 | 0·412 | 0·270 | 0·147 | 0·077 |
| 2·0 | $h_f =$ | 2·91 | 2·83 | 2·65 | 2·44 | 2·44 | 2·44 |
|  | $\varrho_f =$ | 0·727 | 0·566 | 0·378 | 0·222 | 0·116 | 0·060 |
| 3·0 | $h_f =$ | 3·92 | 3·86 | 3·78 | 3·52 | 3·25 | 3·25 |
|  | $\varrho_f =$ | 0·715 | 0·552 | 0·368 | 0·220 | 0·105 | 0·053 |
| 5·0 | $h_f =$ | 5·95 | 5·88 | 5·80 | 5·70 | 5·35 | 5·12 |
|  | $\varrho_f =$ | 0·700 | 0·535 | 0·362 | 0·218 | 0·104 | 0·051 |
| 10 | $h_f =$ | 11·0 | 10·9 | 10·8 | 10·6 | 10·4 | 10·0 |
|  | $\varrho_f =$ | 0·688 | 0·520 | 0·348 | 0·208 | 0·103 | 0·050 |
| 20 | $h_f =$ | 21·0 | 20·8 | 20·7 | 20·6 | 20·2 | 20·1 |
|  | $\varrho_f =$ | 0·677 | 0·507 | 0·339 | 0·203 | 0·100 | 0·050 |
| 30 | $h_f =$ | 31·0 | 30·9 | 30·8 | 30·6 | 30·4 | 30·3 |
|  | $\varrho_f =$ | 0·644 | 0·504 | 0·336 | 0·202 | 0·100 | 0·050 |
| 50 | $h_f =$ | 51·0 | 51·0 | 50·9 | 50·7 | 50·6 | 50·5 |
|  | $\varrho_f =$ | 0·671 | 0·501 | 0·334 | 0·200 | 0·100 | 0·050 |

# APPENDIX E

## Values for the integral expressed per unit area $dx\,dz$

$$\int\limits_{0}^{\infty} \frac{dy}{[x^2 + y^2 + z^2]^{3/2}\,[(x-a)^2 + y^2 + z^2]^{1/2}}$$

| $Z/a$ | $x/a = 0\cdot1$ | $0\cdot2$ | $0\cdot3$ | $0\cdot4$ | $0\cdot5$ |
|---|---|---|---|---|---|
| $0\cdot1$ | 35·3 | 13·2 | 7·65 | 6·34 | 6·63 |
| $0\cdot2$ | 11·0 | 7·56 | 5·76 | 5·26 | 5·60 |
| $0\cdot3$ | 4·87 | 4·43 | 4·08 | 4·05 | 4·38 |
| $0\cdot4$ | 2·75 | 2·83 | 2·88 | 3·02 | 3·28 |
| $0\cdot5$ | 1·79 | 1·94 | 2·07 | 2·22 | 2·41 |
| $0\cdot6$ | 1·27 | 1·40 | 1·52 | 1·64 | 1·78 |
| $0\cdot7$ | 0·940 | 1·04 | 1·14 | 1·23 | 1·32 |
| $0\cdot8$ | 0·721 | 0·796 | 0·867 | 0·934 | 0·996 |
| $0\cdot9$ | 0·565 | 0·620 | 0·672 | 0·720 | 0·762 |
| $1\cdot0$ | 0·451 | 0·491 | 0·529 | 0·563 | 0·592 |
| $1\cdot2$ | 0·299 | 0·321 | 0·342 | 0·359 | 0·373 |
| $1\cdot4$ | 0·207 | 0·220 | 0·231 | 0·240 | 0·248 |
| $1\cdot6$ | 0·149 | 0·156 | 0·162 | 0·168 | 0·172 |
| $1\cdot8$ | 0·110 | 0·114 | 0·118 | 0·121 | 0·124 |
| $2\cdot0$ | 0·0831 | 0·0859 | 0·0883 | 0·0902 | 0·0917 |
| $2\cdot5$ | 0·0452 | 0·0462 | 0·0470 | 0·0476 | 0·0481 |
| $3\cdot0$ | 0·0270 | 0·0274 | 0·0278 | 0·0280 | 0·0282 |
| $3\cdot5$ | 0·0174 | 0·0176 | 0·0177 | 0·0178 | 0·0179 |
| $4\cdot0$ | 0·0118 | 0·0119 | 0·0120 | 0·0120 | 0·0120 |
| $4\cdot5$ | 0·00835 | 0·00840 | 0·00843 | 0·00846 | 0·00847 |
| $5\cdot0$ | 0·00612 | 0·00615 | 0·00617 | 0·00618 | 0·00619 |

continued

| $Z/a$ | $x/a = 0.1$ | 0.6 | 0.7 | 0.8 | 0.9 | 1.0 |
|---|---|---|---|---|---|---|
| 0.1 | 35.5 | 8.14 | 11.5 | 18.8 | 35.4 | 50.8 |
| 0.2 | 11.0 | 6.68 | 8.71 | 12.1 | 16.5 | 18.1 |
| 0.3 | 4.87 | 5.05 | 6.10 | 7.42 | 8.57 | 8.60 |
| 0.4 | 2.75 | 3.68 | 4.18 | 4.69 | 4.99 | 4.85 |
| 0.5 | 1.79 | 2.64 | 2.89 | 3.09 | 3.16 | 3.03 |
| 0.6 | 1.27 | 1.91 | 2.04 | 2.12 | 2.12 | 2.02 |
| 0.7 | 0.940 | 1.40 | 1.47 | 1.50 | 1.49 | 1.42 |
| 0.8 | 0.721 | 1.05 | 1.08 | 1.10 | 1.08 | 1.03 |
| 0.9 | 0.565 | 0.795 | 0.815 | 0.820 | 0.807 | 0.775 |
| 1.0 | 0.451 | 0.614 | 0.626 | 0.628 | 0.617 | 0.595 |
| 1.2 | 0.299 | 0.383 | 0.388 | 0.388 | 0.382 | 0.370 |
| 1.4 | 0.207 | 0.253 | 0.255 | 0.255 | 0.251 | 0.244 |
| 1.6 | 0.149 | 0.175 | 0.176 | 0.175 | 0.173 | 0.169 |
| 1.8 | 0.110 | 0.125 | 0.126 | 0.125 | 0.124 | 0.122 |
| 2.0 | 0.0831 | 0.0925 | 0.0928 | 0.0925 | 0.0916 | 0.0901 |
| 2.5 | 0.0452 | 0.0484 | 0.0484 | 0.0482 | 0.0478 | 0.0473 |
| 3.0 | 0.0270 | 0.0283 | 0.0283 | 0.0282 | 0.0280 | 0.0277 |
| 3.5 | 0.0174 | 0.0179 | 0.0179 | 0.0178 | 0.0177 | 0.0176 |
| 4.0 | 0.0188 | 0.0120 | 0.0120 | 0.0120 | 0.0199 | 0.0118 |
| 4.5 | 0.00835 | 0.00847 | 0.00846 | 0.00844 | 0.00841 | 0.00836 |
| 5.0 | 0.00612 | 0.00619 | 0.00618 | 0.00616 | 0.00614 | 0.00611 |

continued

| $Z/a$ | $x/a = 0\cdot1$ | $1\cdot2$ | $1\cdot4$ | $1\cdot6$ | $1\cdot8$ | $2\cdot0$ |
|---|---|---|---|---|---|---|
| 0·1 | 35·3 | 12·7 | 3·54 | 1·45 | 0·736 | 0·423 |
| 0·2 | 11·0 | 8·24 | 3·00 | 1·34 | 0·700 | 0·409 |
| 0·3 | 4·87 | 5·16 | 2·39 | 1·18 | 0·646 | 0·387 |
| 0·4 | 2·75 | 3·34 | 1·84 | 1·00 | 0·582 | 0·360 |
| 0·5 | 1·79 | 2·26 | 1·41 | 0·842 | 0·515 | 0·330 |
| 0·6 | 1·27 | 1·59 | 1·08 | 0·698 | 0·450 | 0·298 |
| 0·7 | 0·940 | 1·16 | 0·842 | 0·576 | 0·390 | 0·267 |
| 0·8 | 0·721 | 0·869 | 0·661 | 0·476 | 0·336 | 0·238 |
| 0·9 | 0·565 | 0·665 | 0·526 | 0·394 | 0·289 | 0·211 |
| 1·0 | 0·451 | 0·520 | 0·423 | 0·328 | 0·248 | 0·186 |
| 1·2 | 0·299 | 0·332 | 0·283 | 0·232 | 0·184 | 0·145 |
| 1·4 | 0·207 | 0·224 | 0·197 | 0·167 | 0·139 | 0·113 |
| 1·6 | 0·149 | 0·157 | 0·142 | 0·124 | 0·106 | 0·0888 |
| 1·8 | 0·110 | 0·114 | 0·105 | 0·0934 | 0·0817 | 0·0704 |
| 2·0 | 0·0831 | 0·0855 | 0·0792 | 0·0719 | 0·0641 | 0·0564 |
| 2·5 | 0·0452 | 0·0455 | 0·0432 | 0·0403 | 0·0371 | 0·0338 |
| 3·0 | 0·0270 | 0·0269 | 0·0259 | 0·0246 | 0·0231 | 0·0216 |
| 3·5 | 0·0174 | 0·0172 | 0·0167 | 0·0160 | 0·0153 | 0·0145 |
| 4·0 | 0·0118 | 0·0116 | 0·0113 | 0·0110 | 0·0106 | 0·0101 |
| 4·5 | 0·00835 | 0·00823 | 0·00806 | 0·00785 | 0·00761 | 0·00734 |
| 5·0 | 0·00612 | 0·00603 | 0·00593 | 0·00580 | 0·00565 | 0·00548 |

continued

| $Z/a$ | $x/a = 0.1$ | 2·2 | 2·4 | 2·6 | 2·8 | 3·0 |
|---|---|---|---|---|---|---|
| 0·1 | 35·3 | 0·265 | 0·177 | 0·123 | 0·0894 | 0·0667 |
| 0·2 | 11·0 | 0·259 | 0·173 | 0·122 | 0·0883 | 0·0661 |
| 0·3 | 4·87 | 0·249 | 0·168 | 0·119 | 0·0866 | 0·0651 |
| 0·4 | 2·75 | 0·236 | 0·161 | 0·115 | 0·0844 | 0·0637 |
| 0·5 | 1·79 | 0·220 | 0·153 | 0·110 | 0·0816 | 0·0619 |
| 0·6 | 1·27 | 0·204 | 0·144 | 0·105 | 0·0785 | 0·0599 |
| 0·7 | 0·940 | 0·188 | 0·135 | 0·0994 | 0·0750 | 0·0577 |
| 0·8 | 0·721 | 0·171 | 0·125 | 0·0936 | 0·0713 | 0·0553 |
| 0·9 | 0·565 | 0·155 | 0·116 | 0·0876 | 0·0674 | 0·0528 |
| 1·0 | 0·451 | 0·140 | 0·106 | 0·0816 | 0·0636 | 0·0502 |
| 1·2 | 0·299 | 0·113 | 0·0890 | 0·0702 | 0·0559 | 0·0449 |
| 1·4 | 0·207 | 0·0916 | 0·0740 | 0·0599 | 0·0487 | 0·0398 |
| 1·6 | 0·149 | 0·0740 | 0·0613 | 0·0508 | 0·0421 | 0·0350 |
| 1·8 | 0·110 | 0·0600 | 0·0509 | 0·0430 | 0·0363 | 0·0307 |
| 2·0 | 0·0831 | 0·0490 | 0·0424 | 0·0364 | 0·0313 | 0·0268 |
| 2·5 | 0·0452 | 0·0305 | 0·0274 | 0·0244 | 0·0216 | 0·0192 |
| 3·0 | 0·0270 | 0·0199 | 0·0183 | 0·0168 | 0·0152 | 0·0138 |
| 3·5 | 0·0174 | 0·0136 | 0·0127 | 0·0118 | 0·0110 | 0·0101 |
| 4·0 | 0·0118 | 0·00964 | 0·00913 | 0·00861 | 0·00809 | 0·00758 |
| 4·5 | 0·00835 | 0·00705 | 0·00674 | 0·00642 | 0·00610 | 0·00577 |
| 5·0 | 0·00612 | 0·00530 | 0·00510 | 0·00490 | 0·00469 | 0·00448 |

continued

| $Z/a$ | $x/a = 0\cdot1$ | 3·5 | 4·0 | 4·5 | 5·0 | 5·5 |
|---|---|---|---|---|---|---|
| 0·1 | 35·3 | 0·0356 | 0·0212 | 0·0136 | 0·00924 | 0·00658 |
| 0·2 | 11·0 | 0·0354 | 0·0211 | 0·0135 | 0·00922 | 0·00657 |
| 0·3 | 4·87 | 0·0350 | 0·0209 | 0·0135 | 0·00918 | 0·00656 |
| 0·4 | 2·75 | 0·0345 | 0·0207 | 0·0134 | 0·00913 | 0·00652 |
| 0·5 | 1·79 | 0·0339 | 0·0204 | 0·0132 | 0·00906 | 0·00648 |
| 0·6 | 1·27 | 0·0332 | 0·0201 | 0·0131 | 0·00898 | 0·00643 |
| 0·7 | 0·940 | 0·0323 | 0·0198 | 0·0129 | 0·00889 | 0·00638 |
| 0·8 | 0·721 | 0·0314 | 0·0193 | 0·0127 | 0·00878 | 0·00632 |
| 0·9 | 0·565 | 0·0304 | 0·0189 | 0·0125 | 0·00866 | 0·00625 |
| 1·0 | 0·451 | 0·0293 | 0·0184 | 0·0122 | 0·00853 | 0·00617 |
| 1·2 | 0·299 | 0·0271 | 0·0174 | 0·0117 | 0·00824 | 0·00600 |
| 1·4 | 0·207 | 0·0249 | 0·0163 | 0·0112 | 0·00792 | 0·00582 |
| 1·6 | 0·149 | 0·0226 | 0·0152 | 0·0105 | 0·00758 | 0·00561 |
| 1·8 | 0·110 | 0·0205 | 0·0140 | 0·00993 | 0·00722 | 0·00539 |
| 2·0 | 0·0831 | 0·0185 | 0·0130 | 0·00930 | 0·00685 | 0·00516 |
| 2·5 | 0·0452 | 0·0141 | 0·0104 | 0·00780 | 0·00593 | 0·00458 |
| 3·0 | 0·0270 | 0·0107 | 0·00831 | 0·00646 | 0·00506 | 0·00400 |
| 3·5 | 0·0174 | 0·00822 | 0·00662 | 0·00531 | 0·00428 | 0·00346 |
| 4·0 | 0·0118 | 0·00636 | 0·00528 | 0·00436 | 0·00360 | 0·00298 |
| 4·5 | 0·00835 | 0·00498 | 0·00424 | 0·00359 | 0·00303 | 0·00256 |
| 5·0 | 0·00612 | 0·00394 | 0·00343 | 0·00297 | 0·00255 | 0·00219 |

continued

| $Z/a$ | $x/a = 0.1$ | 6·0 | 6·5 | −0·1 | −0·2 | −0·3 |
|---|---|---|---|---|---|---|
| 0·1 | 35·3 | 0·00369 | 0·00287 | 28·4 | 7·95 | 3·10 |
| 0·2 | 11·0 | 0·00368 | 0·00286 | 8·70 | 4·42 | 2·29 |
| 0·3 | 4·87 | 0·00367 | 0·00286 | 3·74 | 2·50 | 1·60 |
| 0·4 | 2·75 | 0·00366 | 0·00285 | 2·08 | 1·57 | 1·14 |
| 0·5 | 1·79 | 0·00365 | 0·00284 | 1·35 | 1·09 | 0·848 |
| 0·6 | 1·27 | 0·00363 | 0·00283 | 0·963 | 0·802 | 0·654 |
| 0·7 | 0·940 | 0·00361 | 0·00282 | 0·726 | 0·618 | 0·519 |
| 0·8 | 0·721 | 0·00359 | 0·00280 | 0·566 | 0·491 | 0·421 |
| 0·9 | 0·565 | 0·00356 | 0·00278 | 0·453 | 0·398 | 0·347 |
| 1·0 | 0·451 | 0·00353 | 0·00276 | 0·368 | 0·328 | 0·290 |
| 1·2 | 0·299 | 0·00346 | 0·00272 | 0·253 | 0·230 | 0·208 |
| 1·4 | 0·207 | 0·00339 | 0·00267 | 0·180 | 0·167 | 0·153 |
| 1·6 | 0·149 | 0·00330 | 0·00261 | 0·132 | 0·124 | 0·116 |
| 1·8 | 0·110 | 0·00322 | 0·00255 | 0·0998 | 0·0944 | 0·0889 |
| 2·0 | 0·0831 | 0·00312 | 0·00249 | 0·0767 | 0·0732 | 0·0695 |
| 2·5 | 0·0452 | 0·00286 | 0·00231 | 0·0428 | 0·0415 | 0·0400 |
| 3·0 | 0·0270 | 0·00259 | 0·00212 | 0·0260 | 0·0255 | 0·0248 |
| 3·5 | 0·0174 | 0·00232 | 0·00193 | 0·0169 | 0·0166 | 0·0163 |
| 4·0 | 0·0118 | 0·00207 | 0·00174 | 0·0116 | 0·0114 | 0·0112 |
| 4·5 | 0·00835 | 0·00183 | 0·00156 | 0·00822 | 0·00814 | 0·00805 |
| 5·0 | 0·00612 | 0·00162 | 0·00139 | 0·00604 | 0·00600 | 0·00594 |

continued

| $Z/a$ | $x/a = 0.1$ | $-0.4$ | $-0.5$ | $-0.6$ | $-0.7$ | $-0.8$ |
|---|---|---|---|---|---|---|
| 0·1 | 35·3 | 1·57 | 0·933 | 0·618 | 0·440 | 0·329 |
| 0·2 | 11·0 | 1·32 | 0·838 | 0·575 | 0·418 | 0·317 |
| 0·3 | 4·87 | 1·05 | 0·719 | 0·516 | 0·386 | 0·298 |
| 0·4 | 2·75 | 0·823 | 0·604 | 0·454 | 0·350 | 0·276 |
| 0·5 | 1·79 | 0·653 | 0·504 | 0·394 | 0·313 | 0·252 |
| 0·6 | 1·27 | 0·526 | 0·423 | 0·341 | 0·277 | 0·228 |
| 0·7 | 0·940 | 0·431 | 0·356 | 0·295 | 0·245 | 0·205 |
| 0·8 | 0·721 | 0·358 | 0·303 | 0·256 | 0·217 | 0·184 |
| 0·9 | 0·565 | 0·300 | 0·259 | 0·222 | 0·191 | 0·165 |
| 1·0 | 0·451 | 0·254 | 0·222 | 0·194 | 0·169 | 0·148 |
| 1·2 | 0·299 | 0·187 | 0·167 | 0·149 | 0·133 | 0·118 |
| 1·4 | 0·207 | 0·140 | 0·128 | 0·116 | 0·105 | 0·0953 |
| 1·6 | 0·149 | 0·107 | 0·0991 | 0·0914 | 0·0840 | 0·0771 |
| 1·8 | 0·110 | 0·0834 | 0·0780 | 0·0727 | 0·0676 | 0·0628 |
| 2·0 | 0·0831 | 0·0658 | 0·0621 | 0·0585 | 0·0549 | 0·0515 |
| 2·5 | 0·0452 | 0·0385 | 0·0370 | 0·0355 | 0·0339 | 0·0323 |
| 3·0 | 0·0270 | 0·0242 | 0·0234 | 0·0227 | 0·0220 | 0·0212 |
| 3·5 | 0·0174 | 0·0160 | 0·0156 | 0·0153 | 0·0149 | 0·0145 |
| 4·0 | 0·0118 | 0·0111 | 0·0109 | 0·0107 | 0·0105 | 0·0102 |
| 4·5 | 0·00835 | 0·00795 | 0·00784 | 0·00772 | 0·00760 | 0·00747 |
| 5·6 | 0·00612 | 0·00588 | 0·00582 | 0·00575 | 0·00568 | 0·00560 |

continued

| $Z/a$ | $x/a = 0.1$ | $-0.9$ | $-1.0$ | $-1.2$ | $-1.4$ | $-1.6$ |
|---|---|---|---|---|---|---|
| 0·1 | 35·3 | 0·256 | 0·204 | 0·138 | 0·0992 | 0·0741 |
| 0·2 | 11·0 | 0·248 | 0·199 | 0·136 | 0·0979 | 0·0734 |
| 0·3 | 4·87 | 0·236 | 0·192 | 0·132 | 0·0959 | 0·0722 |
| 0·4 | 2·75 | 0·222 | 0·182 | 0·127 | 0·0931 | 0·0705 |
| 0·5 | 1·79 | 0·206 | 0·171 | 0·121 | 0·0898 | 0·0685 |
| 0·6 | 1·27 | 0·189 | 0·159 | 0·115 | 0·0861 | 0·0662 |
| 0·7 | 0·940 | 0·173 | 0·147 | 0·108 | 0·0820 | 0·0636 |
| 0·8 | 0·721 | 0·157 | 0·135 | 0·101 | 0·0778 | 0·0609 |
| 0·9 | 0·565 | 0·143 | 0·124 | 0·0945 | 0·0734 | 0·0581 |
| 1·0 | 0·451 | 0·129 | 0·113 | 0·0878 | 0·0691 | 0·0551 |
| 1·2 | 0·299 | 0·106 | 0·0941 | 0·0752 | 0·0606 | 0·0493 |
| 1·4 | 0·207 | 0·0863 | 0·0781 | 0·0640 | 0·0528 | 0·0437 |
| 1·5 | 0·149 | 0·0707 | 0·0648 | 0·0544 | 0·0457 | 0·0385 |
| 1·8 | 0·110 | 0·0582 | 0·0539 | 0·0461 | 0·0394 | 0·0338 |
| 2·0 | 0·0331 | 0·0482 | 0·0450 | 0·0392 | 0·0340 | 0·0296 |
| 2·7 | 0·0452 | 0·0308 | 0·0293 | 0·0264 | 0·0236 | 0·0211 |
| 3·0 | 0·0270 | 0·0204 | 0·0197 | 0·0181 | 0·0167 | 0·0152 |
| 3·5 | 0·0174 | 0·0141 | 0·0137 | 0·0128 | 0·0120 | 0·0112 |
| 4·0 | 0·0118 | 0·0100 | 0·00978 | 0·00930 | 0·00880 | 0·00830 |
| 4·5 | 0·00835 | 0·00737 | 0·00720 | 0·00691 | 0·00660 | 0·00629 |
| 5·0 | 0·00612 | 0·00551 | 0·00543 | 0·00524 | 0·00505 | 0·00485 |

continued

| $Z/a$ | $x/a = 0.1$ | $-1.8$ | $-2.0$ | $-2.2$ | $-2.4$ | $-2.6$ |
|---|---|---|---|---|---|---|
| 0·1 | 35·3 | 0·0571 | 0·0451 | 0·0363 | 0·0297 | 0·0246 |
| 0·2 | 11·0 | 0·0567 | 0·0448 | 0·0361 | 0·0295 | 0·0245 |
| 0·3 | 4·87 | 0·0559 | 0·0443 | 0·0357 | 0·0293 | 0·0243 |
| 0·4 | 2·75 | 0·0549 | 0·0436 | 0·0352 | 0·0289 | 0·0240 |
| 0·5 | 1·79 | 0·0536 | 0·0427 | 0·0346 | 0·0285 | 0·0237 |
| 0·6 | 1·27 | 0·0520 | 0·0417 | 0·0339 | 0·0280 | 0·0234 |
| 0·7 | 0·940 | 0·0504 | 0·0405 | 0·0331 | 0·0274 | 0·0229 |
| 0·8 | 0·721 | 0·0485 | 0·0393 | 0·0322 | 0·0268 | 0·0224 |
| 0·9 | 0·565 | 0·0466 | 0·0379 | 0·0313 | 0·0261 | 0·0219 |
| 1·0 | 0·451 | 0·0446 | 0·0365 | 0·0302 | 0·0253 | 0·0214 |
| 1·2 | 0·299 | 0·0405 | 0·0336 | 0·0281 | 0·0237 | 0·0202 |
| 1·4 | 0·207 | 0·0364 | 0·0306 | 0·0259 | 0·0220 | 0·0189 |
| 1·6 | 0·149 | 0·0326 | 0·0277 | 0·0238 | 0·0203 | 0·0176 |
| 1·8 | 0·110 | 0·0290 | 0·0249 | 0·0125 | 0·0187 | 0·0162 |
| 2·0 | 0·0831 | 0·0257 | 0·0224 | 0·0195 | 0·0170 | 0·0150 |
| 2·5 | 0·0452 | 0·0189 | 0·0168 | 0·0150 | 0·0134 | 0·0120 |
| 3·0 | 0·0270 | 0·0139 | 0·0127 | 0·0115 | 0·0105 | 0·00952 |
| 3·5 | 0·0174 | 0·0104 | 0·00958 | 0·00885 | 0·00817 | 0·00753 |
| 4·0 | 0·0118 | 0·00781 | 0·00732 | 0·00685 | 0·00640 | 0·00597 |
| 4·5 | 0·00835 | 0·00598 | 0·00566 | 0·00536 | 0·00505 | 0·00476 |
| 5·0 | 0·00612 | 0·00465 | 0·00444 | 0·00423 | 0·00403 | 0·00382 |

continued

| $Z/a$ | $x/a = 0.1$ | $-2.8$ | $-3.0$ | $-3.5$ | $-4.0$ | $-4.5$ |
|---|---|---|---|---|---|---|
| 0·1 | 35·3 | 0·0206 | 0·0174 | 0·0119 | 0·00850 | 0·00626 |
| 0·2 | 11·0 | 0·0205 | 0·0174 | 0·0119 | 0·00848 | 0·00625 |
| 0·3 | 4·87 | 0·0204 | 0·0173 | 0·0118 | 0·00845 | 0·00623 |
| 0·4 | 2·75 | 0·0202 | 0·0171 | 0·0118 | 0·00841 | 0·00621 |
| 0·5 | 1·79 | 0·0200 | 0·0170 | 0·0117 | 0·00835 | 0·00617 |
| 0·6 | 1·27 | 0·0197 | 0·0167 | 0·0116 | 0·00829 | 0·00614 |
| 0·7 | 0·940 | 0·0194 | 0·0165 | 0·0114 | 0·00821 | 0·00609 |
| 0·8 | 0·721 | 0·0190 | 0·0162 | 0·0113 | 0·00812 | 0·00604 |
| 0·9 | 0·565 | 0·0186 | 0·0159 | 0·0111 | 0·00803 | 0·00598 |
| 1·0 | 0·451 | 0·0182 | 0·0156 | 0·0109 | 0·00792 | 0·00591 |
| 1·2 | 0·299 | 0·0173 | 0·0149 | 0·0105 | 0·00769 | 0·00577 |
| 1·4 | 0·207 | 0·0163 | 0·0141 | 0·0101 | 0·00742 | 0·00560 |
| 1·6 | 0·149 | 0·0152 | 0·0133 | 0·0961 | 0·00714 | 0·00542 |
| 1·8 | 0·110 | 0·0142 | 0·0124 | 0·00912 | 0·00684 | 0·00523 |
| 2·0 | 0·0831 | 0·0132 | 0·0116 | 0·00862 | 0·00652 | 0·00503 |
| 2·5 | 0·0452 | 0·0107 | 0·00962 | 0·00738 | 0·00572 | 0·00450 |
| 3·0 | 0·0270 | 0·00865 | 0·00786 | 0·00622 | 0·00495 | 0·00397 |
| 3·5 | 0·0174 | 0·00693 | 0·00638 | 0·00519 | 0·00423 | 0·00346 |
| 4·0 | 0·0118 | 0·00556 | 0·00518 | 0·00432 | 0·00360 | 0·00300 |
| 4·5 | 0·00835 | 0·00448 | 0·00421 | 0·00359 | 0·00305 | 0·00259 |
| 5·0 | 0·00612 | 0·00363 | 0·00344 | 0·00298 | 0·00258 | 0·00223 |

continued

| $Z/a$ | $x/a = 0.1$ | $-5.0$ | $-5.5$ | $-6.0$ | $-6.5$ | $-7.0$ |
|---|---|---|---|---|---|---|
| 0·1 | 35·3 | 0·00474 | 0·00367 | 0·00290 | 0·00233 | 0·00190 |
| 0·2 | 11·0 | 0·00473 | 0·00367 | 0·00290 | 0·00233 | 0·00189 |
| 0·3 | 4·87 | 0·00472 | 0·00366 | 0·00289 | 0·00232 | 0·00189 |
| 0·4 | 2·75 | 0·00471 | 0·00365 | 0·00288 | 0·00232 | 0·00188 |
| 0·5 | 1·79 | 0·00469 | 0·00364 | 0·00287 | 0·00231 | 0·00188 |
| 0·6 | 1·27 | 0·00466 | 0·00362 | 0·00286 | 0·00230 | 0·00188 |
| 0·7 | 0·940 | 0·00463 | 0·00360 | 0·00285 | 0·00229 | 0·00187 |
| 0·8 | 0·721 | 0·00460 | 0·00358 | 0·00284 | 0·00228 | 0·00186 |
| 0·9 | 0·565 | 0·00456 | 0·00355 | 0·00282 | 0·00227 | 0·00186 |
| 1·0 | 0·451 | 0·00452 | 0·00353 | 0·00280 | 0·00226 | 0·00185 |
| 1·2 | 0·299 | 0·00443 | 0·00346 | 0·00276 | 0·00223 | 0·00182 |
| 1·4 | 0·207 | 0·00432 | 0·00339 | 0·00271 | 0·00219 | 0·00180 |
| 1·6 | 0·149 | 0·00420 | 0·00331 | 0·00264 | 0·00216 | 0·00178 |
| 1·8 | 0·110 | 0·00408 | 0·00323 | 0·00260 | 0·00211 | 0·00174 |
| 2·0 | 0·0831 | 0·00394 | 0·00314 | 0·00253 | 0·00207 | 0·00171 |
| 2·5 | 0·0452 | 0·00358 | 0·00289 | 0·00236 | 0·00194 | 0·00162 |
| 3·0 | 0·0270 | 0·00322 | 0·00263 | 0·00217 | 0·00180 | 0·00151 |
| 3·5 | 0·0174 | 0·00285 | 0·00237 | 0·00198 | 0·00166 | 0·00141 |
| 4·0 | 0·0118 | 0·00251 | 0·00212 | 0·00179 | 0·00152 | 0·00130 |
| 4·5 | 0·00835 | 0·00220 | 0·00188 | 0·00161 | 0·00138 | 0·00119 |
| 5·0 | 0·00612 | 0·00192 | 0·00166 | 0·00144 | 0·00125 | 0·00108 |

# APPENDIX F

## Magneto-telluric resistivity* (from Jackson, Wait and Walters, 1962)

| $\lambda_1/h_1$ | $\varrho_2/\varrho_1 = 2\cdot1$ | 4·0 | 6·25 | 10·9 | 25 |
|---|---|---|---|---|---|
| 0·25 | 1·002 | 1·004 | 1·005 | 1·006 | 1·007 |
| 0·33 | 0·996 | 0·993 | 0·990 | 0·987 | 0·983 |
| 0·50 | 0·958 | 0·930 | 0·911 | 0·888 | 0·861 |
| 0·67 | 0·958 | 0·927 | 0·902 | 0·874 | 0·850 |
| 0·825 | 0·985 | 0·972 | 0·963 | 0·953 | 0·944 |
| 1·0 | 1·02 | 1·05 | 1·07 | 1·08 | 1·10 |
| 2·5 | 1·40 | 1·89 | 2·26 | 2·78 | 3·54 |
| 5 | 1·67 | 2·66 | 3·56 | 5·06 | 7·72 |
| 10 | 1·84 | 3·24 | 4·66 | 7·33 | 13·1 |
| 20 | 1·94 | 3·61 | 5·40 | 9·00 | 18·0 |
| 25 | 1·96 | 3·68 | 5·56 | 9·35 | 19·0 |
| 33 | 1·98 | 3·76 | 5·72 | 9·75 | 20·5 |
| 50 | 2·00 | 3·86 | 5·89 | 10·2 | 21·8 |
| 100 | 2·02 | 3·92 | 6·10 | 10·6 | 22·9 |

| 100 | 400 | 625 | 1090 | 2500 | 10,000 | ∞ |
|---|---|---|---|---|---|---|
| 1·010 | 1·010 | 1·010 | 1·011 | 1·011 | 1·011 | 1·011 |
| 0·980 | 0·977 | 0·976 | 0·975 | 0·974 | 0·974 | 0·973 |
| 0·832 | 0·818 | 0·814 | 0·810 | 0·806 | 0·802 | 0·798 |
| 0·825 | 0·800 | 0·797 | 0·792 | 0·788 | 0·784 | 0·780 |
| 0·930 | 0·923 | 0·921 | 0·920 | 0·918 | 0·915 | 0·910 |
| 1·13 | 1·14 | 1·14 | 1·14 | 1·14 | 1·15 | 1·15 |
| 4·65 | 5·38 | 5·54 | 5·73 | 5·90 | 6·10 | 6·28 |
| 13·1 | 17·8 | 19·1 | 20·4 | 21·9 | 23·4 | 25 |
| 29·5 | 51·4 | 58·4 | 66·2 | 75·7 | 87·1 | 100 |
| 51·7 | 108 | 145 | 182 | 232 | 303 | 400 |
| 58·2 | 145 | 183 | 239 | 320 | 442 | 625 |
| 66·3 | 182 | 238 | 326 | 467 | 678 | 1090 |
| 75·6 | 232 | 320 | 466 | 733 | 1270 | 2500 |
| 87·0 | 302 | 461 | 678 | 1270 | 2940 | 10,000 |

Appendix F continued

## Phase angle (difference from 45°), in radians

| | | | | |
|---|---|---|---|---|
| 0·25 | − 0.0007 | − 0·0014 | − 0·0018 | − 0·0022 | − 0·0027 |
| 0·33 | − 0·0045 | − 0·0085 | − 0·0110 | − 0·0138 | − 0·0171 |
| 0·50 | 0·0064 | 0·0121 | 0·0156 | 0·0196 | 0·0243 |
| 0·67 | 0·0360 | 0·0681 | 0·0876 | 0·110 | 0·136 |
| 0·825 | 0·0641 | 0·121 | 0·156 | 0·195 | 0·241 |
| 1·0 | 0·0847 | 0·160 | 0·205 | 0·257 | 0·318 |
| 2·5 | 0·108 | 0·207 | 0·270 | 0·347 | 0·442 |
| 5 | 0·0754 | 0·148 | 0·199 | 0·265 | 0·360 |
| 10 | 0·0442 | 0·0888 | 0·121 | 0·167 | 0·240 |
| 20 | 0·0239 | 0·0485 | 0·0670 | 0·0944 | 0·142 |
| 25 | 0·0194 | 0·0395 | 0·0547 | 0·0775 | 0·118 |
| 50 | 0·0010 | 0·0205 | 0·0285 | 0·0408 | 0·0631 |

| | | | | | | |
|---|---|---|---|---|---|---|
| − 0·0034 | − 0·0037 | − 0·0038 | − 0·0038 | − 0·0039 | − 0·0040 | − 0·0041 |
| − 0·0210 | − 0·0232 | − 0·0236 | − 0·0241 | − 0·0246 | − 0·0251 | − 0·0256 |
| 0·0298 | 0·0303 | 0·0337 | 0·0344 | 0·0351 | 0·0358 | 0·0365 |
| 0·167 | 0·185 | 0·189 | 0·192 | 0·196 | 0·200 | 0·204 |
| 0·295 | 0·326 | 0·333 | 0·339 | 0·346 | 0·353 | 0·360 |
| 0·388 | 0·428 | 0·437 | 0·445 | 0·454 | 0·463 | 0·472 |
| 0·566 | 0·642 | 0·659 | 0·677 | 0·694 | 0·713 | 0·732 |
| 0·507 | 0·618 | 0·645 | 0·674 | 0·704 | 0·737 | 0·772 |
| 0·384 | 0·524 | 0·563 | 0·607 | 0·658 | 0·716 | 0·782 |
| 0·252 | 0·390 | 0·437 | 0·494 | 0·567 | 0·661 | 0·785 |
| 0·214 | 0·345 | 0·391 | 0·451 | 0·529 | 0·636 | 0·785 |
| 0·122 | 0·216 | 0·255 | 0·310 | 0·392 | 0·530 | 0·785 |

# APPENDIX G

## Mutual coupling of coplanar loops lying on the surface of the ground

| $\gamma r$ | Magnitude | Phase | Real component | Imaginary component |
|---|---|---|---|---|
| 0·1 | 1·0002 | 0·1279° | 1·0002 | 0·0022 |
| 0·2 | 1·0013 | 0·4864 | 1·0013 | 0·0085 |
| 0·3 | 1·0043 | 0·9963 | 1·0041 | 0·0175 |
| 0·4 | 1·0095 | 1·599 | 1·0091 | 0·0282 |
| 0·5 | 1·0174 | 2·233 | 1·0166 | 0·0396 |
| 0·6 | 1·0279 | 2·844 | 1·0266 | 0·0510 |
| 0·7 | 1·0409 | 3·383 | 1·0391 | 0·0614 |
| 0·8 | 1·0563 | 3·813 | 1·0540 | 0·0702 |
| 0·9 | 1·0737 | 4·105 | 1·0709 | 0·0769 |
| 1·0 | 1·0926 | 4·241 | 1·0896 | 0·0808 |
| 1·2 | 1·1333 | 4·007 | 1·1306 | 0·0792 |
| 1·4 | 1·1751 | 3·101 | 1·1734 | 0·0636 |
| 1·6 | 1·2149 | 1·576 | 1·2144 | 0·0334 |
| 1·8 | 1·2502 | − 0·4841 | 1·2502 | − 0·0106 |
| 2·0 | 1·2779 | − 2·989 | 1·2779 | − 0·0667 |
| 2·2 | 1·3019 | − 5·856 | 1·2951 | − 0·1328 |
| 2·4 | 1·3165 | − 9·010 | 1·3003 | − 0·2062 |
| 2·6 | 1·3234 | − 12·388 | 1·2925 | − 0·2839 |
| 2·8 | 1·3225 | − 15·938 | 1·2717 | − 0·3632 |
| 3·0 | 1·3144 | − 19·616 | 1·2381 | − 0·4412 |
| 3·2 | 1·2994 | − 23·385 | 1·1927 | − 0·5157 |
| 3·4 | 1·2782 | − 27·215 | 1·1367 | − 0·5846 |
| 3·6 | 1·2515 | − 31·081 | 1·0718 | − 0·6461 |
| 3·8 | 1·2198 | − 34·961 | 0·9997 | − 0·6990 |
| 4·0 | 1·1840 | − 38·836 | 0·9223 | − 0·7425 |
| 4·2 | 1·1447 | − 42·689 | 0·8414 | − 0·7761 |
| 4·4 | 1·1025 | − 46·506 | 0·7588 | − 0·7998 |
| 4·6 | 1·0580 | − 50·275 | 0·6762 | − 0·8137 |
| 4·8 | 1·0119 | − 53·982 | 0·5950 | − 0·8185 |
| 5·0 | 0·9647 | − 57·617 | 0·5166 | − 0·8147 |
| 5·5 | 0·8448 | − 66·321 | 0·3393 | − 0·7737 |
| 6·0 | 0·7272 | − 74·351 | 0·1962 | − 0·7003 |
| 6·5 | 0·6167 | − 81·549 | 0·0906 | − 0·6100 |
| 7·0 | 0·5164 | − 87·750 | 0·0203 | − 0·5160 |
| 7·5 | 0·4282 | − 87·217 | − 0·0208 | − 0·4277 |
| 8·0 | 0·3530 | − 83·521 | − 0·0398 | − 0·3508 |
| 8·5 | 0·2910 | − 81·294 | − 0·0440 | − 0·2876 |
| 9·0 | 0·2415 | − 80·570 | − 0·0396 | − 0·2382 |
| 9·5 | 0·2033 | − 81·216 | − 0·0310 | − 0·2009 |
| 10·0 | 0·1749 | − 82·882 | − 0·0217 | − 0·1735 |

## Mutual coupling of coaxial loops on the surface of a homogeneous earth

| $\gamma r$ | Magnitude | Phase | Real component | Imaginary component |
|---|---|---|---|---|
| 0·1 | 1·0000 | 0·0041° | 1·0000 | 0·0001 |
| 0·2 | 0·9997 | 0·0211 | 0·9997 | 0·0004 |
| 0·3 | 0·9990 | 0·0723 | 0·9990 | 0·0013 |
| 0·4 | 0·9980 | 0·1676 | 0·9979 | 0·0029 |
| 0·5 | 0·9964 | 0·3234 | 0·9964 | 0·0056 |
| 0·6 | 0·9945 | 0·5498 | 0·9944 | 0·0095 |
| 0·7 | 0·9923 | 0·8570 | 0·9922 | 0·0148 |
| 0·8 | 0·9899 | 1·252 | 0·9897 | 0·0216 |
| 0·9 | 0·9876 | 1·742 | 0·9872 | 0·0300 |
| 1·0 | 0·9856 | 2·329 | 0·9848 | 0·0401 |
| 1·2 | 0·9832 | 3·794 | 0·9811 | 0·0651 |
| 1·4 | 0·9847 | 5·618 | 0·9800 | 0·0964 |
| 1·6 | 0·9918 | 7·729 | 0·9828 | 0·1334 |
| 1·8 | 1·0060 | 10·016 | 0·9907 | 0·1750 |
| 2·0 | 1·0282 | 12·348 | 1·0044 | 0·2199 |
| 2·2 | 1·0586 | 14·590 | 1·0245 | 0·2667 |
| 2·4 | 1·0969 | 16·628 | 1·0511 | 0·3139 |
| 2·6 | 1·1424 | 18·379 | 1·0841 | 0·3602 |
| 2·8 | 1·1937 | 19·793 | 1·1232 | 0·4042 |
| 3·0 | 1·2495 | 20·855 | 1·1676 | 0·4448 |
| 3·2 | 1·3083 | 21·574 | 1·2166 | 0·4811 |
| 3·4 | 1·3688 | 21·976 | 1·2694 | 0·5122 |
| 3·6 | 1·4298 | 22·095 | 1·3248 | 0·5378 |
| 3·8 | 1·4902 | 21·971 | 1·3819 | 0·5575 |
| 4·0 | 1·5498 | 21·642 | 1·4397 | 0·5712 |
| 4·2 | 1·6054 | 21·143 | 1·4973 | 0·5791 |
| 4·4 | 1·6589 | 20·509 | 1·5538 | 0·5812 |
| 4·6 | 1·7091 | 19·770 | 1·6083 | 0·5781 |
| 4·8 | 1·7556 | 18·951 | 1·6604 | 0·5701 |
| 5·0 | 1·7982 | 18·075 | 1·7094 | 0·5579 |
| 5·5 | 1·8873 | 15·759 | 1·8163 | 0·5126 |
| 6·0 | 1·9522 | 13·438 | 1·8987 | 0·4537 |
| 6·5 | 1·9955 | 11.263 | 1·9571 | 0·3898 |
| 7·0 | 2·0213 | 9·327 | 1·9945 | 0·3276 |
| 7·5 | 2·0336 | 7·6739 | 2·0153 | 0·2715 |
| 8·0 | 2·0365 | 6·3142 | 2·0241 | 0·2240 |
| 10·0 | 2·0147 | 3·2972 | 2·0114 | 0·1159 |
| 12·0 | 2·0012 | 2·2920 | 1·9996 | 0·0800 |
| 14·0 | 2·0000 | 1·7526 | 1·9991 | 0·0612 |
| 16·0 | 2·0000 | 1·3494 | 1·9999 | 0·0471 |

## Mutual coupling of coplanar coils with axes parallel to the ground

| $\gamma r$ | Magnitude | Phase | Real component | Imaginary component |
|------|-----------|-------|----------------|---------------------|
| 0·1 | 1·0001 | 0·1369° | 1·0001 | 0·0024 |
| 0·2 | 1·0007 | 0·5296 | 1·0007 | 0·0093 |
| 0·3 | 1·0024 | 1·1415 | 1·0022 | 0·0200 |
| 0·4 | 1·0056 | 1·9390 | 1·0050 | 0·0340 |
| 0·5 | 1·0107 | 2·887 | 1·0094 | 0·0509 |
| 0·6 | 1·0179 | 3·948 | 1·0155 | 0·0701 |
| 0·7 | 1·0275 | 5·087 | 1·0235 | 0·0911 |
| 0·8 | 1·0396 | 6·269 | 1·0334 | 0·1135 |
| 0·9 | 1·0542 | 7·462 | 1·0453 | 0·1369 |
| 1·0 | 1·0713 | 8·638 | 1·0592 | 0·1609 |
| 1·2 | 1·1126 | 10·844 | 1·0927 | 0·2093 |
| 1·4 | 1·1620 | 12·746 | 1·1333 | 0·2564 |
| 1·6 | 1·3176 | 14·273 | 1·1800 | 0·3002 |
| 1·8 | 1·2774 | 15·407 | 1·2315 | 0·3394 |
| 2·0 | 1·3396 | 16·167 | 1·2866 | 0·3730 |
| 2·2 | 1·4024 | 16·594 | 1·3440 | 0·4005 |
| 2·4 | 1·4644 | 16·734 | 1·4024 | 0·4216 |
| 2·6 | 1·5246 | 16·636 | 1·4608 | 0·4365 |
| 2·8 | 1·5819 | 16·346 | 1·5180 | 0·4452 |
| 3·0 | 1·6359 | 15·906 | 1·5733 | 0·4484 |
| 3·2 | 1·6861 | 15·351 | 1·6260 | 0·4464 |
| 3·4 | 1·7323 | 14·711 | 1·6755 | 0·4399 |
| 3·6 | 1·7742 | 14·012 | 1·7214 | 0·4296 |
| 3·8 | 1·8120 | 13·274 | 1·7636 | 0·4161 |
| 4·0 | 1·8456 | 12·517 | 1·8018 | 0·4000 |
| 4·2 | 1·8753 | 11·754 | 1·8360 | 0·3820 |
| 4·4 | 1·9012 | 10·996 | 1·8663 | 0·3626 |
| 4·6 | 1·9236 | 10·254 | 1·8929 | 0·3424 |
| 4·8 | 1·9427 | 9·535 | 1·9159 | 0·3218 |
| 5·0 | 1·9588 | 8·844 | 1·9355 | 0·3012 |
| 5·5 | 1·9879 | 7·267 | 1·9719 | 0·2515 |
| 6·0 | 2·0043 | 5·929 | 1·9936 | 0·2070 |
| 6·5 | 2·0120 | 4·8333 | 2·0049 | 0·1695 |
| 7·0 | 2·0142 | 3·9624 | 2·0094 | 0·1392 |
| 7·5 | 2·0132 | 3·2862 | 2·0099 | 0·1154 |
| 8·0 | 2·0108 | 2·7697 | 2·0084 | 0·0972 |
| 10·0 | 2·0020 | 1·6673 | 2·0012 | 0·0583 |
| 12·0 | 2·0001 | 1·1838 | 1·9996 | 0·0413 |
| 14·0 | 2·0001 | 0·8789 | 1·9999 | 0·0307 |
| 16·0 | 2·0001 | 0·6721 | 2·0000 | 0·0235 |

# INDEX